The William Lowell Putnam Mathematical Competition

Problems and Solutions: 1938–1964

Originally published by
The Mathematical Association of America, 2006.

ISBN: 978-1-4704-5122-6
LCCN: 2003110417

∞ The paper used in this book is acid-free and falls within the guidelines
established to ensure permanence and durability.
Visit the AMS home page at https://www.ams.org/

10 9 8 7 6 5 4 3 2 24 23 22 21 20 19

AMS/MAA | PROBLEM BOOKS

VOL 32

The William Lowell Putnam Mathematical Competition

Problems and Solutions: 1938–1964

A. M. Gleason

R. E. Greenwood

L. M. Kelly

MAA PRESS

An Imprint of the AMERICAN MATHEMATICAL SOCIETY

Providence, Rhode Island

MAA PROBLEM BOOKS SERIES

Problem Books is a series of the Mathematical Association of America consisting of collections of problems and solutions from annual mathematical competitions; compilations of problems (including unsolved problems) specific to particular branches of mathematics; books on the art and practice of problem solving, etc.

A Friendly Mathematics Competition: 35 Years of Teamwork in Indiana, edited by Rick Gillman

The Inquisitive Problem Solver, Paul Vaderlind, Richard K. Guy, and Loren C. Larson

Mathematical Olympiads 1998–1999: Problems and Solutions From Around the World, edited by Titu Andreescu and Zuming Feng

Mathematical Olympiads 1999–2000: Problems and Solutions From Around the World, edited by Titu Andreescu and Zuming Feng

Mathematical Olympiads 2000–2001: Problems and Solutions From Around the World, edited by Titu Andreescu, Zuming Feng, and George Lee, Jr.

The William Lowell Putnam Mathematical Competition Problems and Solutions: 1938–1964, A. M. Gleason, R. E. Greenwood, L. M. Kelly

The William Lowell Putnam Mathematical Competition Problems and Solutions: 1965–1984, Gerald L. Alexanderson, Leonard F. Klosinski, and Loren C. Larson

The William Lowell Putnam Mathematical Competition 1985–2000: Problems, Solutions, and Commentary, Kiran S. Kedlaya, Bjorn Poonen, Ravi Vakil

USA and International Mathematical Olympiads 2000, edited by Titu Andreescu and Zuming Feng

USA and International Mathematical Olympiads 2001, edited by Titu Andreescu and Zuming Feng

DEDICATED TO

THE PUTNAM CONTESTANTS

PREFACE

The William Lowell Putnam Mathematical Competition, since its inception in 1938, has had a substantial impact on the field of mathematics in the United States and Canada. It rivals in this respect the classic Tripos in Cambridge and the influential Eötvös competition in Hungary. While there have been many different reasons for the remarkable expansion of mathematics during the past forty years, we believe that the challenge provided by the Putnam Competition has led many gifted college students into serious involvement with mathematics, and our profession is the richer for it.

There are some who feel that isolated problems, especially competition problems, present and emphasize an inappropriate view of mathematics. Yet progress in mathematics has often been made by separating problems from their contexts, and curiosity about isolated problems has frequently led to significant mathematical discoveries. Hermann Weyl wrote

> Important though the general concepts and propositions may be with which the modern industrious passion for axiomatizing and generalizing has presented us . . . nevertheless I am convinced that the special problems in all their complexity constitute the stock and core of mathematics; and to master their difficulties requires on the whole the harder labor. (From the Preface to *The Classical Groups*, Princeton University Press, 1939, 2nd ed., 1946.)

Although this book is primarily about the problems, we have included four articles about the competition from the *American Mathematical Monthly*: Garrett Birkhoff's account of the founding of the competition, L. E. Bush's summary of the first twenty-four contests (which we have revised with Professor Bush's permission to include the twenty-fifth), and an important exchange of views between L. J. Mordell and L. M. Kelly. We hope that these articles will give additional perspective on the role of the competition in American mathematics.

The first twenty-five competitions involved a total of three hundred forty-seven problems. They are collected in Part I essentially as they were presented to the contestants. Each problem is reprinted in Part II together with its solution. Sometimes we have given two or more solutions to a problem. In such cases we present first a solution which in our judgment could reasonably have been found by a contestant under examination conditions. Alternative solutions may involve more sophisticated approaches. We have also included references and historical backgrounds for some of the problems.

We found that it is not always easy to decide just what constitutes a solution of a problem. This is particularly true of the problems in geometry. For example, A.M. 3 of the Seventh Competition involves the concept of a polygonal line "crossing" a segment. Without a formal definition of this term a rigorous proof is impossible, yet a complete discussion could hardly be expected of a contestant pressed for time. Obviously, some compromise between rigor and intuition must be made, but it is by no means clear that a compromise appropriate for an examination booklet is equally appropriate for a book whose authors are under no time constraints.

The early competitions always included a mechanics problem, and in this area standards are even less clear. We can attest that an argument in mechanics can satisfy one mathematician and leave another unconvinced.

It is not surprising that some of the problems involve subtleties not envisioned by their proposers. Problem P.M. 12 of the Third Competition, concerning the director sphere of a central quadric, is a case in point. Lacking any instruction to the contrary, the problem is presumably set in euclidean space. In this setting a complete description of the locus is quite involved; in fact, although the problem itself is a classic, we believe that ours is the first complete solution to be published.

Inevitably, there have been some errors in the exams. These have varied from typographical errors of various degrees of seriousness to outright mistakes. Of the latter, the most interesting is problem A.M. 1 of the Seventeenth Competition, in which the desired conclusion is true locally, but not globally.

By and large we feel that the examinations have been well constructed. The continually increasing popularity of the competition is prima facie evidence that the examinations are generally perceived as fair and challenging. Each of us has served at one time or another on the examination committee, so we are well aware how hard it is to produce fresh and interesting problems year after year. Looking back on the considerable effort we have spent in compiling this book of solutions (but with no feeling that we as examiners did any better), we would offer one piece of advice to future examination committees: Spend more time on the question, What constitutes an acceptable solution?

Originally, it was planned that this volume should appear shortly after the Twenty-Fifth Competition was held. We thank the officers of the Mathematical Association of America for their patient understanding of the difficulties we have encountered and we apologize to them and to the interested mathematical community for the long delay in completing this project.

It is hardly necessary to observe that we and everyone else interested in the Putnam Competition owe much to the vision of William Lowell Putnam and to the generosity of his widow, Elizabeth Lowell Putnam. We would like to record here our appreciation of the encouragement given us by

George Putnam, grandson of the donor, who continues to give close personal attention to all aspects of the competition.

We have received advice and assistance from many friends in the mathematical world, but we must take special note of the invaluable contributions of Robert Brooks, Harley Flanders, Alan Grenadir, David Harbater, Fritz Herzog, David Jerison, James McKay, and L. E. Bush. We are particularly indebted to Basil Gordon, Murray Klamkin, and E. G. Strauss. Their criticisms, often trenchant, but always helpful, served to improve the solutions immeasurably. Our thanks are also due to E. F. Beckenbach, chairman of the Editorial Committee, and Raoul Hailpern, Editorial Director of the Mathematical Association of America.

Glendora Milligan typed the manuscript and James R. Holmes prepared the figures. We are especially grateful to Mrs. Milligan for her patience in handling the many, many revisions.

Summer 1978

A. M. GLEASON
R. E. GREENWOOD
L. M. KELLY

CONTENTS

PROBLEMS

THE FIRST WILLIAM LOWELL PUTNAM MATHEMATICAL COMPETITION
April 16, 1938

Morning Session

1. A solid is bounded by two bases in the horizontal planes $z = h/2$ and $z = -h/2$, and by such a surface that the area of every section in a horizontal plane is given by a formula of the sort

$$\text{Area} = a_0 z^3 + a_1 z^2 + a_2 z + a_3$$

(where as special cases some of the coefficients may be 0). Show that the volume is given by the formula

$$V = \frac{1}{6} h[B_1 + B_2 + 4M],$$

where B_1 and B_2 are the areas of the bases, and M is the area of the middle horizontal section. Show that the formulas for the volume of a cone and of a sphere can be included in this formula when $a_0 = 0$. (page 73)*

2. A can buoy is to be made of three pieces, namely, a cylinder and two equal cones, the altitude of each cone being equal to the altitude of the cylinder. For a given area of surface, what shape will have the greatest volume? (page 75)

3. If a particle moves in the plane, we may express its coordinates x and y as functions of the time t. If $x = t^3 - t$ and $y = t^4 + t$, show that the curve has a point of inflection at $t = 0$ and that the velocity of the moving particle has a maximum at $t = 0$. (page 76)

4. A lumberman wishes to cut down a tree whose trunk is cylindrical and whose material is uniform. He will cut a notch, the two sides of which will be planes intersecting at a dihedral angle θ along a horizontal line through the axis of the cylinder. If θ is given, show that the least volume of material is cut out when the plane bisecting the dihedral angle is horizontal. (page 77)

***The page number at the end of each problem indicates where the corresponding solution appears.**

5. Evaluate the following limits:

(i) $$\lim_{n \to \infty} \frac{n^2}{e^n}.$$

(ii) $$\lim_{x \to 0} \frac{1}{x} \int_0^x (1 + \sin 2t)^{1/t}\, dt.$$ (page 80)

6. A swimmer stands at one corner of a square swimming pool and wishes to reach the diagonally opposite corner. If w is his walking speed and s is his swimming speed ($s < w$), find his path for shortest time. [Consider two cases: (i) $w/s < \sqrt{2}$, and (ii) $w/s > \sqrt{2}$.] (page 81)

7. Take either (i) or (ii).

(i) Show that the gravitational attraction exerted by a thin homogeneous spherical shell at an external point is the same as if the material of the shell were concentrated at its center. (page 82)

(ii) Determine all the straight lines which lie upon the surface $z = xy$, and draw a figure to illustrate your result. (page 84)

Afternoon Session

8. Take either (i) or (ii).

(i) Let A_{ik} be the cofactor of a_{ik} in the determinant

$$d = \begin{vmatrix} a_{11} & a_{12} & a_{13} & a_{14} \\ a_{21} & a_{22} & a_{23} & a_{24} \\ a_{31} & a_{32} & a_{33} & a_{34} \\ a_{41} & a_{42} & a_{43} & a_{44} \end{vmatrix}.$$

Let D be the corresponding determinant with a_{ik} replaced by A_{ik}. Prove $D = d^3$. (page 86)

(ii) Let $P(y) = Ay^2 + By + C$ be a quadratic polynomial in y. If the roots of the quadratic equation $P(y) - y = 0$ are a and b ($a \neq b$), show that a and b are roots of the biquadratic equation $P[P(y)] - y = 0$. Hence write down a quadratic equation which will give the other two roots, c and d, of the biquadratic. Apply this result to solving the following biquadratic equation:

$$(y^2 - 3y + 2)^2 - 3(y^2 - 3y + 2) + 2 - y = 0.$$

(page 87)

9. Find all the solutions of the equation

$$yy'' - 2(y')^2 = 0$$

which pass through the point $x = 1, y = 1$. (page 88)

10. A horizontal disc of diameter 3 inches is rotating at 4 revolutions per minute. A light is shining at a distant point in the plane of the disc. An insect is placed at the edge of the disc furthest from the light, facing the light. It at once starts crawling, and crawls so as always to face the light, at 1 inch per second. Set up the differential equation of motion, and find at what point the insect again reaches the edge of the disc. (page 90)

11. Given the parabola $y^2 = 2mx$, what is the length of the shortest chord that is normal to the curve at one end? (page 91)

12. From the center of a rectangular hyperbola a perpendicular is dropped upon a variable tangent. Find the locus of the foot of the perpendicular. Obtain the equation of the locus in polar coordinates, and sketch the curve. (page 92)

13. Find the shortest distance between the plane $Ax + By + Cz + 1 = 0$ and the ellipsoid $x^2/a^2 + y^2/b^2 + z^2/c^2 = 1$. (For brevity, let

$$h = 1/\sqrt{A^2 + B^2 + C^2} \quad \text{and} \quad m = \sqrt{a^2A^2 + b^2B^2 + c^2C^2}.)$$

State algebraically the condition that the plane shall lie outside the ellipsoid.

(page 93)

THE SECOND WILLIAM LOWELL PUTNAM MATHEMATICAL COMPETITION
March 4, 1939

Morning Session

1. Find the length of the curve $y^2 = x^3$ from the origin to the point where the tangent makes an angle of $45°$ with the x-axis. (page 95)

2. A point P is taken on the curve $y = x^3$. The tangent at P meets the curve again at Q. Prove that the slope of the curve at Q is *four* times the slope at P. (page 95)

3. Find the cubic equation whose roots are the cubes of the roots of

$$x^3 + ax^2 + bx + c = 0.$$

(page 96)

4. Find the equations of the *two* straight lines each of which cuts all the *four* straight lines

$$x = 1, y = 0; \qquad y = 1, z = 0; \qquad z = 1, x = 0; \qquad x = y = -6z.$$

(page 98)

5. Take either (i) or (ii).

(i) Solve the system of differential equations

$$\frac{dx}{dt} = x + y - 3,$$

$$\frac{dy}{dt} = -2x + 3y + 1,$$

subject to the conditions $x = y = 0$ for $t = 0$. (page 101)

(ii) A heavy particle is attached to the end A of a light rod AB of length a. The rod is hinged at B so that it can turn freely in a vertical plane. The rod is balanced in the vertical position above the hinge and then slightly disturbed. Prove that the time taken to pass from the horizontal position to the lowest position is

$$\sqrt{\frac{a}{g}}\log_e(1 + \sqrt{2}).$$

(page 103)

6. Take either (i) or (ii).

(i) A circle of radius a rolls on the inner side of the circumference of a circle of radius $3a$. Find the area contained within the closed curve generated by a point on the circumference of the rolling circle. (page 105)

(ii) A shell strikes an airplane flying at a height h above the ground. It is known that the shell was fired from a gun on the ground with a muzzle velocity of magnitude V, but the position of the gun and its angle of elevation are both unknown. Deduce that the gun is situated within a circle whose center lies directly below the airplane and whose radius is

$$\frac{V}{g}\sqrt{V^2 - 2gh}.$$

(Neglect the resistance of the atmosphere.) (page 107)

7. Take either (i) or (ii).

(i) Find the curve touched by all the curves of the family

$$(y - k^2)^2 = x^2(k^2 - x^2).$$

Make a rough sketch showing this curve and two curves of the family.

(page 109)

(ii) If the expansion in powers of x of the function

$$\frac{1}{(1 - ax)(1 - bx)}$$

is given by

$$c_0 + c_1x + c_2x^2 + c_3x^3 + \cdots,$$

prove that the expansion in powers of x of the function

$$\frac{1 + abx}{(1 - abx)(1 - a^2x)(1 - b^2x)}$$

is given by

$$c_0^2 + c_1^2x + c_2^2x^2 + c_3^2x^3 + \cdots.$$

(page 112)

Afternoon Session

8. From the vertex $(0, c)$ of the catenary

$$y = c\cosh\frac{x}{c}$$

a line L is drawn perpendicular to the tangent to the catenary at a point P. Prove that the length of L intercepted by the axes is equal to the ordinate y of the point P. (page 114)

9. Evaluate the definite integrals

$$\text{(i)} \int_1^3 \frac{dx}{\sqrt{(x-1)(3-x)}}, \qquad \text{(ii)} \int_1^\infty \frac{dx}{e^{x+1}+e^{3-x}}.$$

(page 114)

10. Given the power-series

$$a_0 + a_1x + a_2x^2 + \cdots$$

in which

$$a_n = (n^2+1)3^n,$$

show that there is a relation of the form

$$a_n + pa_{n+1} + qa_{n+2} + ra_{n+3} = 0,$$

in which p, q, r are constants independent of n. Find these constants and the sum of the power-series. (page 115)

11. Find the equation of the parabola which touches the x-axis at the point (1, 0) and the y-axis at the point (0, 2). Find the equation of the axis of the parabola and the coordinates of its vertex. (page 117)

12. Take either (i) or (ii).

(i) Prove that

$$\int_1^a [x]f'(x)\,dx = [a]f(a) - \{f(1) + \cdots + f([a])\},$$

where a is greater than 1 and where $[x]$ denotes the greatest of the integers not exceeding x. Obtain a corresponding expression for

$$\int_1^a [x^2]f'(x)\,dx.$$

(page 121)

(ii) A particle moves on a straight line, the only force acting on it being a resistance proportional to the velocity. If it started with a velocity of 1,000 ft. per sec. and had a velocity of 900 ft. per sec. when it had travelled 1,200 ft., calculate to the nearest hundredth of a second the time it took to travel this distance. (page 122)

13. Take either (i) or (ii).

(i) Let $f(x)$ be defined for $a \le x \le b$. Assuming appropriate properties of continuity and derivability, prove for $a < x < b$ that

$$\frac{\dfrac{f(x) - f(a)}{x - a} - \dfrac{f(b) - f(a)}{b - a}}{x - b} = \tfrac{1}{2} f''(\xi),$$

where ξ is some number between a and b. (page 123)

(ii) Calculate the mutual gravitational attraction of two uniform rods, each of mass m and length $2a$, placed parallel to one another and perpendicular to the line joining their centers at a distance b apart.

In your answer let a approach zero, and comment on the form of the result. (page 125)

14. Take either (i) or (ii).

(i) If

$$u = 1 + \frac{x^3}{3!} + \frac{x^6}{6!} + \cdots,$$

$$v = \frac{x}{1!} + \frac{x^4}{4!} + \frac{x^7}{7!} + \cdots,$$

$$w = \frac{x^2}{2!} + \frac{x^5}{5!} + \frac{x^8}{8!} + \cdots,$$

prove that

$$u^3 + v^3 + w^3 - 3uvw = 1.$$

(page 129)

(ii) Consider the central conics

$$(ax^2 + by^2) + 2(px + qy) + c = 0,$$

$$(ax^2 + by^2) + 2\lambda(px + qy) + \lambda^2 c = 0,$$

where λ is a given positive constant.

Show that if all radii from the origin to the first conic are changed in the ratio λ to 1 the tips of these new radii generate the second conic.

Let P be the point with coordinates

$$x = -\frac{p}{a}\frac{2\lambda}{1 + \lambda}, \qquad y = -\frac{q}{b}\frac{2\lambda}{1 + \lambda}.$$

Show that if all radii from P to the first conic are changed in the ratio λ to 1 and then reversed about P the tips of these new radii generate the second conic.

Comment on these results in case $\lambda = 1$. (page 130)

THE THIRD WILLIAM LOWELL PUTNAM MATHEMATICAL COMPETITION
March 2, 1940

Morning Session

1. Prove that if $f(x)$ is a polynomial with integral coefficients, and there exists an integer k such that none of the integers $f(1), f(2), \ldots, f(k)$ is divisible by k, then $f(x)$ has no integral root. (page 133)

2. Let A and B be two fixed points on the curve $y = f(x)$, where $f(x)$ is continuous and has a continuous derivative, and the arc AB is concave to the chord AB. If P is a point of the arc AB for which $AP + PB$ is a maximum, prove that PA and PB are equally inclined to the tangent to the curve $y = f(x)$ at the point P. (page 133)

3. Find $f(x)$ such that

$$\int [f(x)]^n \, dx = \left[\int f(x) \, dx \right]^n,$$

when constants of integration are suitably chosen. (page 135)

4. The parabola $y^2 = -4px$ rolls without slipping around the parabola $y^2 = 4px$. Find the equation of the locus of the vertex of the rolling parabola. (page 137)

5. Prove that the simultaneous equations

$$x^4 - x^2 = y^4 - y^2 = z^4 - z^2$$

are satisfied by the points of four straight lines and six ellipses, and by no other points. (page 139)

6. $f(x)$ is a polynomial of degree n, such that a power of $f(x)$ is divisible by a power of its derivative $f'(x)$; i.e., $[f(x)]^p$ is divisible by $[f'(x)]^q$; p, q, positive integers. Prove that $f(x)$ is divisible by $f'(x)$ and that $f(x)$ has a single root of multiplicity n. (page 140)

7. If $u_1^2 + u_2^2 + \cdots$ and $v_1^2 + v_2^2 + \cdots$ are convergent series of real constants, prove that

$$(u_1 - v_1)^p + (u_2 - v_2)^p + \cdots, \ p \text{ an integer} \geq 2,$$

is convergent. (page 141)

8. A triangle is bounded by the lines

$$A_1x + B_1y + C_1 = 0, \qquad A_2x + B_2y + C_2 = 0,$$
$$A_3x + B_3y + C_3 = 0.$$

Show that the area, disregarding sign, is

$$\frac{\begin{vmatrix} A_1 & B_1 & C_1 \\ A_2 & B_2 & C_2 \\ A_3 & B_3 & C_3 \end{vmatrix}^2}{2\begin{vmatrix} A_2 & B_2 \\ A_3 & B_3 \end{vmatrix} \cdot \begin{vmatrix} A_3 & B_3 \\ A_1 & B_1 \end{vmatrix} \cdot \begin{vmatrix} A_1 & B_1 \\ A_2 & B_2 \end{vmatrix}}.$$

(page 142)

Afternoon Session

9. A projectile, thrown with initial velocity v_0 in a direction making angle α with the horizontal, is acted on by no force except gravity. Find the length of its path until it strikes a horizontal plane through the starting point. Show that the flight is longest when

$$\sin \alpha \log (\sec \alpha + \tan \alpha) = 1.$$

(page 144)

10. A cylindrical hole of radius r is bored through a cylinder of radius R ($r \leq R$) so that the axes intersect at right angles.

(i) Show that the area of the larger cylinder which is inside the smaller can be expressed in the form

$$S = 8r^2 \int_0^1 \frac{1 - v^2}{\sqrt{(1 - v^2)(1 - m^2v^2)}}\, dv \quad \text{where} \quad m = \frac{r}{R}.$$

(ii) If

$$K = \int_0^1 \frac{dv}{\sqrt{(1 - v^2)(1 - m^2v^2)}} \quad \text{and} \quad E = \int_0^1 \sqrt{\frac{1 - m^2v^2}{1 - v^2}}\, dv$$

show that

$$S = 8[R^2E - (R^2 - r^2)K].$$

(page 146)

11. From any point (a, b) in the Cartesian plane, show that (i) three normals, real or imaginary, can be drawn to the parabola $y^2 = 4px$; (ii) these

are real and distinct if $4(2p - a)^3 + 27pb^2 < 0$; (iii) two of them coincide if (a, b) lies on the curve $27py^2 = 4(x - 2p)^3$; (iv) all three coincide only if (a, b) is the point $(2p, 0)$. (page 147)

12. Prove that the locus of the point of intersection of three mutually perpendicular planes tangent to the surface

$$(1) \qquad ax^2 + by^2 + cz^2 = 1 \qquad (abc \neq 0)$$

is the sphere

$$(2) \qquad x^2 + y^2 + z^2 = \frac{1}{a} + \frac{1}{b} + \frac{1}{c}.$$

(page 149)

13. Determine all rational values for which a, b, c are the roots of

$$x^3 + ax^2 + bx + c = 0.$$

(page 157)

14. Prove that

$$\begin{pmatrix} a_1^2 + k & a_1a_2 & a_1a_3 & \ldots & a_1a_n \\ a_2a_1 & a_2^2 + k & a_2a_3 & \ldots & a_2a_n \\ \ldots & \ldots & \ldots & \ldots & \ldots \\ a_na_1 & a_na_2 & a_na_3 & \ldots & a_n^2 + k \end{pmatrix}$$

is divisible by k^{n-1} and find its other factor. (page 158)

15. Which is greater

$$(\sqrt{n})^{\sqrt{n+1}} \quad \text{or} \quad (\sqrt{n+1})^{\sqrt{n}}$$

where $n > 8$? (page 160)

THE FOURTH WILLIAM LOWELL PUTNAM MATHEMATICAL COMPETITION

March 1, 1941

Morning Session

1. Prove that the polynomial

$$(a - x)^6 - 3a(a - x)^5 + \tfrac{5}{2} a^2(a - x)^4 - \tfrac{1}{2} a^4(a - x)^2$$

takes only negative values for $0 < x < a$. (page 161)

2. Find the nth derivative with respect to x of

$$\int_0^x \left[1 + \frac{(x - t)}{1!} + \frac{(x - t)^2}{2!} + \cdots + \frac{(x - t)^{n-1}}{(n - 1)!}\right] e^{nt}\, dt.$$

(page 162)

3. A circle of *radius* a rolls in its plane along the x-axis. Show that the envelope of a diameter is a cycloid, coinciding with the cycloid traced out by a point on the circumference of a circle of *diameter* a, likewise rolling in its plane along the x-axis. (page 163)

4. Let the roots a, b, c of

$$f(x) \equiv x^3 + px^2 + qx + r = 0$$

be real, and let $a \le b \le c$. Prove that, if the interval (b, c) is divided into *six* equal parts, a root of $f'(x) = 0$ will lie in the *fourth* part counting from the end b. What will be the form of $f(x)$ if the root in question of $f'(x) = 0$ falls at either end of the *fourth* part? (page 165)

5. Show that the line which moves parallel to the plane $y = z$ and which intersects the two parabolas $y^2 = 2x$, $z = 0$ and $z^2 = 3x$, $y = 0$ sweeps out the surface

$$x = (y - z)\left(\frac{y}{2} - \frac{z}{3}\right).$$ (page 166)

6. If the x-coordinate $\bar{x}$ of the center of mass of the area lying between the

x-axis and the curve $y = f(x)$, $(f(x) > 0)$, and between the lines $x = 0$ and $x = a$ is given by

$$\bar{x} = g(a),$$

show that

$$f(x) = A \frac{g'(x)}{[x - g(x)]^2} e^{\int dx/(x - g(x))},$$

where A is a positive constant. (page 168)

7. Take either (i) or (ii).

(i) Prove that

$$\begin{vmatrix} 1 + a^2 - b^2 - c^2 & 2(ab + c) & 2(ca - b) \\ 2(ab - c) & 1 + b^2 - c^2 - a^2 & 2(bc + a) \\ 2(ca + b) & 2(bc - a) & 1 + c^2 - a^2 - b^2 \end{vmatrix}$$

$$= (1 + a^2 + b^2 + c^2)^3.$$ (page 169)

(ii) A semi-ellipsoid of revolution is formed by revolving about the x-axis the area lying within the first quadrant of the ellipse

$$\frac{x^2}{a^2} + \frac{y^2}{b^2} = 1.$$

Show that this semi-ellipsoid will balance in stable equilibrium, with its vertex resting on a horizontal plane, when and only when

$$b\sqrt{8} \geq a\sqrt{5}.$$ (page 171)

Afternoon Session

8. A particle (x, y) moves so that its angular velocities about $(1, 0)$ and $(-1, 0)$ are equal in magnitude but opposite in sign. Prove that

$$y(x^2 + y^2 + 1)\,dx = x(x^2 + y^2 - 1)\,dy,$$

and verify that this is the differential equation of the family of rectangular hyperbolas passing through $(1, 0)$ and $(-1, 0)$ and having the origin as center. (page 173)

9. Evaluate the following limits:

$$\lim_{n\to\infty}\left(\frac{1}{\sqrt{n^2+1^2}}+\frac{1}{\sqrt{n^2+2^2}}+\cdots+\frac{1}{\sqrt{n^2+n^2}}\right);$$

$$\lim_{n\to\infty}\left(\frac{1}{\sqrt{n^2+1}}+\frac{1}{\sqrt{n^2+2}}+\cdots+\frac{1}{\sqrt{n^2+n}}\right);$$

$$\lim_{n\to\infty}\left(\frac{1}{\sqrt{n^2+1}}+\frac{1}{\sqrt{n^2+2}}+\cdots+\frac{1}{\sqrt{n^2+n^2}}\right).$$

(page 175)

10. Find the differential equation satisfied by the product z of any two linearly independent integrals of the equation

$$y'' + y'P(x) + yQ(x) = 0.$$ (page 176)

11. Two perpendicular diameters of the ellipse

$$\frac{x^2}{a^2}+\frac{y^2}{b^2}=1$$

are given, and the two diameters conjugate to them are constructed. Show that the rectangular hyperbola passing through the ends of these conjugate diameters passes through the foci of the ellipse. (page 178)

12. A car is being driven so that its wheels, all of radius a feet, have an angular velocity of ω radians per second. A particle is thrown off from the tire of one of these wheels, where it is supposed that $a\omega^2 > g$. Neglecting the resistance of the air, show that the maximum height above the roadway which the particle can reach is

$$\frac{(a\omega + g\omega^{-1})^2}{2g}.$$ (page 179)

13. Assuming that $f(x)$ is continuous in the interval $(0, 1)$, prove that

$$\int_{x=0}^{x=1}\int_{y=x}^{y=1}\int_{z=x}^{z=y} f(x)f(y)f(z)\,dx\,dy\,dz = \frac{1}{3!}\left(\int_{t=0}^{t=1} f(t)\,dt\right)^3.$$

(page 180)

14. Take either (i) or (ii).

(i) Show that any solution $f(t)$ of the functional equation

$$f(x+y)f(x-y) = f(x)f(x) + f(y)f(y) - 1, \quad (x, y, \text{real}),$$

is such that

$$f''(t) = \pm m^2 f(t), \quad (m \text{ constant and } \geq 0),$$

assuming the existence and continuity of the second derivative. Deduce that $f(t)$ is one of the functions

$$\pm \cos mt, \qquad \pm \cosh mt.$$

(page 181)

(ii) With n constant values $a_1, a_2, \ldots, a_n$, supposed all different, let n constant values $b_1, b_2, \ldots, b_n$ be associated, and let a polynomial $P(x)$ be defined by the identity in x

$$\begin{vmatrix} 1 & x & x^2 & \cdots & x^{n-1} & P(x) \\ 1 & a_1 & a_1^2 & \cdots & a_1^{n-1} & b_1 \\ 1 & a_2 & a_2^2 & \cdots & a_2^{n-1} & b_2 \\ \cdots & \cdots & \cdots & \cdots & \cdots & \cdots \\ 1 & a_n & a_n^2 & \cdots & a_n^{n-1} & b_n \end{vmatrix} \equiv 0.$$

Given a polynomial $\phi(t)$, let a polynomial $Q(x)$ be defined by the identity in x obtained on replacing $P(x)$, $b_1, b_2, \ldots, b_n$ of the identity above by $Q(x)$, $\phi(b_1)$, $\phi(b_2)$, $\ldots$, $\phi(b_n)$. Prove that the remainder obtained on dividing $\phi(P(x))$ by $(x - a_1)(x - a_2) \cdots (x - a_n)$ is $Q(x)$.

(page 182)

THE FIFTH WILLIAM LOWELL PUTNAM MATHEMATICAL COMPETITION

March 7, 1942

Morning Session

1. A square of side $2a$, lying always in the first quadrant of the XY plane, moves so that two consecutive vertices are always on the X- and Y-axes respectively. Find the locus of the midpoint of the square. (page 184)

2. If a polynomial $f(x)$ is divided by $(x - a)^2 (x - b)$, where $a \neq b$, derive a formula for the remainder. (page 185)

3. Is the following series convergent or divergent?

$$1 + \frac{1}{2} \cdot \frac{19}{7} + \frac{2!}{3^2}\left(\frac{19}{7}\right)^2 + \frac{3!}{4^3}\left(\frac{19}{7}\right)^3 + \frac{4!}{5^4}\left(\frac{19}{7}\right)^4 + \cdots.$$

(page 186)

4. Find the orthogonal trajectories of the family of conics $(x + 2y)^2 = a(x + y)$. At what angle do the curves of one family cut the curves of the other family at the origin? (page 187)

5. A circle of radius a is revolved through 180° about a line in its plane, distant b from the center of the circle, where $b > a$. For what value of the ratio b/a does the center of gravity of the solid thus generated lie on the surface of the solid? (page 189)

6. Any circle in the XY (horizontal) plane is "represented" by a point on the vertical line through the center of the circle and at a distance "above" the plane of the circle equal to the radius of the circle.

Show that the locus of the representations of all the circles which cut a fixed circle at a constant angle is a (portion of a) one-sheeted hyperboloid.

By consideration of suitable families of circles in the plane, demonstrate the existence of two families of rulings on the hyperboloid. (page 191)

Afternoon Session

7. A square of side $2a$, lying always in the first quadrant of the XY plane, moves so that two consecutive vertices are always on the X- and Y-axes respectively. Prove that a point within or on the boundary of the square will

in general describe a (portion of a) conic. For what points of the square does this locus degenerate? (page 194)

8. For the family of parabolas

$$y = \frac{a^3x^2}{3} + \frac{a^2x}{2} - 2a,$$

(i) find the locus of vertices,
(ii) find the envelope,
(iii) sketch the envelope and two typical curves of the family. (page 195)

9. Given

$$x = \phi(u, v)$$
$$y = \psi(u, v)$$

where ϕ and ψ are solutions of the partial differential equation

$$\frac{\partial\phi}{\partial u}\frac{\partial\psi}{\partial v} - \frac{\partial\phi}{\partial v}\frac{\partial\psi}{\partial u} = 1. \tag{1}$$

By assuming that x and v are the independent variables, show that (1) may be transformed to

$$\frac{\partial y}{\partial v} = \frac{\partial u}{\partial x}. \tag{2}$$

Integrate (2), and show how this effects in general the solution of (1). What other solutions does (1) possess? (page 198)

10. A particle moves under a central force inversely proportional to the kth power of the distance. If the particle describes a circle (the central force proceeding from a point *on the circumference of the circle*), find k.

(page 201)

11. Sketch the curve

$$y = \frac{x}{1 + x^6 \sin^2 x},$$

and show that

$$\int_0^\infty \frac{x\,dx}{1 + x^6 \sin^2 x}$$

exists. (page 203)

THE SIXTH WILLIAM LOWELL PUTNAM MATHEMATICAL COMPETITION

June 1, 1946

Morning Session

1. Suppose that the function $f(x) = ax^2 + bx + c$, where a, b, c are real constants, satisfies the condition $|f(x)| \le 1$ for $|x| \le 1$. Prove that $|f'(x)| \le 4$ for $|x| \le 1$. (page 206)

2. If $a(x)$, $b(x)$, $c(x)$, and $d(x)$ are polynomials in x, show that

$$\int_1^x a(x)c(x)\,dx \cdot \int_1^x b(x)d(x)\,dx - \int_1^x a(x)d(x)\,dx \cdot \int_1^x b(x)c(x)\,dx$$

is divisible by $(x-1)^4$. (page 208)

3. A projectile in flight is observed simultaneously from four radar stations which are situated at the corners of a square of side b. The distances of the projectile from the four stations, taken in order around the square, are found to be R_1, R_2, R_3, R_4. Show that

$$R_1^2 + R_3^2 = R_2^2 + R_4^2.$$

Show also that the height h of the projectile above the ground is given by

$$h^2 = -\frac{1}{2}b^2 + \frac{1}{4}(R_1^2 + R_2^2 + R_3^2 + R_4^2)$$
$$- \frac{1}{8b^2}(R_1^4 + R_2^4 + R_3^4 + R_4^4 - 2R_1^2R_3^2 - 2R_2^2R_4^2).$$

(page 210)

4. Let $g(x)$ be a function that has a continuous first derivative $g'(x)$ for all values of x. Suppose that the following conditions hold for every x: (i) $g(0) = 0$; (ii) $|g'(x)| \le |g(x)|$. Prove that $g(x)$ vanishes identically. (page 211)

5. Find the smallest volume bounded by the coordinate planes and by a tangent plane to the ellipsoid

$$\frac{x^2}{a^2} + \frac{y^2}{b^2} + \frac{z^2}{c^2} = 1.$$ (page 213)

6. A particle of unit mass moves on a straight line under the action of a force which is a function $f(v)$ of the velocity v of the particle, but the form of this function is not known. A motion is observed, and the distance x covered in time t is found to be connected with t by the formula $x = at + bt^2 + ct^3$, where a, b, c have numerical values determined by observation of the motion. Find the function $f(v)$ for the range of v covered by the experiment. (page 214)

Afternoon Session

1. Let K denote the circumference of a circular disc of radius one, and let k denote a circular arc that joins two points a, b on K and lies otherwise in the given circular disc. Suppose that k divides the circular disc into two parts of equal area. Prove that the length of k exceeds 2. (page 215)

2. Let A, B be variable points on a parabola P, such that the tangents at A and B are perpendicular to each other. Show that the locus of the centroid of the triangle formed by A, B and the vertex of P is a parabola P_1. Apply the same process to P_1, obtaining a parabola P_2, and repeat the process, obtaining altogether the sequence of parabolas $P, P_1, P_2, \ldots, P_n$. If the equation of P is $y^2 = mx$, find the equation of P_n. (page 216)

3. In a solid sphere of radius R the density ρ is a function of r, the distance from the center of the sphere. If the magnitude of the gravitational force of attraction due to the sphere at any point inside the sphere is kr^2, where k is a constant, find ρ as a function of r. Find also the magnitude of the force of attraction at a point outside the sphere at a distance r from the center. (Assume that the magnitude of the force of attraction at a point P due to a thin spherical shell is zero if P is inside the shell, and is m/r^2 if P is outside the shell, m being the mass of the shell, and r the distance of P from the center.) (page 218)

4. For each positive integer n, put

$$p_n = (1 + 1/n)^n, \quad P_n = (1 + 1/n)^{n+1}, \quad h_n = \frac{2p_nP_n}{p_n + P_n}.$$

Prove that $h_1 < h_2 < \cdots < h_n < \cdots$. (page 219)

5. Show that the integer next above $(\sqrt{3} + 1)^{2n}$ is divisible by 2^{n+1}. (page 220)

6. A particle moves on a circle with center O, starting from rest at a point P and coming to rest again at a point Q, without coming to rest at any intermediate point. Prove that the acceleration vector of the particle does not vanish at any point between P and Q and that, at some point R between P and Q, the acceleration vector points in along the radius RO. (page 221)

THE SEVENTH WILLIAM LOWELL PUTNAM MATHEMATICAL COMPETITION

May 24, 1947

Morning Session

1. If $\{a_n\}$ is a sequence of numbers such that for $n \geq 1$

$$(2 - a_n)a_{n+1} = 1,$$

prove that $\lim a_n$, as $n \to \infty$, exists and is equal to one. (page 223)

2. A real valued continuous function satisfies for all real x and y the functional equation

$$f(\sqrt{x^2 + y^2}) = f(x)f(y).$$

Prove that

$$f(x) = [f(1)]^{x^2}.$$ (page 225)

3. Given this figure (see p. 23) and any two points Q_1, Q_2 in the plane not lying on any of the segments $s_1, s_2, \ldots, s_6$, show that there does not exist a polygonal line P joining Q_1 and Q_2 such that:

(i) P crosses each s_i, $i = 1, 2, \ldots, 6$, exactly once;

(ii) P does not intersect itself;

(iii) P does not pass through any vertex V_1, V_2, V_3, V_4. (page 227)

4. A coast artillery gun can fire at any angle of elevation between $0°$ and $90°$ in a fixed vertical plane. If air resistance is neglected and the muzzle velocity is constant ($= v_0$), determine the set H of points in the plane and above the horizontal which can be hit. (page 229)

5. a_1, b_1, c_1 are positive numbers whose sum is 1, and for $n = 1, 2, \ldots$ we define

$$a_{n+1} = a_n^2 + 2b_nc_n, \quad b_{n+1} = b_n^2 + 2c_na_n, \quad c_{n+1} = c_n^2 + 2a_nb_n.$$

Show that a_n, b_n, c_n approach limits as $n \to \infty$ and find these limits.

(page 230)

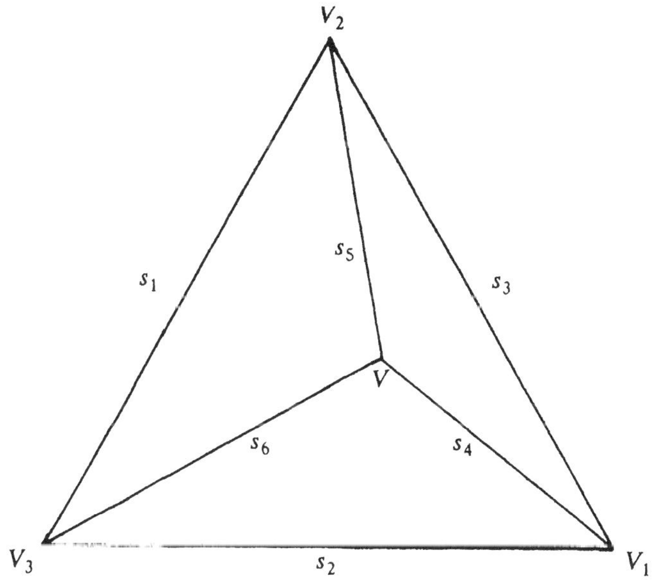

Note. There is an unimportant error in the problem as it was presented to the contestants and as shown above: The "V" of the figure corresponds to the "V_4" of the text.

6. A three-by-three matrix has determinant zero, and has the further property that the cofactor of any element is equal to the square of that element. (The cofactor of a_{ij} is $(-1)^{i+j}$ multiplied by the determinant obtained by striking out the ith row and jth column.) Show that every element in the matrix is zero. (page 234)

Afternoon Session

7. Let $f(x)$ be a function such that $f(1) = 1$ and for $x \geq 1$

$$f'(x) = \frac{1}{x^2 + f^2(x)}.$$

Prove that

$$\lim_{x \to \infty} f(x)$$

exists and is less than $1 + \pi/4$. (page 235)

8. Let $f(x)$ be a differentiable function defined in the closed interval $(0, 1)$ and such that

$$|f'(x)| \leq M, \qquad 0 < x < 1.$$

Prove that

$$\left| \int_0^1 f(x)\,dx - \frac{1}{n}\sum_{k=1}^{n} f\left(\frac{k}{n}\right) \right| \le \frac{M}{n}.$$ (page 236)

9. Let x, y be Cartesian coordinates in the plane. I denotes the line segment $1 \le x \le 3$, $y = 1$. For every point P on I, let P^* denote that point that lies on the segment joining the origin to P and such that the distance PP^* is equal to $1/100$. As P describes I, the corresponding point P^* describes a certain curve C^*. Let $l(I)$, $l(C^*)$ be the lengths of I and C^* respectively. Which one of $l(I)$, $l(C^*)$ is greater? Prove your answer. (page 238)

10. Given $P(z) = z^2 + az + b$, a quadratic polynomial of the complex variable z with complex coefficients a, b. Suppose that $|P(z)| = 1$ for every z such that $|z| = 1$. Prove that $a = b = 0$. (page 238)

11. a, b, c, d are distinct integers such that

$$(x - a)(x - b)(x - c)(x - d) - 4 = 0$$

has an integral root r. Show that $4r = a + b + c + d$. (page 240)

12. C is a fixed point on OZ and U, V are variable points on OX, OY respectively, where OX, OY, OZ are mutually orthogonal lines. Find the locus of a point P such that PU, PV, PC are mutually orthogonal.

(page 241)

THE EIGHTH WILLIAM LOWELL PUTNAM MATHEMATICAL COMPETITION

March 20, 1948

Morning Session

1. What is the maximum of $|z^3 - z + 2|$, where z is a complex number with $|z| = 1$? (page 243)

2. Two spheres in contact have a common tangent cone. These three surfaces divide the space into various parts, only one of which is bounded by all three surfaces; it is "ring-shaped." Being given the radii of the spheres, r and R, find the volume of the "ring-shaped" part. (The desired expression is a rational function of r and R.) (page 243)

3. Let $\{a_n\}$ be a decreasing sequence of positive numbers with limit 0 such that

$$b_n = a_n - 2a_{n+1} + a_{n+2} \geq 0$$

for all n. Prove that

$$\sum_{n=1}^{\infty} nb_n = a_1.$$

(page 247)

4. Let D be a plane region bounded by a circle of radius r. Let (x, y) be a point of D and consider a circle of radius δ and center at (x, y). Denote by $l(x, y)$ the length of that arc of the circle which is outside D. Find

$$\lim_{d \to 0} \frac{1}{\delta^2} \int\int_D l(x, y)\, dx\, dy.$$

(page 248)

Note. This is the text of the problem as published in the *American Mathematical Monthly*, vol. 55 (1948) p. 632. Clearly the "d" under the limit sign should be a "δ." We do not know how the problem was printed on the actual examination.

5. If $x_1, \ldots, x_n$ denote the nth roots of unity, evaluate

$$\pi(x_i - x_j)^2 \qquad (i < j).$$

(page 249)

Note. This is the text of the problem as published in the *American Mathematical Monthly*, vol. 55 (1948) p. 632. The "π" should be a product sign. Again, we do not know how the problem was printed on the actual examination.

6. Answer either (i) or (ii):

(i) A force acts on the element ds of a closed plane curve. The magnitude of this force is $r^{-1}\,ds$ where r is the radius of curvature at the point considered, and the direction of the force is perpendicular to the curve; it points to the convex side. Show that the system of such forces acting on all elements of the curve keep it in equilibrium. (page 252)

(ii) Show that

$$x + \frac{2}{3}x^3 + \frac{2}{3}\frac{4}{5}x^5 + \frac{2}{3}\frac{4}{5}\frac{6}{7}x^7 + \cdots = \frac{\arcsin x}{\sqrt{1-x^2}}.$$

(page 254)

Afternoon Session

1. Let $f(x)$ be a cubic polynomial with roots x_1, x_2, and x_3. Assume that $f(2x)$ is divisible by $f'(x)$ and compute the ratios $x_1 : x_2 : x_3$. (page 255)

2. "A penny in a corner." A circle moves so that it is continually in contact with all three coordinate planes of an ordinary rectangular system. Find the locus of the center of the circle. (page 256)

3. If n is a positive integer, prove that

$$[\sqrt{n} + \sqrt{n+1}] = [\sqrt{4n+2}\,],$$

where $[x]$ denotes as usual the greatest integer not exceeding x. (page 257)

4. Let $\min(x, y)$ denote the smaller of the numbers x and y. For what λ's does the equation

$$\int_0^1 \min(x, y) f(y)\, dy = \lambda f(x)$$

have continuous solutions which do not vanish identically in $(0, 1)$? What are these solutions? (page 258)

5. The pairs of numbers (a, b) such that $|a + bt + t^2| \le 1$ for $0 \le t \le 1$ fill a certain region in the (a, b)-plane. What is the area of this region? (page 259)

6. Answer either (i) or (ii):

(i) Let V_1, V_2, V_3, and V denote four vertices of a cube. V_1, V_2, and V_3 are next neighbors of V, that is, the lines VV_1, VV_2, and VV_3 are edges of the cube. Project the cube orthogonally onto a plane (the z-plane, the Gaussian plane) of which the points are marked with complex numbers. Let the projection of V fall in the origin and the projections of V_1, V_2, and V_3 in points marked with the complex numbers z_1, z_2, and z_3, respectively. Show that $z_1^2 + z_2^2 + z_3^2 = 0$. (page 261)

(ii) Let a_{ij} be a determinant in which each diagonal element exceeds in absolute value the sum of the absolute values of the other elements of its row, that is

$$|a_{ii}| > |a_{i1}| + |a_{i2}| + \cdots + |a_{i,i-1}| + |a_{i,i+1}| + \cdots + |a_{in}|.$$

Show that the determinant is not equal to zero. (Consider the corresponding system of linear homogeneous equations.) (page 263)

THE NINTH WILLIAM LOWELL PUTNAM MATHEMATICAL COMPETITION

March 26, 1949

Morning Session

1. Answer either (i) or (ii):

(i) Three straight lines pass through the three points $(0, -a, a)$, $(a, 0, -a)$, and $(-a, a, 0)$, parallel to the x-axis, y-axis, and z-axis, respectively; $a > 0$. A variable straight line moves so that it has one point in common with each of the three given straight lines. Find the equation of the surface described by the variable line. (page 264)

(ii) Which planes cut the surface $xy + xz + yz = 0$ in (1) circles, (2) parabolas? (page 266)

2. We consider three vectors drawn from the same initial point O, of lengths a, b, and c, respectively. Let E be the parallelepiped with vertex O of which the given vectors are the edges and H the parallelepiped with vertex O of which these vectors are the altitudes. Show that the product of the volumes of E and H equals $(abc)^2$, and generalize the theorem, with proof, to n dimensions. (page 268)

3. Assume that the complex numbers $a_1, a_2, \ldots, a_n, \ldots$ are all different from zero, and that $|a_r - a_s| > 1$ for $r \neq s$. Show that the series

$$\sum_{n=1}^{\infty} \frac{1}{a_n{}^3}$$

converges. (page 269)

4. Given that P is a point inside a tetrahedron with vertices at A, B, C, and D, such that the sum of the distances $PA + PB + PC + PD$ is a minimum, show that the two angles $\angle APB$ and $\angle CPD$ are equal and are bisected by the same straight line. What other pairs of angles must be equal? (page 270)

5. How many roots of the equation $z^6 + 6z + 10 = 0$ lie in each quadrant of the complex plane? (page 272)

6. Prove that for every real or complex x

$$\prod_{k=1}^{\infty} \frac{1 + 2\cos\frac{2x}{3^k}}{3} = \frac{\sin x}{x}.$$ (page 274)

Afternoon Session

1. Each rational number p/q (p, q relatively prime positive integers) of the open interval (0, 1) is covered by a closed interval of length $1/2q^2$, whose center is at p/q. Prove that $\sqrt{2}/2$ is not covered by any of the above closed intervals. (page 275)

2. Answer either (i) or (ii):

(i) Prove that

$$\sum_{n=2}^{\infty} \frac{\cos(\log\log n)}{\log n}$$

diverges. (page 277)

(ii) Assume that $p > 0$, $a > 0$, and $ac - b^2 > 0$, and show that

$$\int_{-\infty}^{\infty}\int_{-\infty}^{\infty} \frac{dx\,dy}{(p + ax^2 + 2bxy + cy^2)^2} = \pi p^{-1}(ac - b^2)^{-1/2}.$$

(page 278)

3. Let K be a closed plane curve such that the distance between any two points of K is always less than 1. Show that K lies inside a circle of radius $1/\sqrt{3}$. (page 279)

4. Show that the coefficients $a_1, a_2, a_3, \ldots$ in the expansion $\frac{1}{4}[1 + x - (1 - 6x + x^2)^{1/2}] = a_1x + a_2x^2 + a_3x^3 + \cdots$ are positive integers. (page 282)

5. Let $a_1, a_2, \ldots, a_n, \ldots$ be an arbitrary sequence of positive numbers. Show that

$$\limsup_{n\to\infty} \left(\frac{a_1 + a_{n+1}}{a_n}\right)^n \geq e.$$ (page 283)

6. Let C be a closed convex curve with a continuously turning tangent and let O be a point inside C. With each point P on C we associate the point $T(P)$ on C which is defined as follows: Draw the tangent to C at P and from O drop the perpendicular to that tangent. $T(P)$ is then the point at which this perpendicular intersects the curve C.

Starting now with a point P_0 on C, we define points P_n by the formula $P_n = T(P_{n-1})$, $n \geq 1$. Prove that the points P_n approach a limit, and characterize those points which can be limits of sequences P_n. (You may consider the facts that T is a continuous transformation and that a convex curve lies on one side of each of its tangents as not requiring proofs.) (page 285)

THE TENTH WILLIAM LOWELL PUTNAM MATHEMATICAL COMPETITION

March 25, 1950

Morning Session

1. For what values of the ratio a/b is the limaçon $r = a - b\cos\theta$ a convex curve? $(a > b > 0)$ (page 288)

2. Answer both (i) and (ii). Test for convergence the series

$$\text{(i)}\quad \frac{1}{\log(2!)} + \frac{1}{\log(3!)} + \frac{1}{\log(4!)} + \cdots + \frac{1}{\log(n!)} + \cdots$$

$$\text{(ii)}\quad \frac{1}{3} + \frac{1}{3\sqrt{3}} + \frac{1}{3\sqrt{3}\sqrt[3]{3}} + \cdots + \frac{1}{3\sqrt{3}\sqrt[3]{3}\cdots\sqrt[n]{3}} + \cdots.$$

(page 290)

3. The sequence $x_0, x_1, x_2, \ldots$ is defined by the conditions

$$x_0 = a,\ x_1 = b,\ x_{n+1} = \frac{x_{n-1} + (2n-1)x_n}{2n} \qquad \text{for } n \geq 1,$$

where a and b are given numbers. Express $\lim_{n\to\infty} x_n$ concisely in terms of a and b. (page 291)

4. Answer either (i) or (ii).

(i) In a right prism with triangular base, given the sum of the areas of three mutually adjacent faces (that is, of two lateral faces and one base), show that these faces are of equal area and perpendicular to each other when the volume attains its maximum. (page 292)

(ii) Show that

$$\frac{\dfrac{x}{1} + \dfrac{x^3}{1\cdot 3} + \dfrac{x^5}{1\cdot 3\cdot 5} + \dfrac{x^7}{1\cdot 3\cdot 5\cdot 7} + \cdots}{1 + \dfrac{x^2}{2} + \dfrac{x^4}{2\cdot 4} + \dfrac{x^6}{2\cdot 4\cdot 6} + \cdots} = \int_0^x e^{-t^2/2}\,dt.$$

(page 293)

5. A function $D(n)$ of the positive integral variable n is defined by the

following properties: $D(1) = 0$, $D(p) = 1$ if p is a prime, $D(uv) = uD(v) + vD(u)$ for any two positive integers u and v. Answer all three parts below.

(i) Show that these properties are compatible and determine uniquely $D(n)$. (Derive a formula for $D(n)/n$, assuming that $n = p_1^{\alpha_1} p_2^{\alpha_2} \cdots p_k^{\alpha_k}$ where $p_1, p_2, \ldots, p_k$ are different primes.)

(ii) For what values of n is $D(n) = n$?

(iii) Define $D^2(n) = D[D(n)]$, etc., and find the limit of $D^m(63)$ as m tends to ∞. (page 294)

6. Each coefficient a_n of the power series

$$a_0 + a_1x + a_2x^2 + a_3x^3 + \cdots = f(x)$$

has either the value 1 or the value 0. Prove the easier of the two assertions:

(i) If $f(0.5)$ is a rational number, $f(x)$ is a rational function.

(ii) If $f(0.5)$ is not a rational number, $f(x)$ is not a rational function. (page 296)

Afternoon Session

1. In each of n houses on a straight street are one or more boys. At what point should all the boys meet so that the sum of the distances that they walk is as small as possible? (page 299)

2. Two obvious approximations to the length of the perimeter of the ellipse with semi-axes a and b are $\pi(a + b)$ and $2\pi(ab)^{1/2}$. Which one comes nearer the truth when the ratio b/a is very close to 1? (page 300)

3. In the Gregorian calendar:

(i) years not divisible by 4 are common years;
(ii) years divisible by 4 but not by 100 are leap years;
(iii) years divisible by 100 but not by 400 are common years;
(iv) years divisible by 400 are leap years;
(v) a leap year contains 366 days; a common year 365 days.

Prove that the probability that Christmas falls on a Wednesday is not 1/7. (page 302)

4. The cross-section of a right cylinder is an ellipse, with semi-axes a and b, where $a > b$. The cylinder is very long, made of very light homogeneous material. The cylinder rests on the horizontal ground which it touches along the straight line joining the lower endpoints of the minor axes of its several cross-sections. Along the upper endpoints of these minor axes lies a very heavy homogeneous wire, straight and just as long as the cylinder. The wire

and the cylinder are rigidly connected. We neglect the weight of the cylinder, the breadth of the wire, and the friction of the ground.

The system described is in equilibrium, because of its symmetry. This equilibrium seems to be stable when the ratio b/a is very small, but unstable when this ratio comes close to 1. Examine this assertion and find the value of the ratio b/a which separates the cases of stable and unstable equilibrium.

(page 303)

5. Answer either (i) or (ii).

(i) Given that the sequence whose nth term is $(s_n + 2s_{n+1})$ converges, show that the sequence $\{s_n\}$ converges also. (page 306)

(ii) A plane varies so that it includes a cone of constant volume equal to $\pi a^3/3$ with the surface the equation of which in rectangular coordinates is $2xy = z^2$. Find the equation of the envelope of the various positions of this plane.

State the result so that it applies to a general cone (that is, conic surface) of the second order. (page 307)

6. Consider the closed plane curves C_i and C_o, their respective lengths $|C_i|$ and $|C_o|$, the closed surfaces S_i and S_o, and their respective areas $|S_i|$ and $|S_o|$. Assume that C_i lies inside C_o and S_i inside S_o. (Subscript i stands for "inner," o for "outer.") Prove the correct assertions among the following four, and disprove the others.

(i) If C_i is convex, $|C_i| \leq |C_o|$.
(ii) If S_i is convex, $|S_i| \leq |S_o|$.
(iii) If C_o is the smallest convex curve containing C_i, then $|C_o| \leq |C_i|$.
(iv) If S_o is the smallest convex surface containing S_i, then $|S_o| \leq |S_i|$.

You may assume that C_i and C_o are polygons and S_i and S_o polyhedra. (Why?) (page 311)

THE ELEVENTH WILLIAM LOWELL PUTNAM MATHEMATICAL COMPETITION

March 31, 1951

Morning Session

1. Show that the determinant:

$$\begin{vmatrix} 0 & a & b & c \\ -a & 0 & d & e \\ -b & -d & 0 & f \\ -c & -e & -f & 0 \end{vmatrix}$$

is non-negative, if its elements a, b, c, etc., are real. (page 317)

2. In the plane, what is the locus of points the sum of the squares of whose distances from n fixed points is a constant? What restrictions, stated in geometric terms, must be put on the constant so that the locus is non-null? (page 319)

3. Find the sum to infinity of the series:

$$1 - \frac{1}{4} + \frac{1}{7} - \frac{1}{10} + \cdots + \frac{(-1)^{n+1}}{3n - 2} + \cdots.$$ (page 319)

4. Trace the curve whose equation is:

$$y^4 - x^4 - 96y^2 + 100x^2 = 0.$$ (page 320)

5. Consider in the plane the network of points having integral coordinates. For lines having rational slope show that:

(i) the line passes through no points of the network or through infinitely many;

(ii) there exists for each line a positive number d having the property that no point of the network, except such as may be on the line, is closer to the line than the distance d. (page 323)

6. Determine the position of a normal chord of a parabola such that it cuts off of the parabola a segment of minimum area. (page 324)

7. Show that if the series $a_1 + a_2 + a_3 + \cdots + a_n + \cdots$ converges, then the series $a_1 + a_2/2 + a_3/3 + \cdots + a_n/n + \cdots$ converges also. (page 325)

Afternoon Session

1. Find the condition that the functions $M(x, y)$ and $N(x, y)$ must satisfy in order that the differential equation $Mdx + Ndy = 0$ shall have an integrating factor of the form $f(xy)$. You may assume that M and N have continuous partial derivatives of all orders. (page 326)

2. Two functions of x are differentiable and not identically zero. Find an example of two such functions having the property that the derivative of their quotient is the quotient of their derivatives. (page 328)

3. Show that if x is positive, then

$$\log_e(1 + 1/x) > 1/(1 + x).$$

(page 329)

4. Investigate, in any way which yields significant results, the existence, in the plane, of the configuration consisting of an ellipse simultaneously tangent to four distinct concentric circles. (page 330)

5. A plane through the center of a torus is tangent to the torus. Prove that the intersection of the plane and the torus consists of two circles. (page 334)

6. Assuming that all the roots of the cubic equation $x^3 + ax^2 + bx + c = 0$ are real, show that the difference between the greatest and the least roots is not less than $(a^2 - 3b)^{1/2}$ or greater than $2(a^2 - 3b)^{1/2}/3^{1/2}$. (page 337)

7. Find the volume of the four-dimensional hypersphere $x^2 + y^2 + z^2 + t^2 = r^2$, and also the hypervolume of its interior $x^2 + y^2 + z^2 + t^2 < r^2$. (page 337)

THE TWELFTH WILLIAM LOWELL PUTNAM MATHEMATICAL COMPETITION

March 22, 1952

Morning Session

1. Let

$$f(x) = \sum_{i=0}^{i=n} a_i x^{n-i}$$

be a polynomial of degree n with integral coefficients. If a_0, a_n, and $f(1)$ are odd, prove that $f(x) = 0$ has no rational roots. (page 340)

2. Show that the equation

$$(9 - x^2)\left(\frac{dy}{dx}\right)^2 = (9 - y^2)$$

characterizes a family of conics touching the four sides of a fixed square. (page 340)

3. Develop necessary and sufficient conditions which ensure that r_1, r_2, r_3 and r_1^2, r_2^2, r_3^2 are simultaneously roots of the equation $x^3 + ax^2 + bx + c = 0$. (page 345)

4. The flag of the United Nations consists of a polar map of the world, with the North Pole as center, extending approximately to 45° South Latitude. The parallels of latitude are concentric circles with radii proportional to their co-latitudes. Australia is near the periphery of the map and is intersected by the parallel of latitude 30° S. In the very close vicinity of this parallel how much are East and West distances exaggerated as compared to North and South distances? (page 349)

5. Let $a_j (j = 1, 2, \ldots, n)$ be entirely arbitrary numbers except that no one is equal to unity. Prove

$$a_1 + \sum_{i=2}^{n} a_i \prod_{j=1}^{i-1} (1 - a_j) = 1 - \prod_{j=1}^{n} (1 - a_j).$$ (page 350)

6. A man has a rectangular block of wood m by n by r inches (m, n, and r

are integers). He paints the entire surface of the block, cuts the block into inch cubes, and notices that exactly half the cubes are completely unpainted. Prove that the number of essentially different blocks with this property is finite. (Do *not* attempt to enumerate them.) (page 351)

7. Directed lines are drawn from the center of a circle, making angles of 0, ± 1, ± 2, ± 3, ... (measured in radians from a prime direction). If these lines meet the circle in points $P_0, P_1, P_{-1}, P_2, P_{-2}, \ldots$, show that there is no interval on the circumference of the circle which does not contain some $P_{\pm i}$. (You may assume that π is irrational.) (page 353)

Afternoon Session

1. A mathematical moron is given two sides and the included angle of a triangle and attempts to use the Law of Cosines: $a^2 = b^2 + c^2 - 2bc \cos A$, to find the third side a. He uses logarithms as follows. He finds log b and doubles it; adds to that the double of log c; subtracts the sum of the logarithms of 2, b, c, and cos A; divides the result by 2; and takes the anti-logarithm. Although his method may be open to suspicion, his computation is accurate. What are the necessary and sufficient conditions on the triangle that this method should yield the correct result? (page 353)

2. Find the surface generated by the solutions of

$$\frac{dx}{yz} = \frac{dy}{zx} = \frac{dz}{xy},$$

which intersects the circle $y^2 + z^2 = 1, x = 0$. (page 354)

3. Develop necessary and sufficient conditions that the equation

$$\begin{vmatrix} 0 & a_1 - x & a_2 - x \\ -a_1 - x & 0 & a_3 - x \\ -a_2 - x & -a_3 - x & 0 \end{vmatrix} = 0 \qquad (a_i \neq 0)$$

shall have a multiple root. (page 356)

4. A homogeneous solid body is made by joining a base of a circular cylinder of height h and radius r, and the base of a hemisphere of radius r. This body is placed with the hemispherical end on a horizontal table, with the axis of the cylinder in a vertical position, and then slightly oscillated. It is intuitively evident that if r is large as compared to h, the equilibrium will be stable; but if r is small as compared to h, the equilibrium will be unstable. What is the critical value of the ratio r/h which enables the body to rest in neutral equilibrium in any position? (page 357)

5. If the terms of a sequence, a_n, are monotonic, and if $\Sigma_1^\infty a_n$ converges, show that $\Sigma_1^\infty n(a_n - a_{n+1})$ converges. (page 359)

6. Prove the necessary and sufficient condition that a triangle inscribed in an ellipse shall have maximum area is that its centroid coincide with the center of the ellipse. (page 360)

7. Given any real number N_0, if $N_{j+1} = \cos N_j$, prove that $\lim_{j\to\infty} N_j$ exists and is independent of N_0. (page 361)

THE THIRTEENTH WILLIAM LOWELL PUTNAM MATHEMATICAL COMPETITION

March 23, 1953

Morning Session

1. Prove that, for every positive integer n,

$$\sqrt{1} + \sqrt{2} + \cdots + \sqrt{n}$$

is more than $\frac{2}{3} n\sqrt{n}$ and less than

$$\frac{4n + 3}{6}\sqrt{n}.$$ (page 364)

2. Six points are in general position in space (no three in a line, no four in a plane). The fifteen line segments joining them in pairs are drawn and then painted, some segments red, some blue. Prove that some triangle has all its sides the same color. (page 365)

3. If x_1, x_2, x_3 are real numbers and the sum of any two is greater than the third, show that

$$\frac{2}{3}\sum_{i=1}^{3} x_i \sum_{i=1}^{3} x_i^2 > \sum_{i=1}^{3} x_i^3 + x_1x_2x_3.$$ (page 366)

4. From the identity

$$\int_0^{\pi/2} \log \sin 2x \, dx = \int_0^{\pi/2} \log \sin x \, dx + \int_0^{\pi/2} \log \cos x \, dx + \int_0^{\pi/2} \log 2 \, dx,$$

deduce the value of

$$\int_0^{\pi/2} \log \sin x \, dx.$$ (page 368)

5. Let P be a point from which three distinct normals can be drawn to a parabola. Show that the sum of the angles which these three normals make

with the axis exceeds by a multiple of π the angle which the line joining P to the focus makes with the axis. (page 369)

6. Show that the sequence

$$\sqrt{7},\ \sqrt{7-\sqrt{7}},\ \sqrt{7-\sqrt{7+\sqrt{7}}},\ \sqrt{7-\sqrt{7+\sqrt{7-\sqrt{7}}}},\ \ldots$$

converges, and evaluate the limit. (page 370)

7. Assuming that the roots of $x^3 + px^2 + qx + r = 0$ are all real and positive, find the relation between p, q, and r which is a necessary and sufficient condition that the roots may be the cosines of the angles of a triangle. (page 372)

Afternoon Session

1. Is the infinite series

$$\sum_{n=1}^{\infty} \frac{1}{n^{(n+1)/n}}$$

convergent? Prove your statement. (page 374)

2. Let $a_0, a_1, \ldots, a_n$ be real numbers and let $f(x) = a_0 + a_1 x + \cdots + a_n x^n$. Suppose that, for every integer i, $f(i)$ is an integer. Prove that $n! \cdot a_k$ is an integer for each k. (page 374)

3. Solve the equations

$$\frac{dy}{dx} = z(y+z)^n \qquad \frac{dz}{dx} = y(y+z)^n,$$

given the initial conditions $y = 1$ and $z = 0$ when $x = 0$. (page 376)

4. Determine the equation of a surface in three dimensional cartesian space which has the following properties: (a) it passes through the point (1, 1, 1); and (b) if the tangent plane be drawn at any point P, and A, B and C are the intersections of this plane with the x, y and z axes respectively, then P is the orthocenter (intersection of the altitudes) of the triangle ABC. (page 377)

5. Show that the roots of $x^4 + ax^3 + bx^2 + cx + d = 0$, if suitably numbered, satisfy the relation $r_1/r_2 = r_3/r_4$, provided $a^2 d = c^2 \neq 0$. (page 378)

6. P and Q are any points inside a circle (C) with center C, such that $CP = CQ$. Determine the location of a point Z on (C) such that $PZ = QZ$ shall be a minimum. (page 379)

Note. This is the text of the problem as it was presented to the contestants. The equals sign in the last sentence is, of course, a misprint. It should read: "Determine the location of a point Z on (C) such that $PZ + QZ$ shall be a minimum."

7. Let w be an irrational number with $0 < w < 1$. Prove that w has a unique convergent expansion of the form

$$w = \frac{1}{p_0} - \frac{1}{p_0 p_1} + \frac{1}{p_0 p_1 p_2} - \frac{1}{p_0 p_1 p_2 p_3} + \cdots,$$

where $p_0, p_1, p_2, \ldots$ are integers and $1 \le p_0 < p_1 < p_2 < \cdots$. If $w = \frac{1}{2}\sqrt{2}$, find p_0, p_1, p_2. (page 384)

THE FOURTEENTH WILLIAM LOWELL PUTNAM MATHEMATICAL COMPETITION

March 6, 1954

Morning Session

1. Let n be an odd integer greater than 1. Let A be an n by n symmetric matrix such that each row and each column of A consists of some permutation of the integers $1, \ldots, n$. Show that each one of the integers $1, \ldots, n$ must appear in the main diagonal of A. (page 387)

2. Consider any five points P_1, P_2, P_3, P_4, P_5 in the interior of a square S of side-length 1. Denote by d_{ij} the distance between the points P_i and P_j. Prove that at least one of the distances d_{ij} is less than $\sqrt{2}/2$. Can $\sqrt{2}/2$ be replaced by a smaller number in this statement? (page 387)

3. Prove that if the family of integral curves of the differential equation

$$\frac{dy}{dx} + p(x)y = q(x) \qquad p(x) \cdot q(x) \neq 0$$

is cut by the line $x = k$, the tangents at the points of intersection are concurrent. (page 388)

4. A uniform rod of length $2k$ and weight w rests with the end A against a smooth vertical wall, while to the lower end B is fastened a string BC of length $2b$ coming from a point C in the wall directly above A. If the system is in equilibrium, determine the angle ABC. (page 389)

5. If $f(x)$ is a real-valued function defined for $0 < x < 1$, then the formula $f(x) = o(x)$ is an abbreviation for the statement that

$$\frac{f(x)}{x} \to 0 \qquad \text{as } x \to 0.$$

Keeping this in mind, prove the following: if

$$\lim_{x \to 0} f(x) = 0 \quad \text{and} \quad f(x) - f\left(\frac{x}{2}\right) = o(x),$$

then $f(x) = o(x)$. (page 390)

6. Suppose that $u_0, u_1, u_2, \ldots$ is a sequence of real numbers such that

$$u_n = \sum_{k=1}^{\infty} u^2_{n+k} \qquad \text{for } n = 0, 1, 2, \ldots .$$

Prove that if Σu_n converges then $u_k = 0$ for all k. (page 391)

7. Prove that there are no integers x and y for which

$$x^2 + 3xy - 2y^2 = 122.$$ (page 392)

Afternoon Session

1. Show that the equation $x^2 - y^2 = a^3$ has always integral solutions for x and y whenever a is a positive integer. (page 392)

2. Assume as known the (true) fact that the alternating harmonic series

$$(1) \qquad 1 - 1/2 + 1/3 - 1/4 + 1/5 - 1/6 + 1/7 - 1/8 + \cdots$$

is convergent, and denote its sum by s. Rearrange the series (1) as follows:

(2)

$$1 + 1/3 - 1/2 + 1/5 + 1/7 - 1/4 + 1/9 + 1/11 - 1/6 + \cdots.$$

Assume as known the (true) fact that the series (2) is also convergent, and denote its sum by S. Denote by s_k, S_k the kth partial sum of the series (1) and (2) respectively. Prove the following statements.

(i) $S_{3n} = s_{4n} + \frac{1}{2} s_{2n}$,

(ii) $S \neq s$. (page 393)

3. Let a and b denote real numbers such that $a < b$. The symbol (a, b) will denote the closed interval with the end points a, b. Let there be given a collection of closed intervals $(a_1, b_1), \ldots, (a_n, b_n)$ such that any two of these closed intervals have at least one point in common. Prove that there exists then a point which is contained in every one of these intervals. (page 394)

4. Given the focus f and the directrix D of a parabola P and a line L, describe (with proof) a Euclidean (i.e., ruler and compass) construction of the point or points of intersection of L and P. Be sure to identify the case for which there are no points of intersection. (page 395)

5. Let $f(x)$ be a real-valued function, defined for $-1 < x < 1$, such that $f'(0)$ exists. Let a_n, b_n be two sequences such that

$$-1 < a_n < 0 < b_n < 1, \quad \lim_{n \to \infty} a_n = 0, \quad \lim_{n \to \infty} b_n = 0.$$

Prove that

$$\lim_{n \to \infty} \frac{f(b_n) - f(a_n)}{b_n - a_n} = f'(0).$$ (page 397)

6. Prove that every positive rational number is the sum of a finite number of distinct terms of the series

$$1 + \frac{1}{2} + \frac{1}{3} + \cdots + \frac{1}{n} + \cdots.$$ (page 398)

7. Show that

$$\lim_{n \to \infty} \sum_{s=1}^{n} \left(\frac{a + s}{n}\right)^n \qquad (a > 0)$$

lies between e^a and e^{a+1}. (page 399)

THE FIFTEENTH WILLIAM LOWELL PUTNAM MATHEMATICAL COMPETITION

March 5, 1955

Morning Session

1. Prove that there is no set of integers m, n, p except 0, 0, 0 for which $m + n\sqrt{2} + p\sqrt{3} = 0$. (page 402)

2. $A_1A_2 \ldots A_n$ is a regular polygon inscribed in a circle of radius r and center O. P is a point on line OA_1 extended beyond A_1. Show that

$$\prod_{i=1}^{n} \overline{PA_i} = \overline{OP}^{\,n} - r^n.$$ (page 403)

3. Suppose that $\sum_{i=1}^{\infty} x_i$ is a convergent series of positive terms which monotonically decrease (that is, $x_1 \geq x_2 \geq x_3 \geq \cdots$). Let P denote the set of all numbers which are sums of some (finite or infinite) subseries of $\sum_{i=1}^{\infty} x_i$. Show that P is an interval if and only if

$$x_n \leq \sum_{i=n+1}^{\infty} x_i \qquad \text{for every integer } n.$$ (page 403)

4. On a circle, n points are selected and the chords joining them in pairs are drawn. Assuming that no three of these chords are concurrent (except at the endpoints), how many points of intersection are there? (page 405)

5. If a parabola is given in the plane, find a geometric construction (ruler and compass) for the focus. (page 405)

6. Find a necessary and sufficient condition on the positive integer n that the equation

$$x^n + (2 + x)^n + (2 - x)^n = 0$$

have a rational root. (page 406)

7. Consider the function f defined by the differential equation

$$f''(x) = (x^3 + ax)f(x)$$

and the initial conditions $f(0) = 1$, $f'(0) = 0$. Prove that the roots of f are bounded above but unbounded below. (page 407)

Afternoon Session

1. A sphere rolls along two intersecting straight lines. Find the locus of its center. (page 408)

2. Suppose that f is a function with two continuous derivatives and $f(0) = 0$. Prove that the function g, defined by $g(0) = f'(0)$, $g(x) = f(x)/x$ for $x \neq 0$, has a continuous derivative. (page 410)

3. Prove that there exists no distance-preserving map of a spherical cap into the plane. (Distances on the sphere are to be measured along great circles on the surface.) (page 411)

4. Do there exist 1,000,000 consecutive integers each of which contains a repeated prime factor? (page 412)

5. Given an infinite sequence of 0's and 1's and a fixed integer k, suppose that there are no more than k distinct blocks of k consecutive terms. Show that the sequence is eventually periodic. (For example, the sequence 11011010101 followed by alternating 0's and 1's indefinitely, which is periodic beginning with the fifth term.) (page 413)

6. Prove: If $f(x) > 0$ for all x and $f(x) \to 0$ as $x \to \infty$, then there exists at most a finite number of solutions of

$$f(m) + f(n) + f(p) = 1$$

in positive integers m, n, and p. (page 414)

7. Four forces acting on a body are in equilibrium. Prove that, if their lines of action are mutually skew, they are rulings of a hyperboloid. (page 415)

THE SIXTEENTH WILLIAM LOWELL PUTNAM MATHEMATICAL COMPETITION

March 3, 1956

Morning Session

1. Evaluate

$$\lim_{x \to \infty} \left[\frac{1}{x} \frac{a^x - 1}{a - 1} \right]^{1/x}$$

where $a > 0$, $a \neq 1$. (page 418)

2. Prove that every positive integer has a multiple whose decimal representation involves all ten digits. (page 419)

3. A particle falls in a vertical plane from rest under the influence of gravity and a force perpendicular to and proportional to its velocity. Obtain the equations of the trajectory and identify the curve. (page 420)

4. Suppose the n times differentiable real function $f(x)$ has at least $n + 1$ distinct zeros in the closed interval $[a, b]$ and that the polynomial $P(z) \equiv z^n + C_{n-1}z^{n-1} + \cdots + C_0$ has only real zeros. Show that $(D^n + C_{n-1}D^{n-1} + \cdots + C_0)f(x)$ has at least one zero in the interval $[a, b]$ where D^n denotes, as usual, d^n/dx^n. (page 421)

5. Given n objects arranged in a row. A subset of these objects is called unfriendly if no two of its elements are consecutive. Show that the number of unfriendly subsets each having k elements is

$$\binom{n - k + 1}{k}.$$

(page 422)

6. (i) A transformation of the plane into itself preserves all rational distances. Prove that it preserves all distances.

(ii) Show that the corresponding theorem for the line is false.

(page 423)

7. Prove that the number of odd binomial coefficients in any finite binomial expansion is a power of 2. (page 424)

Afternoon Session

1. Show that if the differential equation

$$M(x, y)\, dx + N(x, y)\, dy = 0$$

is both homogeneous and exact then the solution $y = f(x)$ satisfies $xM + yN = C$ (constant). (page 425)

2. Suppose that each set X of points in the plane has an associated set $\overline{X}$ of points called its cover. Suppose further that

(1) $\overline{X \cup Y} \supset \overline{\overline{X}} \cup \overline{Y} \cup Y$, where $\cup$ designates point set sum (or union) and $\supset$ denotes set inclusion.

Prove: (i) $\overline{X} \supset X$, (ii) $\overline{\overline{X}} = \overline{X}$, (iii) $X \supset Y$ implies $\overline{X} \supset \overline{Y}$.
Prove conversely that (i), (ii) and (iii) imply (1). (page 426)

3. A sphere is inscribed in a tetrahedron and each point of contact of the sphere with the four faces is joined to the vertices of the face containing the point. Show that the four sets of three angles so formed are identical. (page 427)

4. Prove that if A, B, and C are angles of a triangle measured in radians then $A \cos B + \sin A \cos C > 0$. (page 428)

5. Consider a set of $2n$ points in space, $n > 1$. Suppose they are joined by at least $n^2 + 1$ segments. Show that at least one triangle is formed. Show that for each n it is possible to have $2n$ points joined by n^2 segments without any triangles being formed. (page 429)

6. Given $T_1 = 2$, $T_{n+1} = T_n^2 - T_n + 1$, $n > 0$, Prove:

(i) If $m \neq n$, T_m and T_n have no common factor greater than 1.

(ii) $$\sum_{i=1}^{\infty} \frac{1}{T_i} = 1.$$ (page 430)

7. The polynomials $P(z)$ and $Q(z)$ with complex coefficients have the same set of numbers for their zeros but possibly different multiplicities. The same is true of the polynomials

$$P(z) + 1 \quad \text{and} \quad Q(z) + 1.$$

Prove that $P(z) \equiv Q(z)$. (page 431)

THE SEVENTEENTH WILLIAM LOWELL PUTNAM MATHEMATICAL COMPETITION

March 2, 1957

Morning Session

1. The normals to a surface all intersect a fixed straight line. Show that the surface is a portion of a surface of revolution. (page 433)

2. A uniform wire is bent into a form coinciding with the portion of the curve $y = e^x$, $0 \leq x \leq a$, $a > 1$, and the line segment $a - 1 \leq x \leq a$, $y = e^a$. The wire is then suspended from the point $(a - 1, e^a)$ and a horizontal force F is applied at the point $(0, 1)$ to hold the wire in coincidence with the curve and segment. Assuming the x axis is horizontal, show that the force F is directed to the right. (page 436)

3. A and B are real numbers and k a positive integer. Show that

$$\left|\frac{\cos kB \cos A - \cos kA \cos B}{\cos B - \cos A}\right| < k^2 - 1$$

whenever the left side is defined. (page 438)

4. $P(z)$ is a complex polynomial whose roots (as points in the Argand plane) can be covered by a closed circular disc of radius R. Show that the roots of $nP(z) - kP'(z)$ can be covered by a closed circular disc of radius $R + |k|$, where n is the degree of $P(z)$, k is any complex number, and $P'(z)$ is the derivative of $P(z)$. (page 439)

5. Given n points in the plane, show that the largest distance determined by these points cannot occur more than n times. (page 440)

6. $S_1 = \ln a$ and $S_n = \sum_{i=1}^{n-1} \ln (a - S_i)$, $n > 1$.

Show that

$$\lim_{n \to \infty} S_n = a - 1.$$

(page 441)

7. Each member of a set of circles in the xy plane is tangent to the x axis and no two of the circles intersect. Show that:

(i) the points of tangency can include all the rational points on the axis, but

(ii) the points of tangency cannot include all the irrational points.

(page 443)

Afternoon Session

1. Consider the determinant $|a_{ij}|$ of order 100 with $a_{ij} = i \times j$. Prove that if the absolute value of each of the 100! terms in the expansion of this determinant is divided by 101 then the remainder in each case is 1. (page 444)

2. If facilities for division are not available, it is sometimes convenient in determining the decimal expansion of $1/A$, $A > 0$ to use the iteration $X_{k+1} = X_k(2 - AX_k)$, $k = 0, 1, 2, \ldots$, where X_0 is a selected "starting" value. Find the limitations, if any, on the starting value X_0 in order that the above iteration converges to the desired value $1/A$. (page 444)

3. For $f(x)$ a positive, monotone decreasing function defined in $0 \leq x \leq 1$ prove that

$$\frac{\int_0^1 xf^2(x)\,dx}{\int_0^1 xf(x)\,dx} \leq \frac{\int_0^1 f^2(x)\,dx}{\int_0^1 f(x)\,dx}.$$

(page 446)

4. Let $a(n)$ be the number of representations of the positive integer n as the sums of 1's and 2's taking order into account. For example, since

$$4 = 1 + 1 + 2 = 1 + 2 + 1 = 2 + 1 + 1$$
$$= 2 + 2 = 1 + 1 + 1 + 1,$$

then $a(4) = 5$. Let $b(n)$ be the number of representations of n as the sum of integers greater than 1, again taking order into account and counting the summand n. For example, since $6 = 4 + 2 = 2 + 4 = 3 + 3 = 2 + 2 + 2$, we have $b(6) = 5$. Show that for each n, $a(n) = b(n + 2)$. (page 447)

5. With each subset X of a set is associated a second subset $f(X)$. The association is such that whenever X contains Y then $f(X)$ contains $f(Y)$. Show that for some set A, $f(A) = A$. (page 449)

6. The curve $y = f(x)$ passes through the origin with a slope of 1. It satisfies the differential equation $(x^2 + 9)y'' + (x^2 + 4)y = 0$. Show that it crosses the x axis between

$$x = \frac{3}{2}\pi \quad \text{and} \quad x = \sqrt{\frac{63}{53}}\,\pi.$$

(page 450)

7. Let C be a closed convex planar disc bounded by a regular polygon. Show that for each positive integer n there exists a set of points $S(n)$ in the plane such that each n points of $S(n)$ can be covered by C, but $S(n)$ itself cannot be covered by C. (page 453)

THE EIGHTEENTH WILLIAM LOWELL PUTNAM MATHEMATICAL COMPETITION

February 8, 1958

Morning Session

1. If $a_0, a_1, \ldots, a_n$ are real numbers satisfying

$$\frac{a_0}{1} + \frac{a_1}{2} + \cdots + \frac{a_n}{n+1} = 0,$$

show that the equation $a_0 + a_1 x + a_2 x^2 + \cdots + a_n x^n = 0$ has at least one real root. (page 454)

2. Two uniform solid spheres of equal radii are so placed that one is directly above the other. The bottom sphere is fixed, and the top sphere, initially at rest, rolls off. At what point will contact between the two spheres be "lost"? Assume the coefficient of friction is such that no slipping occurs. (page 454)

3. Real numbers are chosen at random from the interval $(0 \leq x \leq 1)$. If after choosing the nth number the sum of the numbers so chosen first exceeds 1, show that the expected or average value for n is e. (page 457)

4. If $a_1, a_2, \ldots, a_n$ are complex numbers such that

$$|a_1| = |a_2| = \cdots = |a_n| = r \neq 0,$$

and if ${}_nT_s$ denotes the sum of all products of these n numbers taken s at a time, prove that

$$\left|\frac{{}_nT_s}{{}_nT_{n-s}}\right| = r^{2s-n}$$

whenever the denominator of the left-hand side is different from zero. (page 458)

5. Show that the integral equation

$$f(x, y) = 1 + \int_0^x \int_0^y f(u, v)\, du\, dv$$

has at most one solution continuous for $0 \leq x \leq 1, 0 \leq y \leq 1$. (page 459)

6. What is the smallest amount that may be invested at interest rate i, compounded annually, in order that one may withdraw 1 dollar at the end of the first year, 4 dollars at the end of the second year, $\ldots$, n^2 dollars at the end of the nth year, in perpetuity? (page 460)

7. Show that ten equal-sized squares cannot be placed on a plane in such a way that no two have an interior point in common and the first touches each of the others. (page 461)

Afternoon Session

1. (i) Given line segments A, B, C, D, with A the longest, construct a quadrilateral with these sides and with A and B parallel, when possible.

(ii) Given any acute-angled triangle ABC and one altitude AH, select any point D on AH, then draw BD and extend until it intersects AC in E, and draw CD and extend until it intersects AB in F. Prove angle AHE = angle AHF. (page 463)

2. Prove that the product of four consecutive positive integers cannot be a perfect square or cube. (page 466)

3. In a round-robin tournament with n players (each pair of players plays one game) in which there are no draws, the numbers of wins scored by the players are $s_1, s_2, \ldots, s_n$. Prove that a necessary and sufficient condition for the existence of 3 players, A, B, C, such that A beat B, B beat C, and C beat A is

$$s_1^2 + s_2^2 + \cdots + s_n^2 < (n - 1)(n)(2n - 1)/6.$$

(page 467)

4. What is the average straight line distance between two points on a sphere of radius 1? (page 468)

5. Given an infinite number of points in a plane, prove that if all the distances determined between them are integers then the points are all in a straight line. (page 469)

6. A projectile moves in a resisting medium. The resisting force is a function of the velocity and is directed along the velocity vector. The equation $x = f(t)$ gives the horizontal distance in terms of the time t. Show that the vertical distance y is given by

$$y = -gf(t)\int \frac{dt}{f'(t)} + g\int \frac{f(t)}{f'(t)}\,dt + Af(t) + B$$

where A and B are constants and g is the acceleration due to gravity. (page 470)

7. Prove that if $f(x)$ is continuous for $a \le x \le b$ and $\int_a^b x^n f(x)\,dx = 0$ for $n = 0, 1, 2, \ldots$ then $f(x)$ is identically zero on $a \le x \le b$. (page 472)

THE NINETEENTH WILLIAM LOWELL PUTNAM MATHEMATICAL COMPETITION

November 22, 1958

Morning Session

1. Let $f(m, 1) = f(1, n) = 1$ for $m \geq 1$, $n \geq 1$, and let $f(m, n) = f(m - 1, n) + f(m, n - 1) + f(m - 1, n - 1)$ for $m > 1$ and $n > 1$. Also let

$$S(n) = \sum_{a+b=n} f(a, b), \qquad a \geq 1 \text{ and } b \geq 1.$$

Prove that

$$S(n + 2) = S(n) + 2S(n + 1) \qquad \text{for } n \geq 2.$$

(page 476)

2. Let

$$R_1 = 1, \qquad R_{n+1} = 1 + n/R_n, \qquad n \geq 1.$$

Show that for $n \geq 1$,

$$\sqrt{n} \leq R_n \leq \sqrt{n} + 1.$$

(page 477)

3. Under the assumption that the following set of relations has a unique solution for $u(t)$, determine it.

$$\frac{du(t)}{dt} = u(t) + \int_0^1 u(s)\, ds,$$

$$u(0) = 1.$$

(page 478)

4. In assigning dormitory rooms, a college gives preference to pairs of students in this order:

$$AA,\ AB,\ AC,\ BB,\ BC,\ AD,\ CC,\ BD,\ CD,\ DD,$$

in which AA means two seniors, AB means a senior and a junior, etc. Determine numerical values to assign to A, B, C, D so that the set of numbers A

$+ A, A + B, A + C, B + B$, etc., corresponding to the order above will be in descending magnitude. Find the general solution and the solution in least positive integers. (page 480)

5. Show that the number of non-zero terms in the expansion of the nth order determinant having zeros in the main diagonal and ones elsewhere is

$$n!\left[1 - \frac{1}{1!} + \frac{1}{2!} - \frac{1}{3!} + \cdots + \frac{(-1)^n}{n!}\right].$$ (page 481)

6. Let $a(x)$ and $b(x)$ be continuous functions on $0 \le x \le 1$ and let $0 \le a(x) \le a < 1$ on that range. Under what other conditions (if any) is the solution of the equation for u,

$$u = \underset{0 \le x \le 1}{\text{maximum}} \, [b(x) + a(x) \cdot u],$$

given by

$$u = \underset{0 \le x \le 1}{\text{maximum}} \left[\frac{b(x)}{1 - a(x)}\right]?$$ (page 485)

7. Let a and b be relatively prime positive integers, b even. For each positive integer q let $p = p(q)$ be chosen so that

$$\left|\frac{p}{q} - \frac{a}{b}\right|$$

is a minimum. Prove that

$$\lim_{n \to \infty} \sum_{q=1}^{n} \frac{q\left|\frac{p}{q} - \frac{a}{b}\right|}{n} = \frac{1}{4}.$$ (page 486)

Afternoon Session

1. Given

$$b_n = \sum_{k=0}^{n} \binom{n}{k}^{-1}, \qquad n \ge 1,$$

prove that

$$b_n = \frac{n+1}{2n} b_{n-1} + 1, \qquad n \ge 2,$$

and hence, as a corollary,

$$\lim_{n \to \infty} b_n = 2.$$ (page 487)

2. Given a set of $n + 1$ positive integers, none of which exceeds $2n$, show that at least one member of the set must divide another member of the set. (page 489)

3. If a square of unit side be partitioned into two sets, then the diameter (least upper bound of the distances between pairs of points) of one of the sets is not less than $\sqrt{5}/2$. Show also that no larger number will do. (page 489)

4. Let C be a real number, and let f be a function such that

$$\lim_{x \to \infty} f(x) = C, \qquad \lim_{x \to \infty} f'''(x) = 0.$$

Prove that

$$\lim_{x \to \infty} f'(x) = 0 \quad \text{and} \quad \lim_{x \to \infty} f''(x) = 0,$$

where superscripts denote derivatives. (page 490)

5. The lengths of successive segments of a broken line are represented by the successive terms of the harmonic progression $1, 1/2, 1/3, \ldots, 1/n, \ldots$. Each segment makes with the preceding segment a given angle θ. What is the distance and what is the direction of the limiting point (if there is one) from the initial point of the first segment? (page 491)

6. Let a complete oriented graph on n points be given, i.e., a set of n points $1, 2, 3, \ldots, n$, and between any two points i and j a direction, $i \to j$. Show that there exists a permutation of the points, $[a_1, a_2, a_3, \ldots, a_n]$, such that $a_1 \to a_2 \to a_3 \to \cdots \to a_n$. (page 494)

7. Let $a_1, a_2, \ldots, a_n$ be a permutation of the integers $1, 2, \ldots, n$. Call a_i a "big" integer if $a_i > a_j$ for all $j > i$. Find the mean number of "big" integers over all permutations on the first n positive integers. (page 495)

THE TWENTIETH WILLIAM LOWELL PUTNAM MATHEMATICAL COMPETITION

November 21, 1959

Morning Session

1. Let n be a positive integer. Prove that $x^n - (1/x^n)$ is expressible as a polynomial in $x - (1/x)$ with real coefficients if and only if n is odd. (page 497)

2. Prove that if the points in the complex plane corresponding to two distinct complex numbers z_1 and z_2 are two vertices of an equilateral triangle, then the third vertex corresponds to $-\omega z_1 - \omega^2 z_2$, where ω is an imaginary cube root of unity. (page 498)

3. Find all complex-valued functions f of a complex variable such that $f(z) + zf(1 - z) = 1 + z$ for all z. (page 499)

4. If f and g are real-valued functions of one real variable, show that there exist numbers x and y such that $0 \leq x \leq 1$, $0 \leq y \leq 1$, and $|xy - f(x) - g(y)| \geq 1/4$. (page 499)

5. A sparrow, flying horizontally in a straight line, is 50 feet directly below an eagle and 100 feet directly above a hawk. Both hawk and eagle fly directly toward the sparrow, reaching it simultaneously. The hawk flies twice as fast as the sparrow. How far does each bird fly? At what rate does the eagle fly? (page 500)

6. Let m and n be integers greater than 1, and $a_1, \ldots, a_{m+1}$ real numbers. Prove that there exist real n by n matrices $A_1, \ldots, A_m$ such that (i) $\operatorname{Det}(A_j) = a_j$ for $j = 1, \ldots, m$, and (ii) $\operatorname{Det}(A_1 + \cdots + A_m) = a_{m+1}$. (page 503)

7. If f is a real-valued function of one real variable which has a continuous derivative on the closed interval $[a, b]$ and for which there is no $x \in [a, b]$ such that $f(x) = f'(x) = 0$, then prove that there is a function g with continuous first derivative on $[a, b]$ such that $fg' - f'g$ is positive on $[a, b]$. (page 504)

Afternoon Session

1. Let each of m distinct points on the positive part of the X-axis be joined to n distinct points on the positive part of the Y-axis. Obtain a formula for

the number of intersection points of these segments (exclusive of endpoints), assuming that no three of the segments are concurrent. (page 505)

2. Let c be a positive real number. Prove that c can be expressed in infinitely many ways as a sum of infinitely many distinct terms selected from the sequence

$$1/10,\ 1/20,\ \ldots,\ 1/10n,\ \ldots.$$ (page 506)

3. Give an example of a continuous real-valued function f from $[0, 1]$ to $[0, 1]$ which takes on every value in $[0, 1]$ an infinite number of times. (page 507)

4. Given the following matrix of 25 elements

$$\begin{pmatrix} 11 & 17 & 25 & 19 & 16 \\ 24 & 10 & 13 & 15 & 3 \\ 12 & 5 & 14 & 2 & 18 \\ 23 & 4 & 1 & 8 & 22 \\ 6 & 20 & 7 & 21 & 9 \end{pmatrix},$$

choose five of these elements, no two coming from the same row or column, in such a way that the minimum of these five elements is as large as possible. Prove that your answer is correct. (page 510)

5. Find the equation of the smallest sphere which is tangent to both of the lines: (i) $x - t + 1, y - 2t + 4, z = -3t + 5$, and (ii) $x = 4t \quad 12, y = -t + 8, z = t + 17$. (page 511)

6. Prove that, if x and y are positive irrationals such that $1/x + 1/y = 1$, then the sequences $[x], [2x], \ldots, [nx], \ldots$ and $[y], [2y], \ldots, [ny], \ldots$ together include every positive integer exactly once. (The notation $[x]$ means the largest integer not exceeding x.) (page 513)

7. For each positive integer n, let f_n be a real-valued symmetric function of n real variables. Suppose that for all n and for all real numbers $x_1, \ldots, x_{n+1}, y$, it is true that

(1) $f_n(x_1 + y, \ldots, x_n + y) = f_n(x_1, \ldots, x_n) + y$,

(2) $f_n(-x_1, \ldots, -x_n) = -f_n(x_1, \ldots x_n)$,

(3) $f_{n+1}(f_n(x_1, \ldots, x_n), \ldots, f_n(x_1, \ldots, x_n), x_{n+1}) = f_{n+1}(x_1, \ldots, x_{n+1})$.

Prove that

(4) $f_n(x_1, \ldots, x_n) = (x_1 + \cdots + x_n)/n$. (page 515)

THE TWENTY-FIRST WILLIAM LOWELL PUTNAM MATHEMATICAL COMPETITION

December 3, 1960

Morning Session

1. Let n be a given positive integer. How many solutions are there in ordered positive integer pairs (x, y) to the equation

$$\frac{xy}{x+y} = n?$$

(page 516)

2. Show that if three points are inside a closed square of unit side, then some pair of them are within $\sqrt{6} - \sqrt{2}$ units apart. (page 516)

3. Show that if t_1, t_2, t_3, t_4, t_5 are real numbers, then

$$\sum_{j=1}^{5} (1 - t_j) \exp\left(\sum_{k=1}^{j} t_k\right) \le e^{e^{e^e}}.$$

(page 518)

4. Given two points in the plane, P and Q, at fixed distances from a line L, and on the same side of the line, as indicated, the problem is to find a third point R so that $PR + RQ + RS$ is a minimum, where RS is perpendicular to L. Consider all cases.

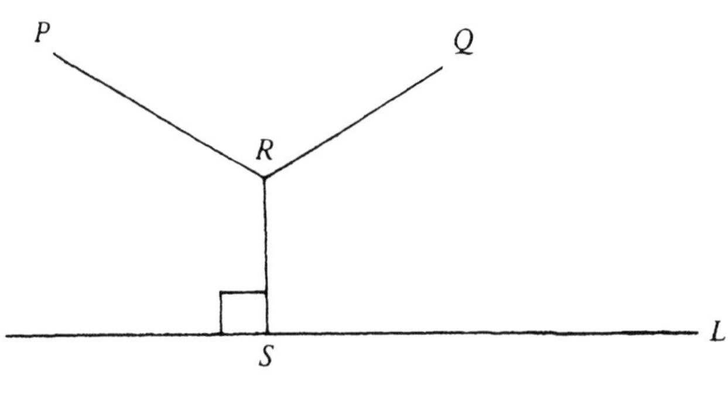

(page 519)

5. Consider a polynomial $f(x)$ with real coefficients having the property $f(g(x)) = g(f(x))$ for every polynomial $g(x)$ with real coefficients. Determine and prove the nature of $f(x)$. (page 521)

6. A player throwing a die scores as many points as on the top face of the die and is to play until his score reaches or passes a total n. Denote by $p(n)$ the probability of making exactly the total n, and find the value of $\lim_{n\to\infty} p(n)$. (page 522)

7. Let $N(n)$ denote the smallest positive integer N such that $x^N = 1$ for every permutation x on n symbols, where 1 denotes the identity permutation. Prove that if $n > 1$,

$$\frac{N(n)}{N(n-1)} = 1 \text{ if } n \text{ is divisible by 2 distinct primes,}$$

$$= p \text{ if } n \text{ is a power of a prime } p.$$ (page 526)

Afternoon Session

1. Find all solutions of $n^m = m^n$ in integers n and m ($x \neq m$). Prove that you have obtained all of them. (page 527)

2. Evaluate the double series

$$\sum_{j=0}^{\infty} \sum_{k=0}^{\infty} 2^{-3k-j-(k+j)^2}.$$ (page 528)

3. The motion of the particles of a fluid in the plane is specified by the following components of velocity

$$(1) \qquad \frac{dx}{dt} = y + 2x(1 - x^2 - y^2),$$

$$\frac{dy}{dt} = -x.$$

Sketch the shape of the trajectories near the origin. Discuss what happens to an individual particle as $t \to +\infty$, and justify your conclusion. (page 529)

Note. The second y in the first equation above was broken and looked like a v on the examination as it was distributed to the contestants.

4. Consider the arithmetic progression $a, a + d, a + 2d, \ldots$, where a and d are positive integers. For any positive integer k, prove that the progression has either no exact kth powers or infinitely many. (page 532)

5. Define a sequence as follows:

$$a_0 = 0$$

$$a_1 = 1 + \sin(-1)$$

$$\vdots$$

$$a_n = 1 + \sin(a_{n-1} - 1)$$

$$\vdots$$

Evaluate

$$\lim_{n\to\infty} \frac{1}{n} \sum_{k=1}^{n} a_k. \qquad \text{(page 534)}$$

6. Any positive integer may be written in the form $n = 2^k(2l + 1)$. Let $a_n = e^{-k}$ and $b_n = a_1 a_2 a_3 \cdots a_n$. Prove that Σb_n converges. (page 535)

7. Let $g(t)$ and $h(t)$ be real, continuous functions for $t \geq 0$. Show that any function $v(t)$ satisfying the differential inequality

$$\frac{dv}{dt} + g(t)v \geq h(t), \qquad v(0) = c,$$

satisfies the further inequality $v(t) \geq u(t)$ where

$$\frac{du}{dt} + g(t)u = h(t), \qquad u(0) = c.$$

From this, conclude that for sufficiently small $t > 0$, the solution of

$$\frac{dv}{dt} + g(t)v = v^2, \qquad v(0) = c_1$$

may be written

$$v = \max_w \left[c_1 e^{-\int_0^t [g(s) - 2w(s)]\,ds} - \int_0^t e^{-\int_0^t [g(s_1) - 2w(s_1)]\,ds_1} w^2(s)\,ds \right]$$

where the maximization is over all continuous functions $w(t)$ defined over some t-interval $[0, t_0]$. (page 536)

THE TWENTY-SECOND WILLIAM LOWELL PUTNAM MATHEMATICAL COMPETITION

December 2, 1961

Morning Session

1. The graph of the equation $x^y = y^x$ in the first quadrant (i.e., the region where $x > 0$ and $y > 0$) consists of a straight line and a curve. Find the coordinates of the intersection point of the line and the curve. (page 538)

2. For a real-valued function $f(x, y)$ of the two positive real variables x and y, define $f(x, y)$ to be *linearly bounded* if and only if there exists a positive number K such that $|f(x, y)| < (x + y)K$ for all positive x and y. Find necessary and sufficient conditions on the real numbers α and β such that $x^\alpha y^\beta$ is linearly bounded. (page 540)

3. Evaluate

$$\lim_{n \to \infty} \sum_{j=1}^{n^2} \frac{n}{n^2 + j^2}.$$

(page 541)

4. Define a function f over the domain of positive integers as follows: $f(1) = 1$, and for $n > 1, f(n) = (-1)^k$ where k is the total number of prime factors of n. For example $f(9) = (-1)^2, f(20) = (-1)^3$. Define $F(n)$ as $\sum f(d)$ where the sum ranges over all positive integer divisors of n. Prove that for every positive integer n, $F(n) = 0$ or $F(n) = 1$. For which integers n is $F(n) = 1$? (page 542)

5. Let Ω be a set of n points, where $n > 2$. Let Σ be a nonempty subcollection of the 2^n subsets of Ω that is closed with respect to unions, intersections, and complements (that is, if A and B are members of Σ, then so are $A \cup B$, $A \cap B$, $\Omega - A$ and $\Omega - B$, where $\Omega - B$ denotes all points in Ω but not in B). If k is the number of members of Σ, what are the possible values of k? Give a proof. (page 543)

6. If $J_2 = \{0, 1\}$ is the field of integers modulo 2, and if $J_2[x]$ is the integral domain of polynomials in one indeterminate with coefficients in J_2, prove that $p(x) = 1 + x + x^2 + \cdots + x^n$ is reducible (factorable) in case $n + 1$ is composite. Is the converse true? That is, if $n + 1$ is prime, is $p(x)$ irreducible? (page 544)

7. Let S be a nonempty closed set in the Euclidean plane for which there is a closed disk D (a circle together with its interior) containing S such that D is a subset of *every* closed disk that contains S. Prove that every point inside D is the midpoint of a segment joining two points of S. (page 545)

Afternoon Session

1. Let $\alpha_1, \alpha_2, \alpha_3, \ldots$ be a sequence of positive real numbers; define s_n as $(\alpha_1 + \alpha_2 + \cdots + \alpha_n)/n$ and r_n as $(\alpha_1^{-1} + \alpha_2^{-1} + \cdots + \alpha_n^{-1})/n$. Given that $\lim s_n$ and $\lim r_n$ exist as $n \to \infty$, prove that the product of these limits is not less than 1. (page 546)

2. Let α and β be given positive real numbers, with $\alpha < \beta$. If two points are selected at random from a straight line segment of length β, what is the probability that the distance between them is at least α? (page 546)

3. Consider four points in a plane, no three of which are collinear, and such that the circle through three of them does not pass through the fourth. Prove that one of the four points can be selected having the property that it lies inside the circle determined by the other three. (page 547)

4. For a fixed positive integer n let $x_1, x_2, \ldots, x_n$ be real numbers satisfying $0 \le x_k \le 1$ for $k = 1, 2, \ldots, n$. Determine the maximum value, as a function of n, of the sum of the $n(n-1)/2$ terms:

$$\sum_{\substack{i,j=1 \\ i<j}}^{n} |x_i - x_j|.$$

(page 549)

5. Let k be a positive integer, and n a positive integer greater than 2. Define

$$f_1(n) = n,\ f_2(n) = n^{f_1(n)},\ \ldots,\ f_{j+1}(n) = n^{f_j(n)},\ \text{etc.}$$

Prove either part of the inequality

$$f_k(n) < n!!!\cdots! < f_{k+1}(n),$$

where the middle term has k factorial symbols. (page 550)

6. Consider the function $y(x)$ satisfying the differential equation $y'' = -(1 + \sqrt{x})y$ with $y(0) = 1$ and $y'(0) = 0$. Prove that $y(x)$ vanishes exactly once on the interval $0 < x < \pi/2$, and find a positive lower bound for the zero. (page 551)

7. Given a sequence $\{a_n\}$ of non-negative real numbers such that $a_{n+m} \le a_n a_m$ for all pairs of positive integers, m and n, prove that the sequence $\{\sqrt[n]{a_n}\}$ has a limit as $n \to \infty$. (page 552)

THE TWENTY-THIRD WILLIAM LOWELL PUTNAM MATHEMATICAL COMPETITION

December 1, 1962

Morning Session

1. Given five points in a plane, no three of which lie on a straight line, show that some four of these points form the vertices of a convex quadrilateral. (page 554)

2. Find every real-valued function f whose domain is an interval I (finite or infinite) having 0 as a left-hand endpoint, such that for every positive member x of I the average of f over the closed interval $[0, x]$ is equal to the geometric mean of the numbers $f(0)$ and $f(x)$. (page 555)

3. In a triangle ABC in the Euclidean plane, let A' be a point on the segment from B to C, B' a point on the segment from C to A, and C' a point on the segment from A to B such that

$$\frac{AB'}{B'C} = \frac{BC'}{C'A} = \frac{CA'}{A'B} = k,$$

where k is a positive constant. Let Δ be the triangle formed by parts of the segments obtained by joining A and A', B and B', and C and C'. Prove that the areas of the triangles Δ and ABC are in the ratio.

$$\frac{(k-1)^2}{k^2+k+1}.$$

(page 556)

4. Assume that $|f(x)| \leq 1$ and $|f''(x)| \leq 1$ for all x on an interval of length at least 2. Show that $|f'(x)| \leq 2$ on the interval. (page 559)

5. Evaluate in closed form

$$\sum_{k=1}^{n} \binom{n}{k} k^2.$$

(page 560)

Note:

$$\binom{n}{k} = \frac{n(n-1)\cdots(n-k+1)}{1\cdot 2\cdots k}.$$

6. Let S be a set of rational numbers such that whenever a and b are members of S, so are $a + b$ and ab, and having the property that for every rational number r exactly one of the following three statements is true:

$$r \in S, \quad -r \in S, \quad r = 0.$$

Prove that S is the set of all positive rational numbers. (page 561)

Afternoon Session

1. Let $x^{(n)} = x(x-1)\cdots(x-n+1)$ for n a positive integer and let $x^{(0)} = 1$. Prove that

$$(x+y)^{(n)} = \sum_{k=0}^{n} \binom{n}{k} x^{(k)} y^{(n-k)}.$$

(page 562)

Note:

$$\binom{n}{k} = \frac{n(n-1)\cdots(n-k+1)}{1\cdot 2\cdots k}.$$

2. Let R be the set of all real numbers and S the set of all subsets of the positive integers. Construct a function f whose domain is R and whose range is in S, such that $f(a)$ is a proper subset of $f(b)$ whenever $a < b$. (page 563)

3. Let S be a convex region in the Euclidean plane containing the origin. Assume that every ray (that is, half-line) from the origin has at least one point outside S. Prove that S is bounded. (A region in the plane is defined to be convex if and only if the line segment joining every pair of its points lies entirely within the region.) (page 564)

4. The Euclidean plane is divided into regions by drawing a finite number of circles. Show that it is possible to color each of these regions either red or blue in such a way that no two adjacent regions have the same color. (Two such regions are said to be adjacent if and only if their boundaries have an arc of a circle in common.) (page 566)

5. Prove that for every integer n greater than 1:

$$\frac{3n+1}{2n+2} < \left(\frac{1}{n}\right)^n + \left(\frac{2}{n}\right)^n + \cdots + \left(\frac{n}{n}\right)^n < 2.$$

(page 566)

6. Let

$$f(x) = \sum_{k=0}^{n} a_k \sin kx + b_k \cos kx,$$

where a_k and b_k are constants. Show that, if $|f(x)| \leq 1$ for $0 \leq x \leq 2\pi$ and $|f(x_i)| = 1$ for $0 \leq x_1 < x_2 < \cdots < x_{2n} < 2\pi$, then $f(x) = \cos(nx + a)$ for some constant a. (page 567)

THE TWENTY-FOURTH WILLIAM LOWELL PUTNAM MATHEMATICAL COMPETITION

December 7, 1963

Morning Session

1. (i) Show that a regular hexagon, six squares, and six equilateral triangles can be assembled without overlapping to form a regular dodecagon.

(ii) Let $P_1, P_2, \ldots, P_{12}$ be the successive vertices of a regular dodecagon. Explain how the three diagonals P_1P_9, P_2P_{11}, and P_4P_{12} intersect.

(page 569)

2. Let $\{f(n)\}$ be a strictly increasing sequence of positive integers such that $f(2) = 2$ and $f(mn) = f(m)f(n)$ for every relatively prime pair of positive integers m and n (the greatest common divisor of m and n is equal to 1). Prove that $f(n) = n$ for every positive integer n. (page 570)

3. Find an integral formula for the solution of the differential equation

$$\delta(\delta - 1)(\delta - 2)\cdots(\delta - n + 1)y = f(x), \qquad x \geq 1,$$

for y as a function of x satisfying the initial conditions $y(1) = y'(1) = \cdots = y^{(n-1)}(1) = 0$, where f is continuous and

$$\delta \equiv x\frac{d}{dx}.$$

(page 571)

4. Let $\{a_n\}$ be a sequence of positive real numbers. Show that

$$\limsup_{n\to\infty} n\left(\frac{1 + a_{n+1}}{a_n} - 1\right) \geq 1.$$

Show that the number 1 on the right-hand side of this inequality cannot be replaced by any larger number. (The symbol lim sup is sometimes written $\overline{\lim}$.)

(page 572)

5. (i) Prove that if a function f is continuous on the closed interval $[0, \pi]$ and if

$$\int_0^\pi f(\theta)\cos\theta\, d\theta = \int_0^\pi f(\theta)\sin\theta\, d\theta = 0$$

then there exist points α and β such that

$$0 < \alpha < \beta < \pi \quad \text{and} \quad f(\alpha) = f(\beta) = 0.$$ (page 574)

(ii) Let R be any bounded convex open region in the Euclidean plane (that is, R is a connected open set contained in some circular disk, and the line segment joining any two points of R lies entirely in R). Prove with the help of part (i) that the centroid (center of gravity) of R bisects at least three distinct chords of the boundary of R. (page 574)

6. Let U and V be any two distinct points on an ellipse, let M be the midpoint of the chord UV, and let AB and CD be any two other chords through M. If the line UV meets the line AC in the point P and the line BD in the point Q, prove that M is the midpoint of the segment PQ. (page 575)

Afternoon Session

1. For what integer a does $x^2 - x + a$ divide $x^{13} + x + 90$? (page 577)

2. Let S be the set of all numbers of the form $2^m 3^n$, where m and n are integers, and let P be the set of all positive real numbers. Is S dense in P? (page 578)

3. Find every twice-differentiable real-valued function f with domain the set of all real numbers and satisfying the functional equation

$$(f(x))^2 - (f(y))^2 = f(x+y)f(x-y)$$

for all real numbers x and y. (page 579)

4. Let C be a closed plane curve that has a continuously turning tangent and bounds a convex region. If T is a triangle inscribed in C with maximum perimeter, show that the normal to C at each vertex of T bisects the angle of T at that vertex. If a triangle T has the property just described, does it necessarily have maximum perimeter? What is the situation if C is a circle? (A convex region is a connected open set such that the line segment joining any two points of the set lies entirely in the set.) (page 580)

5. Let $\{a_n\}$ be a sequence of real numbers satisfying the inequalities

$$0 \le a_k \le 100a_n \text{ for } n \le k \le 2n \quad \text{and} \quad n = 1, 2, \ldots,$$

and such that the series

$$\sum_{n=0}^{\infty} a_n$$

converges. Prove that

$$\lim_{n \to \infty} na_n = 0.$$ (page 582)

6. Let E be a Euclidean space of at most three dimensions. If A is a nonempty subset of E, define $S(A)$ to be the set of all points that lie on closed segments joining pairs of points of A. For a given nonempty set A_0, define $A_n \equiv S(A_{n-1})$ for $n = 1, 2, \ldots$. Prove that $A_2 = A_3 = \cdots$. (A one-point set should be considered to be a special case of a closed segment.)

(page 583)

THE TWENTY-FIFTH WILLIAM LOWELL PUTNAM MATHEMATICAL COMPETITION

December 5, 1964

Morning Session

1. Given a set of 6 points in the plane, prove that the ratio of the longest distance between any pair to the shortest is at least $\sqrt{3}$. (page 586)

2. Find all continuous positive functions $f(x)$, for $0 \leq x \leq 1$, such that

$$\int_0^1 f(x)\, dx = 1$$

$$\int_0^1 f(x)x\, dx = \alpha$$

$$\int_0^1 f(x)x^2\, dx = \alpha^2$$

where α is a given real number. (page 587)

3. Let $P_1, P_2, \ldots$ be a sequence of distinct points which is dense in the interval (0, 1). The points $P_1, P_2, \ldots, P_{n-1}$ decompose the interval into n parts, and P_n decomposes one of these into two parts. Let a_n and b_n be the lengths of these two intervals. Prove that

$$\sum_{n=1}^{\infty} a_n b_n (a_n + b_n) = 1/3.$$

(A sequence of points in an interval is said to be dense when every subinterval contains at least one point of the sequence.) (page 588)

4. Let p_n $(n = 1, 2, \ldots)$ be a bounded sequence of integers which satisfies the recursion

$$p_n = \frac{p_{n-1} + p_{n-2} + p_{n-3}p_{n-4}}{p_{n-1}p_{n-2} + p_{n-3} + p_{n-4}}.$$

Show that the sequence eventually becomes periodic. (page 589)

5. Prove that there is a constant K such that the following inequality holds for any sequence of positive numbers $a_1, a_2, a_3, \ldots$:

$$\sum_{n=1}^{\infty} \frac{n}{a_1 + a_2 + \cdots + a_n} \leq K \sum_{n=1}^{\infty} \frac{1}{a_n}.$$ (page 589)

6. Let S be a finite subset of a straight line. Say that S has the repeated distance property when every value of the distance between pairs of points of S (except for the longest) occurs at least twice. Show that if S has the repeated distance property then the ratio of any two distances between two points of S is a rational number. (page 592)

Afternoon Session

1. Let u_k $(k = 1, 2, \ldots)$ be a sequence of integers, and let V_n be the number of those which are less than or equal to n. Show that if

$$\sum_{k=1}^{\infty} 1/u_k < \infty,$$

then

$$\lim_{n \to \infty} V_n / n = 0.$$ (page 594)

2. Let S be a set of $n > 0$ elements, and let $A_1, A_2, \ldots, A_k$ be a family of distinct subsets, with the property that any two of these subsets meet. Assume that no other subset of S meets all of the A_i.

Prove that $k = 2^{n-1}$ (page 595)

3. Let $f(x)$ be a real continuous function defined for all real x. Assume that for every $\epsilon > 0$

$$\lim_{n \to \infty} f(n\epsilon) = 0, \qquad \text{(where } n \text{ is a positive integer).}$$

Prove that

$$\lim_{x \to \infty} f(x) = 0.$$ (page 595)

4. Into how many regions do n great circles (no three concurrent) decompose the surface of the sphere on which they lie? (page 597)

5. Let u_n $(n = 1, 2, 3, \ldots)$ denote the least common multiple of the first n

terms of a strictly increasing sequence of positive integers (for example, the sequence 1, 2, 3, 4, 5, 6, 10, 12, ...). Prove that the series

$$\sum_{n=1}^{\infty} 1/u_n$$

is convergent. (page 598)

6. Show that the unit disk in the plane cannot be partitioned into two disjoint congruent subsets. (page 599)

SOLUTIONS

THE FIRST WILLIAM LOWELL PUTNAM MATHEMATICAL COMPETITION
April 16, 1938

Morning Session

1. A solid is bounded by two bases in the horizontal planes $z = h/2$ and $z = -h/2$, and by such a surface that the area of every section in a horizontal plane is given by a formula of the sort

$$\text{Area} = a_0 z^3 + a_1 z^2 + a_2 z + a_3$$

(where as special cases some of the coefficients may be 0). Show that the volume is given by the formula

$$V = \frac{1}{6} h[B_1 + B_2 + 4M],$$

where B_1 and B_2 are the areas of the bases, and M is the area of the middle horizontal section. Show that the formulas for the volume of a cone and of a sphere can be included in this formula when $a_0 = 0$.

Solution. The volume in question is given by

$$V = \int_{-h/2}^{h/2} (a_0 z^3 + a_1 z^2 + a_2 z + a_3)dz$$

$$= \frac{a_1 h^3}{12} + a_3 h.$$

On the other hand, the base areas and M are given by

$$B_1 = \frac{a_0 h^3}{8} + \frac{a_1 h^2}{4} + \frac{a_2 h}{2} + a_3$$

$$B_2 = -\frac{a_0 h^3}{8} + \frac{a_1 h^2}{4} - \frac{a_2 h}{2} + a_3$$

$$M = a_3,$$

so that the suggested expression $(1/6)h[B_1 + B_2 + 4M]$ works out to be

$$\frac{1}{6}h\left(\frac{a_1h^2}{2} + 6a_3\right) = \frac{a_1h^3}{12} + a_3h = V$$

as required.

The formula $V = (1/6)h(B_1 + B_2 + 4M)$ is known in solid geometry as the prismoidal formula. It is closely related to Simpson's rule in numerical integration.

Indeed, for functions of class C^4 it can be proved that

$$\int_{-h/2}^{h/2} f(z)dz = (1/6)h[f(-h/2) + 4f(0) + f(h/2)] + E$$

where $E = -(1/2880)\ h^5f^{(4)}(\xi)$ for some ξ lying in $(-h/2, h/2)$. See Kunz, *Numerical Analysis*, McGraw-Hill, 1957, p. 146, or any other book on numerical integration.

In many cases the error term E is very small and therefore we may use the approximate relation

$$\int_{-h/2}^{h/2} f(z)dz \simeq (h/6)[f(-h/2) + 4f(0) + f(h/2)].$$

This approximation is known as Simpson's rule. In particular, when f is a polynomial of degree at most three, $E = 0$ and the result is exact.

For the special cases of the cone and the sphere, we can proceed as follows. For the cone, let the vertex be in the plane $z = h/2$ and the base in the plane $z = -h/2$. Then the area of a cross-section at level z is given by $A = (B/h^2)(z - (h/2))^2$ where B is the area of the base. Since the expression for A is a polynomial of degree two,

$$V = (h/6)(B + 4\,(B/4) + 0) = (1/3)Bh,$$

a well-known result.

For the sphere of radius $r = h/2$, included between the two planes $z = -h/2$ and $z = h/2$, the cross-sectional area at level z is given by $A = \pi(r^2 - z^2)$. This expression for A is also a polynomial of degree 2, and we get

$$V = (h/6)(4\pi r^2) = \frac{4}{3}\pi r^3.$$

For both the sphere and the cone, the coefficient a_0 of z^3 in the cross-section area formula is zero.

2. A can buoy is to be made of three pieces, namely, a cylinder and two equal cones, the altitude of each cone being equal to the altitude of the cylinder. For a given area of surface, what shape will have the greatest volume?

Solution. Let r be the radius of the cylinder, and h its altitude. The given condition is

$$S = 2\pi rh + 2(\pi r\sqrt{h^2 + r^2}) = \text{constant}, \tag{1}$$

and the volume of the buoy is

$$V = \pi r^2 h + \frac{2\pi r^2 h}{3} = \frac{5\pi r^2 h}{3}. \tag{2}$$

The required problem is to find the maximum value of V subject to condition (1). This can be done by the method of Lagrange multipliers, but in this particular problem it is easier to solve (1) for h and express V as a function of r. We have

$$(S - 2\pi rh)^2 = 4\pi^2 r^2 (h^2 + r^2),$$

whence

$$h = \frac{S^2 - 4\pi^2 r^4}{4\pi r S}, \tag{3}$$

and the expression for V becomes

$$V = \frac{5r}{12S}(S^2 - 4\pi^2 r^4). \tag{4}$$

Since r and V must be positive, the domain of interest is given by

$$0 < r < \sqrt[4]{S^2/4\pi^2}.$$

We compute the derivative and equate it to zero to get

$$\frac{dV}{dr} = \frac{5S}{12} - \frac{100\,\pi^2 r^4}{12S} = 0.$$

The only critical value is

$$r_0 = \sqrt[4]{\frac{S^2}{20\pi^2}}.$$

Since $V \to 0$ as $r \to 0$ or as $r \to \sqrt[4]{S^2/4\pi^2}$, and is positive in between, the critical value r_0 yields a maximum for V.

The corresponding value of h is found from (3) to be $h_0 = \frac{2}{5}\sqrt{5}\, r_0$. The shape of the buoy is completely determined by the ratio

$$\frac{h_0}{r_0} = \frac{2}{5}\sqrt{5}.$$

---

3. If a particle moves in the plane, we may express its coordinates x and y as functions of the time t. If $x = t^3 - t$ and $y = t^4 + t$, show that the curve has a point of inflection at $t = 0$ and that the velocity of the moving particle has a maximum at $t = 0$.

REMARK. This problem was in the 1938 competition. There is considerable difference in notation now (over a third of a century later). A similar problem now would distinguish between velocity (a vector function) and the speed (its norm and hence a scalar function). The obvious intention of the problem is to show that the speed has a *local* maximum at $t = 0$.

First Solution. If the velocity vector at time t is of length v and has direction θ, then $\dot{x} = v\cos\theta$, $\dot{y} = v\sin\theta$, and $\ddot{x} = \dot{v}\cos\theta - v\dot{\theta}\sin\theta$, $\ddot{y} = \dot{v}\sin\theta + v\dot{\theta}\cos\theta$. Thus $\dot{x}\ddot{y} - \dot{y}\ddot{x} = v^2\dot{\theta}$ and $v^2 = \dot{x}^2 + \dot{y}^2$. From the given parametric data,

$$v^2\dot{\theta} = (3t^2 - 1)12t^2 - (4t^3 + 1)6t = 6t(2t^3 - 2t - 1)$$

$$v^2 = 16t^6 + 9t^4 + 8t^3 - 6t^2 + 2.$$

From the $v^2\dot{\theta}$ relation, $\dot{\theta}$ changes sign as t passes through 0. To rule out the possibility of a cusp point, one notes that $v^2 \neq 0$ at $t = 0$. Hence the curve has an inflection point at $t = 0$. Also

$$\frac{d(v^2)}{dt} = 96t^5 + 36t^4 + 24t^2 - 12t$$

$$\left.\frac{d(v^2)}{dt}\right]_{t=0} = 0$$

and

$$\left.\frac{d^2(v^2)}{dt^2}\right]_{t=0} = -12.$$

So v^2 has a (local) maximum at $t = 0$. Hence the speed v also has a local maximum.

Second Solution. Since dx/dt does not vanish for $t = 0$, we can solve for t in terms of x near $t = 0$ and write y as a function of x. Then we have

$$\frac{dy}{dx} = \frac{dy}{dt}\bigg/\frac{dx}{dt} = \frac{4t^3 + 1}{3t^2 - 1} = -1 - 3t^2 - \cdots.$$

Therefore, dy/dx has a (local) maximum at $t = 0$, so the curve has an inflection point at $t = 0$.

The magnitude v of the velocity is given by

$$v^2 = \left(\frac{dx}{dt}\right)^2 + \left(\frac{dy}{dt}\right)^2 = (3t^2 - 1)^2 + (4t^3 + 1)^2 = 2 - 6t^2 + \cdots.$$

Hence v has a (local) maximum at $t = 0$.

---

4. A lumberman wishes to cut down a tree whose trunk is cylindrical and whose material is uniform. He will cut a notch, the two sides of which will be planes intersecting at a dihedral angle θ along a horizontal line through the axis of the cylinder. If θ is given, show that the least volume of material is cut out when the plane bisecting the dihedral angle is horizontal.

First Solution. Suppose $0 \le \alpha_1 < \alpha_2 < \alpha_2 + \theta < \pi/2$; then the wedge-shaped solid between the planes at angles α_1 and $\alpha_1 + \theta$ is smaller than the wedge between α_2 and $\alpha_2 + \theta$, because a simple rotation of the former through an angle $\alpha_2 - \alpha_1$ makes it a proper subset of the latter.

Consider now any asymmetrical wedge of angle θ with cross-section AOB. If A and B are on the same side of the horizontal through O, then the above argument shows that the wedge does not have minimal volume.

Suppose then that A is below the horizontal, and B above it. By symmetry we can assume that AOB lies below the symmetrical wedge SOT of angle θ, as shown. The wedge AOS is congruent by symmetry with the

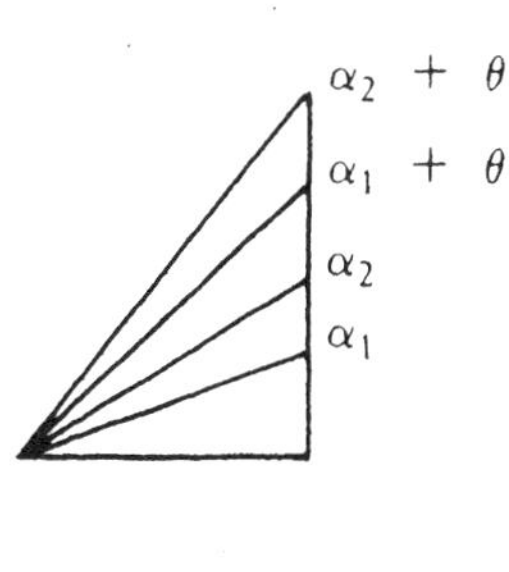

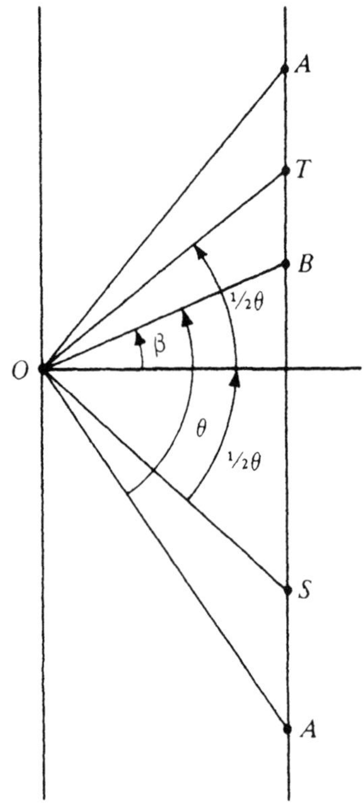

wedge $A'OT$, which is, in turn, larger than wedge BOT (as shown above). Hence

$$\begin{aligned} \text{wedge } AOB &= \text{wedge } AOS + \text{wedge } SOB \\ &> \text{wedge } SOB + \text{wedge } BOT \\ &= \text{wedge } SOT. \end{aligned}$$

Thus the symmetrical wedge is a strict minimum.

Second Solution. Let a be the radius of the cylindrical tree, and let α and β be the angles between the planes of the cut and the horizontal;

$$\beta = \alpha + \theta.$$

The volume of the wedge is

$$\begin{aligned} V &= \int_0^a 2x(\tan\beta - \tan\alpha)\sqrt{a^2 - x^2}\,dx \\ &= A(\tan\beta - \tan\alpha). \end{aligned}$$

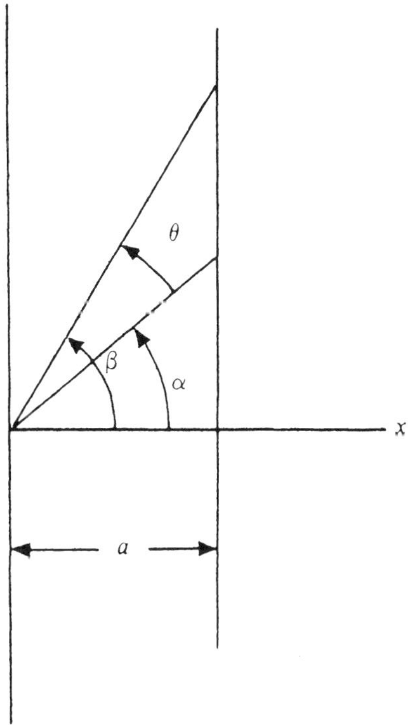

(It is easy, but unnecessary, to evaluate the integral; in fact, $A = 2a^3/3$.) We seek to minimize V by choice of α. This is equivalent to minimizing

$$W = \tan(\alpha + \theta) - \tan \alpha$$

for $-\pi/2 < \alpha < \pi/2 - \theta$, since $-\pi/2 < \alpha$ and $\beta < \pi/2$. The critical points are found by solving

$$\frac{dW}{d\alpha} = \sec^2(\alpha + \theta) - \sec^2 \alpha = 0.$$

Since both $\sec(\alpha + \theta)$ and $\sec \alpha$ are positive through the interval in question, $\sec(\alpha + \theta) = \sec \alpha$, whence $\alpha + \theta = \pm\alpha$. Since θ is not zero, the only critical point is given by $\alpha = -\theta/2$. It is easily seen to correspond to a minimum. When $\alpha = -\theta/2$, $\beta = \theta/2$ and the horizontal plane bisects the wedge.

5. Evaluate the following limits:

(i) $$\lim_{n\to\infty} \frac{n^2}{e^n}.$$

(ii) $$\lim_{x\to 0} \frac{1}{x} \int_0^x (1 + \sin 2t)^{1/t}\, dt.$$

Solution. (i) It follows from L'Hospital's rule that

$$\lim_{x\to\infty} \frac{x^2}{e^x} = \lim_{x\to\infty} \frac{2x}{e^x} = \lim_{x\to\infty} \frac{2}{e^x} = 0,$$

whence the desired limit is zero.

Alternatively, one could use the fact that for $x > 0$,

$$e^x = \sum_{n=0}^{\infty} \frac{x^n}{n!} > \frac{x^3}{6}, \qquad \text{whence } 0 < \frac{x^2}{e^x} < \frac{6}{x}$$

and, as $x \to \infty$, $\lim_{x\to\infty} 6/x = 0$, whence the desired limit is zero.

(ii) By L'Hospital's rule,

$$\lim_{x\to 0} \frac{1}{x} \int_0^x (1 + \sin 2t)^{1/t}\, dt = \lim_{x\to 0} (1 + \sin 2x)^{1/x},$$

provided the latter limit exists. Let $\phi(x) = (1 + \sin 2x)^{1/x}$. Then

$$\log \phi(x) = \frac{\log(1 + \sin 2x)}{x}.$$

Again using L'Hospital's rule,

$$\lim_{x\to 0} \log \phi(x) = \lim_{x\to 0} \frac{2\cos 2x}{1 + \sin 2x} = 2.$$

Since the exponential function is continuous,

$$\lim_{x\to 0} \phi(x) = e^2.$$

Therefore,

$$\lim_{x\to 0} \frac{1}{x} \int_0^x (1 + \sin 2t)^{1/t}\, dt = e^2.$$

6. A swimmer stands at one corner of a square swimming pool and wishes to reach the diagonally opposite corner. If w is his walking speed and s is his swimming speed ($s < w$), find his path for shortest time. [Consider two cases: (i) $w/s < \sqrt{2}$, and (ii) $w/s > \sqrt{2}$.]

Solution. Let the square pool be denoted by $ABCD$, with the swimmer initially at A and desirous of reaching C. The path of least time can evidently be described as follows. The swimmer walks from A to E (a point on side AB), swims from E to F where F is on BC, and then walks from F to C. Note that a path like $AGHC$ is time equivalent to a path of the type described with $F = C$.

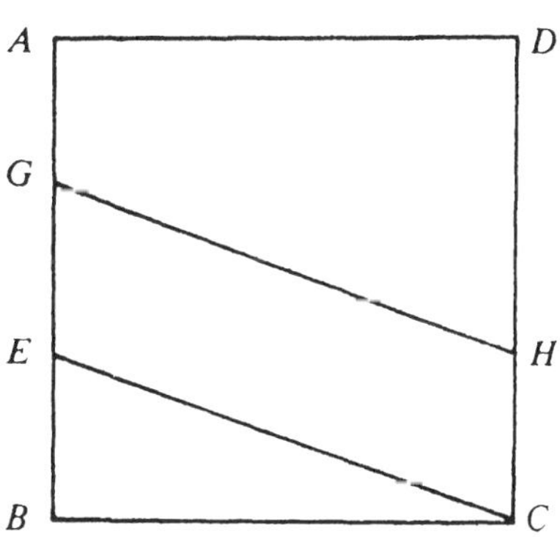

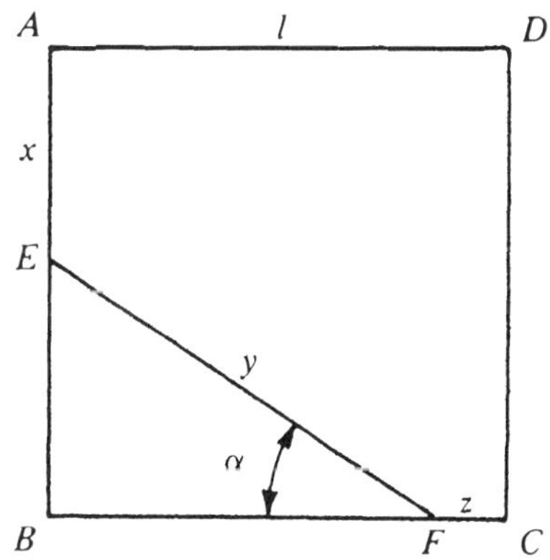

Let $\overline{AE} = x$, $\overline{EF} = y$, $\overline{FC} = z$. Then the time T is given by $T = (x + z)/w + (y/s)$. If the sum $x + z$ is fixed, then the sum $y \sin \alpha + y \cos \alpha$ is also fixed, and y is minimal when $(\sin \alpha + \cos \alpha)$ is maximal. This maximum is attained for $\alpha = 45°$.

Thus for a minimal time path, $x = z$ and $y = \sqrt{2}\,(l - x)$, where l is the length of a side of the pool. Accordingly, we have to minimize $T = (2x/w) + \sqrt{2}\,(l - x)/s$ for $0 \leq x \leq l$.

But T is a linear function of x, so its maximum occurs at an endpoint of the interval. If $x = 0$, $T = \sqrt{2l}/s$, and if $x = l$, $T = 2l/w$.

If $\sqrt{2}l/s < 2l/w$ then $w/s < \sqrt{2}$, and conversely. Hence, if $w/s < \sqrt{2}$ the minimal path is unique and the swimmer should swim diagonally across the pool from A to C. If $w/s > \sqrt{2}$, he should walk from A to B to C. Finally, if $w/s = \sqrt{2}$, T is independent of x and there are infinitely many minimizing paths, in fact any path $AEFC$ for which $\alpha = 45°$.

7. (i) Show that the gravitational attraction exerted by a thin homogeneous spherical shell at an external point is the same as if the material of the shell were concentrated at its center.

First Solution. A thin homogeneous spherical shell is, of course, the surface of a sphere with constant uniform density, say σ. Let the shell have radius a and let its center be the origin of a rectangular coordinate system. Consider a particle P of mass m at the point $(R, 0, 0)$, where $R > a$ so that P is outside the shell. The figure shows the cross-section in the x-y plane. We shall calculate the gravitational attraction between P and the shell.

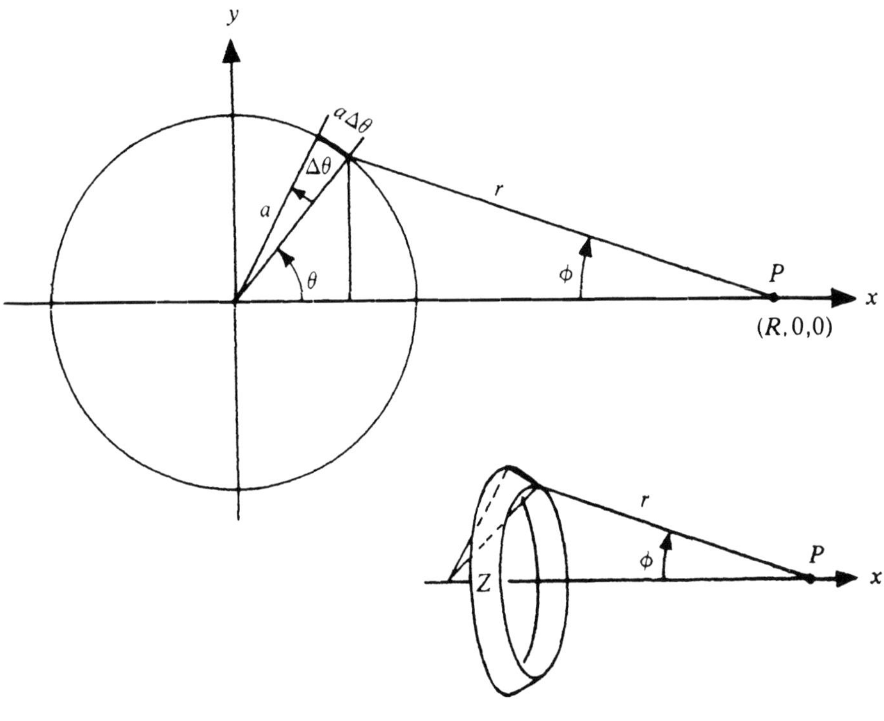

Because of rotational symmetry the resultant force is directed along the x-axis; hence we need compute only the x-component of the force. The short arc of length $a\Delta\theta$ shown in the figure generates (on rotation about the x-axis) a zone Z on the shell of (approximate) area $2\pi a^2 \sin\theta\Delta\theta$ and mass $2\pi a^2\sigma \sin\theta\Delta\theta$. Let r and ϕ be as shown in the figure. Then every point of Z is (essentially) at distance r from P and acts along a line making angle ϕ with the x-axis. Hence the x-component of the gravitational attraction between Z and P has the approximate magnitude

$$\frac{Gm \cdot 2\pi a^2 \sigma \sin\theta\Delta\theta}{r^2} \cos\phi,$$

where G is the constant of gravitation. Therefore the entire attractive force between the shell and P is

$$F = 2\pi Gma^2\sigma \int_0^\pi \frac{\cos\phi}{r^2}\sin\theta d\theta.$$

Here ϕ and r are functions of θ.

We have $r\cos\phi + a\cos\theta = R$, so the integral becomes

$$2\pi Gma^2\sigma \int_0^\pi \frac{R - a\cos\theta}{r^3}\sin\theta d\theta.$$

This apparently complicated integral can be easily evaluated if we change the variable of integration to r. By the law of cosines we have

$$r^2 = R^2 + a^2 - 2aR\cos\theta.$$

Therefore

$$R - a\cos\theta = \frac{1}{2R}(R^2 - a^2 + r^2)$$

and $rdr = aR\sin\theta d\theta$. Hence

$$\text{(1)} \qquad F = \frac{\pi Gma\sigma}{R^2}\int_{R-a}^{R+a}\frac{R^2 - a^2 + r^2}{r^2}dr$$

$$= \frac{\pi Gma\sigma}{R^2}\left[(R^2 - a^2)\left[\frac{1}{R-a} - \frac{1}{R+a}\right] + 2a\right]$$

$$= \frac{4\pi Gma^2\sigma}{R^2} = \frac{GmM}{R^2},$$

where $M = 4\pi a^2\sigma$ is the mass of the spherical shell. Thus the force is the same as it would be if all the mass of the shell were concentrated at the center.

REMARK 1. It follows from this result that any spherically homogeneous mass distribution within a sphere S has the same gravitational field at points outside S as it would have if all the mass were concentrated at the center of S.

REMARK 2. The same computations handle the case in which P is within the shell except that $R < a$ and the limits of integration in (1) are from $a - R$ to $a + R$. We find then $F = 0$. Thus the attraction of a homogeneous spherical shell on a particle inside it is always zero.

Second Solution. Gauss's theorem states that if a closed surface S contains a mass M in its interior, and if $\vec{F}$ is the gravitational field due to the mass, then the total flux through S is

$$\int_S\int \vec{F}\cdot\vec{n}dA = -4G\pi M$$

where $\vec{n}$ is the outward-drawn unit normal to S and dA is the element of surface area.

Now let S be a sphere of radius R external to the given shell and concentric with it. By symmetry,

$$\vec{F} = -F(R)\vec{n} \text{ on } S,$$

where $F = F(R)$ is a constant on S. Thus,

$$-4G\pi M = \int_S\int -F(R)\vec{n}\cdot\vec{n}dA = -F(R)\int_S\int dA = -4\pi R^2F(R)$$

whence $F(R) = GM/R^2$, and as a vector $\vec{F} = -(GM/R^2)\vec{n}$.

---

7. (ii) Determine all the straight lines which lie upon the surface $z = xy$, and draw a figure to illustrate your result.

Solution. Suppose that L is a line through (x_0, y_0, z_0) which lies in the surface $z = xy$. L can be represented parametrically by

$$x = x_0 + \alpha t$$
$$y = y_0 + \beta t$$
$$z = z_0 + \gamma t$$

where α, β, γ are not all zero. In order that L lie in the given surface it is necessary and sufficient that

$$z_0 + \gamma t = (x_0 + \alpha t)(y_0 + \beta t) = x_0y_0 + (\alpha y_0 + \beta x_0)t + \alpha\beta t^2$$

for all values of t. This implies that $\alpha\beta$, the coefficient of t^2, is zero; hence either $\alpha = 0$ or $\beta = 0$.

If $\alpha = 0$, then $\gamma = \beta x_0$, while if $\beta = 0$, then $\gamma = \alpha y_0$. We cannot have both α and β zero, since that would imply also $\gamma = 0$. Hence we can normalize whichever of these is non-zero to be one. Therefore the parametric equations of L can be put in one of the two forms

$$\begin{aligned} x &= x_0 \\ y &= y_0 + t \\ z &= z_0 + x_0 t \end{aligned} \qquad \text{or} \qquad \begin{aligned} x &= x_0 + t \\ y &= y_0 \\ z &= z_0 + y_0 t \end{aligned}$$

with corresponding non-parametric forms

$$\begin{aligned} x &= x_0 \\ z &= x_0 y \end{aligned} \qquad \text{or} \qquad \begin{aligned} y &= y_0 \\ z &= y_0 x . \end{aligned}$$

Conversely, it is obvious that any such line lies entirely on the surface.

An alternative procedure depends on the following fact: If a line L through a point P lies entirely on a surface S, then L lies in the tangent plane to S at P.

For the given surface

$$z = xy \tag{1}$$

the tangent plane at (x_0, y_0, z_0) is

$$(x - x_0) y_0 + (y - y_0) x_0 + (z - z_0)(-1) = 0. \tag{2}$$

Eliminating z between the last two equations we find

$$(x - x_0)(y - y_0) = 0 \tag{3}$$

and this is the equation of the projection on the xy-plane of the intersection of the surface (1) with its tangent plane (2). The projection consists of two lines

$$x = x_0 \qquad \text{and} \qquad y = y_0$$

and these come from two lines

$$\begin{aligned} x &= x_0 \\ z &= x_0 y \end{aligned} \qquad \text{and} \qquad \begin{aligned} y &= y_0 \\ z &= y_0 x \end{aligned}$$

lying on (1) and (2).

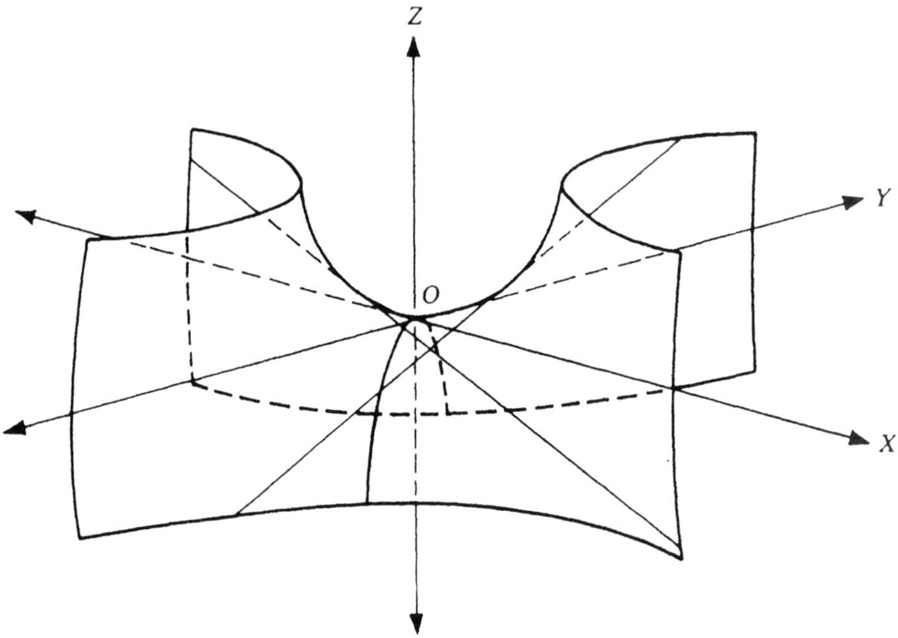

REMARKS. Such lines are called rulings. In this case there are two families of rulings, such that one member of each family passes through every point. Hyperboloids of one sheet and hyperbolic paraboloids both have two such families of rulings, and they are, in effect, the only curved surfaces with two such families.

Afternoon Session

8. (i) Let A_{ik} be the cofactor of a_{ik} in the determinant

$$d = \begin{vmatrix} a_{11} & a_{12} & a_{13} & a_{14} \\ a_{21} & a_{22} & a_{23} & a_{24} \\ a_{31} & a_{32} & a_{33} & a_{34} \\ a_{41} & a_{42} & a_{43} & a_{44} \end{vmatrix}.$$

Let D be the corresponding determinant with a_{ik} replaced by A_{ik}. Prove $D = d^3$.

Solution. Let α be the matrix of the given determinant with elements a_{ik} and let β be the matrix of the cofactors A_{ik}, and let γ be the transpose of β. Then the product matrix $\alpha\gamma$ is a diagonal matrix with all entries on the main diagonal equal to d.

Thus $\det(\alpha\gamma) = d^4 = (\det\alpha)(\det\gamma) = (\det\alpha)(\det\beta) = dD$. The equation

$$dD = d^4. \tag{1}$$

is an identity between polynomials in the 16 matrix entries regarded as independent indeterminates. Since there certainly exists a 4×4 matrix whose determinant is not zero, d is non-zero in the polynomial ring. Since the polynomial ring is an integral domain, the result

$$D = d^3$$

follows from (1).

REMARK. The result can obviously be generalized to matrices of any dimension. The determinant of the matrix of cofactors of an $n \times n$ matrix is d^{n-1}, where d is the determinant of the original matrix.

---

8. (ii) Let $P(y) = Ay^2 + By + C$ be a quadratic polynomial in y. If the roots of the quadratic equation $P(y) - y = 0$ are a and $b (a \neq b)$, show that a and b are roots of the biquadratic equation $P[P(y)] - y = 0$. Hence write down a quadratic equation which will give the other two roots, c and d, of the biquadratic. Apply this result to solving the following biquadratic equation:

$$(y^2 - 3y + 2)^2 - 3(y^2 - 3y + 2) + 2 - y = 0.$$

Solution. Since a is a root of $P(y) - y = 0$, we have $P(a) = a$. Then $P(P(a)) = P(a) = a$, so a is a root of $P(P(y)) - y = 0$. Similarly, b is a root of this biquadratic.

Let $Q(y) = P(P(y)) - y$. To find the other zeros of Q, note that $P(y) - y = Ay^2 + (B - 1)y + C = A(y - a)(y - b)$, whence $A(a + b) = 1 - B$. Then

$$\begin{aligned} Q(y) &= P(P(y)) - P(y) + P(y) - y \\ &= A\{P(y) - a\}\{P(y) - b\} + A(y - a)(y - b) \\ &= A\{A(y - a)(y - b) + y - a\}\{A(y - a)(y - b) + y - b\} \\ &\qquad + A(y - a)(y - b) \\ &= A(y - a)(y - b)R(y), \end{aligned}$$

where

$$\begin{aligned} R(y) &= \{A(y-b)+1\}\{A(y-a)+1\}+1 \\ &= AP(y)+Ay-A(a+b)+2 \\ &= A^2y^2+A(B+1)y+AC+B+1. \end{aligned}$$

The roots c and d are the zeros of R, so the required quadratic equation for c and d is

$$A^2y^2+A(B+1)y+AC+B+1=0.$$

In the special case given, $A=1$, $B=-3$, $C=2$, and $R(y)=y^2-2y$. The zeros of $P(y)-y$ are $2\pm\sqrt{2}$, so the zeros of Q are $2\pm\sqrt{2}$, 0, and 2.

---

9. Find all the solutions of the equation

$$yy''-2(y')^2=0$$

which pass through the point $x=1, y=1$.

First Solution. $1/y^3$ is an integrating factor since

$$\frac{d}{dx}\left(\frac{y'}{y^2}\right)=\frac{yy''-2(y')^2}{y^3}=0.$$

Therefore $y'/y^2=C$ and $-1/y=Cx+D$ for appropriate constants C and D. In order that the solution pass through (1, 1), we require that $C+D=-1$. Hence

$$y=\frac{1}{1+C(1-x)}. \tag{1}$$

Conversely, any function of this form satisfies the equation and the initial conditions. If $C=0$, this is a constant function and its domain may be taken as $(-\infty,+\infty)$. If $C\neq 0$, the right member of (1) becomes infinite for $x=(1+C)/C$, so the domain of (1) must be restricted to

$$\left(-\infty,\frac{1+C}{C}\right) \quad \text{if} \quad C>0$$

$$\left(\frac{1+C}{C},\infty\right) \quad \text{if} \quad C<0.$$

Second Solution. Since x does not appear explicitly in the given differential equation, the substitution $v = y'$, $y'' = v\, dv/dy$ leads to a first-order equation

$$v\left(y \frac{dv}{dy} - 2v\right) = 0.$$

Hence either $v = 0$ and y is constant, or

$$y \frac{dv}{dy} - 2v = 0.$$

In this case the variables are separable and we obtain

$$v = Cy^2,$$

that is,

$$y' = Cy^2,$$

which is again separable. We get

$$-1/y = Cx + D,$$

and the solution proceeds as before.

Note that the special solution $y =$ constant is subsumed in this general case.

If the original equation is solved for the highest derivative

$$y'' = \frac{2(y')^2}{y},$$

it becomes clear that this differential equation is regular in the upper and lower half-planes but may be singular along the x-axis. It is obvious that all solutions (1) are maximal solutions, since they cannot be extended continuously to any larger connected domain. Since none of these solutions passes through any point where the equation might be singular, we are assured that we have found all of the solutions passing through (1,1).

---

10. A horizontal disc of diameter 3 inches is rotating at 4 revolutions per minute. A light is shining at a distant point in the plane of the disc. An insect is placed at the edge of the disc furthest from the light, facing the light. It at once starts crawling, and crawls so as always to face the light, at 1 inch per

second. Set up the differential equation of motion, and find at what point the insect again reaches the edge of the disc.

Solution. Choose both rectangular and polar coordinate systems so that the origin is at the center of the disc, the insect is initially at $(3/2, 0)$, the distant light at $(-\infty, 0)$, and the disc rotates counterclockwise. Suppose that at time t the insect's position is (x, y) in cartesian coordinates, and (r, θ) in polar coordinates.

Then as long as the insect is on the disc, the horizontal and vertical components of its velocity are respectively

$$(1) \qquad V_x = \frac{dx}{dt} = -1 - \frac{2\pi r}{15}\sin\theta = -1 - \frac{2\pi}{15}y,$$

$$(2) \qquad V_y = \frac{dy}{dt} = \frac{2\pi r}{15}\cos\theta = \frac{2\pi}{15}x.$$

Differentiating (1) and using (2) we get

$$(3) \qquad \frac{d^2x}{dt^2} = -\frac{2\pi}{15}\frac{dy}{dt} = -\left(\frac{2\pi}{15}\right)^2 x,$$

whence the differential equation governing x is

$$\frac{d^2x}{dt^2} + \left(\frac{2\pi}{15}\right)^2 x = 0.$$

The solution to (3) is

$$x = A\cos\left(\frac{2\pi}{15}t - \phi\right),$$

and from (1)

$$y = A\sin\left(\frac{2\pi}{15}t - \phi\right) - \frac{15}{2\pi}.$$

Therefore the motion is uniform circular motion along the circle

$$x^2 + \left(y + \frac{15}{2\pi}\right)^2 = A^2$$

which has center at $(0, -15/2\pi)$ and radius A. Here A can be evaluated from the initial conditions

$$x = \frac{3}{2}, y = 0 \quad \text{when} \quad t = 0,$$

giving $A^2 = (3/2)^2 + (15/2\pi)^2$.

By symmetry this circle cuts the boundary of the disc again at $(-3/2, 0)$, so the insect will leave the disc at that point.

---

11. Given the parabola $y^2 = 2mx$, what is the length of the shortest chord that is normal to the curve at one end?

Solution. Any point on the parabola has coordinates of the form $(2mt^2, 2mt)$. Let AB be a chord normal to the parabola at A. Say $A = (2mt^2, 2mt)$ and $B = (2ms^2, 2ms)$. The slope of AB is $1/(s + t)$, and the slope of the tangent at A is $1/(2t)$. Hence $s + t = -1/(2t)$.

The length L of AB is given by

$$L^2 = 4m^2[(s^2 - t^2)^2 + (s - t)^2] = 4m^2(s - t)^2[(s + t)^2 + 1].$$

Substituting $s = -t - 1/(2t)$ we have

$$(1) \qquad L^2 = 4m^2\left(\frac{4t^2 + 1}{2t}\right)^2 \frac{1 + 4t^2}{4t^2} = \frac{m^2}{4} \frac{(4t^2 + 1)^3}{t^4}.$$

We seek the value of t which minimizes L, so we may just as well choose t to minimize

$$\frac{4t^2 + 1}{t^{4/3}} = 4t^{2/3} + t^{-4/3}.$$

Setting the derivative equal to zero, we find two critical points, $t = \pm\sqrt{2}/2$. Since $L \to \infty$ as $t \to 0, \pm\infty$, these two critical values both give minima. Either of the two shortest chords is of length $3\sqrt{3}\,|m|$, from (1).

---

12. From the center of a rectangular hyperbola a perpendicular is dropped upon a variable tangent. Find the locus of the foot of the perpendicular. Obtain the equation of the locus in polar coordinates, and sketch the curve.

Solution. Let the axes be the asymptotes, so that $xy = a^2$ is the equation of the given hyperbola. Let the point (h, k) be on the hyperbola. Then $hk = a^2$ and the equation of the tangent line at (h, k) is $hy + kx - 2hk = 0$.

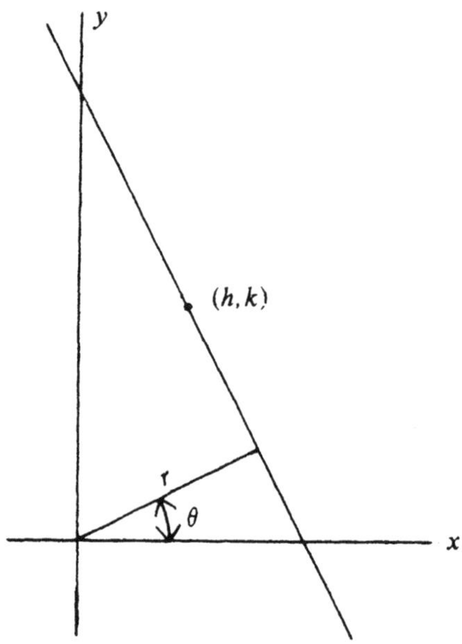

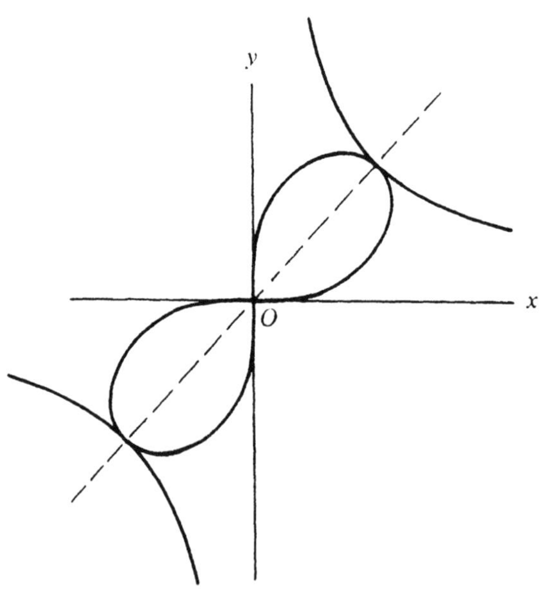

The x and y intercepts of this tangent line are $2h$ and $2k$ respectively. Let (r, θ) be the polar coordinates of the foot of the perpendicular from the origin to the tangent line. Then $2h \cos \theta = r$ and $2k \sin \theta = r$, and hence $r^2 = 4hk \sin \theta \cos \theta$ or

$$r^2 = 2a^2 \sin 2\theta. \tag{1}$$

This is the polar equation of the desired locus. We have shown that the foot of every perpendicular lies on (1).

Conversely, every point satisfying (1), except the pole, is the foot of some perpendicular: Given such a point, $P = (r, \theta)$, which must be in either the first or third quadrant, the equations $2h \cos \theta = r$ and $2k \sin \theta = r$ determine a point (h, k) on the hyperbola and P is the foot of the perpendicular on the tangent at (h, k).

The locus is the well-known lemniscate of Bernoulli.

---

13. Find the shortest distance between the plane $Ax + By + Cz + 1 = 0$ and the ellipsoid $x^2/a^2 + y^2/b^2 + z^2/c^2 = 1$. (For brevity, let

$$h = 1/\sqrt{A^2 + B^2 + C^2} \quad \text{and} \quad m = \sqrt{a^2A^2 + b^2B^2 + c^2C^2}.)$$

State algebraically the condition that the plane shall lie outside the ellipsoid.

Solution. If the given plane intersects the ellipsoid, then the minimum distance is zero. If the plane fails to intersect the ellipsoid, then the shortest distance is the distance between the given plane and the nearer of the two tangent planes to the ellipsoid that are parallel to the given plane.

The tangent plane to the ellipsoid at the point (x_0, y_0, z_0) is

$$\frac{x_0x}{a^2} + \frac{y_0y}{b^2} + \frac{z_0z}{c^2} = 1.$$

If this plane is parallel to $Ax + By + Cz + 1 = 0$, then

$$\frac{x_0}{a^2} = kA, \qquad \frac{y_0}{b^2} = kB, \qquad \text{and} \qquad \frac{z_0}{c^2} = kC,$$

where k is a constant. Since

$$1 = \frac{x_0^2}{a^2} + \frac{y_0^2}{b^2} + \frac{z_0^2}{c^2} = k^2[a^2A^2 + b^2B^2 + c^2C^2],$$

we get $|k| = 1/m$.

THE SECOND WILLIAM LOWELL PUTNAM MATHEMATICAL COMPETITION

March 4, 1939

Morning Session

1. Find the length of the curve $y^2 = x^3$ from the origin to the point where the tangent makes an angle of $45°$ with the x-axis.

Solution. The arc in the first quadrant is represented by the equation $y = x^{3/2}$, and its slope is $\frac{3}{2}x^{1/2}$. The point $P(x_0, y_0)$ where the tangent makes an angle of $45°$ is determined from the relation $\frac{3}{2}x_0^{1/2} = 1$, whence $x_0 = \frac{4}{9}$ The desired length is therefore

$$\int_0^{4/9} \sqrt{1 + \frac{9x}{4}}\, dx = \frac{8}{27}\left(1 + \frac{9x}{4}\right)^{3/2}\Bigg]_0^{4/9} = \frac{8}{27}(2\sqrt{2} - 1).$$

---

2. A point P is taken on the curve $y = x^3$. The tangent at P meets the curve again at Q. Prove that the slope of the curve at Q is *four* times the slope at P.

Solution. Let P have coordinates (x_0, y_0); then the slope at P is $3x_0^2$. The equation of the tangent at P is $y = 3x_0^2(x - x_0) + x_0^3$. The points of intersection of the tangent and the original curve are determined by the relation

$$x^3 = 3x_0^2(x - x_0) + x_0^3,$$

which is equivalent to

$$(x - x_0)^2(x + 2x_0) = 0.$$

Hence the second point of intersection is $(-2x_0, -8x_0^3)$. The slope at this point is $12x_0^2$, which is four times the slope at P, as was to be proved.

If $x_0 = 0$, the tangent does not really meet the curve again. However, since the tangent in this case has a triple point of intersection with the curve, instead of the usual double point of intersection, it is reasonable to say that it meets the curve "again" at (0,0).

---

3. Find the cubic equation whose roots are the cubes of the roots of

$$x^3 + ax^2 + bx + c = 0.$$

First Solution. Let the roots of the given cubic equation be x_1, x_2, x_3. Then the roots of the desired equation are x_1^3, x_2^3, x_3^3. From

$$x^3 + ax^2 + bx + c = (x - x_1)(x - x_2)(x - x_3),$$

it follows that

$$x_1 + x_2 + x_3 = -a, \qquad x_1x_2 + x_2x_3 + x_3x_1 = b, \qquad x_1x_2x_3 = -c.$$

Let the desired cubic equation be

$$x^3 + Ax^2 + Bx + C = (x - x_1^3)(x - x_2^3)(x - x_3^3) = 0.$$

Then we have

$$(x_1 + x_2 + x_3)^3 = x_1^3 + x_2^3 + x_3^3 + 3(x_1 + x_2 + x_3)(x_1x_2 + x_2x_3 + x_3x_1) - 3x_1x_2x_3$$

whence

$$A = -(x_1^3 + x_2^3 + x_3^3) = a^3 - 3ab + 3c.$$

Also,

$$(x_1x_2 + x_2x_3 + x_3x_1)^3 = x_1^3x_2^3 + x_2^3x_3^3 + x_3^3x_1^3 + 3abc - 3c^2$$

and hence

$$B = x_1^3x_2^3 + x_2^3x_3^3 + x_3^3x_1^3 = b^3 - 3abc + 3c^2.$$

Finally $C = -x_1^3x_2^3x_3^3 = c^3$. Thus the desired cubic equation is

$$x^3 + (a^3 - 3ab + 3c)x^2 + (b^3 - 3abc + 3c^2)x + c^3 = 0.$$

Other Solutions. A number of alternative solutions can be given. The method given above, namely, to calculate the symmetric functions of the desired roots, is straightforward and will suffice to find a polynomial whose roots are $F(x_1)$, $F(x_2)$, and $F(x_3)$ for any polynomial function F.

Another approach that is also general depends on the following theorem: If M is a matrix with characteristic roots $\lambda_1, \lambda_2, \ldots, \lambda_n$ and F is any polynomial, then $F(M)$ has characteristic roots $F(\lambda_1)$, $F(\lambda_2)$, $\ldots$, $F(\lambda_n)$. For this problem we take M to be a matrix whose characteristic polynomial is $x^3 + ax^2 + bx + c$. Then M^3 has the required polynomial as its characteristic polynomial. We can take M to be the companion matrix

$$\begin{pmatrix} 0 & 0 & -c \\ 1 & 0 & -b \\ 0 & 1 & -a \end{pmatrix}.$$

Finding M^3 and its characteristic polynomial is tedious, however.

Still another general method is elimination. We eliminate x between the two equations

$$\begin{aligned} y - x^3 &= 0 \\ c + bx + ax^2 + x^3 &= 0 \end{aligned} \tag{1}$$

to obtain the required equation for y. The standard method for accomplishing this [see, for example, G. Salmon, *Modern Higher Algebra*, Dublin, 1876, 71 ff.] is to multiply both equations through by x and by x^2 to obtain six equations that can be written in the matrix form

$$\begin{pmatrix} y & 0 & 0 & -1 & 0 & 0 \\ 0 & y & 0 & 0 & -1 & 0 \\ 0 & 0 & y & 0 & 0 & -1 \\ c & b & a & 1 & 0 & 0 \\ 0 & c & b & a & 1 & 0 \\ 0 & 0 & c & b & a & 1 \end{pmatrix} \begin{pmatrix} 1 \\ x \\ x^2 \\ x^3 \\ x^4 \\ x^5 \end{pmatrix} = 0.$$

If y is the cube of a root of the given equation, then this matrix annihilates a non-zero vector; hence its determinant vanishes. Since this determinant is a polynomial of degree three in y, it must be the required polynomial. Using the first three rows to eliminate the entries in the 3×3 submatrix in the lower right corner, this determinant is seen to be

$$\det \begin{pmatrix} y + c & b & a \\ ay & y + c & b \\ by & ay & y + c \end{pmatrix}$$
$$= (y + c)^3 + a^3y^2 + b^3y - 3aby(y + c).$$

The elimination of x between the equations (1) can also be carried out directly. We have

$$-(y+c) = x(ax+b).$$

Hence

$$\begin{aligned}-(y+c)^3 &= x^3(a^3x^3 + b^3 + 3abx(ax+b)\\ &= y(a^3y + b^3 - 3ab(y+c)).\end{aligned}$$

A method which generalizes easily to find the polynomial for other powers of the roots is as follows: Let P be the given polynomial, and let Q be the required polynomial. Then

$$Q(x^3) = (x^3 - x_1{}^3)(x^3 - x_2{}^3)(x^3 - x_3{}^3).$$

Since $(x^3 - x_1{}^3) = (x - x_1)(\omega x - x_1)(\omega^2 x - x_1)$, where ω and ω^2 are the complex cube roots of unity, we have

$$Q(x^3) = P(x)P(\omega x)P(\omega^2 x).$$

Since we know P, we can multiply this out to obtain $Q(x^3)$ and hence $Q(x)$. This can be done very easily if we recall the identity

$$(u+v+w)(u+\omega v+\omega^2 w)(u+\omega^2 v+\omega w) = u^3 + v^3 + w^3 - 3uvw.$$

---

4. Find the equations of the *two* straight lines each of which cuts all the *four* straight lines

$$x = 1,\ y = 0; \qquad y = 1,\ z = 0; \qquad z = 1,\ x = 0; \qquad x = y = -6z.$$

First Solution. Suppose the required line L meets the given lines in the points A, B, C, and D respectively. Then

$$A = (1, 0, a),\ B = (b, 1, 0),\ C = (0, c, 1), \text{ and } D = (6d, 6d, -d)$$

for some numbers a, b, c, and d. Treat A, B, C, and D as vectors. The condition that they be collinear is that the vectors

$$\begin{aligned}B - A &= (b-1, 1, -a)\\ C - A &= (-1, c, 1-a)\\ D - A &= (6d-1, 6d, -d-a)\end{aligned}$$

be proportional.

The proportionality of the first two tells us that

$$c = \frac{1}{1-b} = \frac{a-1}{a} \tag{1}$$

while the first and third give

$$6d = \frac{1-6d}{1-b} = \frac{a+d}{a}.$$

Rewrite the middle member here using (1)

$$6d = (1-6d)\frac{a-1}{a} = \frac{a+d}{a}.$$

Clearing fractions

$$6ad = a + d$$
$$a + 6d - 1 - 6ad = a + d.$$

Adding these equations, we find $4d = a + 1$, so

$$6a(a+1) = 24ad = 4(a+d) = 5a + 1.$$

The quadratic equation $6a(a+1) = 5a + 1$ has roots

$$a = \frac{1}{3}, -\frac{1}{2},$$

and the corresponding values of the other unknowns are

$$b = \frac{3}{2}, \frac{2}{3}$$
$$c = -2, 3$$
$$d = \frac{1}{3}, \frac{1}{8}.$$

The direction vectors of the lines (proportional to $B - A$, $C - A$, and $D - A$) in the two cases are $(3, 6, -2)$ and $(-2, 6, 3)$, respectively. The two lines are given parametrically by

$$L_1 : s \mapsto \left(1, 0, \frac{1}{3}\right) + s(3, 6, -2)$$

and

$$L_2 : t \mapsto \left(1, 0, -\frac{1}{2}\right) + t(-2, 6, 3).$$

These lines cross the given lines (in order) for

$$s = 0, +\frac{1}{6}, -\frac{1}{3}, \frac{1}{3} \quad \text{and} \quad t = 0, \frac{1}{6}, \frac{1}{2}, \frac{1}{8}.$$

In non-parametric form L_1 is given by

$$y = 2(x - 1) = 1 - 3z$$

and L_2 is given by

$$y = 3(1 - x) = 2z + 1.$$

Second Solution. Denote the given lines in order by M_1, M_2, M_3, and M_4. Then the equation of the plane of the required line L and M_1 has the form

$$y = \lambda(x - 1).$$

The equation of the plane of L and M_2 has the form

$$z = \mu(y - 1)$$

and the plane of L and M_3 is given by

$$x = \nu(z - 1).$$

Any two of these equations determine the line L, so, if we eliminate y from the first two of these equations, we must obtain an equation equivalent to the third. Therefore $\lambda\mu\nu = 1$.

Let the point of intersection of L and M_4 be $(6d, 6d, -d)$. It lies on all of the planes considered above, so

$$6d = \lambda(6d - 1)$$
$$-d = \mu(6d - 1)$$
$$6d = \nu(-d - 1).$$

Multiply these equations and use $\lambda\mu\nu = 1$ to get

$$-36d^3 = -(6d - 1)^2(d + 1).$$

This simplifies to $24d^2 - 11d + 1 = 0$, so $d = \frac{1}{3}$ or $\frac{1}{8}$. The corresponding values of λ, μ, ν are $2, -\frac{1}{3}, -\frac{3}{2}$ or $-3, +\frac{1}{2}, -\frac{2}{3}$, and we obtain non-parametric equations for L_1 and L_2 as before.

For a general treatment of this problem, see D. M. Y. Sommerville, *Analytic Geometry of Three Dimensions*, Cambridge, 1934, page 184.

---

5. (i) Solve the system of differential equations

$$\frac{dx}{dt} = x + y - 3,$$

$$\frac{dy}{dt} = -2x + 3y + 1,$$

subject to the conditions $x = y = 0$ for $t = 0$.

First Solution. General existence theorems for linear differential equations assure us that there is a unique solution to the given system satisfying the initial conditions and that this solution is infinitely differentiable.

Solve the first equation for y and then differentiate:

(1) $$y = \frac{dx}{dt} - x + 3$$

(2) $$\frac{dy}{dt} = \frac{d^2x}{dt^2} - \frac{dx}{dt}.$$

Now eliminate y from the second of the original equations:

(3) $$\frac{d^2x}{dt^2} - \frac{dx}{dt} = -2x + 3\left(\frac{dx}{dt} - x + 3\right) + 1$$

whence

(4) $$\frac{d^2x}{dt^2} - 4\frac{dx}{dt} + 5x = 10.$$

This equation has the obvious particular solution $x = 2$ and the general solution

(5) $$x = e^{2t}(A\cos t + B\sin t) + 2.$$

The constants A and B can be evaluated from the initial conditions $x = 0$ and $dx/dt = -3$ (derived from the given conditions and from the first of the original equations) when $t = 0$. We find $A = -2$, $B = +1$, and therefore

$$x = e^{2t}(-2\cos t + \sin t) + 2. \tag{6}$$

From (1), we now obtain

$$y = e^{2t}(-\cos t + 3\sin t) + 1. \tag{7}$$

It is easy to verify that equations (6) and (7) define solutions of the given system.

Second Solution. Treat the given system as a single differential equation in vectors

$$\mathbf{x}' = A\mathbf{x} + \mathbf{b}$$

where

$$\mathbf{x} = \begin{pmatrix} x \\ y \end{pmatrix}, \quad A = \begin{pmatrix} 1 & 1 \\ -2 & 3 \end{pmatrix} \quad \text{and} \quad \mathbf{b} = \begin{pmatrix} -3 \\ 1 \end{pmatrix}.$$

Solving the equation $A\mathbf{x} + \mathbf{b} = 0$, we find the constant particular solution $\binom{2}{1}$. The general solution is therefore

$$\mathbf{x} = (\exp At)\mathbf{c} + \begin{pmatrix} 2 \\ 1 \end{pmatrix}$$

where $\mathbf{c}$ is an arbitrary constant vector. Using the initial condition $x = 0$ when $t = 0$, we find

$$\mathbf{c} = -\begin{pmatrix} 2 \\ 1 \end{pmatrix}.$$

The characteristic polynomial of A is

$$\det(A - uI) = \begin{vmatrix} 1-u & 1 \\ -2 & 3-u \end{vmatrix} = (u-2)^2 + 1,$$

where I is the identity matrix. Hence

$$(A - 2I)^2 = -I$$

$$\exp(A - 2I)t = 1 + (A - 2I)t + \frac{(A - 2I)^2}{2!}t^2 + \frac{(A - 2I)^3}{3!}t^3 + \cdots$$

$$= (\cos t)I + (\sin t)(A - 2I),$$

and

$$\exp At = e^{2t}((\cos t)I + (\sin t)(A - 2I)),$$

giving finally

$$\mathbf{x} = e^{2t}\left\{(\cos t)\begin{pmatrix}-2\\-1\end{pmatrix} + (\sin t)\begin{pmatrix}1\\3\end{pmatrix}\right\} + \begin{pmatrix}2\\1\end{pmatrix},$$

which is equivalent to the two equations (6) and (7).

---

5. (ii) A heavy particle is attached to the end A of a light rod AB of length a. The rod is hinged at B so that it can turn freely in a vertical plane. The rod is balanced in the vertical position above the hinge and then slightly disturbed. Prove that the time taken to pass from the horizontal position to the lowest position is

$$\sqrt{\frac{a}{g}}\log_e(1 + \sqrt{2}).$$

Solution. Let m be the mass of the particle, and let θ be the angular position of the rod, measured from the vertical, at time t. The force of gravity mg can be resolved into two components, $mg\cos\theta$ acting along the rod, and $mg\sin\theta$ acting perpendicular to the rod. The former is counterbalanced by the tension (or compression) in the rod and the latter accelerates the particle along the circle of radius a. By Newton's third law we have

$$mg\sin\theta = ma\frac{d^2\theta}{dt^2}.$$

Multiply through by $\dfrac{2}{m}\dfrac{d\theta}{dt}$ to get

$$2g\sin\theta\frac{d\theta}{dt} = 2a\frac{d\theta}{dt}\frac{d^2\theta}{dt^2}.$$

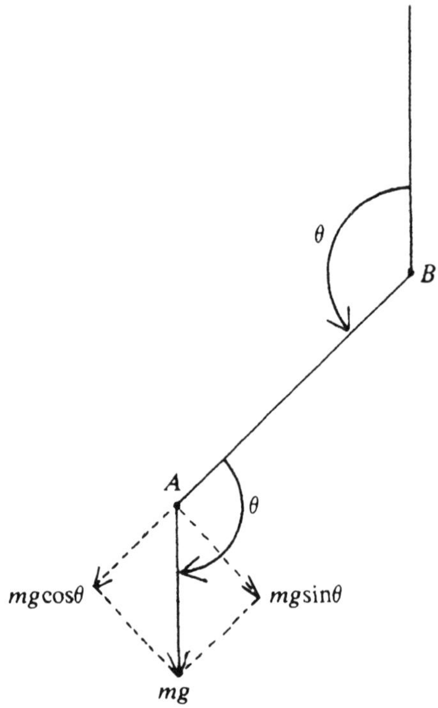

This can be integrated to give

$$-2g \cos \theta + k = a \left(\frac{d\theta}{dt}\right)^2. \tag{1}$$

From the initial conditions, $\theta = d\theta/dt = 0$ when $t = 0$, we find $k = 2g$. Thus we have

$$a\left(\frac{d\theta}{dt}\right)^2 = 2g(1 - \cos \theta) = 4g \sin^2 (\theta/2), \tag{2}$$

whence

$$\frac{d\theta}{dt} = 2\sqrt{(g/a)} \sin (\theta/2).$$

We have chosen the positive square root because $d\theta/dt$ is positive for $0 < \theta \le \pi$.

The time required for the passage from $\theta = \pi/2$ to $\theta = \pi$ is given by

$$\int_{\pi/2}^{\pi} \frac{dt}{d\theta}\, d\theta = \int_{\pi/2}^{\pi} \frac{1}{2}\sqrt{\frac{a}{g}} \csc (\theta/2)\, d\theta$$

$$= \sqrt{\frac{a}{g}}\,[-\log(\csc(\theta/2) + \cot(\theta/2))]_{\pi/2}^{\pi}$$

$$= \sqrt{\frac{a}{g}}\log(\sqrt{2} + 1).$$

First Remark. By using the fact that the kinetic energy of the particle is equal to its loss of potential energy, we could start with the equation

$$\frac{1}{2}m\left(a\frac{d\theta}{dt}\right)^2 = mga(1 - \cos\theta)$$

and obtain (2) directly.

Second Remark. We evaluated the constant of integration in (1) as if the particle fell from the very top of the circle, but actually no such motion is possible, as we can easily see by noting that the time required to fall from $\theta = \epsilon$ to $\theta = \pi$ approaches infinity as $\epsilon \to 0$. The precise result is that if $T(\epsilon)$ is the time required to pass from the horizontal position to the lowest position when the particle starts at $\theta = \epsilon$, then $\lim_{\epsilon \to 0} T(\epsilon) = \sqrt{a/g}\,\log(\sqrt{2} + 1)$. This follows because we can pass to the limit under the sign of integration in the formula

$$T(\epsilon) = \int_{\pi/2}^{\pi} \sqrt{\frac{a}{2g}}\frac{d\theta}{\sqrt{\cos\epsilon - \cos\theta}}.$$

---

6. (i) A circle of radius a rolls on the inner side of the circumference of a circle of radius $3a$. Find the area contained within the closed curve generated by a point on the circumference of the rolling circle.

Solution. Take rectangular coordinates with the origin at the center of the large circle so that the generating point P is in contact with the large circle at $A = (3a, 0)$. It can be seen from the diagram that when the small circle has rolled until the line of centers makes an angle θ with OA, the coordinates of P are

$$x = 2a\cos\theta + a\cos 2\theta$$

$$y = 2a\sin\theta - a\sin 2\theta.$$

These are, then, parametric equations for the path of P.

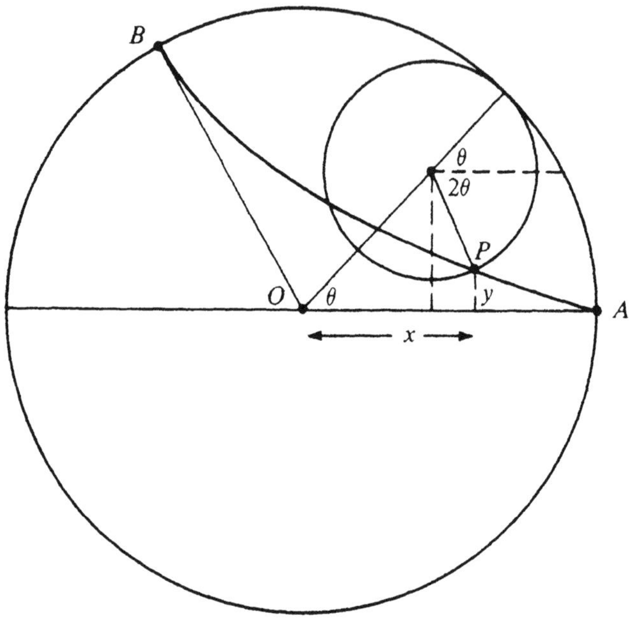

The area is given by

$$A = \frac{1}{2}\oint (x\,dy - y\,dx).$$

$$= \frac{a^2}{2}\int_0^{2\pi} \{[2\cos\theta + \cos 2\theta][2\cos\theta - 2\cos 2\theta]$$

$$- [2\sin\theta - \sin 2\theta][-2\sin\theta - 2\sin 2\theta]\}\,d\theta$$

$$= \frac{a^2}{2}\int_0^{2\pi} (2 - 2\cos 3\theta)\,d\theta = 2\pi\,a^2.$$

REMARKS. There is a vast literature on the properties of cycloids, epicycloids, and hypocycloids. See E. H. Lockwood, *A Book of Curves,* Cambridge University Press, 1961.

The special curve under consideration here is a three-cusped hypocycloid, also known as a deltoid. It is used in studying the properties of the Simson Line of a triangle. See D. C. Kay, *College Geometry,* Holt, Reinhart and Winston, 1969, pp. 248–263; or Lockwood, pp. 73–79.

In 1917, Kakeya proposed the problem of finding the region of least area in which a unit segment can be turned around. For several years it was believed that the region bounded by the deltoid was the smallest such region. However, in 1927, Besicovitch proved that there are regions of arbitrarily small area in which a segment can be turned around. The proof of this surprising fact is presented by Besicovitch in a lecture recorded on film, *The Kakeya Problem*, MAA Films 194X1043, distributed by Ward's—Modern Learning Aids Division, Rochester, New York.

---

6. (ii) A shell strikes an airplane flying at a height h above the ground. It is known that the shell was fired from a gun on the ground with a muzzle velocity of magnitude V, but the position of the gun and its angle of elevation are both unknown. Deduce that the gun is situated within a circle whose center lies directly below the airplane and whose radius is

$$\frac{V}{g}\sqrt{V^2 - 2gh}.$$

(Neglect the resistance of the atmosphere.)

Solution. Choose rectangular coordinates with the y-axis vertical, the origin at the position of the gun, and the airplane over a point of the positive x-axis. Then the coordinates of the airplane are (u, h) where $u \geq 0$.

If the gun is fired at time $t = 0$ with muzzle velocity V and elevation angle α, then (neglecting air resistance) the shell's position at time t is given by

$$x = Vt\cos\alpha$$

$$y = Vt\sin\alpha - \frac{1}{2}gt^2.$$

Since it is given that the shell strikes the airplane, we have

$$(1) \qquad \begin{aligned} u &= Vt\cos\alpha \\ h &= Vt\sin\alpha - \frac{1}{2}gt^2 \end{aligned}$$

for some t and α. Hence

$$u^2 + \left(h + \frac{1}{2}gt^2\right)^2 = V^2t^2,$$

so that

$$(2) \qquad \frac{1}{4}g^2t^4 + (gh - V^2)t^2 + h^2 + u^2 = 0.$$

In order that (2) have a real root t, it is necessary that

$$(gh - V^2)^2 \geq g^2(h^2 + u^2),$$

and therefore that

$$g^2u^2 \leq V^2(V^2 - 2gh).$$

Thus it is necessary that

$$V^2 \geq 2gh$$

and

$$u \leq \frac{V}{g}\sqrt{V^2 - 2gh}. \tag{3}$$

This shows that the gun is within distance $(V/g)\sqrt{V^2 - 2gh}$ from the point directly below the airplane when it was hit.

REMARK. Condition (3) is also sufficient that the airplane be within range of the gun, for when it is satisfied, the gunner can solve (2), obtaining in general two positive values of t, and then use (1) to determine the elevation at which to fire. Suppose for example $u = V^2/2g$, $h = V^2/4g$. Then (2) becomes

$$\frac{1}{4}g^2t^4 - \frac{3}{4}V^2t^2 + \frac{5V^4}{16g^2} = 0,$$

which has positive roots $\sqrt{\frac{1}{2}}\, V/g$ and $\sqrt{\frac{5}{2}}\, V/g$. From (1) we obtain the corresponding angles of elevation arctan 1 and arctan 3.

When the airplane is at the extreme limit of the range, i.e., when

$$g^2u^2 = V^2(V^2 - 2gh) \tag{4}$$

the two values of t coalesce, and so do the trajectories. In this case the unique trajectory is tangent to the boundary of the critical region at the point where the airplane is, because the trajectory does not leave the critical region. Hence, if we think of u and h as coordinates in the vertical plane, (4) is the equation of the envelope of the trajectories.

The sketch shows a portion of the envelope, and the two trajectories through the point

$$P = \left(\frac{1}{2}\frac{V^2}{g}, \frac{1}{4}\frac{V^2}{g}\right).$$

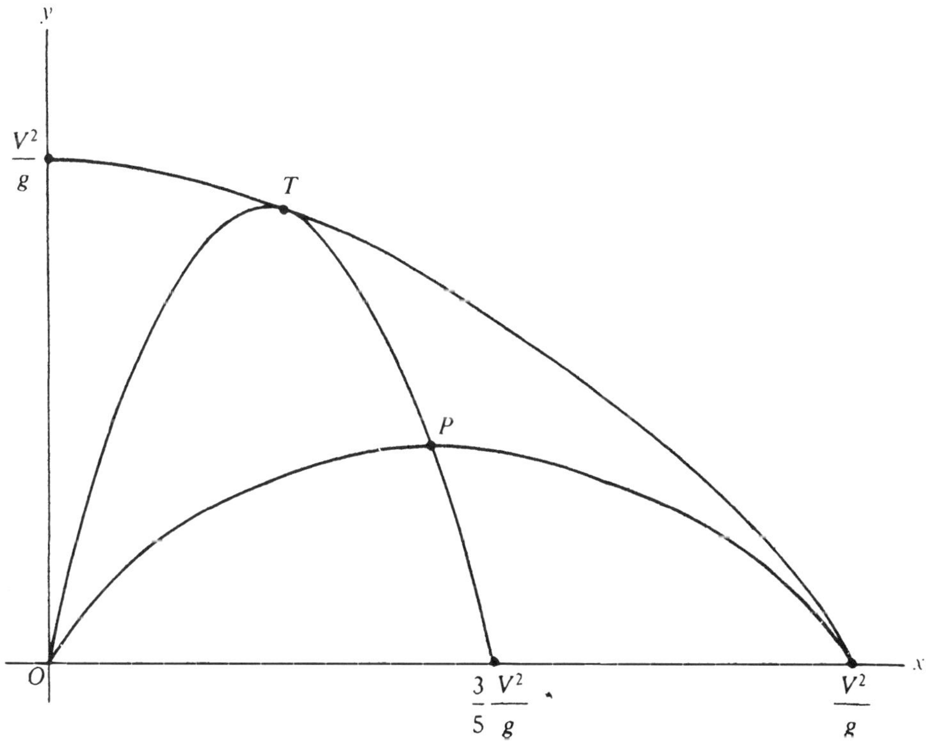

The trajectory (1) is tangent to the envelope at

$$T = \left(\frac{V^2}{g}\cot\alpha, \frac{V^2}{2g}\,(1 - \cot^2\alpha)\right).$$

It follows that the line of fire (the tangent to the trajectory at 0) bisects the angle between OT and the vertical.

---

7. (i) Find the curve touched by all the curves of the family

$$(y - k^2)^2 = x^2(k^2 - x^2).$$

Make a rough sketch showing this curve and two curves of the family.

Solution. We may use the graphs of

$$y = x^2(k^2 - x^2)$$

and

$$y^2 = x^2(k^2 - x^2)$$

as aids in sketching the family of curves:

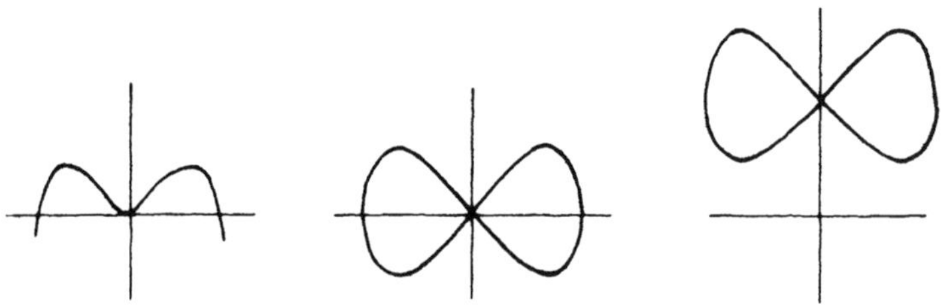

$y = x^2(k^2 - x^2)$ $\quad$ $y^2 = x^2(k^2 - x^2)$ $\quad$ $(y - k^2)^2 = x^2(k^2 - x^2)$

The function $x^2(k^2 - x^2)$ assumes its maximum when $x^2 = k^2 - x^2$; i.e., when $x = \pm k/\sqrt{2}$. Hence the graph of the curve

$$f(x, y, k) = (y - k^2)^2 - x^2(k^2 - x^2) = 0$$

has lower horizontal tangents at $(\pm k/\sqrt{2}, k^2/2)$ and upper horizontal tangents at $(\pm k/\sqrt{2}, 3k^2/2)$. Because f depends on k^2, we need only consider $k \geq 0$. The curve degenerates to a point for $k = 0$, so assume k positive. Clearly, the curve is contained in the strip $-k \leq x \leq k$. There are vertical tangents at $(\pm k, k^2)$. We can check this formally by noting that $\partial f/\partial y$ vanishes at this point, but $\partial f/\partial x$ does not. The curve has a double point at $(0, k^2)$ because both $\partial f/\partial x$ and $\partial f/\partial y$ vanish here. At this point the curve resembles a pair of crossed lines since dropping the terms of degree higher than two in x and $y - k^2$ gives

$$(y - k^2)^2 - k^2x^2 = 0,$$

whose graph is the union of the two lines $y - k^2 = \pm kx$.

To obtain the equation of the envelope, we eliminate k from the two equations

$$f = (y - k^2)^2 - x^2(k^2 - x^2) = 0$$

and

$$\frac{\partial f}{\partial k} = -4k(y - k^2) - 2kx^2 = 0.$$

From the second equation we have written either $k = 0$ or $k^2 = y + \frac{1}{2}x^2$.

The first alternative leads to $y^2 = -x^4$, which is just the origin. The second gives

$$x^2(3x^2 - 4y) = 0$$

which represents the union of the line $x = 0$ and the parabola $4y = 3x^2$.

Although the y-axis meets each curve in a double point, it is not tangent to any curve of the family, so it is not part of the envelope. The parabola $4y = 3x^2$, however, is tangent to the curve $f(x, y, k) = 0$ at each of the points $(\pm(2/\sqrt{5})k, (3/5)k^2)$. Hence this parabola is the envelope, provided the one-point "curve" corresponding to $k = 0$ is regarded as tangent to it; otherwise, the envelope is the parabola less the origin.

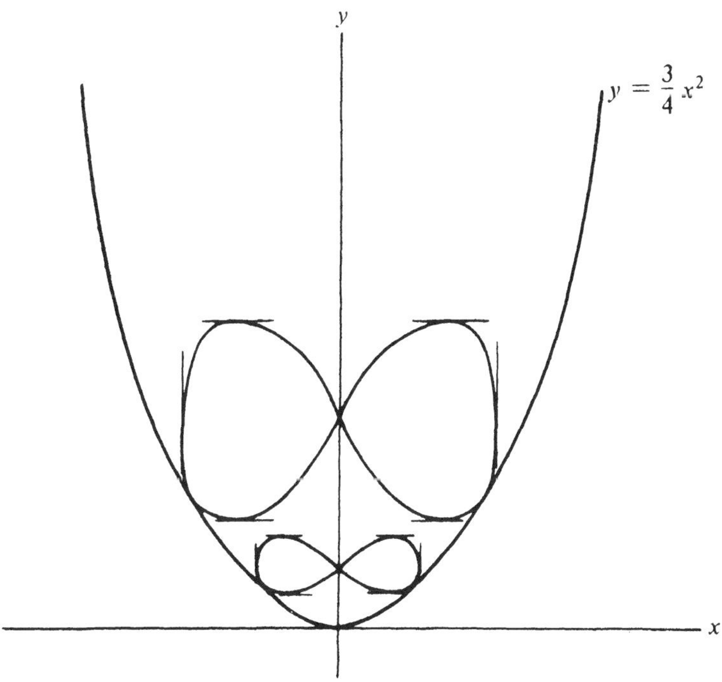

REMARK. One can see, without calculus, that the required curve is a parabola, because the given family of curves is invariant under the transformation $x' = \lambda x$, $y' = \lambda^2 y$, $k' = \lambda k$. Hence if (α, β) lies on the envelope, so does $(\lambda\alpha, \lambda^2\beta)$, and the equation of the envelope must be of the form $\alpha^2 y = \beta x^2$.

7. (ii) If the expansion in powers of x of the function

$$\frac{1}{(1 - ax)(1 - bx)}$$

is given by

$$c_0 + c_1 x + c_2 x^2 + c_3 x^3 + \cdots,$$

prove that the expansion in powers of x of the function

$$\frac{1 + abx}{(1 - abx)(1 - a^2 x)(1 - b^2 x)}$$

is given by

$$c_0^2 + c_1^2 x + c_2^2 x^2 + c_3^2 x^3 + \cdots.$$

Solution. First we obtain an explicit formula for the coefficients $\{c_i\}$. Using partial fractions, and assuming $a \neq b$, we get

$$\frac{1}{(1 - ax)(1 - bx)} = \frac{1}{b - a}\left(\frac{-a}{1 - ax} + \frac{b}{1 - bx}\right)$$

$$= \frac{1}{b - a}\left(-a \sum_0^\infty a^n x^n + b \sum_0^\infty b^n x^n\right)$$

and therefore

$$c_n = \frac{b^{n+1} - a^{n+1}}{b - a}.$$

Then we have

$$\sum_{n=0}^\infty c_n^2 x^n = \frac{1}{(a - b)^2}\left[a^2 \sum_{n=0}^\infty a^{2n} x^n - 2ab \sum_{n=0}^\infty a^n b^n x^n + b^2 \sum_{n=0}^\infty b^{2n} x^n\right]$$

$$= \frac{1}{(a - b)^2}\left[\frac{a^2}{1 - a^2 x} - \frac{2ab}{1 - abx} + \frac{b^2}{1 - b^2 x}\right]$$

$$= \frac{1 + abx}{(1 - a^2 x)(1 - abx)(1 - b^2 x)}.$$

SPECIAL CASE. If $a = b$, then

$$\frac{1}{(1-ax)(1-bx)}=\frac{1}{(1-ax)^2}=\sum_{n=0}^{\infty}(n+1)\,a^n x^n.$$

So, in this case, $c_n = (n+1)\,a^n$. For this value of c_n, we get

$$\sum_{n=0}^{\infty} c_n{}^2 x^n = \sum_{n=0}^{\infty}(n^2+2n+1)\,a^{2n}x^n$$

$$=\sum_{n=0}^{\infty}(n+1)(n+2)\,a^{2n}x^n - \sum_{n=0}^{\infty}(n+1)\,a^{2n}x^n$$

$$=\frac{2}{(1-a^2x)^3}-\frac{1}{(1-a^2x)^2}=\frac{1+a^2x}{(1-a^2x)^3}$$

which is the desired result when $a = b$.

REMARK. The power series involved here all converge for $|x| < \min\{|a|^{-1}, |b|^{-1}\}$. Therefore, the formal manipulations can all be justified provided neither a nor b is zero. If either is zero, the whole problem is easy.

The special case can be seen as a limiting case, for

$$\lim_{b\to a} c_n(a, b) = \lim_{b\to a}\frac{b^{n+1}-a^{n+1}}{b-a} = (n+1)\,a^n.$$

However, the problem is really an algebra problem and convergence is not important here, because we can consider all calculations as taking place in the ring of formal power series in x with coefficients in the field $Q(a, b)$, where a and b are indeterminates. In this field, $a \neq b$, so the special case is unnecessary. When we find that c_n is, in fact, a polynomial in a and b (the denominator $b - a$ divides out), it is automatic that our calculations remain valid if we replace b by a. It is, of course, easy to see from the beginning that the coefficients in either expansion will be polynomials in a and b.

Afternoon Session

8. From the vertex $(0, c)$ of the catenary

$$y = c\cosh\frac{x}{c}$$

a line L is drawn perpendicular to the tangent to the catenary at a point P. Prove that the length of L intercepted by the axes is equal to the ordinate y of the point P.

Solution. At the point $(x_1, c \cosh(x_1/c))$ the slope of the given catenary is $\sinh(x_1/c)$. Hence the equation of the line L is

$$y - c = \frac{-x}{\sinh(x_1/c)},$$

and this line intersects the x-axis at $(c \sinh(x_1/c), 0)$. Therefore the length of the segment of L between the axes is $\sqrt{c^2 \sinh^2(x_1/c) + c^2} = c \cosh(x_1/c)$, which is indeed the ordinate of the point P.

---

9. Evaluate the definite integrals

$$\text{(i)} \int_1^3 \frac{dx}{\sqrt{(x-1)(3-x)}}, \qquad \text{(ii)} \int_1^\infty \frac{dx}{e^{x+1} + e^{3-x}}.$$

Solution. Part (i). Since the integrand is not defined at either bound of integration, one should write

$$\int_1^3 \frac{dx}{\sqrt{(x-1)(3-x)}} = \lim_{\substack{\epsilon \to 0+ \\ \delta \to 0+}} \int_{1+\epsilon}^{3-\delta} \frac{dx}{\sqrt{(x-1)(3-x)}}$$

$$= \lim_{\substack{\epsilon \to 0+ \\ \delta \to 0+}} \int_{1+\epsilon}^{3-\delta} \frac{dx}{\sqrt{1-(x-2)^2}} = \lim_{\substack{\epsilon \to 0+ \\ \delta \to 0+}} \arcsin(x-2)\Big]_{1+\epsilon}^{3-\delta}$$

$$= \lim_{\substack{\epsilon \to 0+ \\ \delta \to 0+}} [\arcsin(1-\delta) - \arcsin(\epsilon - 1)] = \frac{\pi}{2} + \frac{\pi}{2} = \pi.$$

Part (ii). The difficulty here is with the infinite interval of integration. Let $y = x - 1$; then

$$\int \frac{dx}{e^{x+1} + e^{3-x}} = \frac{1}{e^2} \int \frac{dx}{e^{x-1} + e^{1-x}} = \frac{1}{e^2} \int \frac{dy}{e^y + e^{-y}}$$

$$= \frac{1}{e^2} \int \frac{e^y\, dy}{e^{2y} + 1} = \frac{1}{e^2} \arctan e^y + c.$$

Hence

$$\int_1^\infty \frac{dx}{e^{x+1} + e^{3-x}} = \lim_{N\to\infty} \int_1^N \frac{dx}{e^{x+1} + e^{3-x}}$$

$$= \lim_{N\to\infty} \frac{1}{e^2} [\arctan e^{N-1} - \arctan e^0]$$

$$= \frac{1}{e^2}\left[\frac{\pi}{2} - \frac{\pi}{4}\right] = \frac{\pi}{4e^2}.$$

---

10. Given the power-series

$$a_0 + a_1 x + a_2 x^2 + \cdots$$

in which

$$a_n = (n^2 + 1)3^n,$$

show that there is a relation of the form

$$a_n + p a_{n+1} + q a_{n+2} + r a_{n+3} = 0,$$

in which p, q, r are constants independent of n. Find these constants and the sum of the power-series.

First Solution. The desired relation is

$$(n^2 + 1)3^n + p((n + 1)^2 + 1)3^{n+1}$$
$$+ q((n + 2)^2 + 1)3^{n+2} + r((n + 3)^2 + 1)3^{n+3} = 0,$$

which is equivalent to

(1)
$$n^2(1 + 3p + 9q + 27r) + n(6p + 36q + 162r)$$
$$+ (1 + 6p + 45q + 270r) = 0.$$

Equation (1) holds for all n if and only if

$$1 + 3p + 9q + 27r = 0$$
$$p + 6q + 27r = 0$$
$$1 + 6p + 45q + 270r = 0.$$

These linear equations have the solution $p = -1$, $q = \frac{1}{3}$, $r = -\frac{1}{27}$, so

$$a_n - a_{n+1} + \frac{1}{3} a_{n+2} - \frac{1}{27} a_{n+3} = 0.$$

Let $S(x) = a_0 + a_1 x + a_2 x^2 + \cdots$. Proceeding formally, we have

$$\begin{aligned}
x^3 S(x) &= a_0 x^3 + a_1 x^4 + \cdots + a_{n-3} x^n + \cdots \\
px^2 S(x) &= pa_0 x^2 + pa_1 x^3 + pa_2 x^4 + \cdots + pa_{n-2} x^n + \cdots \\
qxS(x) &= qa_0 x + qa_1 x^2 + qa_2 x^3 + qa_3 x^4 + \cdots + qa_{n-1} x^n + \cdots \\
rS(x) &= ra_0 + ra_1 x + ra_2 x^2 + ra_3 x^3 + ra_4 x^4 + \cdots + ra_n x^n + \cdots .
\end{aligned}$$

When we sum these we get

$$S(x)[x^3 + px^2 + qx + r] = (pa_0 + qa_1 + ra_2)x^2 + (qa_0 + ra_1)x + ra_0.$$

Multiplying through by -27, we obtain

$$S(x)[1 - 9x + 27x^2 - 27x^3] = 1 - 3x + 18x^2,$$

and therefore

$$S(x) = \frac{1 - 3x + 18x^2}{(1 - 3x)^3} .$$

Using the ratio test we conclude that the series converges for $|x| < \frac{1}{3}$; hence the formal manipulations above are valid for these values of x.

Second Solution. Let $b_n = a_n/3^n = n^2 + 1$. Then

$$\Delta b_n = b_{n+1} - b_n = 2n + 1, \qquad \Delta^2 b_n = b_{n+2} - 2b_{n+1} + b_n = 2$$

$$\Delta^3 b_n = b_{n+3} - 3b_{n+2} + 3b_{n+1} - b_n = 0.$$

So

$$\frac{a_{n+3}}{3^{n+3}} - \frac{3a_{n+2}}{3^{n+2}} + \frac{3a_{n+1}}{3^{n+1}} - \frac{a_n}{3^n} = 0,$$

whence

$$a_n - a_{n+1} + (1/3)\, a_{n+2} - (1/27)\, a_{n+3} = 0.$$

Since

$$n^2 + 1 = (n + 1)(n + 2) - 3(n + 1) + 2,$$

we have

$$\Sigma b_n y^n = \Sigma(n+1)(n+2)y^n - 3\,\Sigma(n+1)y^n + 2\,\Sigma y^n$$

$$= \frac{2}{(1-y)^3} - \frac{3}{(1-y)^2} + \frac{2}{1-y}$$

$$= \frac{1-y+2y^2}{(1-y)^3},$$

provided $|y| < 1$. Replace y by $3x$.

$$\Sigma a_n x^n = \frac{1-3x+18x^2}{(1-3x)^3},$$

provided $|x| < \frac{1}{3}$.

REMARK. We could assume that the problem is concerned with the ring of formal power series. In that case, $(1 - 3x)$ has an inverse in the ring and our result is that

$$\Sigma a_n x^n = (1-3x+18x^2)(1-3x)^{-3}.$$

---

11. Find the equation of the parabola which touches the x-axis at the point $(1, 0)$ and the y-axis at the point $(0, 2)$. Find the equation of the axis of the parabola and the coordinates of its vertex.

First Solution. Clearly the required parabola does not pass through the origin, and any conic not passing through the origin has an equation of the form

$$Ax^2 + Bxy + Cy^2 + Dx + Ey + 1 = 0.$$

In order that this conic be tangent to the x-axis at $(1, 0)$, the equation obtained by setting $y = 0$ must have a double root at $x = 1$. Hence $A = 1$ and $D = -2$. In order that it be tangent to the y axis at $(0, 2)$, we must have $C = \frac{1}{4}$, $E = -1$. In order that the conic be a parabola we must have $B^2 = 4AC$. This leads to two possibilities

$$x^2 + xy + \frac{1}{4}y^2 - 2x - y + 1 = 0 \tag{1}$$

$$x^2 - xy + \frac{1}{4}y^2 - 2x - y + 1 = 0. \tag{2}$$

Since (1) can be written as $(x + \frac{1}{2}y - 1)^2 = 0$, we see that it represents a degenerate conic, the double line through the two given points. Hence (2) is the equation of the required parabola.

It is convenient to multiply equation (2) by 4 to eliminate the fraction. We obtain

$$4x^2 - 4xy + y^2 - 8x - 4y + 4 = 0. \tag{3}$$

Since the quadratic terms in (3) can be written in the form $(2x - y)^2$, a transformation is suggested. The orthogonal (but not scale-preserving) transformation

$$u = 2x - y$$
$$v = x + 2y$$

with inverse

$$x = \frac{1}{5}(2u + v)$$

$$y = \frac{1}{5}(-u + 2v)$$

transforms (3) into

$$u^2 - \frac{12}{5}u - \frac{16}{5}v + 4 = 0. \tag{4}$$

This has the standard form

$$\left(u - \frac{6}{5}\right)^2 - \frac{16}{5}\left(v - \frac{4}{5}\right) = 0.$$

In this form the axis of the parabola is the line $u = \frac{6}{5}$, and the vertex has uv-coordinates $(\frac{6}{5}, \frac{4}{5})$. In terms of the original coordinates, the axis has the equation $2x - y = \frac{6}{5}$ and the vertex is at $(\frac{16}{25}, \frac{2}{25})$.

There is another way to handle the first part of this solution.

Suppose two distinct conics are given by the quadratic equations $f(x, y) = 0$ and $g(x, y) = 0$. These conics meet in four points (in the complex projective place, counting multiolicities) and any other conic passing through these four points has an equation $\lambda f + \mu g = 0$ for suitable choice of λ and μ (only the ratio $\lambda{:}\mu$ matters, of course). Conics tangent to the x-axis at P and tangent to the y-axis at Q form such a family because P and Q are both double points in this case. Two degenerate conics in this family are $xy = 0$ (the union of the coordinate axes) and $(2x + y - 2)^2 = 0$ (the double line PQ); consequently all conics of the family have equations of the form

$$\lambda xy + \mu(2x + y - 2)^2 = 0.$$

Now we determine the parabola in this family by the condition that its discriminant vanishes. We find $\lambda:\mu = 0$ or $\lambda:\mu = -8$. The former gives back the double line; and the latter gives the desired parabola.

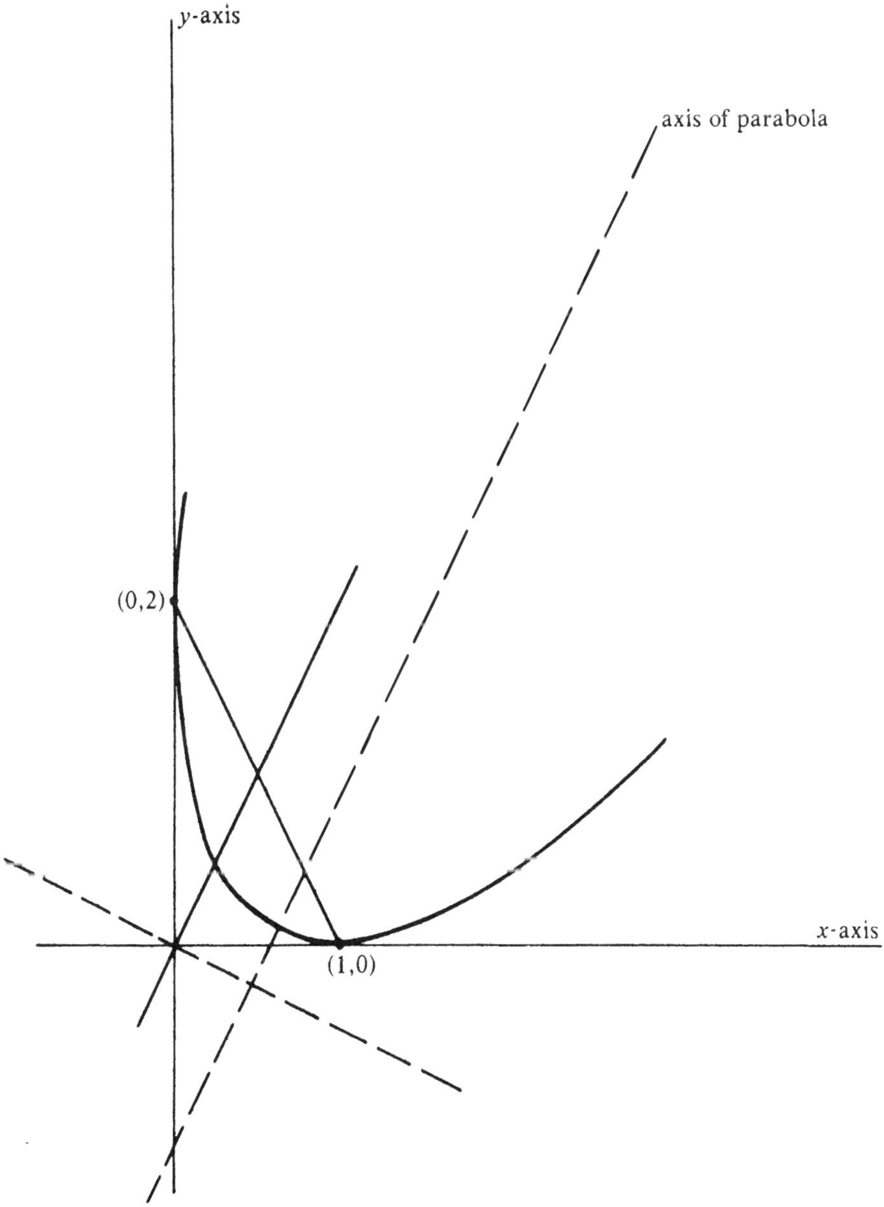

Second Solution. Let $\mathcal{P}$ be a parabola with focus F and directrix d, and suppose two perpendicular lines through O are tangent to $\mathcal{P}$ at P and Q, respectively. Using the well-known reflection property of the parabola, we shall prove synthetically that O is on d and F is the foot of the perpendicular from O on PQ.

Let G and H be the reflections of F in the tangents OP and OQ, respectively. Then triangles POF and POG are congruent, as are triangles QOF and QOH. Because $\angle FPO = \angle GPO$, the ray $\overrightarrow{FP}$ on reflection from the parabola at P has the direction opposite to $\overrightarrow{PG}$. From the reflection property it follows that PG is parallel to the axis of $\mathcal{P}$ and perpendicular to d. Therefore, the distance from P to d is measured along PG. Since P is equidistant from F and d and $PF = PG$, we conclude that G is on d. Similarly, H is on d.

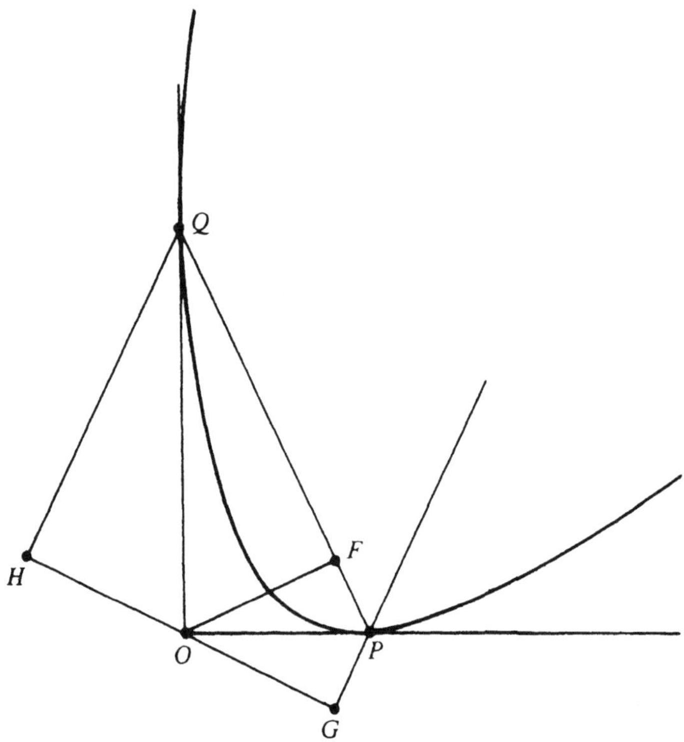

Now $\angle GOF = 2\angle POF$ and $\angle FOH = 2\angle FOQ$, so $\angle GOH = 2\angle POQ = 2$ right angles, since OP and OQ are perpendicular. Hence O lies on $GH = d$. Therefore $\angle OGP$ and $\angle OHQ$ are right angles. From the original congruent triangles it follows that $\angle OFP$ and $\angle OFQ$ are right angles. Hence $\angle PFQ$ is a straight angle, so F is at the foot of the perpendicular from O on PQ, as claimed.

In the problem at hand O is the origin and P and Q are the points $(1, 0)$

and (0, 2). The line PQ has equation $2x + y = 2$ and the perpendicular from O on PQ has slope $\frac{1}{2}$. The point F is found to be $(\frac{4}{5}, \frac{2}{5})$. The direction of the axis is found by reflecting the line PQ in either tangent, so the slope of the axis is 2. Since it passes through F the equation of the axis is

$$5y = 10x - 6.$$

The directrix d has slope $-\frac{1}{2}$ and passes through O; hence its equation is $2y + x = 0$. The directrix and axis meet at $(\frac{12}{25}, \frac{-6}{25})$, and the vertex is halfway between this point and the focus; that is, at $(\frac{16}{25}, \frac{2}{25})$.

If (x, y) is any point on the parabola, it is equidistant from the directrix and the focus. Thus, using the squares of these distances

$$\left(\frac{2y + x}{\sqrt{5}}\right)^2 = \left(x - \frac{4}{5}\right)^2 + \left(y - \frac{2}{5}\right)^2.$$

This simplifies to

$$4x^2 - 4xy + y^2 - 8x - 4y + 4 = 0,$$

the equation of the parabola.

---

12. (i) Prove that

$$\int_1^a [x]f'(x)\,dx = [a]f(a) - \{f(1) + \cdots + f([a])\},$$

where a is greater than 1 and where $[x]$ denotes the greatest of the integers not exceeding x. Obtain a corresponding expression for

$$\int_1^a [x^2]f'(x)\,dx.$$

Solution. We have

$$\int_1^a [x]f'(x)dx = \int_1^2 1\cdot f'(x)dx + \int_2^3 2\cdot f'(x)dx + \cdots + \int_{[a]}^a [a]\cdot f'(x)dx$$

$$= f(2) - f(1) + 2(f(3) - f(2)) + \cdots + [a](f(a) - f([a]))$$

$$= [a]f(a) - \{f(1) + f(2) + \cdots + f([a])\}.$$

For the second part, we have

$$\int_1^a [x^2]f'(x)dx = \int_1^{\sqrt{2}} 1\cdot f'(x)dx + \int_{\sqrt{2}}^{\sqrt{3}} 2f'(x)dx + \cdots + \int_{\sqrt{[a^2]}}^{a} [a^2]f'(x)dx$$

$$= (f(\sqrt{2}) - f(1)) + 2(f(\sqrt{3}) - f(\sqrt{2})) + \cdots$$

$$+ [a^2](f(a) - f(\sqrt{[a^2]}))$$

$$= [a^2]f(a) - \{f(1) + f(\sqrt{2}) + \cdots + f(\sqrt{[a^2]})\}.$$

REMARK. These formulas result from integration by parts applied to Stieltjes integrals; for example:

$$\int_1^a [x]f'(x)dx = \int_{1/2}^a [x]f'(x)dx = [x]f(x)\Big|_{1/2}^a - \int_{1/2}^a f(x)d[x]$$

$$= [a]f(a) - (f(1) + f(2) + \cdots + f([a])).$$

---

12. (ii) A particle moves on a straight line, the only force acting on it being a resistance proportional to the velocity. If it started with a velocity of 1,000 ft. per sec. and had a velocity of 900 ft. per sec. when it had travelled 1,200 ft., calculate to the nearest hundredth of a second the time it took to travel this distance.

Solution. The differential equation governing the motion is

$$m\frac{d^2x}{dt^2} = -k\frac{dx}{dt},$$

and the boundary conditions are

$$x = 0, \qquad \frac{dx}{dt} = 1000, \qquad \text{when } t = 0$$

$$x = 1200, \qquad \frac{dx}{dt} = 900, \qquad \text{when } t = T,$$

where T is the time required.

Let $b = k/m$. Then $d^2x/dt^2 = -b\,dx/dt$, which implies

$$(1) \qquad \frac{dx}{dt} = -bx + c.$$

The boundary conditions give

$$1000 = c$$
$$900 = -1200b + c,$$

whence $b = \frac{1}{12}$. Using these values and (1), we have

$$T = \int_0^{1200} \frac{dt}{dx}\,dx = \int_0^{1200} \frac{dx}{1000 - x/12} = 12 \log \frac{10}{9}.$$

To evaluate this it is convenient to write $T = -12 \log \frac{9}{10}$ and use the series expansion

$$-\log(1 - x) = x + \frac{1}{2}x^2 + \frac{1}{3}x^3 + \frac{1}{4}x^4 + \cdots.$$

Taking $x = \frac{1}{10}$, we have

$$\frac{1}{10} + \frac{1}{200} + \frac{1}{3000} < -\log\frac{9}{10} < \frac{1}{10} + \frac{1}{200} + \frac{1}{3}\sum_{n=3}^{\infty}\left|\frac{1}{10}\right|^n.$$

The lower bound exceeds $.1 + .005 + .0003 = .1053$, and the upper bound is $.1 + .005 + 1/2700 < .1054$. Therefore,

$$1.2636 < -12 \log \frac{9}{10} < 1.2648$$

so, to the nearest hundreth of a second, $T \simeq 1.26$ sec.

---

13. (i) Let $f(x)$ be defined for $a \le x \le b$. Assuming appropriate properties of continuity and derivability, prove for $a < x < b$ that

$$\frac{\dfrac{f(x) - f(a)}{x - a} - \dfrac{f(b) - f(a)}{b - a}}{x - b} = \tfrac{1}{2} f''(\xi),$$

where ξ is some number between a and b.

First Solution. Assume f is continuous on $[a, b]$ and has a second derivative at each point of (a, b).

Let x be fixed with $a < x < b$. Set

$$g(t) = \begin{vmatrix} f(t) & t^2 & t & 1 \\ f(x) & x^2 & x & 1 \\ f(a) & a^2 & a & 1 \\ f(b) & b^2 & b & 1 \end{vmatrix}.$$

Then $g(a) = g(x) = g(b) = 0$. By the mean value theorem there exist numbers α and β such that $a < \alpha < x < \beta < b$ and $g'(\alpha) = g'(\beta) = 0$. Applying the mean value theorem once again, we see that there is a number ξ such that $\alpha < \xi < \beta$ and $g''(\xi) = 0$. Evidently, ξ is between a and b.

Now

$$g''(\xi) = \begin{vmatrix} f''(\xi) & 2 & 0 & 0 \\ f(x) & x^2 & x & 1 \\ f(a) & a^2 & a & 1 \\ f(b) & b^2 & b & 1 \end{vmatrix} = 0.$$

Expanding the determinant by minors of the first row, we find

$$\frac{1}{2}f''(\xi) = \frac{\begin{vmatrix} f(x) & x & 1 \\ f(a) & a & 1 \\ f(b) & b & 1 \end{vmatrix}}{\begin{vmatrix} x^2 & x & 1 \\ a^2 & a & 1 \\ b^2 & b & 1 \end{vmatrix}}$$

$$= \frac{\dfrac{f(x) - f(a)}{x - a} - \dfrac{f(a) - f(b)}{a - b}}{x - b},$$

as required.

Second Solution. Assume f is continuous on $[a, b]$ and has a second derivative at each point of (a, b). For a fixed x with $a < x < b$, let

$$\lambda = \frac{\dfrac{f(x) - f(a)}{x - a} - \dfrac{f(a) - f(b)}{b - a}}{x - b}.$$

Then

$$f(x) = f(a) + \frac{f(b) - f(a)}{b - a}(x - a) + \lambda(x - a)(x - b).$$

Define

$$h(t) = f(t) - \left\{ f(a) + \frac{f(b) - f(a)}{b - a}(t - a) + \lambda(t - a)(t - b) \right\}.$$

Then $h(a) = h(x) = h(b) = 0$, so by the mean value theorem there are numbers α and β such that $a < \alpha < x < \beta < b$ and $h'(\alpha) = h'(\beta) = 0$. Then h'' vanishes for some number ξ in (α, β) and hence between a and b. We have $h''(\xi) = f''(\xi) - 2\lambda = 0$, and therefore

$$\lambda = \frac{1}{2} f''(\xi),$$

as required.

REMARK. The auxiliary functions g and h in the two solutions are related by $g(t) = (x - a)(x - b)(a - b)h(t)$.

---

13. (ii) Calculate the mutual gravitational attraction of two uniform rods, each of mass m and length $2a$, placed parallel to one another and perpendicular to the line joining their centers at a distance b apart.

In your answer let a approach zero, and comment on the form of the result.

First Solution. We first find the vertical component of the force of attraction between a particle P of mass μ situated at the point (h, b) and a uniform rod of mass m lying along the x-axis from $(0, 0)$ to $(0, 2a)$.

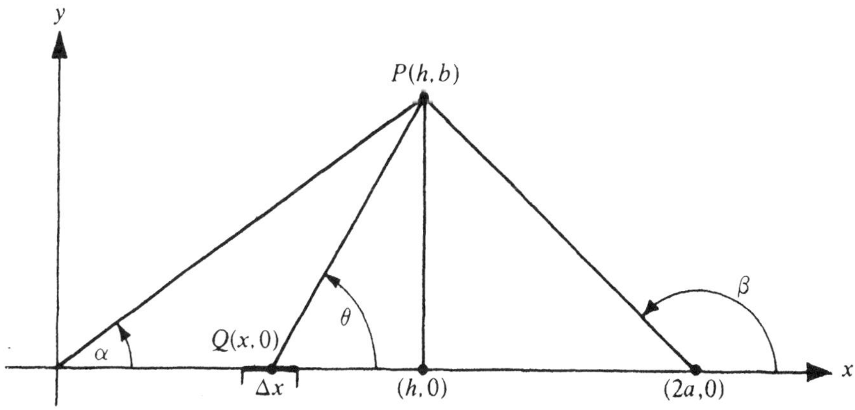

Consider a short segment S of the rod of length Δx and center at $Q = (x, 0)$. Let α, β, θ be the angles shown in the diagram. The mass of S is $m \cdot \Delta x/2a$. If the mass of S were concentrated at Q, the attractive force between S and P would be

$$G\mu \cdot \frac{m}{2a} \Delta x \cdot \frac{1}{b^2 \operatorname{cosec}^2 \theta}$$

and its vertical component would be

$$G\mu \cdot \frac{m}{2a} \Delta x \cdot \frac{1}{b^2 \operatorname{cosec}^2 \theta} \cdot \sin \theta,$$

where G is the constant of gravitation.

The vertical component of the total attractive force between P and the rod is therefore

$$f_y = \frac{G\mu m}{2ab^2} \int_0^{2a} \sin^3 \theta \, dx .$$

Now x and θ are related by $x + b \cot \theta = h$, so if we change the variable of integration to θ, we have

$$\begin{aligned} f_y &= \frac{G\mu m}{2ab} \int_\alpha^\beta \sin \theta \, d\theta \\ &= \frac{G\mu m}{2ab} (\cos \alpha - \cos \beta) \\ &= \frac{G\mu m}{2ab} \left(\frac{h}{\sqrt{h^2 + b^2}} - \frac{h - 2a}{\sqrt{(h - 2a)^2 + b^2}} \right). \end{aligned}$$

Consider now the two parallel rods, the second running from $(0, b)$ to $(2a, b)$. A short segment of the upper rod of length Δh may be considered a particle of mass $m\Delta h/2a$, and it follows as above that the vertical component of the force of attraction between the rods is

$$\begin{aligned} F_y &= \frac{Gm^2}{4a^2 b} \int_0^{2a} \left(\frac{h}{\sqrt{h^2 + b^2}} - \frac{h - 2a}{\sqrt{(h - 2a)^2 + b^2}} \right) dh \\ &= \frac{Gm^2}{4a^2 b} \left[\sqrt{h^2 + b^2} - \sqrt{(h - 2a)^2 + b^2} \right]_0^{2a} \\ &= \frac{Gm^2}{2a^2 b} (\sqrt{4a^2 + b^2} - b). \end{aligned}$$

It is clear from symmetry that the entire force acts along the line of centers, so the force is given by its vertical component.

Second Solution. Imagine the rods placed as shown, and consider the attraction between an element of length Δx at position x on the first rod and an element of length Δw at position w on the second rod.

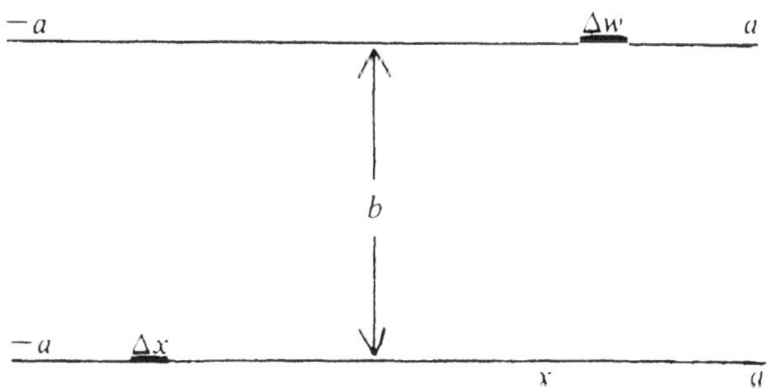

Since the masses of these elements are $m\,\Delta x/2a$ and $m\,\Delta w/2a$, the magnitude of the force is

$$\frac{Gm^2}{4a^2}\,\frac{\Delta w\cdot\Delta x}{b^2+(w-x)^2}$$

where G is the gravitational constant.

The component in the vertical direction is

$$\frac{Gm^2 b}{4a^2}\,\frac{\Delta w\cdot\Delta x}{[b^2+(w-x)^2]^{3/2}}.$$

The total force in the vertical direction is therefore

$$F_y=\frac{Gm^2 b}{4a^2}\iint\frac{dw\,dx}{\sqrt{b^2+(w-x)^2}}$$

where the integration is over the square $[-a, a]\times[-a, a]$.

Make the change of variables

$$w=\frac{1}{2}(u-v),\qquad x=\frac{1}{2}(u+v),$$

with Jacobian

$$\frac{\partial(w, x)}{\partial(u, v)}=\frac{1}{2}.$$

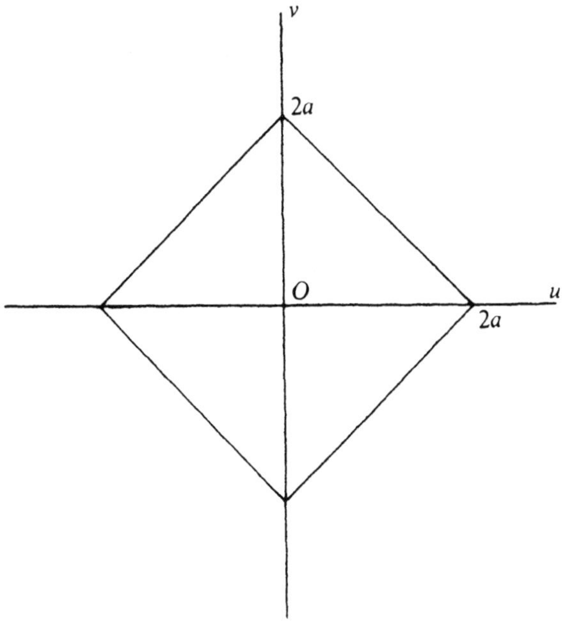

Then the integral becomes

$$\frac{Gm^2b}{8a^2}\iint_R \frac{du\,dv}{(b^2+v^2)^{3/2}},$$

where R is the diamond-shaped region pictured. The integral is symmetric over the four quadrants, so

$$F_y = \frac{Gm^2b}{2a^2}\int_{v=0}^{2a}\int_{u=0}^{2a-v}\frac{du\,dv}{(b^2+v^2)^{3/2}}$$

$$= \frac{Gm^2b}{2a^2}\int_0^{2a}\frac{2a-v}{(b^2+v^2)^{3/2}}\,dv.$$

The substitution $v = b\tan\theta$ reduces this integral to a tractable form, and after some calculation we find an indefinite integral. We have

$$F_y = \frac{Gm^2b}{2a^2}\left[\frac{2av}{b^2\sqrt{b^2+v^2}} + \frac{1}{\sqrt{b^2+v^2}}\right]_0^{2a}$$

$$= \frac{Gm^2b}{2a^2}\left[\frac{4a^2}{b^2\sqrt{b^2+4a^2}} + \frac{1}{\sqrt{b^2+4a^2}} - \frac{1}{b}\right]$$

$$= \frac{Gm^2}{2ba^2}\left[\sqrt{b^2+4a^2} - b\right].$$

The horizontal component of the force is evidently zero.

FIRST REMARK. If m is held fixed and a approaches zero, one expects to get the gravitational attraction between two particles each of mass m separated by a distance b. Indeed, using L'Hospital's rule we find that

$$\lim_{a \to 0} \frac{Gm^2}{2b} \frac{\sqrt{4a^2 + b^2} - b}{a^2} = \lim_{a \to 0} \frac{Gm^2}{2b} \frac{4a(4a^2 + b^2)^{-1/2}}{2a} = \frac{Gm^2}{b^2}.$$

SECOND REMARK. In both solutions we have treated the rods as limits of arrays of particles. By routine calculations with Riemann sums we could prove rigorously that the integral in (1) is the limit of the vertical components of the forces of attraction between the particle P and the arrays of particles that "approximate" the rod. That this limit is the vertical component of the actual force between P and the rod is an essential postulate of mechanics. For a discussion of this point, see Kellogg, *Foundations of Potential Theory,* Ungar Publishing Company, 1929, chapter 1.

---

14. (i) If

$$u = 1 + \frac{x^3}{3!} + \frac{x^6}{6!} + \cdots,$$

$$v = \frac{x}{1!} + \frac{x^4}{4!} + \frac{x^7}{7!} + \cdots,$$

$$w = \frac{x^2}{2!} + \frac{x^5}{5!} + \frac{x^8}{8!} + \cdots,$$

prove that

$$u^3 + v^3 + w^3 - 3uvw = 1.$$

First Solution. The power series for u, v, and w converge for all x, and

$$\frac{du}{dx} = w, \qquad \frac{dv}{dx} = u, \qquad \frac{dw}{dx} = v,$$

as we see by differentiating them. Letting $f = u^3 + v^3 + w^3 - 3uvw$, we have

$$\begin{aligned} f' &= 3u^2u' + 3v^2v' + 3w^2w' - 3uvw' - 3uv'w - 3u'vw \\ &= 3u^2w + 3v^2u + 3w^2v - 3uv^2 - 3u^2w - 3vw^2 = 0. \end{aligned}$$

Thus $f =$ constant. But $f(0) = [u(0)]^3 = 1$, so $f(x) = 1$ for all x.

Second Solution. Let $\omega = \exp(2\pi i/3)$, a primitive cube root of unity. Then $1 + \omega + \omega^2 = 0$. Also

$$\begin{aligned} u^3 + v^3 + w^2 - 3uvw &= (u + v + w)(u + \omega v + \omega^2 w)(u + \omega^2 v + \omega w) \\ &= (\exp x)(\exp \omega x)(\exp \omega^2 x) \\ &= \exp[(1 + \omega + \omega^2)x] = \exp 0 = 1. \end{aligned}$$

---

14. (ii) Consider the central conics

$$(ax^2 + by^2) + 2(px + qy) + c = 0,$$
$$(ax^2 + by^2) + 2\lambda(px + qy) + \lambda^2 c = 0,$$

where λ is a given positive constant.

Show that if all radii from the origin to the first conic are changed in the ratio λ to 1 the tips of these new radii generate the second conic.

Let P be the point with coordinates

$$x = -\frac{p}{a}\,\frac{2\lambda}{1+\lambda}, \qquad y = -\frac{q}{b}\,\frac{2\lambda}{1+\lambda}.$$

Show that if all radii from P to the first conic are changed in the ratio λ to 1 and then reversed about P the tips of these new radii generate the second conic.

Comment on these results in case $\lambda = 1$.

Solution. Call the two conics C and D,

$$C\colon (ax^2 + by^2) + 2(px + qy) + c = 0,$$
$$D\colon (ax^2 + by^2) + 2\lambda(px + qy) + \lambda^2 c = 0.$$

Suppose a point (x_0, y_0) is on C, and is transformed to (x_1, y_1) as described, with $(x_1, y_1) = (\lambda x_0, \lambda y_0)$. Then substituting (x_1, y_1) in the equation of

D we get

$$\begin{aligned}&ax_1^2 + by_1^2 + 2\lambda(px_1 + qy_1) + \lambda^2 c\\ &= \lambda^2[ax_0^2 + by_0^2 + 2(px_0 + qy_0) + c] = 0\end{aligned}$$

since (x_0, y_0) is on C. So (x_1, y_1) is indeed on D. Conversely, if (x_1, y_1) is on D the same equation shows that (x_0, y_0) is on C. [Here we need to divide by λ^2 and use the hypothesis $\lambda \neq 0$.]

Under the second transformation a point (x_0, y_0) becomes the point (x_2, y_2) where

$$x_2 = -\lambda\left(x_0 + \frac{p}{a}\frac{2\lambda}{1+\lambda}\right) - \frac{p}{a}\frac{2\lambda}{1+\lambda} = -\lambda\left(x_0 + \frac{2p}{a}\right)$$

$$y_2 = -\lambda\left(y_0 + \frac{q}{b}\frac{2\lambda}{1+\lambda}\right) - \frac{q}{b}\frac{2\lambda}{1+\lambda} = -\lambda\left(y_0 + \frac{2q}{b}\right).$$

Then if (x_0, y_0) is on C, we have

$$\begin{aligned}&ax_2^2 + by_2^2 + 2\lambda(px_2 + qy_2) + \lambda^2 c\\ &= \lambda^2\left[ax_0^2 + 4px_0 + \frac{4p^2}{a} + by_0^2 + 4qy_0 + \frac{4q^2}{b}\right.\\ &\quad\left. - 2\left(px_0 + \frac{2p^2}{a} + qy_0 + \frac{2q^2}{b}\right) + c\right]\\ &= \lambda^2[ax_0^2 + by_0^2 + 2(px_0 + qy_0) + c] = 0\end{aligned}$$

and hence (x_2, y_2) is on D. As before, the same procedure shows that if (x_2, y_2) is on D, (x_0, y_0) is on C.

If $\lambda = 1$, the two conics are the same. This conic has central symmetry about $(-p/a, -q/a)$, as we see by writing its equation in the form

$$a\left(x + \frac{p}{a}\right)^2 + b\left(y + \frac{q}{b}\right)^2 + c' = 0.$$

In this case, the first transformation is the identity and the second transformation is the central symmetry of the conic.

REMARK. If two figures E and F in Euclidean space of any dimension are homothetic about a point O with ratio $\lambda \neq -1$, and E has central symmetry about a point $Q \neq O$, then E and F are also homothetic about another point P with ratio $-\lambda$. In fact, the point P is determined by

$$\overrightarrow{QP} = \frac{\lambda - 1}{\lambda + 1}\overrightarrow{QO}.$$

This is equivalent to

$$\overrightarrow{OP} = \frac{2\lambda}{1 + \lambda}\overrightarrow{OQ},$$

and from this form the genesis of the problem is obvious.

The most familiar form of this result concerns circles in the plane. If two circles have different centers and different radii, they have two centers of similitude.

When $\lambda = 1$, E and F are the same, P coincides with Q, and the homothety about P becomes the central symmetry of E.

When $\lambda = -1$, but $Q \neq O$, the homothety about another point degenerates into a translation ("P is at infinity") and the conclusion becomes: F is a translate of E. This is because the product of two central symmetries with different centers is a translation. E is mapped onto itself by the central symmetry about Q and then onto F by the central symmetry about O; the effect of the two mappings is to map E onto F by a translation.

THE THIRD WILLIAM LOWELL PUTNAM MATHEMATICAL COMPETITION

March 2, 1940

Morning Session

1. Prove that if $f(x)$ is a polynomial with integral coefficients, and there exists an integer k such that none of the integers $f(1), f(2), \ldots, f(k)$ is divisible by k, then $f(x)$ has no integral root.

Solution. Suppose f has an integral root r. Then $f(x) = (x - r)g(x)$ where $g(x)$ is also a polynomial with integral coefficients. Then there are integers p and q such that $r = p + kq$ and $1 \le p \le k$. But $f(p) = (p - r)g(p) = -kq\, g(p)$ and hence $f(p)$ is divisible by k contrary to the hypothesis. This contradiction shows that $f(x)$ has no integral root.

---

2. Let A and B be two fixed points on the curve $y = f(x)$, where $f(x)$ is continuous and has a continuous derivative, and the arc AB is concave to the chord AB. If P is a point of the arc AB for which $AP + PB$ is a maximum, prove that PA and PB are equally inclined to the tangent to the curve $y = f(x)$ at the point P.

Solution. The theorem of the problem is a corollary to the following lemma which proves much more.

LEMMA. *Suppose S is a subset of a plane π, and A and B are points of π. Let P be a point of S such that*

$$AP + PB \ge AX + XB$$

for all X in S. Then the line t through P perpendicular to the bisector of $\angle APB$ is a support line of S (i.e., S is contained in one of the closed half-planes with edge t). Moreover, if P is not on the segment AB, P is the only point of $S \cap t$.

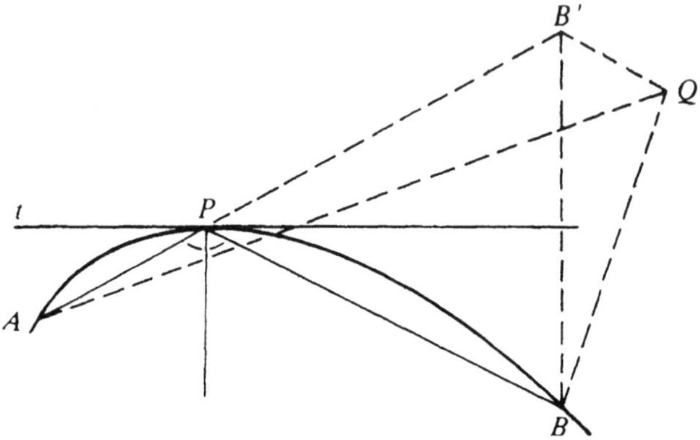

Proof. Suppose P is not on the segment AB. Then A and B are on the same side of the line t. Let B' be the reflection of B in the line t. Then A, P, B' are collinear.

Let Q be any point in the open half-plane containing B'. Then $QB > QB'$ and we have

$$AQ + QB > AQ + QB' \geq AB' = AP + PB' = AP + PB,$$

so $Q \notin S$. If Q is a point of t other than P, we have

$$AQ + QB = AQ + QB' > AB' = AP + PB,$$

so again $Q \notin S$. Thus, except for the point P itself, S lies in the other open half-plane.

If P is on the segment AB, it is clear that all of S lies on AB and hence in both of the closed half-planes with edge t. ■

Applying this result to the problem at hand, we see that PA and PB are equally inclined to a line of support of the arc AB of the differentiable curve $y = f(x)$. But if a differentiable arc has a line of support at a point P, other than an endpoint—it is clear that P is not A or B—then that line is the tangent line at P.

The last statement is clear, but details can be supplied as follows. If the equation of the line of support is $y = mx + b$ (it cannot be vertical) and $P = (x_0, y_0)$, then the differentiable function

$$g(x) = f(x) - mx - b$$

has either a maximum or a minimum at x_0, so $g'(x_0) = 0$. Hence $f'(x_0) = m$ and the tangent to the curve is $y = mx + b$.

---

3. Find $f(x)$ such that

$$\int [f(x)]^n \, dx = \left| \int f(x) \, dx \right|^n,$$

when constants of integration are suitably chosen.

Solution. We assume that only real-valued continuous functions f defined on an interval are to be considered. If we put $g(x) = \int f(x)^n \, dx$ and $h(x) = \int f(x) \, dx$, we are asked to find all pairs of C^1-functions g and h defined on an interval such that

$$g(x) = h(x)^n \tag{1}$$

and

$$g'(x) = h'(x)^n. \tag{2}$$

If $n = 1$, then obviously any continuous function f and corresponding functions g and h solve the problem, so we assume from now on that $n \neq 1$.

We proceed formally. Differentiate (1) to get

$$g'(x) = nh(x)^{n-1}h'(x) \tag{3}$$

whence

$$h'(x)^n = nh(x)^{n-1}h'(x) \tag{4}$$

$$h'(x)^{n-1} = nh(x)^{n-1} \tag{5}$$

$$h'(x) = Ah(x) \tag{6}$$

where $A = n^{1/(n-1)}$. Hence

$$h(x) = ce^{Ax} \tag{7}$$

for some constant c. Finally

$$f(x) = h'(x) = cAe^{Ax}. \tag{8}$$

Now let us examine this formal work critically. Since the step from (4) to (5) requires $h'(x) \neq 0$, we shall restrict ourselves temporarily to an open interval I on which neither h nor h' vanishes. There is a problem about the meaning of the exponent so we are obliged to consider several cases.

CASE 1. Suppose n cannot be represented as the quotient of two integers $n = p/q$ where q is odd; i.e., n is either irrational or a rational number having even denominator when written in lowest terms. In this case we have no interpretation of b^n if $b < 0$. Hence h and h' must both be positive throughout I and therefore (4) is clearly impossible if $n < 0$. So we must have $n > 0$. Then the solution proceeds as written with the proviso that the constant of integration c must be positive.

CASE 2. $n = p/q$ where p and q are integers, q odd. Then $b^n = (\sqrt[q]{b})^p$ makes sense for both positive and negative b. Equations (3), (4), and (5) follow from (1) and (2). The step to (6) requires that we subdivide the case.

CASE 2*a*. p is odd. Then $n - 1 = (p - q)/q$ with even numerator, so $h(x)^{n-1}$ and $h'(x)^{n-1}$ are positive. So again we must have $n > 0$ from (5). Since we have excluded the possibility $n - 1 = 0$, (6) follows, except that we may have instead

$$h'(x) = -Ah(x) \tag{6'}$$

leading to

$$h(x) = ce^{-Ax} \tag{7'}$$

$$f(x) = -cAe^{-Ax}. \tag{8'}$$

In both (7) and (7′), c may be negative; we need only $c \neq 0$.

CASE 2*b*. p is even. Then $h(x)^{n-1}$ and $h'(x)^{n-1}$ have the signs of $h(x)$ and $h'(x)$, respectively, so n can be either positive or negative. (It cannot be zero, however, under our hypothesis on h and h'.) The formal solution is then correct, $n^{1/(n-1)}$ being well defined, while the constant c can be either positive or negative.

Now we examine the role of our hypothesis on h and h'. Suppose we had a solution h defined on an interval J such that either h or h' vanished at some point of J but not throughout all of J. Then there would be an open subinterval I of J on which neither h nor h' vanishes but with one of h and h' vanishing at an endpoint of I. On I, h and h' are given by exponential functions as we have shown (we are still assuming $n \neq 1$) and these functions do not have limit zero at an endpoint of I. Hence there are no such solutions, and the only solutions not covered above are those identically zero on an interval. Evidently these functions are solutions if $n > 0$.

Note that there are never solutions if $n = 0$ since in that case (1) and (2) become $g(x) = 1$, $g'(x) = 1$.

In summary we have found that all solutions are given as follows:

If $n = 1$, any continuous function f is a solution.

If $n > 0$, $n \neq 1$, then (8) gives solutions with $c \geq 0$. If, furthermore, $n = p/q$ with q odd, the constant c may be taken negative. And if also p is odd we have additional solutions (8′) with arbitrary c.

If $n = 0$, there are no solutions.

If $n < 0$, there are no solutions unless $n = p/q$ where p is even and q is odd, in which case (8) is a solution for $c \neq 0$.

---

4. The parabola $y^2 = -4px$ rolls without slipping around the parabola $y^2 = 4px$. Find the equation of the locus of the vertex of the rolling parabola.

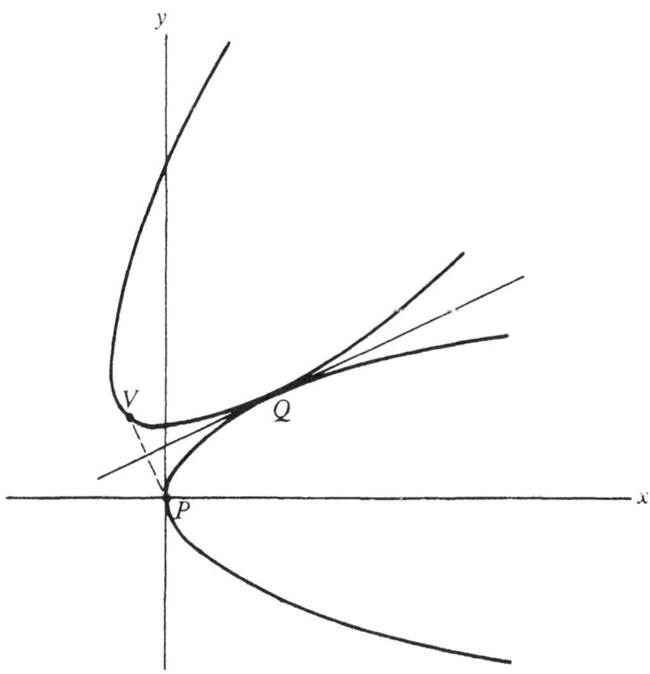

Solution. If the rolling parabola and the fixed parabola are tangent at the point Q, it is obvious from symmetry that the vertex V of the rolling parabola is the reflection of the origin (the vertex of the fixed parabola) in the tangent line at Q.

In the sketch, we have tacitly assumed $p > 0$. Suppose that Q is the point $(4pt^2, 4pt)$. [Any point on the fixed parabola will have this form for a unique t.] The slope of the tangent at Q is $1/(2t)$ and the equation of the tangent is

$$y = \frac{1}{2t}x + 2pt.$$

The perpendicular on this line through the origin has the equation

$$y = -2tx.$$

These lines intersect at

$$\left(\frac{-4pt^2}{1+4t^2}, \frac{8pt^3}{1+4t^2}\right)$$

so the vertex V is at

$$\left(\frac{-8pt^2}{1+4t^2}, \frac{16pt^3}{1+4t^2}\right).$$

This gives us the parametric equations

$$x = \frac{-8pt^2}{1+4t^2}, \qquad y = \frac{16pt^3}{1+4t^2}$$

for the path of the point V. Since $-2t = y/x$ (if $x \neq 0$), the elimination of t gives

$$x = \frac{-2p\left(\dfrac{y}{x}\right)^2}{1+\left(\dfrac{y}{x}\right)^2}$$

so that

$$(x^2 + y^2)x + 2py^2 = 0 \tag{1}$$

is an equation satisfied by all points of the locus (including (0, 0)). Conversely, any point (x, y) other than the origin satisfying (1) leads to a value of $t(= -y/2x)$ and is therefore on the locus. Thus (1) describes the locus precisely.

---

5. Prove that the simultaneous equations

$$x^4 - x^2 = y^4 - y^2 = z^4 - z^2$$

are satisfied by the points of four straight lines and six ellipses, and by no other points.

Solution. Let L denote the locus of the given equations. Then a point is on L if and only if its coordinates (x, y, z) satisfy

$$(x^2 + y^2 - 1)(x^2 - y^2) = 0 \tag{1}$$

$$(y^2 + z^2 - 1)(y^2 - z^2) = 0 \tag{2}$$

$$(z^2 + x^2 - 1)(z^2 - x^2) = 0. \tag{3}$$

Consider the loci A, B, C, D defined as follows:

$$A: \begin{cases} x^2 + y^2 - 1 = 0 \\ y^2 - z^2 \qquad = 0, \end{cases}$$

$$B: \begin{cases} y^2 + z^2 - 1 = 0 \\ z^2 - x^2 \qquad = 0, \end{cases}$$

$$C: \begin{cases} z^2 + x^2 - 1 = 0 \\ x^2 - y^2 \qquad = 0, \end{cases}$$

$$D: \quad x^2 = y^2 = z^2.$$

A is the intersection of a right circular cylinder with the union of the planes $z = y$ and $z = -y$. Hence A is the union of two ellipses. Similarly B and C are each the union of two ellipses. D, on the other hand, is the union of four straight lines, namely:

$$x = y = z, \qquad x = y = -z, \qquad x = -y = z \qquad \text{and} \quad x = -y = -z.$$

Any point common to A and B is in fact also on D, so the ellipses of A and B are different. Similarly for B and C and for C and A. Hence $A \cup B \cup C \cup D$ consists of the union of 6 (distinct) ellipses and 4 (distinct) lines.

We now show that $L = A \cup B \cup C \cup D$. If $(x, y, z) \in A$, then evidently (x, y, z) satisfies (1), (2), and (3), the latter because

$$z^2 + x^2 - 1 = (x^2 + y^2 - 1) + (z^2 - y^2) = 0.$$

Thus $A \subseteq L$. Similarly $B \subseteq L$ and $C \subseteq L$. It is immediate that $D \subseteq L$. So $A \cup B \cup C \cup D \subseteq L$.

Now consider a point $p(x, y, z)$ of L that is not on D. Assume, therefore,

$x^2 \neq y^2$ and $x^2 \neq z^2$. Since p satisfies (1) and (3), we have

$$x^2 + y^2 - 1 = 0$$
$$x^2 + z^2 - 1 = 0$$

and therefore $y^2 = z^2$, so $p \in A$. The other cases of inequalities lead to $p \in B$ or $p \in C$ by the same argument. Hence $L \subseteq A \cup B \cup C \cup D$ and indeed $L = A \cup B \cup C \cup D$ is the union of 4 lines and 6 ellipses.

---

6. $f(x)$ is a polynomial of degree n, such that a power of $f(x)$ is divisible by a power of its derivative $f'(x)$; i.e., $[f(x)]^p$ is divisible by $[f'(x)]^q$; p, q, positive integers. Prove that $f(x)$ is divisible by $f'(x)$ and that $f(x)$ has a single root of multiplicity n.

Solution. Let the factorization of f be

$$f = \alpha p_1^{e_1} p_2^{e_2} \cdots p_k^{e_k}$$

where α is a scalar and $p_1, p_2, \ldots, p_k$ are distinct monic irreducible polynomials, and $e_1, e_2, \ldots, e_k$ are positive integers. Using the product rule for differentiation,

$$f' = \alpha \sum_{i=1}^{k} e_i p_1^{e_1} \cdots p_{i-1}^{e_{i-1}} p_i^{e_i - 1} p_{i+1}^{e_{i+1}} \cdots p_k^{e_k} \cdot p_i'.$$

Since $p_j^{e_j - 1}$ divides each term of this sum and $p_j^{e_j}$ divides all terms but one which it definitely does not divide, one sees that

(1) $$f' = p_1^{e_1 - 1} p_2^{e_2 - 1} \cdots p_k^{e_k - 1} \cdot g$$

where g is a polynomial not divisible by any of the p_i's. Since some power of f is divisible by a power of f', any irreducible factor of g divides a power of f. By the unique factorization theorem for polynomials this is impossible, so one concludes that g has no irreducible factors; hence

(2) g has degree 0.

Now the degree of f is $n = \sum_{i=1}^{k} e_i d_i$ where d_i is the degree of p_i. By (1) and (2) the degree of f' is $n - 1 = \sum_{i=1}^{k} (e_i - 1) d_i$, and subtracting we obtain $1 = \sum_{i=1}^{k} d_i$. Since the d's are positive integers, we conclude that $k = 1$, $d_1 = 1$. Hence

$$f = \alpha p_1^{e_1} = \alpha p_1^n$$

where p_1 is a linear polynomial. Therefore f has a simple root of multiplicity n, and f is indeed divisible by $f' = n\alpha p_1^{n-1}$.

The proof could be made less formalistic if we assume that the polynomials are defined over the complex field. We could then start off with

$$f(x) = \alpha(x - a_1)^{e_1}(x - a_2)^{e_2} \cdots (x - a_k)^{e_k}.$$

REMARK. The proof just given uses the ring of formal polynomials over an arbitrary field, with differentiation being formally defined. The proof remains valid for arbitrary fields of characteristic zero (this hypothesis is used at the point where it is asserted that there is a term in the sum not divisible by $p_i^{e_i}$). The result is false for fields of characteristic $q \neq 0$. Consider the polynomial $x(x^q - 1)$. Its derivative is $x^q - 1$ which divides f, but f does not have a single root of multiplicity $q + 1$.

---

7. If $u_1^2 + u_2^2 + \cdots$ and $v_1^2 + v_2^2 + \cdots$ are convergent series of real constants, prove that

$$(u_1 - v_1)^p + (u_2 - v_2)^p + \cdots, \; p \text{ an integer } \geq 2,$$

is convergent.

Solution. Let $A = u_1^2 + u_2^2 + \cdots$ and $B = v_1^2 + v_2^2 + \cdots$. Since

$$(u_i + v_i)^2 + (u_i - v_i)^2 = 2u_i^2 + 2v_i^2$$

we have, for any positive integer n,

$$\sum_{i=1}^{n} (u_i - v_i)^2 \leq 2 \sum_{i=1}^{n} u_i^2 + 2 \sum_{i=1}^{n} v_i^2 \leq 2A + 2B.$$

Since the terms are all non-negative, it follows that

$$\sum_{i=1}^{\infty} (u_i - v_i)^2 \text{ is convergent.}$$

Therefore, the terms approach zero, so there exists an integer k such that

$$(u_i - v_i)^2 < 1 \quad \text{for all} \quad i \geq k.$$

If p is an integer and $p \geq 2$, then $|u_i - v_i|^p \leq (u_i - v_i)^2$ for all $i \geq k$, so the series

$$\sum_{i=1}^{\infty} (u_i - v_i)^p$$

is absolutely convergent, and therefore convergent.

---

8. A triangle is bounded by the lines

$$A_1x + B_1y + C_1 = 0, \qquad A_2x + B_2y + C_2 = 0,$$
$$A_3x + B_3y + C_3 = 0.$$

Show that the area, disregarding sign, is

$$\frac{\begin{vmatrix} A_1 & B_1 & C_1 \\ A_2 & B_2 & C_2 \\ A_3 & B_3 & C_3 \end{vmatrix}^2}{2\begin{vmatrix} A_2 & B_2 \\ A_3 & B_3 \end{vmatrix} \cdot \begin{vmatrix} A_3 & B_3 \\ A_1 & B_1 \end{vmatrix} \cdot \begin{vmatrix} A_1 & B_1 \\ A_2 & B_2 \end{vmatrix}}.$$

Solution. Let L_i be the line with equation $A_ix + B_iy + C_i = 0$, and let (x_i, y_i) be the vertex of the triangle opposite L_i. Let

$$M = \begin{pmatrix} A_1 & B_1 & C_1 \\ A_2 & B_2 & C_2 \\ A_3 & B_3 & C_3 \end{pmatrix}$$

and

$$X = \begin{pmatrix} x_1 & x_2 & x_3 \\ y_1 & y_2 & y_3 \\ 1 & 1 & 1 \end{pmatrix}.$$

Then the area of the triangle is $\frac{1}{2}|\det X|$.

Since the point (x_i, y_i) is on the side L_j for $i \neq j$, we find

(1) $$MX = \begin{pmatrix} d_1 & 0 & 0 \\ 0 & d_2 & 0 \\ 0 & 0 & d_3 \end{pmatrix}$$

where the diagonal entries remain to be determined.

If we solve the system of equations

$$M \begin{pmatrix} x \\ y \\ z \end{pmatrix} = \begin{pmatrix} d_1 \\ 0 \\ 0 \end{pmatrix}$$

by Cramer's rule, we find

$$z \det M = d_1 \begin{vmatrix} A_2 & B_2 \\ A_3 & B_3 \end{vmatrix}.$$

But from (1) we know that the solution is

$$\begin{pmatrix} x_1 \\ y_1 \\ 1 \end{pmatrix}.$$

Hence,

$$d_1 = \frac{\det M}{\begin{vmatrix} A_2 & B_2 \\ A_3 & B_3 \end{vmatrix}}.$$

Similarly,

$$d_2 = \frac{\det M}{\begin{vmatrix} A_3 & B_3 \\ A_1 & B_1 \end{vmatrix}}$$

$$d_3 = \frac{\det M}{\begin{vmatrix} A_1 & B_1 \\ A_2 & B_2 \end{vmatrix}}.$$

Therefore, from (1),

$$(\det M)(\det X) = \frac{(\det M)^3}{\begin{vmatrix} A_2 & B_2 \\ A_3 & B_3 \end{vmatrix} \cdot \begin{vmatrix} A_3 & B_3 \\ A_1 & B_1 \end{vmatrix} \cdot \begin{vmatrix} A_1 & B_1 \\ A_2 & B_2 \end{vmatrix}}.$$

Now $\det M \neq 0$ is precisely the condition that the given lines form a triangle, so

$$\text{Area of triangle} = \frac{1}{2} |\det X|$$

$$= \pm \frac{(\det M)^2}{2 \begin{vmatrix} A_2 & B_2 \\ A_3 & B_3 \end{vmatrix} \cdot \begin{vmatrix} A_3 & B_3 \\ A_1 & B_1 \end{vmatrix} \cdot \begin{vmatrix} A_1 & B_1 \\ A_2 & B_2 \end{vmatrix}}.$$

Afternoon Session

9. A projectile, thrown with initial velocity v_0 in a direction making angle α with the horizontal, is acted on by no force except gravity. Find the length of its path until it strikes a horizontal plane through the starting point. Show that the flight is longest when

$$\sin \alpha \log (\sec \alpha + \tan \alpha) = 1.$$

Solution. The differential equations of the motion (using x for the horizontal coordinate and y for the vertical coordinate and taking the origin at the initial point) are

$$\frac{d^2x}{dt^2} = 0, \qquad \frac{d^2y}{dt^2} = -g,$$

where g is the acceleration due to gravity. Using the given initial conditions these can be solved to get

$$x = v_0 t \cos \alpha, \qquad y = v_0 t \sin \alpha - \frac{1}{2} g t^2.$$

The flight lasts from time $t = 0$ to $t = T = (2v_0 \sin \alpha)/g$. The length of the trajectory is given by

$$S(\alpha) = \int_0^T \sqrt{(v_0 \sin \alpha - gt)^2 + (v_0 \cos \alpha)^2}\, dt.$$

Putting $w = v_0 \sin \alpha - gt$ and $u = v_0 \cos \alpha$ this becomes

$$S(\alpha) = -\frac{1}{g}\int_{v_0 \sin \alpha}^{-v_0 \sin \alpha} \sqrt{w^2 + u^2}\, dw = \frac{2}{g}\int_0^{v_0 \sin \alpha} \sqrt{w^2 + u^2}\, dw.$$

Bearing in mind that u depends on α, we differentiate this with respect to α and obtain

$$S'(\alpha) = \frac{2}{g}\sqrt{v_0^2 \sin^2 \alpha + u^2}\cdot v_0 \cos \alpha + \frac{2}{g}\int_0^{v_0 \sin \alpha} \frac{u\, dw}{\sqrt{w^2 + u^2}}\cdot \frac{du}{d\alpha}$$

$$= \frac{2v_0^2 \cos \alpha}{g}\left(1 - \sin \alpha \int_0^{v_0 \sin \alpha} \frac{dw}{\sqrt{w^2 + u^2}}\right).$$

$$= \frac{2v_0^2 \cos \alpha}{g}\left(1 - \sin \alpha \left[\log (w + \sqrt{w^2 + u^2})\right]_0^{v_0 \sin \alpha}\right)$$

$$= \frac{2v_0^2 \cos \alpha}{g}(1 - \sin \alpha \log (\sec \alpha + \tan \alpha)).$$

Now $\sin \alpha$ increases from 0 to 1 as α varies from 0 to $\pi/2$ and $\log (\sec \alpha + \tan \alpha)$ increases from 0 to $+\infty$, while $\cos \alpha$ is positive except for $\alpha = \pi/2$. It follows that $\sin \alpha \log (\sec \alpha + \tan \alpha) = 1$ for a unique value $\alpha = \alpha_0 \in (0, \pi/2)$ and that $S'(\alpha) > 0$ for $0 < \alpha < \alpha_0$, $S'(\alpha) < 0$ for $\alpha_0 < \alpha < \pi/2$. Since S is obviously a continuous function on $[0, \pi/2]$, it has a unique maximum on this interval at the point α_0; i.e., the flight is longest for $\alpha = \alpha_0$. Calculation shows that $\alpha_0 = 56°28'$ approximately.

---

10. A cylindrical hole of radius r is bored through a cylinder of radius R ($r \leq R$) so that the axes intersect at right angles.

(i) Show that the area of the larger cylinder which is inside the smaller can be expressed in the form

$$S = 8r^2 \int_0^1 \frac{1 - v^2}{\sqrt{(1 - v^2)(1 - m^2v^2)}}\, dv \quad \text{where} \quad m = \frac{r}{R}.$$

(ii) If

$$K = \int_0^1 \frac{dv}{\sqrt{(1 - v^2)(1 - m^2v^2)}} \quad \text{and} \quad E = \int_0^1 \sqrt{\frac{1 - m^2v^2}{1 - v^2}}\, dv$$

show that

$$S = 8[R^2E - (R^2 - r^2)K].$$

Solution. Consider the sketch.

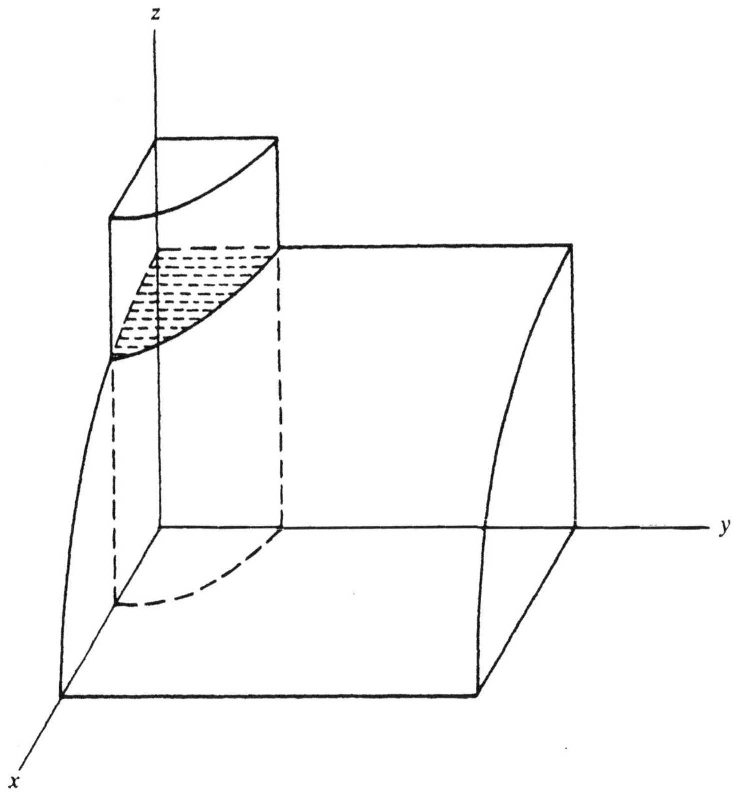

Let the two cylindrical surfaces be $x^2 + z^2 = R^2$ and $x^2 + y^2 = r^2$, where $r \leq R$. The shaded area shown in the diagram is the part of the required area that lies in one octant. The equation of this surface is

$$z = \sqrt{R^2 - x^2}.$$

The required area is

$$S = 8 \iint \sqrt{1 + \left(\frac{\partial z}{\partial y}\right)^2 + \left(\frac{\partial z}{\partial x}\right)^2}\, dy\, dx$$

where the double integral is over the region

$$x^2 + y^2 \leq r^2, \qquad x \geq 0, \qquad y \geq 0.$$

Converted to an iterated integral, this becomes

$$S = 8\int_0^r \left(\int_0^{\sqrt{r^2-x^2}} \frac{R}{\sqrt{R^2 - x^2}}\, dy\right) dx$$

$$= 8R \int_0^r \sqrt{\frac{r^2 - x^2}{R^2 - x^2}}\, dx.$$

Now let $x/r = v$ and $r/R = m$ and simplify further to get

$$S = 8r^2 \int_0^1 \frac{1 - v^2}{\sqrt{(1 - v^2)(1 - m^2v^2)}}\, dv,$$

which completes part (i).

To obtain (ii), write $r^2(1 - v^2) = R^2(1 - m^2v^2) - (R^2 - r^2)$ and substitute to get

$$S = 8\int_0^1 \frac{R^2(1 - m^2v^2)}{\sqrt{(1 - v^2)(1 - m^2v^2)}}\, dv - 8\int_0^1 \frac{(R^2 - r^2)}{\sqrt{(1 - v^2)(1 - m^2v^2)}}\, dv$$

$$= 8[R^2E - (R^2 - r^2)K].$$

REMARK. The integrals K and E are known as the complete elliptic integrals of the first and second kinds, respectively. Their values have been tabulated in terms of the parameter m.

---

11. From any point (a, b) in the Cartesian plane, show that (i) three normals, real or imaginary, can be drawn to the parabola $y^2 = 4px$; (ii) these are real and distinct if $4(2p - a)^3 + 27pb^2 < 0$; (iii) two of them coincide if (a, b) lies on the curve $27py^2 = 4(x - 2p)^3$; (iv) all three coincide only if (a, b) is the point $(2p, 0)$.

Solution. The slope of the parabola at the point (x, y) is $2p/y$. Hence the line joining (a, b) to (x, y) is normal to the parabola at (x, y) if and only if

$$(y - b) = -\frac{y}{2p}(x - a) = -\frac{y}{2p}\left(\frac{y^2}{4p} - a\right).$$

(The case $y = 0$ is also covered by this equation.) This reduces to

$$y^3 + 4p(2p - a)y - 8p^2b = 0. \tag{1}$$

This cubic will have three roots if counted with multiplicity, the roots may be real or complex. Since (real or complex) values of y correspond one to one with (real or imaginary) points on the parabola, part (i) of the problem has been established.

A real cubic equation of the form $y^3 + Ay + B = 0$ will have all real roots if and only if $\Delta = 27B^2 + 4A^3 \le 0$. Two roots will coincide if and only if $\Delta = 0$, and all three roots will coincide if and only if $A = B = 0$.

In the present case, $A = 4p(2p - a)$, $B = -8p^2b$. So $\Delta = 64p^3[4(2p - a)^3 + 27pb^2]$. The problem apparently intends that we take $p > 0$ (as is usual with the standard form of the parabola). With this assumption, Δ has the same sign as

$$\Delta' = 4(2p - a)^3 + 27pb^2.$$

The roots of (1), and hence the points of the parabola, will be real and distinct if and only if $\Delta' < 0$. Two normals will coincide if and only if $\Delta' = 0$; that is, (a, b) lies on the curve $27py^2 = 4(x - 2p)^3$. Finally, all three normals will coincide if and only if $4p(2p - a) = 0$ and $8p^2b = 0$, i.e., if and only if $(a, b) = (2p, 0)$.

The algebraic facts stated above for the roots of $y^3 + Ay + B = 0$ can be derived by the methods of calculus as follows.

The graph of $y^3 + Ay$ will be increasing with strictly positive slope if $A > 0$, so any equation $y^3 + Ay = -B$ will have just one real root and therefore two complex roots. Thus the roots are distinct if $A > 0$.

If $A = 0$, the equation will have one real and two complex roots for $B \neq 0$, and a triple root at 0 if $B = 0$ also.

For $A < 0$, the graph has turning points at

$$\left(-\sqrt{-\frac{A}{3}}, -\frac{2}{3}A\sqrt{-\frac{A}{3}}\right) \quad \text{and} \quad \left(+\sqrt{-\frac{A}{3}}, \frac{2}{3}A\sqrt{-\frac{A}{3}}\right).$$

It is then clear that the equation will have three distinct real roots if

$$|B| < \left|\frac{2}{3}A\sqrt{-\frac{A}{3}}\right|,$$

one double and one single root if

$$|B| = \left|\frac{2}{3}A\sqrt{-\frac{A}{3}}\right|,$$

and one real and two complex roots if

$$|B| > \left|\frac{2}{3} A \sqrt{-\frac{A}{3}}\right| .$$

Squaring these relations we can write them in terms of $\Delta = 27B^2 + 4A^3$, and we have

$$\begin{aligned} \Delta < 0 &\Rightarrow \text{three distinct real roots,} \\ \Delta = 0 &\Rightarrow \text{(at least) two equal real roots,} \\ \Delta > 0 &\Rightarrow \text{one real and two complex roots,} \\ A = B = 0 &\Leftrightarrow \text{a triple root.} \end{aligned}$$

These facts are often derived algebraically by showing that $-\Delta$, which is called the discriminant of the cubic, is the product of the squares of the differences of the roots. See any text on the theory of equations.

---

12. Prove that the locus of the point of intersection of three mutually perpendicular planes tangent to the surface

$$(1) \qquad ax^2 + by^2 + cz^2 = 1 \qquad (abc \neq 0)$$

is the sphere

$$(2) \qquad x^2 + y^2 + z^2 = \frac{1}{a} + \frac{1}{b} + \frac{1}{c}.$$

Solution. We first find the conditions on the coefficients in order that the plane

$$(3) \qquad \alpha x + \beta y + \gamma z = \delta$$

be tangent to the quadric surface $\mathfrak{Q}$ given by (1).

The tangent plane to $\mathfrak{Q}$ at the point (x_1, y_1, z_1) has the equation

$$(4) \qquad ax_1x + by_1y + cz_1z = 1.$$

Suppose (3) and (4) are the same plane. Then $\delta \neq 0$ and

$$(5) \qquad x_1 = \frac{\alpha}{a\delta}, \qquad y_1 = \frac{\beta}{b\delta}, \qquad z_1 = \frac{\gamma}{c\delta}.$$

Since (x_1, y_1, z_1) is a point of $\mathcal{Q}$ we must have

$$\frac{1}{\delta^2}\left(\frac{\alpha^2}{a}+\frac{\beta^2}{b}+\frac{\gamma^2}{c}\right)=1.$$

Hence

$$\frac{\alpha^2}{a}+\frac{\beta^2}{b}+\frac{\gamma^2}{c}=\delta^2 \tag{6}$$

with $\delta \neq 0$ is a necessary condition in order that the plane (3) be tangent to $\mathcal{Q}$. Conversely, if (6) holds and $\delta \neq 0$, then (5) determines a point $P = (x_1, y_1, z_1)$ of $\mathcal{Q}$ and the tangent to $\mathcal{Q}$ and P has equation (4) which is equivalent to (3). So our condition is both necessary and sufficient.

Later we shall have to know what happens if (6) holds with $\delta = 0$. (We assume, of course, that α, β, and γ are not all zero.) Then the plane (3) is asymptotic to $\mathcal{Q}$. In the context of projective geometry, (6) is necessary and sufficient that (3) be projectively tangent to $\mathcal{Q}$, the point of tangency having homogeneous coordinates $(\alpha/a, \beta/b, \gamma/c, \delta)$. When $\delta \neq 0$, this is an ordinary point with Cartesian coordinates given by (5). When $\delta = 0$, this is a point at infinity on $\mathcal{Q}$. (A projective point having homogeneous coordinates (x, y, z, t) is on $\mathcal{Q}$ if and only if $ax^2 + by^2 + cz^2 = t^2$. This equation is derived from the equation for $\mathcal{Q}$ by multiplying each term by a power of t to bring its degree up to two.) An asymptotic plane is thus "tangent to $\mathcal{Q}$ at infinity." We may also view it as the limit of a sequence of tangent planes whose points of tangency approach some point at infinity on $\mathcal{Q}$.

Suppose $(\alpha_1, \beta_1, \gamma_1)$, $(\alpha_2, \beta_2, \gamma_2)$, and $(\alpha_3, \beta_3, \gamma_3)$ are mutually orthogonal unit vectors. Then the matrix

$$\begin{pmatrix} \alpha_1 & \beta_1 & \gamma_1 \\ \alpha_2 & \beta_2 & \gamma_2 \\ \alpha_3 & \beta_3 & \gamma_3 \end{pmatrix} \tag{7}$$

is orthogonal, so its columns are also mutually orthogonal unit vectors. Hence

$$\Sigma\, \alpha_i^2 = \Sigma\, \beta_i^2 = \Sigma\, \gamma_i^2 = 1.$$

If we have planes tangent to $\mathcal{Q}$ with these vectors as normals, these planes are mutually perpendicular; their equations are

$$\alpha_i x + \beta_i y + \gamma_i z = \delta_i, \qquad i = 1, 2, 3; \tag{8}$$

where

$$\frac{\alpha_i^2}{a} + \frac{\beta_i^2}{b} + \frac{\gamma_i^2}{c} = \delta_i^2.$$

Since $|\delta_i|$ is the distance from the origin to the ith plane, the Pythagorean Theorem shows that the distance from the origin O to the point P where these planes intersect is given by

$$OP^2 = \delta_1^2 + \delta_2^2 + \delta_3^2 = \frac{1}{a}\Sigma\,\alpha_i^2 + \frac{1}{b}\Sigma\,\beta_i^2 + \frac{1}{c}\Sigma\,\gamma_i^2 = \frac{1}{a} + \frac{1}{b} + \frac{1}{c}.$$

This can also be derived analytically as follows: If x, y, and z are the coordinates of P, then all of the equations (8) hold. Square these equations and add, using the orthogonality of the columns of (7). This gives

$$x^2 + y^2 + z^2 = \delta_1^2 + \delta_2^2 + \delta_3^2 = \frac{1}{a} + \frac{1}{b} + \frac{1}{c}.$$

Either way we have proved: If P is the point of intersection of three mutually perpendicular planes all tangent to $\mathcal{Q}$, then P is on the sphere $\mathcal{S}$ given by (2). This remains true if the tangencies are allowed to be projective.

The sphere $\mathcal{S}$ is known as the *director sphere* of $\mathcal{Q}$.

With minor changes our argument remains valid in any dimension. Thus an ellipse or hyperbola in the plane has a director circle which contains the point of intersection of any two perpendicular tangents, and for higher dimensional central quadrics there are director spheres of corresponding dimension.

We come now to the converse part of the question, whether every point of the director sphere is the intersection of some three mutually perpendicular tangent planes. This is more difficult.

Consider for a moment the two-dimensional problem. Suppose we start with an ellipse $\mathcal{E}$: $ax^2 + by^2 = 1$, a and b positive. The director circle $\mathcal{C}$: $x^2 + y^2 = (1/a) + (1/b)$ is easily seen to surround $\mathcal{E}$, so if P is any point of $\mathcal{C}$, there is a line l through P tangent to $\mathcal{E}$. Moreover, there are two other lines, m and m', tangent to $\mathcal{E}$ and perpendicular to l (an ellipse obviously has two tangents in every direction). Now $l \cap m$ and $l \cap m'$ are both on $\mathcal{C} \cap l$ (by the analog of what we have proved), but the latter set consists of just two points, one of which is P. Hence P is either $l \cap m$ or $l \cap m'$, so it is possible to draw two perpendicular tangents from P.

In the case of a hyperbola $\mathcal{H}$: $ax^2 + by^2 = 1$, $ab < 0$, it may happen that $(1/a) + (1/b) < 0$. Then it is impossible to find two perpendicular tangent lines. Or $(1/a) + (1/b)$ may be zero, then the director circle

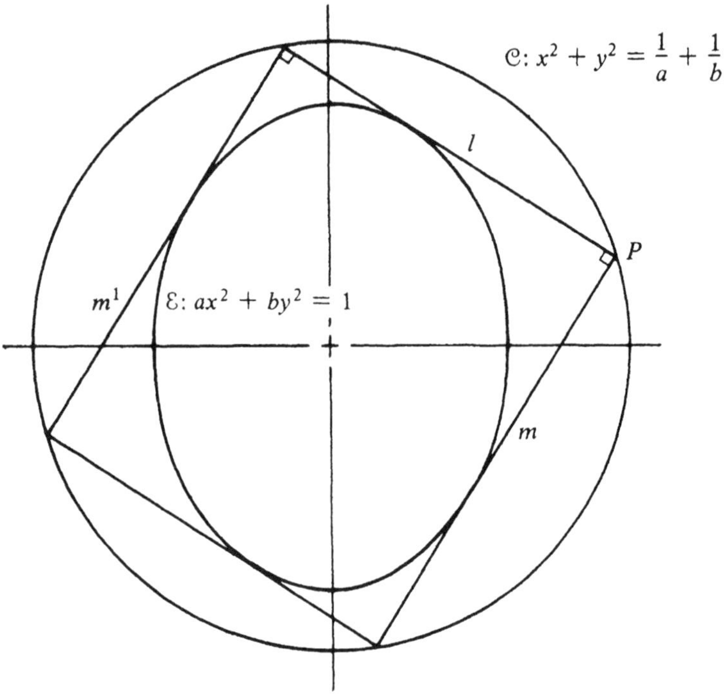

shrinks to a point, the origin. From the origin it is impossible to draw a line tangent to $\mathcal{H}$, but it is possible to draw two asymptotes, and sure enough, they are perpendicular, since $\mathcal{H}$ is equilateral in this case and $a = -b$. If $(1/a) + (1/b) > 0$, then it can be checked that from any point of the director circle just two lines can be drawn that are either tangent or asymptotic to $\mathcal{H}$ and these lines are perpendicular. At the four points where $\mathcal{C}$ crosses the asymptotes, however, we cannot draw two perpendicular tangents.

We shall see that from any point P of the director sphere of $\mathcal{Q}$ it is possible to draw three mutually perpendicular planes, each of which is projectively tangent to $\mathcal{Q}$. Moreover, this can be done in infinitely many ways. Hence, we expect to find a triple involving only true tangencies; this is true with a few rare exceptions. One exception we can see immediately. If $(1/a) + (1/b) + (1/c) = 0$, $\mathcal{S}$ degenerates to a point, the origin; there are no planes through the origin properly tangent to $\mathcal{Q}$, so the origin is an exceptional point.

To prove the converse result, we shall need the following theorem (only in the case $n = 3$).

THEOREM. *Let M be an $n \times n$ matrix. A necessary and sufficient condition that there exist n mutually orthogonal unit vectors* $\mathbf{u}_1, \mathbf{u}_2, \ldots, \mathbf{u}_n$

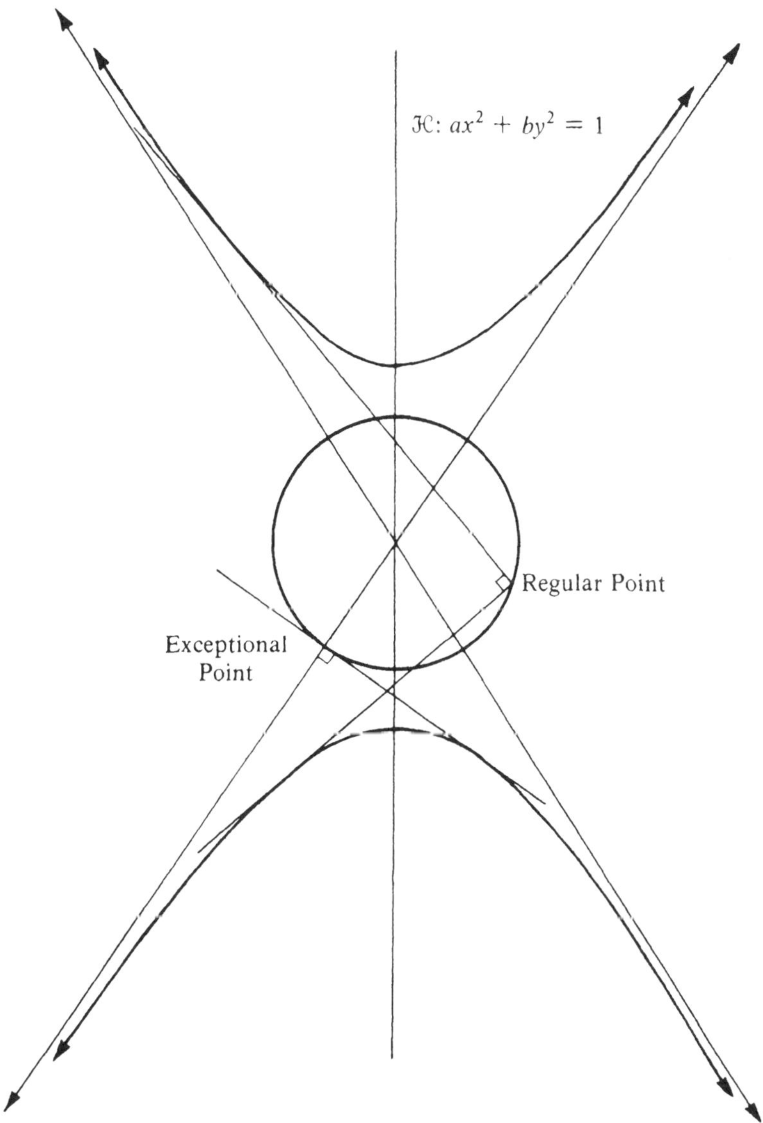

in $\mathbf{R}^n$ *such that* $\mathbf{u}_i{}^T M\, \mathbf{u}_i = 0$ *for* $i = 1, 2, \ldots, n$ *is that the trace of* M *be zero.*

Proof. Let F be a quadratic form defined on an inner product space V over $\mathbf{R}$. The trace of F on V, denoted $\mathrm{tr}_V F$, is the sum $F(\mathbf{v}_1) + F(\mathbf{v}_2) + \cdots + F(\mathbf{v}_n)$, where $\mathbf{v}_1, \mathbf{v}_2, \ldots, \mathbf{v}_n$ is any orthonormal basis of V. (This is the trace of the matrix representing F with respect to the basis $\mathbf{v}_1, \mathbf{v}_2, \ldots, \mathbf{v}_n$. If C and D are matrices for F with respect to different orthonormal bases, then $C = SDS^T = SDS^{-1}$ for some orthogonal matrices

S; hence tr C = tr D. Thus it doesn't matter what orthonormal basis is used.) Now F also has a trace on any linear subspace W of V and the trace is additive on orthogonal subspaces; that is, if X and Y are orthogonal, $\mathrm{tr}_{X \oplus Y} F = \mathrm{tr}_X F + \mathrm{tr}_Y F$. (In choosing an orthonormal basis of $X \oplus Y$ to compute the trace, we may as well coalesce an orthonormal basis of X with an orthonormal basis of Y; then this equation is obvious.)

Suppose $\mathrm{tr}_V F = 0$ where V has positive dimension n. Then F is certainly neither positive definite nor negative definite, so there is a non-zero vector $\mathbf{u}_1$ with $F(\mathbf{u}_1) = 0$ and we can arrange that $\mathbf{u}_1$ is a unit vector. Let V_1 be the subspace orthogonal to $\mathbf{u}_1$. Then $\mathrm{tr}_{V_1} F = 0$. Assuming V_1 has positive dimension we can find a unit vector $\mathbf{u}_2 \in V_1$ such that $F(\mathbf{u}_2) = 0$. Let V_2 be the subspace orthogonal to both $\mathbf{u}_1$ and $\mathbf{u}_2$. Then $\mathrm{tr}_{V_2} F = 0$. Continuing this argument inductively we see that:

A necessary and sufficient condition that there exist an orthonormal basis $\mathbf{u}_1, \mathbf{u}_2, \ldots, \mathbf{u}_n$ *of* V *such that* $F(\mathbf{u}_1) = F(\mathbf{u}_2) = \cdots = F(\mathbf{u}_n) = 0$ *is that* $\mathrm{tr}_V F = 0$.

Now if M is any $n \times n$ matrix, define a quadratic form F on R^n by $F(\mathbf{x}) = \mathbf{x}^T M\mathbf{x}$. Then $\mathrm{tr}_V F$ is the trace of the matrix M, so the theorem follows immediately. ■

Now we can attack the converse problem. Suppose three mutually orthogonal unit vectors $\mathbf{u}_i = (\alpha_i, \beta_i, \gamma_i)$ are given as before. The planes through the point $P = (r, s, t)$ having these vectors as normals are given by

$$\alpha_i x + \beta_i y + \gamma_i z = \alpha_i r + \beta_i s + \gamma_i t$$

and they will all be projectively tangent to $\mathcal{Q}$ if and only if

$$\frac{\alpha_i^2}{a} + \frac{\beta_i^2}{b} + \frac{\gamma_i^2}{c} - (\alpha_i r + \beta_i s + \gamma_i t)^2 = 0 \tag{9}$$

for $i = 1, 2, 3$. If M is the matrix

$$\begin{pmatrix} \frac{1}{a} - r^2 & -rs & -rt \\ -rs & \frac{1}{b} - s^2 & -st \\ -rt & -st & \frac{1}{c} - t^2 \end{pmatrix}$$

then (9) is just $\mathbf{u}_i^T M \mathbf{u}_i = 0$, $i = 1, 2, 3$. According to our theorem, given P and hence M, we can find such mutually orthogonal unit vectors if and only if

$$\operatorname{tr} M = \frac{1}{a} + \frac{1}{b} + \frac{1}{c} - r^2 - s^2 - t^2 = 0.$$

In other words, there are three mutually perpendicular planes through P each projectively tangent to $\mathfrak{Q}$ if and only if P lies on the director sphere.

Finally we should investigate whether the three planes through P can be found properly tangent to $\mathfrak{Q}$. We have already observed that they cannot if $(1/a) + (1/b) + (1/c) = 0$, so assume from now on that $(1/a) + (1/b) + (1/c) > 0$.

If we look back at the proof of the theorem we see that $\mathbf{u}_1$ can be taken as any unit vector such that $\mathbf{u}_1{}^T M \mathbf{u}_1 = 0$. This means that, in finding the mutually perpendicular planes through a point P of $\mathfrak{S}$, we can start with any plane through P that is projectively tangent to $\mathfrak{Q}$. Suppose P is on $\mathfrak{S}$ but not on $\mathfrak{Q}$. Then if there are any planes tangent to $\mathfrak{Q}$ through P there are infinitely many. They can be found as follows. If $P = (r, s, t)$ then the plane $arx + bsy + ctz = 1$ is called the *polar* of P. It intersects $\mathfrak{Q}$ in a certain conic $\mathfrak{D}$. Every line tangent to $\mathfrak{D}$ determines with P a plane tangent to $\mathfrak{Q}$ and this gives all tangent planes through P. $\mathfrak{D}$ may have one or two points at infinity, in which case there are one or two planes through P asymptotic to $\mathfrak{Q}$. After the first tangent plane π_1 is chosen the other two must contain the line perpendicular to π_1 through P. There can be only two planes projectively tangent to $\mathfrak{Q}$ containing this line since it is not a ruling of $\mathfrak{Q}$, so there is just one set of three mutually perpendicular tangent planes through P including π_1. Therefore, the infinitely many planes through P and projectively tangent to $\mathfrak{Q}$ are divided into disjoint sets of three mutually perpendicular planes. At most two of these sets can contain an asymptotic plane, so there must be infinitely many sets consisting of three properly tangent planes.

Now suppose P is on both $\mathfrak{Q}$ and $\mathfrak{S}$. Then $\mathfrak{Q}$ must be a hyperboloid of one sheet, for if $\mathfrak{Q}$ is either an ellipsoid or a hyperboloid of two sheets and P is on $\mathfrak{Q}$, there is only one plane through P tangent to $\mathfrak{Q}$, even projectively. Thus $\mathfrak{Q}$ is a ruled quadric and any plane containing either of the two rulings through P is tangent to $\mathfrak{Q}$ at least projectively, and all tangent planes through P are found in this way. Of the planes containing a fixed ruling r, just one is asymptotic, namely, the plane determined by r and the origin. Now if three projectively tangent planes through P are mutually perpendicular, they do not share the same ruling, so two contain one ruling and the third contains the other. This third plane is perpendicular to the line of intersection of the first two, which is the ruling. It follows that the two rulings through P must be perpendicular. Conversely, if P is any point of $\mathfrak{Q}$ at which the rulings are perpendicular, then the plane perpendicular to either ruling and any two perpendicular planes containing that ruling form a mutually perpendicular triple of planes each projectively tangent to $\mathfrak{Q}$. Thus we see that $\mathfrak{Q} \cap \mathfrak{S}$ consists of just those points of $\mathfrak{Q}$

at which the rulings are perpendicular. Moreover, it is clear that we can choose the three planes all properly tangent to $\mathcal{Q}$ unless both of the planes π_1 and π_2 perpendicular to the rulings at P are asymptotic. Since the asymptotic planes pass through the origin, this awkward possibility occurs precisely when $\pi_1 \cap \pi_2$, that is the normal to $\mathcal{Q}$ at P, passes through the origin. To identify these points, we note that for $\mathcal{Q}$ to be a hyperboloid of one sheet we must have just one of the numbers a, b, and c negative, say $a > 0$, $b > 0 > c$. Then points of $\mathcal{Q}$ at which the normal passes through the origin are just four $(\pm a^{-1/2}, 0, 0)$ and $(0, \pm b^{-1/2}, 0)$, unless $a = b$, in which case $\mathcal{Q}$ is rotationally symmetric about the z-axis and such points fill the entire circle in which $\mathcal{Q}$ meets the x-y plane. Taking into account the fact that our exceptional point must be on $\mathcal{S}$ we see that there are

two exceptional points $(\pm a^{-1/2}, 0, 0)$,	if $a \neq b = -c$,
two exceptional points $(0, \pm b^{-1/2}, 0)$,	if $b \neq a = -c$,
a circle of exceptional points,	if $a = b = -c$,

and no exceptional points in any other case.

Those interested in pursuing the techniques used in this solution will find the following problem attractive: Any ellipsoid or hyperboloid can be put in the form (1) by translation and rotation of axes. Hence the problem of three perpendicular tangent planes has been solved for all of these. A paraboloid, however, can be put in the canonical form

$$z = ax^2 + by^2, \qquad ab \neq 0.$$

(This is an elliptic paraboloid if $ab > 0$, a hyperbolic paraboloid if $ab < 0$.) Show that the locus of points of intersection of three mutually orthogonal tangent planes to this paraboloid has the equation

$$z = -\frac{1}{4}\left(\frac{1}{a} + \frac{1}{b}\right)$$

and discuss the existence of exceptional points. (The analog in the plane of this result is the well-known theorem that the directrix is the locus of points of intersection of perpendicular tangents to a parabola.)

REMARK. There is a reason why the converse argument is so much longer than the first part. Suppose we were restricted to calculation in the rational field; i.e., we consider only equations with rational coefficients and points with rational coordinates. Then our first result would remain valid: the intersection of any three mutually perpendicular tangent planes is on $\mathcal{S}$. But there may be points of $\mathcal{S}$ (with all coordinates rational) from which it is impossible to find three mutually orthogonal tangent planes having rational equations. *Example*: Suppose $\mathcal{Q}$ is given by $x^2 + y^2 + \frac{1}{2}z^2 = 1$. Then $P = (0, 0, 2)$ is on $\mathcal{S}$, but no three mutually perpendicular planes tangent to $\mathcal{Q}$ with rational equations pass through P. (The proof of this is an interesting problem in itself.) The argument for the converse must therefore use some property of the real field not possessed by the rational field. Where?

---

13. Determine all rational values for which a, b, c are the roots of

$$x^3 + ax^2 + bx + c = 0.$$

Solution. The conditions on the roots are equivalent to

$$a + b + c = -a, \tag{1}$$

$$ab + bc + ca = b, \tag{2}$$

$$abc = -c. \tag{3}$$

If $c = 0$, then $ab = b$ and $2a + b = 0$, so either $b = 0$, $a = 0$, or $a = 1$, $b = -2$.

If $c \neq 0$, then $ab = -1$. If $a + b = 0$, then (2) becomes $ab = b$ so that $a = 1$, $b = -1$, $c = -1$. If $a + b \neq 0$, then

$$c = \frac{b+1}{a+b} = \frac{a(b+1)}{a(a+b)} = \frac{-1+a}{a^2-1} = \frac{1}{a+1}$$

and (1) becomes

$$2a - \frac{1}{a} + \frac{1}{a+1} = 0$$

whence

$$2a^3 + 2a^2 - 1 = 0.$$

This equation has no rational roots, since the only possibilities are ± 1, $\pm 1/2$, and these are not roots. There are therefore three solutions

a	b	c	corresponding to
0	0	0	$x^3 = 0$
+1	−2	0	$x^3 + x^2 - 2x = 0$
+1	−1	−1	$x^3 + x^2 - x - 1 = (x^2 - 1)(x + 1) = 0.$

---

14. Prove that

$$\begin{pmatrix} a_1^2+k & a_1a_2 & a_1a_3 & \cdots & a_1a_n \\ a_2a_1 & a_2^2+k & a_2a_3 & \cdots & a_2a_n \\ \cdots & \cdots & \cdots & \cdots & \cdots \\ a_na_1 & a_na_2 & a_na_3 & \cdots & a_n^2+k \end{pmatrix}$$

is divisible by k^{n-1} and find its other factor.

First Solution. Let B be the matrix

$$\begin{pmatrix} a_1^2 & a_1a_2 & a_1a_3 & \cdots & a_1a_n \\ a_2a_1 & a_2^2 & a_2a_3 & \cdots & a_2a_n \\ \cdots & \cdots & \cdots & \cdots & \cdots \\ a_na_1 & a_na_2 & a_na_3 & \cdots & a_n^2 \end{pmatrix}.$$

B has rank at most one, since any two rows (or columns) are clearly dependent. So there are $(n-1)$ zeros among the eigenvalues of B. Therefore the characteristic polynomial of B is divisible by x^{n-1}. Hence

$$\begin{aligned}\det(x\cdot I - B) &= x^n - (\text{trace } B)x^{n-1}\\ &= x^{n-1}(x - a_1^2 - a_2^2 - \cdots - a_n^2)\end{aligned}$$

so

$$\begin{aligned}\det(B+kI) &= (-1)^n \det(-kI - B)\\ &= k^{n-1}(k + a_1^2 + a_2^2 + \cdots + a_n^2),\end{aligned}$$

and the other factor is $(k + a_1^2 + a_2^2 + \cdots + a_n^2)$.

Second Solution. Assume for a moment that none of the a's are zero, and let

$$B_n = \begin{pmatrix} a_1^2+k & a_1a_2 & a_1a_3 & \cdots & a_1a_n \\ a_2a_1 & a_2^2+k & a_2a_3 & \cdots & a_2a_n \\ \cdots & \cdots & \cdots & \cdots & \cdots \\ a_na_1 & a_na_2 & a_na_3 & \cdots & a_n^2+k \end{pmatrix}$$

Since the determinant is linear in the last row, we find

$$\det B_n = \det \left(\begin{array}{ccc|c} & B_{n-1} & & \begin{matrix} a_1 a_n \\ \vdots \\ a_{n-1}a_n \end{matrix} \\ \hline a_n a_1 & \cdots & a_n a_{n-1} & a_n^2 \end{array}\right) + \det \left(\begin{array}{ccc|c} & B_{n-1} & & \begin{matrix} a_1 a_n \\ \vdots \\ a_{n-1}a_n \end{matrix} \\ \hline 0 & \cdots & 0 & k \end{array}\right).$$

Now in the first of these new determinants, subtract multiples of the last rows from the others to get

$$\det \left(\begin{array}{ccc|c} k & & 0 & 0 \\ & k & & \vdots \\ 0 & & k & 0 \\ \hline a_n a_1 & \cdots & a_n a_{n-1} & a_n^2 \end{array}\right).$$

Then we have

$$\det B_n = k^{n-1} a_n^2 + k \det B_{n-1}$$

Since $\det B_1 = k + a_1^2$, the relation

(1) $$\det B_n = k^{n-1}(k + a_1^2 + \cdots + a_n^2)$$

follows easily by induction.

Although this derivation depends on the assumption that the a's are not zero, the result remains valid for the case where some of the a's are zero, since D_n is evidently some polynomial in k and the a's which agrees with $k^{n-1}(k + a_1^2 + \cdots + a_n^2)$ as long as none of the a's are zero. Therefore (1) must be a polynomial identity.

Alternatively, we can regard the computation as taking place in the field $Q(k, a_1, a_2, \ldots, a_n)$ where k and the a_i are independent indeterminates. In this field the condition $a_i \neq 0$ is satisfied.

For a discussion of such fields, see I. N. Herstein, *Topics in Algebra*, Blaisdell, Waltham, Mass., 1964.

---

15. Which is greater

$$(\sqrt{n})^{\sqrt{n+1}} \quad \text{or} \quad (\sqrt{n+1})^{\sqrt{n}}$$

where $n > 8$?

Solution. $(\sqrt{n})^{\sqrt{n+1}}$ is greater than $(\sqrt{n+1})^{\sqrt{n}}$ for $n > 8$. Consider the function $f(x) = (\log x)/x$ for $x > 0$. Its derivative is $(1 - \log x)/x^2$ which is negative for $x > e$.

Hence, if $e \le x < y$ we have $f(x) > f(y)$, and

$$xy\left(\frac{\log x}{x}\right) > xy\left(\frac{\log y}{y}\right).$$

Taking exponentials we get

$$e^{y \log x} > e^{x \log y},$$

that is,

$$x^y > y^x \tag{1}$$

provided $e \le x < y$.

If $n \ge 8$, then $e < \sqrt{n} < \sqrt{n+1}$, so

$$(\sqrt{n})^{\sqrt{n+1}} > (\sqrt{n+1})^{\sqrt{n}}.$$

REMARK. A number of interesting problems are based on the inequality (1). See for example *Journal of Recreational Mathematics*, vol. 2, no. 4 (October 1969), pages 255–256, where the inequality $e^\pi > \pi^e$ and some generalizations are discussed. The inequality is related to Problem P.M. 1 of Competition 21 and Problem A.M. 1 of Competition 22.

THE FOURTH WILLIAM LOWELL PUTNAM MATHEMATICAL COMPETITION

March 1, 1941

Morning Session

1. Prove that the polynomial

$$(a - x)^6 - 3a(a - x)^5 + \tfrac{5}{2}a^2(a - x)^4 - \tfrac{1}{2}a^4(a - x)^2$$

takes only negative values for $0 < x < a$.

Solution. Make the substitution $x = a(1 - y)$. Then the given polynomial becomes

$$a^6y^2\left(y^4 - 3y^3 + \frac{5}{2}y^2 - \frac{1}{2}\right).$$

Since a^6y^2 is surely positive, it suffices to prove that

$$g(y) = y^4 - 3y^3 + \frac{5}{2}y^2 - \frac{1}{2} < 0$$

for $0 < y < 1$.

Since $g'(y) = 4y^3 - 9y^2 + 5y = y(y - 1)(4y - 5)$, the critical values for g are 0, 1, 5/4. Between consecutive critical values g is strictly monotonic. Therefore, since $g(0) = -\frac{1}{2}$ and $g(1) = 0$, we have

$$-\frac{1}{2} < g(y) < 0$$

for $0 < y < 1$.

---

2. Find the nth derivative with respect to x of

$$\int_0^x \left[1 + \frac{(x - t)}{1!} + \frac{(x - t)^2}{2!} + \cdots + \frac{(x - t)^{n-1}}{(n - 1)!}\right] e^{nt}\, dt.$$

First Solution. Let

$$\phi_k(x) = \int_0^x \frac{(x-t)^k}{k!} e^{nt}\, dt.$$

Then, for $k > 0$,

$$\phi_k{}'(x) = \phi_{k-1}(x).$$

Also

$$\phi_0(x) = \int_0^x e^{nt}\, dt = \frac{e^{nx}-1}{n}.$$

Therefore

$$\left(\frac{d}{dx}\right)^n \phi_k(x) = \left(\frac{d}{dx}\right)^{n-k} \phi_0(x) = n^{n-k-1} e^{nx} \quad \text{for} \quad n > k.$$

Accordingly, the nth derivative of the given function is

$$\left(\frac{d}{dx}\right)^n [\phi_0(x) + \phi_1(x) + \cdots + \phi_{n-1}(x)]$$

$$= [n^{n-1} + n^{n-2} + \cdots + 1]\, e^{nx}$$

$$= \begin{cases} \dfrac{n^n - 1}{n-1} e^{nx} & \text{for} \quad n \neq 1, \\ e^x & \text{for} \quad n = 1. \end{cases}$$

Second Solution. Let

$$f(x) = \int_0^x \left(1 + \frac{(x-t)}{1!} + \frac{(x-t)^2}{2!} + \cdots + \frac{(x-t)^{n-1}}{(n-1)!}\right) e^{nt}\, dt.$$

Substituting $t = x - u$, we get $f(x) = e^{nx}\,\psi(x)$, where

$$\psi(x) = \int_0^x \left(1 + \frac{u}{1!} + \frac{u^2}{2!} + \cdots + \frac{u^{n-1}}{(n-1)!}\right) e^{-nu}\, du.$$

Then

$$\psi'(x) = \left(1 + \frac{x}{1!} + \frac{x^2}{2!} + \cdots + \frac{x^{n-1}}{(n-1)!}\right) e^{-nx}$$

and

$$f'(x) - nf(x) = e^{nx}\psi'(x), \tag{1}$$

where the right member is a polynomial of degree $n - 1$.

The general solution of the differential equation (1) is

$$f(x) = Ce^{nx} + P,$$

where P is another polynomial of degree $n - 1$. It follows that

$$f^{(n)}(x) = Cn^n e^{nx},$$

and it remains to find C.

We have

$$\begin{aligned} C &= \lim_{x\to\infty} f(x)e^{-nx} = \lim_{x\to\infty} \psi(x) \\ &= \int_0^\infty \left(1 + \frac{u}{1!} + \frac{u^2}{2!} + \cdots + \frac{u^{n-1}}{(n-1)!}\right) e^{-nu}\, du \\ &= \frac{1}{n} + \frac{1}{n^2} + \cdots + \frac{1}{n^n}. \end{aligned}$$

Therefore

$$\begin{aligned} f^{(n)}(x) &= (n^{n-1} + n^{n-2} + \cdots + 1)\, e^{nx} \\ &= \begin{cases} \dfrac{n^n - 1}{n - 1}\, e^{nx}, & \text{if } n > 1, \\ e^x, & \text{if } n = 1. \end{cases} \end{aligned}$$

3. A circle of *radius a* rolls in its plane along the x-axis. Show that the envelope of a diameter is a cycloid, coinciding with the cycloid traced out by a point on the circumference of a circle of *diameter a*, likewise rolling in its plane along the x-axis.

First Solution. Let the two circles roll along the upper side of the x-axis so that at time t they both touch the axis at $(at,\ 0)$. Let P and Q be the points that start in contact with the origin O, and let D and E be the centers of the larger and smaller circles, respectively. Let P' be the point opposite P on the larger circle. We may suppose that PP' is the moving diameter of the problem, since we can choose the origin along the x-axis at our convenience.

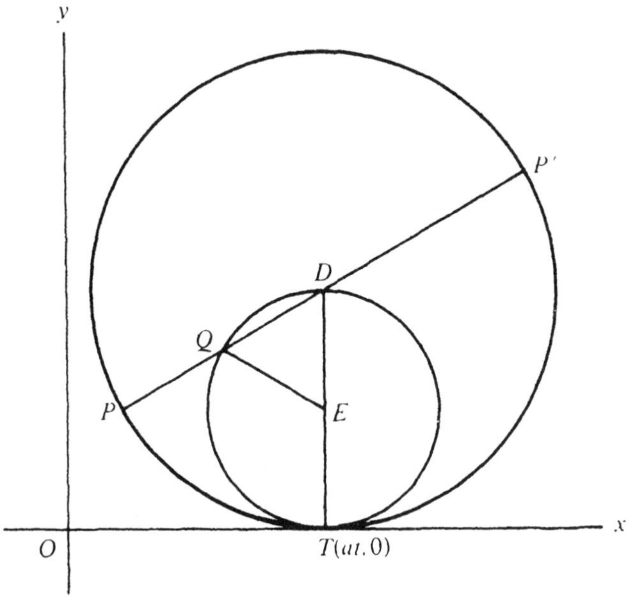

The point Q traces a cycloid C, and we shall prove that PP' is always tangent to C at Q. This will prove that the envelope of PP' is C.

Since arc PT = arc $QT = OT = at$, while the circles are of radius a and $a/2$, we see that $\angle PDT = t$, $\angle QET = 2t$. Hence at time t, P is at $(at - a \sin t, a - a \cos t)$, P' is at $(at + a \sin t, a + a \cos t)$, and the coordinates of Q are

$$x = at - \frac{a}{2} \sin 2t$$

$$y = \frac{a}{2} - \frac{a}{2} \cos 2t.$$

These are parametric equations for the motion of Q, that is, for the cycloid C. This parametrization is non-singular when t is not a multiple of π and singular when it is (for then both dx/dt and dy/dt are zero). The tangent to C at a point corresponding to a non-singular t has direction vector

$$(a - a \cos 2t, a \sin 2t)$$

so the tangent to C at Q has the representation

$$(1) \qquad \begin{aligned} x &= at - \frac{a}{2}\sin 2t + u(a - a\cos 2t) \\ y &= \frac{a}{2} - \frac{a}{2}\cos 2t + u(a \sin 2t) \end{aligned}$$

in terms of the parameter u. Putting

$$u = \frac{\cos t - 1}{2 \sin t} \qquad \text{and} \qquad u = \frac{\cos t + 1}{2 \sin t}$$

in (1), we see that P and P' are both on the tangent line. This proves that PP' is tangent to C at Q for t not a multiple of π.

When t is a multiple of π, Q is on the x-axis, the curve C has a cusp at Q with a vertical tangent, and PP' stands vertical with either P or P' on the x-axis. Hence PP' is tangent to C in this case too.

Second Solution. A purely synthetic argument can be given as follows: As before, we see that $\angle QET = 2\angle PDT$. Since $\angle QDT$ is measured by half of arc QT, we have $\frac{1}{2}\angle QET = \angle QDT$. It follows that $\angle QDT = \angle PDT$, so Q is on PP'.

As the small circle rolls, the point T is instantaneously at rest; hence all other points of the moving circle are instantaneously in rotation about T. This implies that the tangent at Q to the orbit of Q is perpendicular to QT. But $QD = PP'$ is this perpendicular since $\angle TQD$ is inscribed in a semicircle.

As is often the case with such synthetic arguments, a slightly different analysis is required after Q attains its highest point.

---

4. Let the roots a, b, c of

$$f(x) \equiv x^3 + px^2 + qx + r = 0$$

be real, and let $a \leq b \leq c$. Prove that, if the interval (b, c) is divided into *six* equal parts, a root of $f'(x) = 0$ will lie in the *fourth* part counting from the end b. What will be the form of $f(x)$ if the root in question of $f'(x) = 0$ falls at either end of the *fourth* part?

Solution. The proposition is valid for $f(x)$ if and only if it is valid for $f(x + b)$ so we can translate all the roots by $-b$ and thus arrange that the middle root is zero. It is no loss of generality, therefore, to assume that $b = 0$ to begin with. Hence we consider

$$f(x) = (x - a)x(x - c) = x^3 - (a + c)x^2 + acx$$

where $a \le 0 \le c$. The fourth subinterval referred to in the problem is $[c/2, 2c/3]$.

From $f'(x) = 3x^2 - 2(a + c)x + ac$ we find $f'(c/2) = -\frac{1}{4}c^2 \le 0$ and $f'(2c/3) = -\frac{1}{3}ac \ge 0$. Hence, since f' is continuous, there is a root of $f'(x) = 0$ on $[c/2, 2c/3]$.

A root occurs at the left endpoint $c/2$ if and only if $c = 0$; that is, the two largest roots coincide. In this event, $f(x) = (x - a)x^2$.

A root occurs at the right endpoint $2c/3$ if and only if $a = 0$ or $c = 0$. If $c = 0$ we have the previous case, and the interval in question has degenerated to a single point. If $c \neq 0$, then $a = 0$ and the two smallest roots coincide. In this case, $f(x) = x^2(x - c)$.

To answer the second part of the question in terms of the *original* a, b, c: the zero of f' occurs at the left endpoint iff $f(x) = (x - a)(x - b)^2$ and at the right endpoint iff $f(x) = (x - b)^2(x - c)$ or $f(x) = (x - a)(x - b)^2$.

---

5. Show that the line which moves parallel to the plane $y = z$ and which intersects the two parabolas $y^2 = 2x$, $z = 0$ and $z^2 = 3x$, $y = 0$ sweeps out the surface

$$x = (y - z)\left(\frac{y}{2} - \frac{z}{3}\right).$$

Solution. Suppose that the line meets the first parabola at $(18s^2, 6s, 0)$ and the second parabola at a distinct point $(12t^2, 0, 6t)$. The condition that the line be parallel to the plane $y = z$ is that the direction vector of the line, namely

$$(18s^2 - 12t^2, 6s, -6t),$$

should lie in that plane, i.e., $6s = -6t$. Hence the two points are $(18t^2, -6t, 0)$ and $(12t^2, 0, 6t)$, where $t \neq 0$.

This line has the parametric representation

$$(18t^2, -6t, 0) + u(6t^2, -6t, -6t), \tag{1}$$

where u is the parameter. So we have the two-parameter representation

$$x = 18t^2 + 6ut^2$$
$$y = -6t - 6ut$$
$$z = -6tu,$$

where $t \neq 0$, for the surface generated by the moving line.

We eliminate the parameters by noting that

$$(2) \qquad y - z = -6t, \quad \text{and} \qquad \frac{y}{2} - \frac{z}{3} = -t(3 + u).$$

So we get

$$(3) \qquad x = 6t(3t + ut) = (y - z)\left(\frac{y}{2} - \frac{z}{3}\right).$$

This proves that any point on any line parallel to the plane $y = z$ which meets the two given parabolas at distinct points lies on the surface (3).

Conversely, if (x_1, y_1, z_1) is a point on surface (3), with $y_1 \neq z_1$, the numbers t and u can be determined from (2) so that (x_1, y_1, z_1) has the form (1). Points on the plane $y = z$, however, are special because the two given parabolas intersect the plane at the point $(0, 0, 0)$. So every point of the plane is on some line which is parallel to (in fact contained in) the plane and which intersects both parabolas. If these lines are to be taken into account, then the locus (3) must be supplemented by adjoining the whole plane $y = z$. The surface (3) itself meets the plane $y = z$ in just one line, $x = 0, y = z$.

---

6. If the x-coordinate $\bar{x}$ of the center of mass of the area lying between the x-axis and the curve $y = f(x)$, $(f(x) > 0)$, and between the lines $x = 0$ and $x = a$ is given by

$$\bar{x} = g(a),$$

show that

$$f(x) = A \frac{g'(x)}{[x - g(x)]^2} e^{\int dx/(x - g(x))},$$

where A is a positive constant.

Solution. By the definition of centroid,

$$g(x) = \frac{\int_0^x tf(t)dt}{\int_0^x f(t)dt}, \qquad x \neq 0. \tag{1}$$

We confine our attention to positive values of x. Put

$$F(x) = \int_0^x f(t)dt; \qquad \text{then } F' = f.$$

Write (1) in the form

$$F(x)g(x) = \int_0^x tf(t)dt,$$

and differentiate to get

$$F'(x)g(x) + F(x)g'(x) = xf(x) = xF'(x).$$

Thus F satisfies the linear differential equation

$$F'(x) = \frac{g'(x)}{x - g(x)} F(x). \tag{2}$$

Hence

$$F(x) = Ae^{\psi(x)}, \tag{3}$$

where A is a positive constant and

$$\psi(x) = \int^x \frac{g'(t)}{t - g(t)} dt = \int^x \frac{dt}{t - g(t)} - \log(x - g(x)). \tag{4}$$

Thus

$$F(x) = \frac{A}{x - g(x)} \exp \int^x \frac{dt}{t - g(t)} \tag{5}$$

Substituting (5) into (2) gives us

$$f(x) = A \frac{g'(x)}{(x - g(x))^2} \exp \int^x \frac{dt}{t - g(t)} \tag{6}$$

as required.

Under the conditions of the problem, $x > g(x)$ for all positive x, so the denominator $x - g(x)$ causes no singularities for positive x. For $x < 0$, we have $g(x) > x$, so we must replace $\log(x - g(x))$ by $\log(g(x) - x)$ in (4). Then (6) becomes

$$f(x) = \frac{-Ag'(x)}{(x - g(x))^2} \exp \int^x \frac{dt}{t - g(t)}.$$

However, in this case the constant A in (3) must be negative so we again have the required form.

---

7. (i) Prove that

$$\begin{vmatrix} 1 + a^2 - b^2 - c^2 & 2(ab + c) & 2(ca - b) \\ 2(ab - c) & 1 + b^2 - c^2 - a^2 & 2(bc + a) \\ 2(ca + b) & 2(bc - a) & 1 + c^2 - a^2 - b^2 \end{vmatrix} = (1 + a^2 + b^2 + c^2)^3.$$

First Solution. In the determinant add b times row 3 and subtract c times row 2 from row 1 to get

$$\begin{vmatrix} 1 + a^2 + b^2 + c^2 & c(1 + a^2 + b^2 + c^2) & -b(1 + a^2 + b^2 + c^2) \\ 2ab - 2c & 1 + b^2 - c^2 - a^2 & 2bc + 2a \\ 2ac + 2b & 2bc - 2a & 1 + c^2 - a^2 - b^2 \end{vmatrix}$$

$$= (1 + a^2 + b^2 + c^2) \begin{vmatrix} 1 & c & -b \\ 2ab - 2c & 1 + b^2 - c^2 - a^2 & 2bc + 2a \\ 2ac + 2b & 2bc - 2a & 1 + c^2 - a^2 - b^2 \end{vmatrix}$$

$$= (1 + a^2 + b^2 + c^2) \begin{vmatrix} 1 & 0 & 0 \\ 2ab - 2c & 1 + b^2 + c^2 - a^2 - 2abc & 2ab^2 + 2a \\ 2ac + 2b & -2a - 2ac^2 & 1 + c^2 - a^2 + b^2 + 2abc \end{vmatrix}$$

$$= (1 + a^2 + b^2 + c^2)[(1 + b^2 + c^2 - a^2)^2 - 4a^2b^2c^2 + 4a^2(b^2 + 1)(c^2 + 1)]$$

$$= (1 + a^2 + b^2 + c^2)^3.$$

Second Solution. Let

$$M = \begin{pmatrix} 0 & c & -b \\ -c & 0 & a \\ b & -a & 0 \end{pmatrix}$$

$$M^2 = \begin{pmatrix} -b^2 - c^2 & ab & ac \\ ab & -c^2 - a^2 & bc \\ ac & bc & -a^2 - b^2 \end{pmatrix}$$

and we are to find the determinant of X where

$$X = (1 + a^2 + b^2 + c^2)I + 2M^2 + 2M.$$

The characteristic polynomial of M is $x^3 + (a^2 + b^2 + c^2)x$, and its eigenvalues are 0, $\pm ui$ where $u^2 = a^2 + b^2 + c^2$. Hence the eigenvalues of X are

$$1 + u^2,\ 1 + u^2 - 2u^2 \pm 2ui = 1 - u^2 \pm 2ui.$$

The determinant of a matrix is the product of its eigenvalues, so

$$\det X = (1 + u^2)(1 - u^2 + 2ui)(1 - u^2 - 2ui)$$
$$= (1 + u^2)^3 = (1 + a^2 + b^2 + c^2)^3.$$

---

7. (ii) A semi-ellipsoid of revolution is formed by revolving about the x-axis the area lying within the first quadrant of the ellipse

$$\frac{x^2}{a^2} + \frac{y^2}{b^2} = 1.$$

Show that this semi-ellipsoid will balance in stable equilibrium, with its vertex resting on a horizontal plane, when and only when

$$b\sqrt{8} \geq a\sqrt{5}.$$

First Solution. Let C be the center of gravity of the solid semi-ellipsoid S, and let V be its vertex. Consider the sphere with center C and radius CV.

Suppose that near V the sphere lies strictly inside S (except for the point V, of course). Then if S rests on a horizontal plane with point of contact V any small displacement raises the center of gravity, and therefore S is stably balanced.

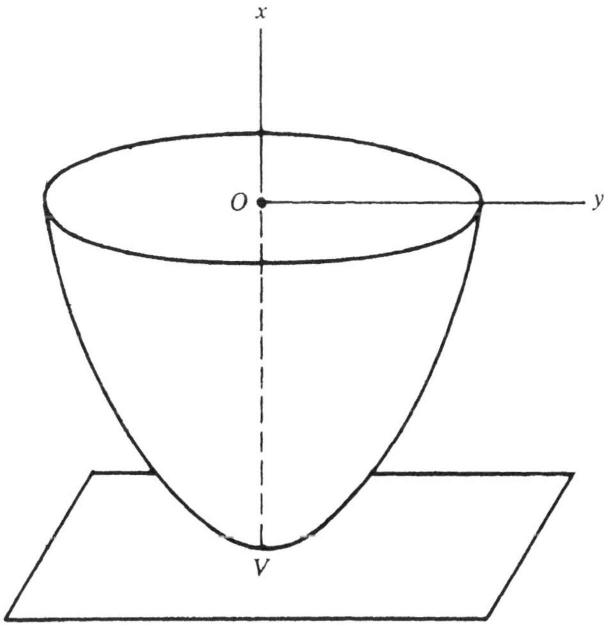

On the other hand, suppose that near V the sphere lies strictly outside S (again except for V itself). Then if S rests on a horizontal plane with point of contact V, any small displacement lowers the center of gravity, so S is unstable.

Consider therefore the function $f(P) = CP$, the distance from C to a variable point P on the surface of the ellipsoid. If this function has a strict local minimum at V, the balance will be stable; if it has a strict local maximum at P, the balance will be unstable. We may as well consider $f(P)^2$ instead of $f(P)$.

From the circular symmetry of the problem it is clear that C is at $(c, 0, 0)$ for some $c > 0$; moreover we may restrict ourselves to considering the function $f(P)^2$ where P varies along the generating ellipse $(x^2/a^2) + (y^2/b^2) = 1$ instead of the whole surface. If $P = (x, y)$ we have

$$\begin{aligned} f(P)^2 = (x - c)^2 + y^2 &= (x - c)^2 + b^2\left(1 - \frac{x^2}{a^2}\right) \\ &= \left(1 - \frac{b^2}{a^2}\right) x^2 - 2cx + b^2 + c^2. \end{aligned}$$

We want to determine whether $x = a$ is a local minimum for this polynomial relative to the interval $[0, a]$. Considering this polynomial along the whole positive x-axis we see that it is strictly decreasing if $b^2 \geq a^2$. If $b^2 < a^2$, then it decreases from $x = 0$ to $x = ca^2/(a^2 - b^2)$ and then increases. Hence $V(=(a, 0)$ in the two-dimensional coordinate system) is a strict local minimum point if $b^2 \geq a^2$ or if $b^2 < a^2$ and $a \leq ca^2/(a^2 - b^2)$; this is equivalent to $b^2 \geq a^2 - ac$. V is a strict local maximum if $b^2 < a(a - c)$.

Now we locate the center of gravity. Since nothing is said about the mass distribution, it is presumably uniform, so the center of gravity coincides with the centroid. Hence

$$c = \int_0^a x \cdot \pi y^2 dx \bigg/ \int_0^a \pi y^2 dx$$

$$= \int_0^a x\left(1 - \frac{x^2}{a^2}\right) dx \bigg/ \int_0^a \left(1 - \frac{x^2}{a^2}\right) dx = \frac{3}{8} a,$$

and our condition for stability becomes $b^2 \geq \frac{5}{8} a^2$; that is, $\sqrt{8}\, b \geq \sqrt{5}\, a$ as required.

Remark. With minor changes the argument for stability applies in much more general circumstances. If S is any convex solid and C is its center of gravity, it will rest stably on the surface point V if and only if the distance from a surface point to C has a strict local minimum at V.

Second Solution. From the circular symmetry of the ellipsoid it follows that every normal to the surface passes through the x-axis. If the solid is rocked slightly away from V so that it rests momentarily on the point Q, the force of support acts upward through Q. The force of gravity acts downward through C. Hence if the normal at Q meets the axis above C (i.e., at a point farther from V than C), the two forces produce a couple tending to restore the solid to its original position, so the balance is stable. Conversely, if the normal at Q meets the axis below C, the couple tends to accentuate the displacement, and the original balance is unstable. Hence the condition for stability is that the normals to the surface at all points Q near V (but not V) meet the axis above C.

Again we may compute with the generating ellipse instead of the whole ellipsoid. If Q is the point $(a \cos \theta, b \sin \theta)$, $\theta \neq 0, \pi$, the normal to the ellipse at Q crosses the x-axis at

$$a\left(1 - \frac{b^2}{a^2}\right) \cos \theta.$$

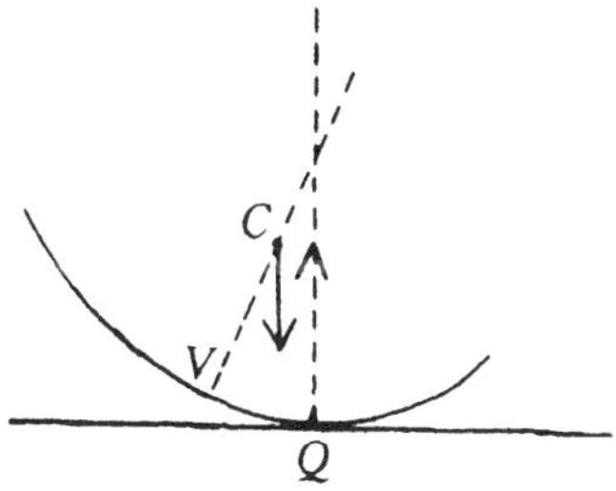

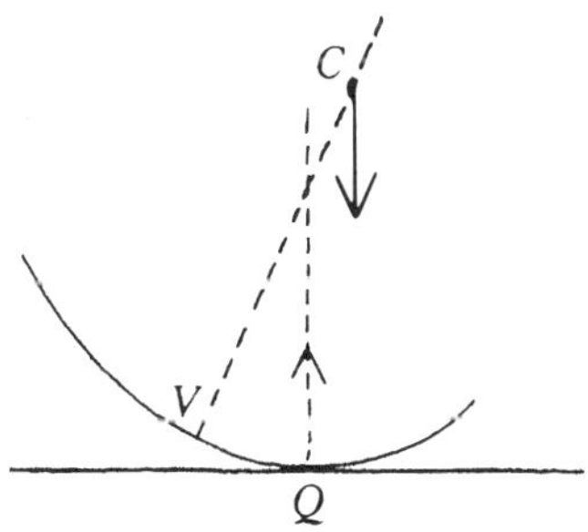

This is farther from V than C for all small non-zero θ if and only if $a(1 - (b^2/a^2)) \leq c$; that is, $b^2 \geq a^2 - ac$, as before.

Afternoon Session

8. A particle (x, y) moves so that its angular velocities about $(1, 0)$ and $(-1, 0)$ are equal in magnitude but opposite in sign. Prove that

$$y(x^2 + y^2 + 1)\,dx = x(x^2 + y^2 - 1)\,dy,$$

and verify that this is the differential equation of the family of rectangular hyperbolas passing through $(1, 0)$ and $(-1, 0)$ and having the origin as center.

Solution. The angular velocity about the origin of a point moving according to the parametric equations

$$x = x(t), \qquad y = y(t)$$

is

$$\frac{x\dot{y} - y\dot{x}}{x^2 + y^2},$$

where the dots indicate differentiation with respect to t. Translating the center of reference, first to $(1, 0)$ and then to $(-1, 0)$, we can express the condition as

$$\frac{(x-1)\dot{y} - y\dot{x}}{(x-1)^2 + y^2} + \frac{(x+1)\dot{y} - y\dot{x}}{(x+1)^2 + y^2} = 0.$$

From this we get

$$(x\dot{y} - y\dot{x} - \dot{y})(x^2 + y^2 + 1 + 2x)$$
$$+ (x\dot{y} - y\dot{x} + \dot{y})(x^2 + y^2 + 1 - 2x) = 0,$$

which simplifies to

(1) $$x(x^2 + y^2 - 1)\dot{y} = y(x^2 + y^2 + 1)\dot{x}.$$

This is equivalent to the required differential equation.

A central rectangular hyperbola has an equation of the form

$$A(x^2 - y^2) + Bxy = 1.$$

It passes through $(-1, 0)$ and $(1, 0)$ if and only if $A = 1$. So the suggested family is given by

(2) $$x^2 - y^2 + Bxy = 1.$$

Differentiation with respect to t yields

(3) $$2x\dot{x} - 2y\dot{y} + B(x\dot{y} + \dot{x}y) = 0.$$

Eliminating B between (2) and (3) we get

$$xy(2x\dot{x} - 2y\dot{y}) + (x\dot{y} + \dot{x}y)(1 - x^2 + y^2) = 0.$$

This is the differential equation of the suggested family of hyperbolas and it simplifies to (1) as required.

---

9. Evaluate the following limits:

$$\lim_{n\to\infty}\left(\frac{1}{\sqrt{n^2+1^2}} + \frac{1}{\sqrt{n^2+2^2}} + \cdots + \frac{1}{\sqrt{n^2+n^2}}\right);$$

$$\lim_{n\to\infty}\left(\frac{1}{\sqrt{n^2+1}}+\frac{1}{\sqrt{n^2+2}}+\cdots+\frac{1}{\sqrt{n^2+n}}\right);$$

$$\lim_{n\to\infty}\left(\frac{1}{\sqrt{n^2+1}}+\frac{1}{\sqrt{n^2+2}}+\cdots+\frac{1}{\sqrt{n^2+n^2}}\right).$$

Solution. (i) For the first sum,

$$\frac{1}{\sqrt{n^2+1}}+\frac{1}{\sqrt{n^2+2^2}}+\cdots+\frac{1}{\sqrt{n^2+n^2}}=$$

$$\frac{1}{n}\left[\frac{1}{\sqrt{1+\left(\frac{1}{n}\right)^2}}+\frac{1}{\sqrt{1+\left(\frac{2}{n}\right)^2}}+\cdots+\frac{1}{\sqrt{1+\left(\frac{n}{n}\right)^2}}\right].$$

This latter form is the lower Riemann sum for

$$\int_0^1\frac{dx}{\sqrt{1+x^2}}$$

corresponding to the subdivision points

$$\frac{1}{n},\frac{2}{n},\ldots,\frac{n-1}{n}.$$

Therefore its limit as $n\to\infty$ is

$$\int_0^1\frac{dx}{\sqrt{1+x^2}}=\log\left(x+\sqrt{1+x^2}\right)\Big|_0^1=\log\left(1+\sqrt{2}\right).$$

(ii) For the second sum, an individual term $1/\sqrt{n^2+i}$ satisfies

$$\frac{1}{\sqrt{n^2+\mathrm{n}}}\le\frac{1}{\sqrt{n^2+i}}\le\frac{1}{\sqrt{n^2+1}},\qquad i=1,2,\ldots,n,$$

and hence

$$\frac{n}{\sqrt{n^2+n}}\le\sum_{i=1}^{n}\frac{1}{\sqrt{n^2+i}}\le\frac{n}{\sqrt{n^2+1}}.$$

Now as $n\to\infty$ both extremes have the limit 1; hence

$$\lim_{n\to\infty}\left[\sum_{i=1}^{n}\frac{1}{\sqrt{n^2+i}}\right]=1.$$

(iii) For the third sum,

$$\frac{1}{\sqrt{n^2+i}} \geq \frac{1}{\sqrt{n^2+n^2}} = \frac{1}{n\sqrt{2}}, \qquad i = 1, 2, \ldots, n^2.$$

Hence

$$\sum_{i=1}^{n^2} \frac{1}{\sqrt{n^2+i}} \geq \frac{n}{\sqrt{2}}.$$

Therefore

$$\lim_{n\to\infty} \sum_{i=1}^{n^2} \frac{1}{\sqrt{n^2+i}} = \infty.$$

---

10. Find the differential equation satisfied by the product z of any two linearly independent integrals of the equation

$$y'' + y'P(x) + yQ(x) = 0.$$

Solution. Suppose y_1 and y_2 are two linearly independent solutions of the given differential equation. Then any two solutions have the form $u = ay_1 + by_2$ and $v = cy_1 + dy_2$.

Since uv falls in the linear space spanned by y_1^2, y_1y_2, y_2^2, we expect to find that it satisfies a linear differential equation of the third order.

Letting $z = uv$ we have

(1) $$u'' + Pu' + Qu = 0$$

(2) $$v'' + Pv' + Qv = 0$$

(3) $$z' = uv' + u'v$$

(4) $$z'' = uv'' + 2u'v' + vu''.$$

We find

(5)

$$\begin{aligned} z'' + Pz' + 2Qz &= u(v'' + Pv' + Qv) + v(u'' + Pu' + Qu) + 2u'v' \\ &= 2u'v'. \end{aligned}$$

Differentiating (5), we get

$$z''' + Pz'' + (P' + 2Q)z' + 2Q'z = 2u'v'' + 2v'u''. \tag{6}$$

Next multiply (1) by $2v'$ and (2) by $2u'$ and add to obtain.

$$2u'v'' + 2v'u'' + 4Pu'v' + 2Qz' = 0. \tag{7}$$

Multiply (5) by $2P$ to get

$$2Pz'' + 2P^2z' + 4PQz = 4Pu'v'. \tag{8}$$

Add (6), (7), and (8), cancelling the terms that appear on the right, to get

$$z''' + 3Pz'' + (2P^2 + P' + 4Q)z' + (4PQ + 2Q')z = 0, \tag{9}$$

a third-order differential equation satisfied by the product z of any two solutions of the original equation.

REMARK 1. The Wronskian of the three functions y_1^2, y_1y_2, and y_2^2 is twice the cube of the Wronskian of y_1 and y_2. So if y_1 and y_2 are linearly independent solutions of the original differential equation, then y_1^2, y_1y_2, and y_2^2 are linearly independent solutions of (9).

REMARK 2. If $z = Ay_1^2 + By_1y_2 + Cy_2^2$, where A, B, and C are constants, is differentiated three times, we may regard the four equations for z, z', z'', z''' as a system of four homogeneous linear equations in 1, A, B, C. Then the determinant of the system must be zero. The evaluation of this determinant leads to the desired differential equation (9).

Historical Note. The problem was first treated by Appell, *Comptes Rendus*, vol. 91 (1880), pp. 211-214. Recently, Bellman, *Bolletino della Unione Matematica Italiana*, series 3, vol. 12 (1957), pp. 12-15, has given a matrix method for finding the linear differential equation satisfied by the product of the solutions of two given linear differential equations.

---

11. Two perpendicular diameters of the ellipse

$$\frac{x^2}{a^2} + \frac{y^2}{b^2} = 1$$

are given, and the two diameters conjugate to them are constructed. Show that the rectangular hyperbola passing through the ends of these conjugate diameters passes through the foci of the ellipse.

Solution. Let E be the given ellipse. We assume that $a^2 > b^2$. Then the foci are at $(c, 0)$ and $(-c, 0)$, where $c = \sqrt{a^2 - b^2}$.

Suppose the given diameters have equations $y = mx$ and $-my = x$. Recall that one diameter of an ellipse is conjugate to another if it is parallel to the tangents at the ends of the other. The conjugate diameters are easily found to have the equations

$$b^2x + ma^2y = 0 \quad \text{and} \quad mb^2x - a^2y = 0,$$

respectively. Hence

$$(b^2x + ma^2y)(mb^2x - a^2y) = mb^4x^2 - ma^4y^2 + (m^2 - 1)a^2b^2xy = 0$$

is the equation of the degenerate conic D consisting of the lines of the two conjugate diameters. Any conic (except E) passing through the four points where D meets E (i.e., the ends of the conjugate diameters) has an equation of the form

$$(1) \quad \lambda\left(\frac{x^2}{a^2} + \frac{y^2}{b^2} - 1\right) + mb^4x^2 - ma^4y^2 + (m^2 - 1)a^2b^2xy = 0.$$

This conic is a rectangular hyperbola (possibly degenerate) if and only if the sum of the coefficients of x^2 and y^2 is zero, that is,

$$\lambda\left(\frac{1}{a^2} + \frac{1}{b^2}\right) + mb^4 - ma^4 = 0,$$

i.e., $\lambda = ma^2b^2c^2$. With this value of λ, the conic (1) passes through $(c, 0)$ and $(-c, 0)$, as required.

The rectangular hyperbola, (1) with $\lambda = ma^2b^2c^2$, is degenerate if and only if the constant term λ is zero. If the given ellipse is a genuine ellipse (i.e., $a^2 \neq b^2$), this occurs only for $m = 0$, that is, when the original diameters are the axes of the ellipse. In this case the conjugate diameters are also the axes and there is no proper rectangular hyperbola through the four points. The union of the two axes is then a degenerate rectangular hyperbola fulfilling the conditions.

If the given ellipse is actually a circle, $c = 0$, so λ is always zero. For a circle, perpendicular diameters are always conjugate and the union of any two such diameters is a degenerate rectangular hyperbola which passes through the foci, which coincide at the center.

12. A car is being driven so that its wheels, all of radius a feet, have an angular velocity of ω radians per second. A particle is thrown off from the tire of one of these wheels, where it is supposed that $a\omega^2 > g$. Neglecting the resistance of the air, show that the maximum height above the roadway which the particle can reach is

$$\frac{(a\omega + g\omega^{-1})^2}{2g}.$$

Solution. If a particle is thrown into motion in a gravitational field starting at height h and with upward component of velocity v, it will rise to the height $h + (v^2/2g)$. [The horizontal components of the motion do not influence the maximum height].

As long as the particle remains attached to the tire, it follows the path of a cycloid, and we may take its equations of motion as

$$x = a\omega t - a \sin \omega t$$
$$y = a(1 - \cos \omega t).$$

If the particle leaves the tire when $\omega t = \theta$, then it starts into gravitational motion with height $a(1 - \cos \theta)$ and upward velocity component $y' = a\omega \sin \theta$. Hence it reaches the height

$$H = a(1 - \cos \theta) + \frac{a^2\omega^2}{2g} \sin^2 \theta, \tag{1}$$

provided $0 \leq \theta \leq \pi$ (to ensure that the particle starts upward).

We are asked to maximize H by choice of θ. We set

$$\frac{dH}{d\theta} = a \sin \theta + \frac{a^2\omega^2}{g} \sin \theta \cos \theta = 0$$

and find that the critical points are 0, π, θ_0, where $\theta_0 = \arccos(-g/a\omega^2)$. (Since $a\omega^2 > g$, there is such a θ_0.) The corresponding values of H are 0, $2a$, and

$$H_0 = a\left(1 + \frac{g}{a\omega^2}\right) + \frac{a^2\omega^2}{2g}\left(1 - \frac{g^2}{a^2\omega^4}\right) = \frac{1}{2g}(a\omega + g\omega^{-1})^2.$$

Since $(a\omega + g\omega^{-1})^2 > 4ag$, $H_0 > 2a$, so the maximum value of H is H_0.

We can also find the maximum value of H by writing (1) in the form

$$H = H_0 - \frac{a^2\omega^2}{2g}\left(\cos \theta + \frac{g}{a\omega^2}\right)^2.$$

REMARK. If $a\omega^2 \leq g$, the maximum value of H will be $2a$, attained for $\theta = \pi$, which means that particles flying off never go higher than the top of the wheel.

---

13. Assuming that $f(x)$ is continuous in the interval (0, 1), prove that

$$\int_{x=0}^{x=1} \int_{y=x}^{y=1} \int_{z=x}^{z=y} f(x)f(y)f(z)\, dx\, dy\, dz = \frac{1}{3!}\left(\int_{t=0}^{t=1} f(t)\, dt\right)^3.$$

First Solution. Let $F(u) = \int_0^u f(t)\, dt$. Then $F'(u) = f(u)$. The right member of the desired equation is $\frac{1}{6} F(1)^3$. The left member can be integrated in successive steps. We get

$$\begin{aligned}
&\int_{x=0}^{x=1} f(x)\left(\int_x^1 f(y)(F(y) - F(x))\, dy\right) dx \\
&\qquad = \int_0^1 f(x)\left[\frac{1}{2}(F(y) - F(x))^2\right]_{y=x}^{y=1} dx \\
&\qquad = \frac{1}{2}\int_0^1 f(x)(F(1) - F(x))^2\, dx \\
&\qquad = -\frac{1}{6}(F(1) - F(x))^3 \Big|_0^1 = \frac{1}{6} F(1)^3,
\end{aligned}$$

as required.

Second Solution. Consider the unit cube in the positive octant. Points (x, y, z) of this unit cube can be divided into six subsets according to the ordering of x, y, z. (Note that the set of points having two or more coordinates the same is negligible.) Symmetry shows that the integral of $f(x)f(y)f(z)$ is the same over any of these sets. The required integral is $\iiint f(x)f(y)f(z)\, dx\, dy\, dz$ over the region

$$\{(x, y, z) : x < z < y\}.$$

It is therefore one sixth of the integral over the entire cube. The integral over the entire cube is

$$\int_0^1 f(x)\, dx \int_0^1 f(y)\, dy \int_0^1 f(z)\, dz = \left(\int_0^1 f(t)\, dt\right)^3$$

and the required equation follows.

---

14. (i) Show that any solution $f(t)$ of the functional equation

$$f(x+y)f(x-y) = f(x)f(x) + f(y)f(y) - 1, \quad (x, y, \text{ real}),$$

is such that

$$f''(t) = \pm m^2 f(t), \quad (m \text{ constant and } \geq 0),$$

assuming the existence and continuity of the second derivative. Deduce that $f(t)$ is one of the functions

$$\pm \cos mt, \qquad \pm \cosh mt.$$

Solution. Starting with the given functional equation

$$(1) \qquad f(x+y)f(x-y) = f(x)^2 + f(y)^2 - 1,$$

we differentiate with respect to y to get

$$(2) \qquad f'(x+y)f(x-y) - f(x+y)f'(x-y) = 2f(y)f'(y),$$

and then with respect to x to get

$$(3) \qquad f''(x+y)f(x-y) - f(x+y)f''(x-y) = 0.$$

Setting $x = y = 0$ in (1) and (2), we obtain $f(0)^2 = 2f(0)^2 - 1$ and $2f(0)f'(0) = 0$, whence

$$(4) \qquad f(0) = \pm 1, \qquad f'(0) = 0.$$

Now for any given number t, put $x = y = t/2$ in (3) to get

$$f''(t)f(0) - f(t)f''(0) = 0,$$

which is equivalent to

$$(5) \qquad f''(t) \pm m^2 f(t) = 0,$$

where $m = |f''(0)|^{1/2}$.

Integrating (5) using the initial conditions (4), we obtain

$$f(t) = \pm \cos mt \quad \text{or} \quad f(t) = \pm \cosh mt,$$

depending on whether the sign in (5) is plus or minus. If $m = 0$, these solutions are constant.

Conversely, either of the above solutions satisfies the given functional equation for any value of m.

REMARK. The continuity of the second derivative was not used in this proof. In fact, the result can be proved under far weaker hypotheses; for example, it is enough to assume that f itself is continuous. See J. Aczèl, *Lectures on Functional Equations and Their Applications,* Academic Press, New York, 1966.

---

14. (ii) With n constant values $a_1, a_2, \ldots, a_n$, supposed all different, let n constant values $b_1, b_2, \ldots, b_n$ be associated, and let a polynomial $P(x)$ be defined by the identity in x

$$\begin{vmatrix} 1 & x & x^2 & \cdots & x^{n-1} & P(x) \\ 1 & a_1 & a_1^2 & \cdots & a_1^{n-1} & b_1 \\ 1 & a_2 & a_2^2 & \cdots & a_2^{n-1} & b_2 \\ \cdots & \cdots & \cdots & \cdots & \cdots & \cdots \\ 1 & a_n & a_n^2 & \cdots & a_n^{n-1} & b_n \end{vmatrix} \equiv 0.$$

Given a polynomial $\phi(t)$, let a polynomial $Q(x)$ be defined by the identity in x obtained on replacing $P(x), b_1, b_2, \ldots, b_n$ of the identity above by $Q(x)$, $\phi(b_1), \phi(b_2), \ldots, \phi(b_n)$. Prove that the remainder obtained on dividing $\phi(P(x))$ by $(x - a_1)(x - a_2) \cdots (x - a_n)$ is $Q(x)$.

Solution. We use the fact that the Vandermonde determinant

$$V = \begin{vmatrix} 1 & a_1 & a_1^2 & \ldots & a_1^{n-1} \\ 1 & a_2 & a_2^2 & \ldots & a_2^{n-1} \\ \cdot & \cdot & \cdot & \ldots & \cdot \\ \cdot & \cdot & \cdot & \ldots & \cdot \\ 1 & a_n & a_n^2 & \ldots & a_n^{n-1} \end{vmatrix} = \prod_{i>j} (a_i - a_j)$$

is not zero.

If we replace x by a_1 in the determinant defining $P(x)$, subtract the second row from the first, and expand by minors of the first row, we see that

$$V(P(a_1) - b_1) = 0.$$

Hence, $P(a_1) = b_1$. Similarly, $P(a_i) = b_i$ for $i = 2, 3, \ldots, n$.

A similar argument using the determinant for Q shows that $Q(a_i) = \phi(b_i)$ for $i = 1, 2, \ldots, n$. Moreover, it is evident that Q is a polynomial of degree less than n.

Suppose the polynomial $\phi(P(x))$ is divided by $(x - a_1)(x - a_2)\cdots(x - a_n)$ leaving a remainder $R(x)$ of degree less than n. Then

$$\phi(P(x)) = (x - a_1)(x - a_2)\cdots(x - a_n)S(x) + R(x),$$

where $S(x)$ is a polynomial. Substituting $x = a_i$, we get

$$R(a_i) = \phi(P(a_i)) = \phi(b_i) = Q(a_i).$$

Thus the polynomial $Q(x) - R(x)$ is zero for n distinct values of x. But its degree is less than n, so $Q(x) - R(x)$ is identically zero, and the remainder is $Q(x)$, as required.

REMARK. The determinant given provides a useful expression for the unique polynomial of degree less than n with values prescribed at $a_1, a_2, \ldots, a_n$.

THE FIFTH WILLIAM LOWELL PUTNAM MATHEMATICAL COMPETITION

March 7, 1942

Morning Session

1. A square of side $2a$, lying always in the first quadrant of the XY plane, moves so that two consecutive vertices are always on the X- and Y-axes respectively. Find the locus of the midpoint of the square.

Solution. Let A and B be two consecutive vertices of the square lying on the X- and Y-axes, respectively. Let C be the center of the square and D and E the feet of perpendiculars from C to the X- and Y-axes, respectively. Let O be the origin.

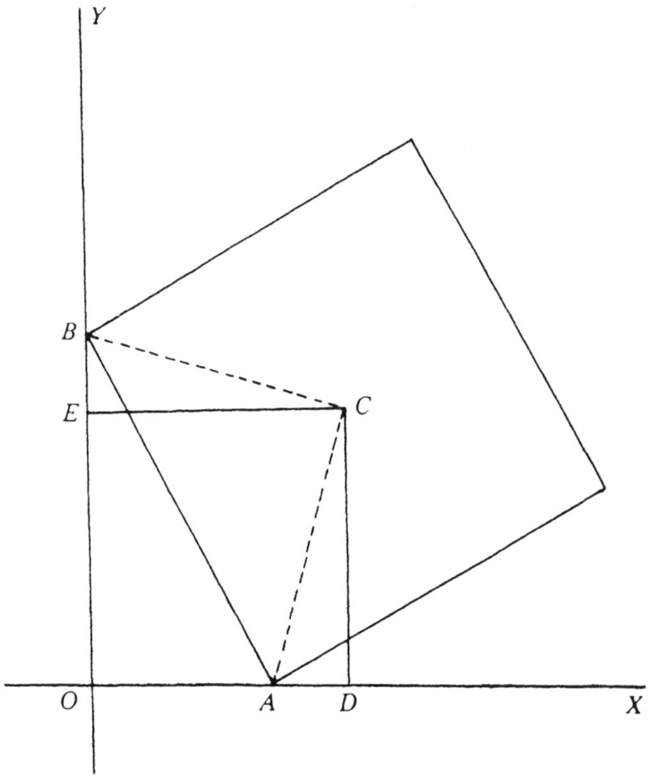

Then $\angle ACB$ is right angle, and $\angle DCE$ is a right angle and $\triangle ACD$ is congruent to $\triangle BCE$ (hypotenuse and acute angle). Hence the x and y coordinates of C are equal, and $OECD$ is a square. As the given square moves under the specified constraint, the center C moves back and forth along the segment from (a, a) to $(\sqrt{2}\,a, \sqrt{2}\,a)$.

REMARK. A generalization of this problem appears as Problem P.M. 7 in this examination.

---

2. If a polynomial $f(x)$ is divided by $(x - a)^2(x - b)$, where $a \neq b$, derive a formula for the remainder.

First Solution. Since $f(x)$ is divided by a cubic polynomial, the remainder $R(x)$ will be of degree at most two in x, say $R(x) = Ax^2 + Bx + C$. Then

$$f(x) = (x - a)^2(x - b)Q(x) + Ax^2 + Bx + C$$

and

$$f'(x) = 2(x - a)(x - b)Q(x) + (x - a)^2Q(x) + (x - a)^2(x - b)Q'(x) + 2Ax + B.$$

From these relations one gets

$$f(a) = Aa^2 + Ba + C$$
$$f(b) = Ab^2 + Bb + C$$
$$f'(a) = 2Aa + B.$$

Solving for A, B, C one gets

$$A = \frac{1}{(b - a)^2}[f(b) - f(a) - (b - a)f'(a)]$$

$$B = \frac{-1}{(b - a)^2}[2a(f(b) - f(a)) - (b^2 - a^2)f'(a)]$$

$$C = \frac{1}{(b - a)^2}[(b - a)^2 f(a) + a^2(f(b) - f(a)) + ab(a - b)f'(a)].$$

Hence

$$R(x) = \frac{1}{(b-a)^2}\{(f(b) - f(a) - (b-a)f'(a))x^2$$

$$- (2a(f(b) - f(a)) + (b^2 - a^2)f'(a))x$$

$$+ ((b-a)^2 f(a) + a^2(f(b) - f(a)) + ab(a-b)f'(a))\}.$$

This is easier to check if written in the form

$$R(x) = f(a) + \frac{f(b) - f(a)}{(b-a)^2}(x-a)^2 - \frac{f'(a)}{b-a}(x-a)(x-b).$$

Second Solution. We can write

$$\text{(1)} \qquad f(x) = f(a) + (x-a)f'(a) + (x-a)^2 g(x)$$

where g is a polynomial, and $g(x) = g(b) + (x-b)h(x)$ where h is a polynomial. Then

$$f(x) = f(a) + (x-a)f'(a) + (x-a)^2 g(b) + (x-a)^2(x-b)h(x)$$

and the desired remainder is

$$f(a) + (x-a)f'(a) + (x-a)^2 g(b).$$

Substituting b for x in (1) gives us $g(b)$ and the remainder is

$$f(a) + (x-a)f'(a) + \frac{(x-a)^2}{(b-a)^2}(f(b) - f(a) - (b-a)f'(a)).$$

---

3. Is the following series convergent or divergent?

$$1 + \frac{1}{2}\cdot\frac{19}{7} + \frac{2!}{3^2}\left(\frac{19}{7}\right)^2 + \frac{3!}{4^3}\left(\frac{19}{7}\right)^3 + \frac{4!}{5^4}\left(\frac{19}{7}\right)^4 + \cdots.$$

Solution. Use the ratio test. Let

$$a_n = \frac{(n-1)!}{n^{n-1}}\left(\frac{19}{7}\right)^{n-1}, \qquad a_{n+1} = \frac{n!}{(n+1)^n}\left(\frac{19}{7}\right)^n.$$

Then

$$R_n = \frac{a_{n+1}}{a_n} = \frac{n^n}{(n+1)^n}\frac{19}{7} = \frac{1}{\left(1+\frac{1}{n}\right)^n}\frac{19}{7},$$

and

$$\lim_{n\to\infty} R_n = \frac{19}{7}\lim\frac{1}{\left(1+\frac{1}{n}\right)^n} = \frac{19}{7}\frac{1}{e}.$$

Since $19/7 < 2.715$ and $e > 2.718$, $19/7e < 1$ and the series converges.

---

4. Find the orthogonal trajectories of the family of conics $(x + 2y)^2 = a(x + y)$. At what angle do the curves of one family cut the curves of the other family at the origin?

Solution. The given family is a family of parabolas all tangent to the line $x + y = 0$ at the origin. For $a = 0$ the parabola degenerates to the double line $(x + 2y)^2 = 0$ which should be viewed as two degenerate parabolas, the ray in the fourth quadrant being the limiting case as a goes to zero through positive values and the ray in the second quadrant being the limiting parabola as a goes to zero through negative values.

To find the differential equation of the family we differentiate the given equation and eliminate a between the original equation and its derivative.

(1) $$(x + 2y)^2 = a(x + y),$$

(2) $$2(x + 2y)(1 + 2y') = a(1 + y').$$

We get

$$2(x + y)(x + 2y)(1 + 2y') = (x + 2y)^2(1 + y'),$$

which simplifies to

(3) $$(3x + 2y)y' + x = 0.$$

The factor $x + 2y$ that was cancelled reflects the degeneracy along the line $x + 2y = 0$.

This differential equation is defined along the line $x + y = 0$ (where the original family of parabolas has no members), so in effect the line $x + y = 0$ is another degenerate member of the family corresponding to the case $a = \infty$.

The orthogonal trajectories are obtained by integrating the differential equation

$$xy' = 3x + 2y. \tag{4}$$

We write this as

$$\frac{d}{dx}(y + 3x) = \frac{2(y + 3x)}{x}.$$

The solution is

$$y + 3x = kx^2 \tag{5}$$

where k is an arbitrary constant. This is a new family of parabolas, with a unique member through every point of the plane except for points on the y-axis. The y-axis is an integral curve of differential equation (4) rewritten in the form

$$x = (3x + 2y)\frac{dx}{dy}$$

so that the y-axis also belongs to the family of orthogonal trajectories.

All the curves in the new family (5) are tangent to the line $y + 3x = 0$ at the origin (except the degenerate double parabola made by the y-axis).

The angle between the two families at the origin is then the angle θ between the two lines $x + y = 0$ and $3x + y = 0$.

Using the slopes we get

$$\tan\theta = \frac{m_1 - m_2}{1 + m_1 m_2} = \frac{-1 - (-3)}{(1 + (-1)(-3))} = \frac{2}{4} = \frac{1}{2}.$$

Hence $\theta = \arctan\left(\frac{1}{2}\right)$.

REMARK. If the degenerate cases are allowed, the answer is not unique and there will be other angles. For example, the angle between the y-axis (degenerate member of the orthogonal family) and $x + 2y = 0$ (degenerate member of the second family) is $\arctan 2$.

5. A circle of radius a is revolved through $180°$ about a line in its plane, distant b from the center of the circle, where $b > a$. For what value of the ratio b/a does the center of gravity of the solid thus generated lie on the surface of the solid?

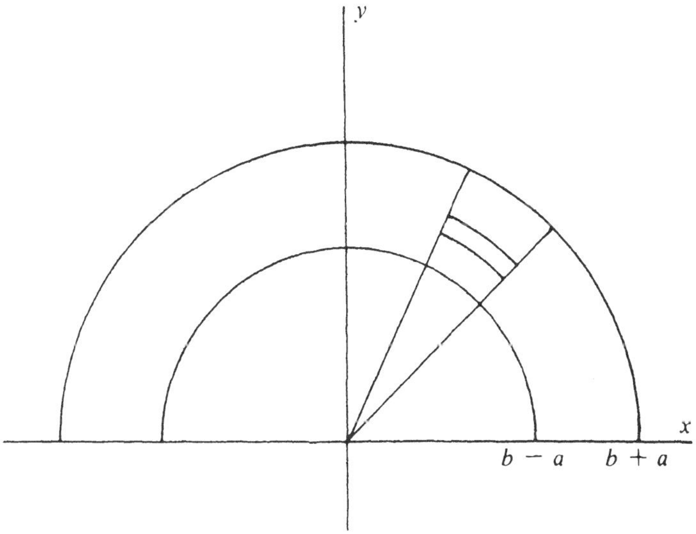

Solution. We choose axes so that the generating circle starts in the x-z plane and is revolved about the z-axis. The generated solid is half of a toroid (i.e., a solid bounded by a torus).

It is clear from symmetry that the centroid lies at a point $(0, \bar{y}, 0)$ on the y-axis, and the requirement of the problem is that $\bar{y} = b - a$.

To find the centroid we introduce polar coordinates in the x-y plane. Corresponding to the element of area $r\,dr\,d\theta$ in the plane there is the element of volume

$$2\sqrt{a^2 - (r - b)^2}\, r\, dr\, d\theta$$

which contributes

$$2r \sin \theta \sqrt{a^2 - (r - b)^2}\, r\, dr\, d\theta$$

to the moment M_y of the solid in the y direction. We have $\bar{y} = M_y/V$ where V is the volume of the semi-toroid.

$$V = \int_0^\pi \int_{b-a}^{b+a} 2\sqrt{a^2 - (r-b)^2}\; r\,dr\,d\theta$$

$$= 2\pi \int_{b-a}^{b+a} \sqrt{a^2 - (r-b)^2}\; r\,dr$$

$$= 2\pi a^2 \int_{-\pi/2}^{\pi/2} \cos\phi\,(b + a\sin\phi)\cos\phi\,d\phi$$

$$= \pi^2 a^2 b.$$

$$M_y = \int_0^\pi \int_{b-a}^{b+a} 2\sqrt{a^2 - (r-b)^2}\; r^2 dr \sin\theta\,d\theta$$

$$= 4 \int_{b-a}^{b+a} \sqrt{a^2 - (r-b)^2}\; r^2 dr$$

$$= 4a^2 \int_{-\pi/2}^{\pi/2} \cos\phi\,(b + a\sin\phi)^2 \cos\phi\,d\phi$$

$$= 2\pi\, a^2 b^2 + \frac{\pi}{2} a^4.$$

In both integrals we used the substitution $r = b + a\sin\phi$.

Hence

$$\bar{y} = \frac{M_y}{V} = \frac{a^2 + 4b^2}{2\pi b}.$$

But we require $\bar{y} = b - a$, so

$$2\pi b^2 - 2\pi ab = a^2 + 4b^2.$$

If $c = b/a$, then

$$(2\pi - 4)c^2 - 2\pi c - 1 = 0$$

and

$$c = \frac{\pi + \sqrt{\pi^2 + 2\pi - 4}}{2\pi - 4}.$$

We chose the positive sign since c must be positive.

Remarks. The volume of the semi-toroid could have been obtained from Pappus's Theorem that the volume of a solid of revolution is the product of the area of the generating region times the length of the circle traversed by the centroid of the area.

The critical ratio is about 2.91.

---

6. Any circle in the XY (horizontal) plane is "represented" by a point on the vertical line through the center of the circle and at a distance "above" the plane of the circle equal to the radius of the circle.

Show that the locus of the representations of all the circles which cut a fixed circle at a constant angle is a (portion of a) one-sheeted hyperboloid.

By consideration of suitable families of circles in the plane, demonstrate the existence of two families of rulings on the hyperboloid.

Solution. Denote the fixed circle by C and suppose it has center O and radius a. Take coordinates in the XY plane with origin at O and introduce a third coordinate z as usual. The angle α between two intersecting smooth curves is the smaller of the two angles between their tangents at the point of intersection. Hence $0 \le \alpha \le \pi/2$.

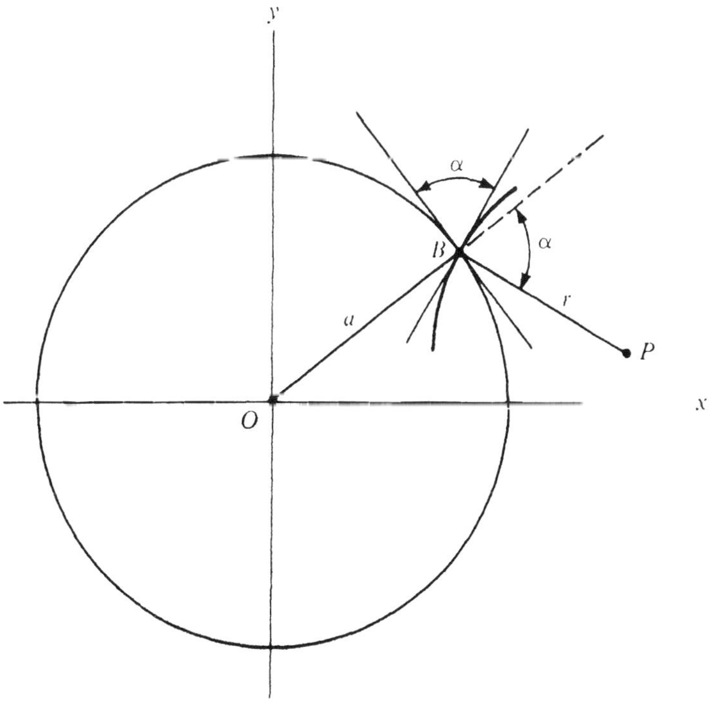

Fix such an α. Suppose a circle C' with center P and radius r cuts the fixed circle C at B making an angle α. Then

$$(OP)^2 = a^2 + r^2 \pm 2ar\cos\alpha$$

the sign being + if $\angle OBP$ is obtuse and − if $\angle OBP$ is acute. If the plane coordinates of P are (x_0, y_0) then C' will be represented by the point (x_0, y_0, r). Hence all such representative points lie on the set $\mathcal{L} = \mathcal{L}_1 \cup \mathcal{L}_2$ where $\mathcal{L}_1$ is determined by the conditions

$$x^2 + y^2 = a^2 + z^2 - 2az\cos\alpha, \qquad z > 0$$

and $\mathcal{L}_2$ by the conditions

$$x^2 + y^2 = a^2 + z^2 + 2az\cos\alpha, \qquad z > 0.$$

Conversely, given any point (x_0, y_0, r) of $\mathcal{L}$ the circle C' of radius r and center (x_0, y_0) cuts C at two points (just one if $\alpha = 0$). To see this, note that

$$(r - a)^2 \le a^2 + r^2 \pm 2ar\cos\alpha = x_0{}^2 + y_0{}^2 \le (r + a)^2$$

and therefore the distance from the center of C to the center of C' is between $|r - a|$ and $r + a$.

Except in the cases $\alpha = 0, \pi/2$, the sets $\mathcal{L}_1$ and $\mathcal{L}_2$ are portions of two distinct hyperboloids obtained by rotating about the z-axis the hyperbolas

$$x^2 = (z \pm a\cos\alpha)^2 + a^2\sin^2\alpha.$$

If $\alpha = \pi/2$, these hyperbolas coincide, as do $\mathcal{L}_1$ and $\mathcal{L}_2$. If $\alpha = 0$, the hyperbolas degenerate into cones (and the phrase "cut at angle α" becomes "tangent to").

Let B be a fixed point of C and let R be one of the four rays (assuming $0 < \alpha < \pi/2$) starting from B which makes an angle α with $\overleftrightarrow{AB}$. All circles with centers on R and passing through B cut C at angle α and so are represented by a point on $\mathcal{L}$. Since the radii of these circles will increase in proportion to the change in their x- (or y-) coordinate, the points representing them form a ray in space, and the ray is a ruling of $\mathcal{L}$. The two rays at B that meet C again generate rays lying in $\mathcal{L}_1$.

The other two rays at B generate rays in $\mathcal{L}_2$. If the ruling corresponding to one of these rays is rotated about the z-axis it generates a whole family of rulings on $\mathcal{L}$. Thus $\mathcal{L}$ has four families of rulings, two on $\mathcal{L}_1$ and two on $\mathcal{L}_2$. If $\alpha = \pi/2$ there are just two families of rulings, since there are only two rays from B perpendicular to OB, but in this case $\mathcal{L}_1 = \mathcal{L}_2$.

Each point Q of $\mathcal{L}$ corresponds to a circle with center P which cuts C twice, say at B and B' (we are here assuming $0 < \alpha \leq \pi/2$). Then the rulings corresponding to $\overrightarrow{BP}$ and $\overrightarrow{B'P}$ are two different rulings through Q and they come from different families, since $\overrightarrow{BP}$ does not coincide with $\overrightarrow{B'P}$ rotated. Thus there are two rulings from different families through every point of $\mathcal{L}$.

If $\alpha = 0$ there is only one ruling through each point of $\mathcal{L}$ except the point $(0, 0, a)$, corresponding to C, through which pass all the rulings of $\mathcal{L}$.

REMARK. The framers of this problem seem to have overlooked the fact that in general two hyperboloids are involved. It is interesting to verify that, if all circles and angles are considered directed, a circle described in the negative direction is regarded as having a negative radius, and point circles are accepted, then the locus $\mathcal{L}$ is all of a single hyperboloid (cone, if $\alpha = 0$ or π).

Afternoon Session

7. A square of side $2a$, lying always in the first quadrant of the XY plane, moves so that two consecutive vertices are always on the X- and Y-axes respectively. Prove that a point within or on the boundary of the square will in general describe a (portion of a) conic. For what points of the square does this locus degenerate?

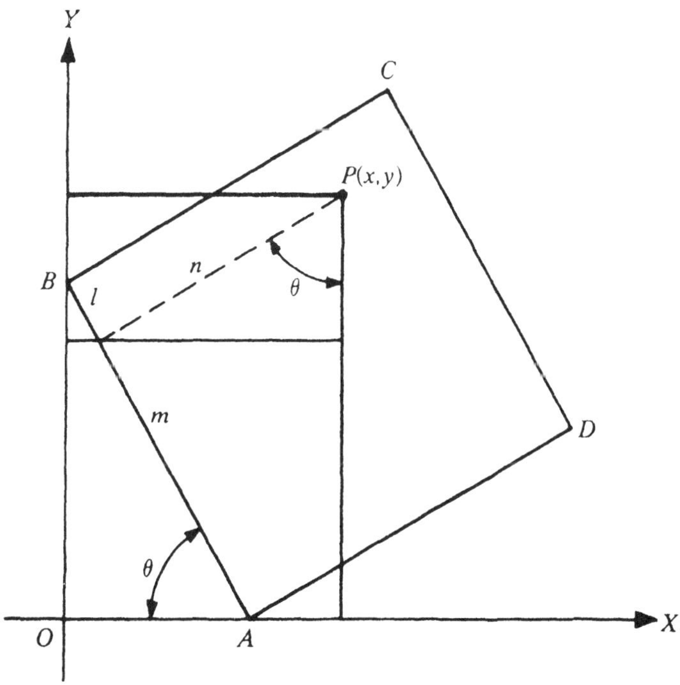

Solution. Let A and B be two consecutive vertices of the square lying on the X- and Y-axes, respectively, and let $P(x, y)$ be a specified point in the square. The coordinates of P are given by

$$x = n \sin \theta + l \cos \theta$$
(1)
$$y = m \sin \theta + n \cos \theta$$

where n, l, m, and θ are defined implicitly in the diagram.

We solve these equations for $\sin \theta$ and $\cos \theta$:

$$(lm - n^2) \cos \theta = mx - ny$$

$$(lm - n^2) \sin \theta = -nx + ly.$$

Hence

(2) $$(lm - n^2)^2 = (mx - ny)^2 + (-nx + ly)^2$$

is an equation satisfied by the coordinates of P for any value of θ. Now (2) reduces to

(3) $$(m^2 + n^2)x^2 - 2n(l + m)xy + (l^2 + n^2)y^2 = (lm - n^2)^2,$$

which in general represents a central conic. Its discriminant is

$$\Delta = 4n^2(l + m)^2 - 4(m^2 + n^2)(l^2 + n^2) = -4(lm - n^2)^2.$$

Evidently, Δ cannot be positive. If $\Delta < 0$, then (3) represents an ellipse or circle. It will be a circle if and only if $m^2 = l^2$ and $n(l + m) = 0$. Since $l + m = AB$ cannot be zero, a circle occurs if and only if $n = 0$ and $l = m$; i.e., P is the midpoint of AB.

If $\Delta = 0$, the right member of (3) is also zero and (3) reduces to

$$\frac{l + m}{m}(mx - ny)^2 = 0$$

which is equivalent to $y = (m/n)x$, so P moves along a straight line.

The geometrical meaning of $\Delta = 0$, or $lm = n^2$ is that P is on a semicircle of which AB is a diameter. [Note that n is the mean proportional between l and m.] Notice that $l = m = n$ makes P the center of the given square and the locus is a portion of the line $y = x$, as already found in Problem 1 of this competition.

If $\Delta \neq 0$, i.e., if $lm \neq n^2$, the parametrization (1) is non-singular (that is, $dx/d\theta$ and $dy/d\theta$ do not vanish together) so the point P traverses its elliptical path smoothly as θ goes from 0 to $\pi/2$.

If $\Delta = 0$, then as noted above, P lies on a semicircle with diameter AB. The other half of the circle will pass through the origin O. Hence as θ varies from 0 to $\pi/2$, a point P on the semicircle will move away from the origin until PO is a diameter of this moving circle and then move back. To see this analytically, note that reversal can occur only when x has a critical value as a function of θ. But $dx/d\theta = 0$ requires $\tan\theta = n/l$, which means that PB is perpendicular to the y-axis and therefore PO is a diameter of the moving circle.

REMARKS. The problem could be interpreted as meaning that after A gets to the origin, it continues up the y-axis while D moves toward the origin along the x-axis, etc. With this interpretation, a point will, in general, describe portions of four different conics. This interpretation makes no difference if P is the center of the square, since all four partial conics degenerate to the same segment.

This problem, the first problem of the afternoon session, is a nice generalization of the first problem of the morning session.

---

8. For the family of parabolas

$$y = \frac{a^3x^2}{3} + \frac{a^2x}{2} - 2a,$$

(i) find the locus of vertices,
(ii) find the envelope,
(iii) sketch the envelope and two typical curves of the family.

Solution. (i) The given equation can be written in standard form as

$$y + \frac{35}{16}a = \frac{a^3}{3}\left(x + \frac{3}{4a}\right)^2,$$

whence a typical vertex is

$$\left(-\frac{3}{4a}, -\frac{35}{16}a\right).$$

If $a = 0$, then the given curve is a straight line, not a parabola, and therefore it has no vertex. The vertices of the parabolas in the system all lie on the hyperbola $xy = 105/64$.

Conversely every point of this hyperbola is the vertex of a unique member of this family, since if (x_0, y_0) is on the hyperbola, then $x_0 = -3/4a$, $y_0 = -35a/16$ can be solved uniquely for a.

(ii) Let

$$f(x, y, a) = \frac{a^3x^2}{3} + \frac{a^2x}{2} - 2a - y.$$

Then

$$\frac{\partial f}{\partial a}(x, y, a) = (ax + 2)(ax - 1).$$

To find the envelope of the family, we eliminate a between $\partial f/\partial a = 0$ and $f = 0$. This gives

$$xy = \frac{(ax)^3}{3} + \frac{(ax)^2}{2} - 2ax = \frac{1}{3} + \frac{1}{2} - 2 = \frac{-7}{6}$$

or

$$= \frac{-8}{3} + \frac{4}{2} + 4 = \frac{10}{3}$$

depending on which of the two factors of $\partial f/\partial a$ we choose to equate to zero.

We can readily verify that the parabola corresponding to the parameter a is tangent to the hyperbola

$$xy = -7/6 \text{ at } \left(\frac{1}{a}, -\frac{7a}{6}\right)$$

and tangent to the hyperbola

$$xy = 10/3 \text{ at } \left(-\frac{2}{a}, -\frac{5}{3}a\right).$$

Hence the envelope is the union of the two hyperbolas.

(iii) Sketch.

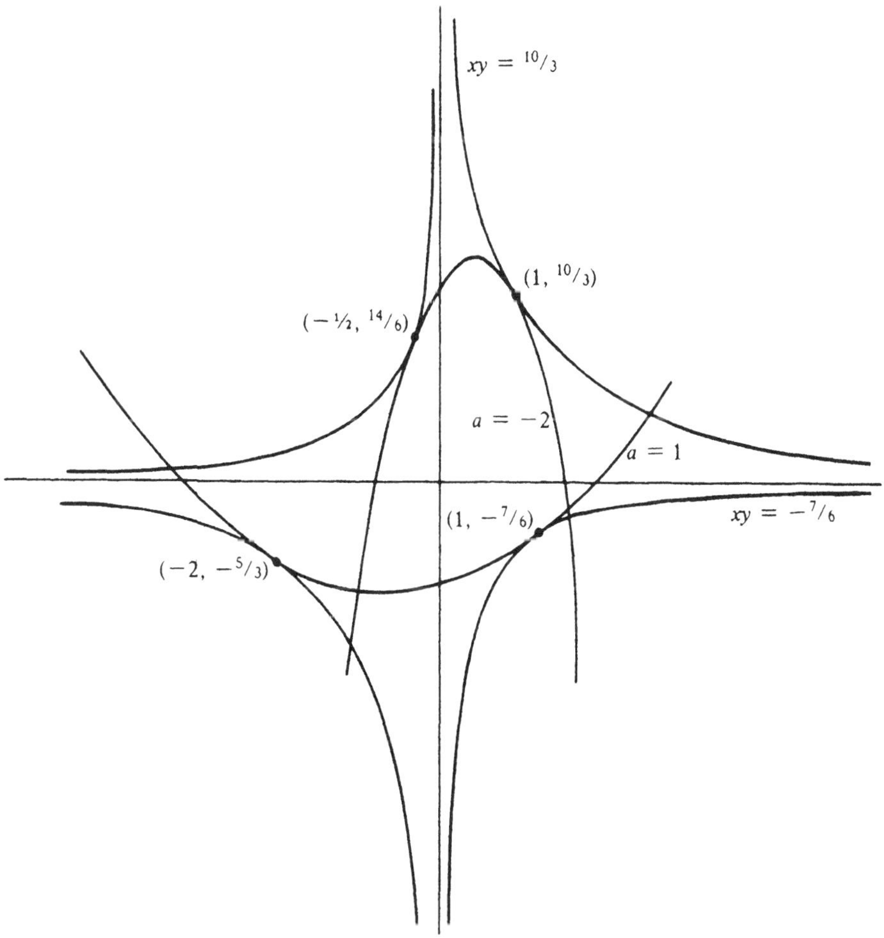

9. Given

$$x = \phi(u, v)$$
$$y = \psi(u, v)$$

where ϕ and ψ are solutions of the partial differential equation

$$\frac{\partial \phi}{\partial u}\frac{\partial \psi}{\partial v} - \frac{\partial \phi}{\partial v}\frac{\partial \psi}{\partial u} = 1. \tag{1}$$

By assuming that x and v are the independent variables, show that (1) may be transformed to

$$\frac{\partial y}{\partial v} = \frac{\partial u}{\partial x}. \tag{2}$$

Integrate (2), and show how this effects in general the solution of (1). What other solutions does (1) possess?

Solution. The statement of the problem implies, of course, that for each x and v there exist unique u and y such that $x = \phi(u, v)$ and $y = \psi(u, v)$, that is, there are functions α and β such that $u = \alpha(x, v)$ and $y = \beta(x, v)$. We assume that these functions have continuous first partial derivatives. Later we shall discuss the differentiability assumptions more carefully.

To reduce confusion in the notation we let $\phi_1, \phi_2, \psi_1, \psi_2$ be the partial derivatives of ϕ and ψ with respect to their first and second arguments, respectively. In this notation equation (1) is

$$\phi_1\psi_2 - \phi_2\psi_1 = 1.$$

Let

$$\frac{\partial y}{\partial x}, \frac{\partial y}{\partial v}, \frac{\partial u}{\partial x}, \text{ and } \frac{\partial u}{\partial v}$$

be the partial derivatives of y and u when x and v are taken as independent variables. Then

$$\phi_1 \frac{\partial u}{\partial v} + \phi_2 = 0, \tag{3}$$

$$\phi_1 \frac{\partial u}{\partial x} = 1, \tag{4}$$

$$\psi_1 \frac{\partial u}{\partial v} + \psi_2 = \frac{\partial y}{\partial v}, \tag{5}$$

$$\psi_1 \frac{\partial u}{\partial x} = \frac{\partial y}{\partial x}. \tag{6}$$

From (5), (3), and (1) we get

$$\phi_1 \frac{\partial y}{\partial v} = \psi_1\phi_1 \frac{\partial u}{\partial v} + \psi_2\phi_1 = -\psi_1\phi_2 + \phi_1\psi_2 = 1.$$

Multiply by $\partial u/\partial x$ and use (4) to get

$$\frac{\partial y}{\partial v} = \frac{\partial u}{\partial x}$$

which is the required equation (2).

Suppose now that

$$y = \int^{v} f(x, \eta)d\eta + g(x)$$

$$u = \int^{x} f(\xi, v)d\xi + h(v)$$

where f, g and h are continuous functions. Clearly

$$\frac{\partial y}{\partial v} = f(x, v) = \frac{\partial u}{\partial x},$$

and we have a wide class of solutions of (2).

Suppose

$$y = \alpha(x, v) \tag{7}$$

$$u = \beta(x, v) \tag{8}$$

give a solution of (2), that is, $\alpha_2 = \beta_1$; and suppose moreover that β_1 is never zero. Then (8) can be solved for x in terms of u and v, say $x = \phi(u, v)$, and the result substituted into (7) to express y in terms of u and v, say $y = \psi(u, v)$. Then considering u and v as independent variables, we have

$$\beta_1 \frac{\partial x}{\partial u} = 1 \tag{9}$$

$$\beta_1 \frac{\partial x}{\partial v} + \beta_2 = 0 \tag{10}$$

$$\alpha_1 \frac{\partial x}{\partial u} = \frac{\partial y}{\partial u} \tag{11}$$

$$\alpha_1 \frac{\partial x}{\partial v} + \alpha_2 = \frac{\partial y}{\partial v}. \tag{12}$$

From (9) and (11) $\alpha_1 = \beta_1 \, \partial y/\partial u$, so

$$\beta_1 \left[\frac{\partial x}{\partial u}\frac{\partial y}{\partial v} - \frac{\partial x}{\partial v}\frac{\partial y}{\partial u} \right] = \frac{\partial y}{\partial v} - \alpha_1 \frac{\partial x}{\partial v} = \alpha_2. \tag{13}$$

Since we are assuming $\alpha_2 = \beta_1$ and β_1 is never zero, we obtain

$$\frac{\partial x}{\partial u} \cdot \frac{\partial y}{\partial v} - \frac{\partial x}{\partial v} \frac{\partial y}{\partial u} = 1$$

which is (1). Thus solutions of (2) for which β_1 does not vanish give rise to solutions of (1).

The equivalence of (1) and (2) was established under the hypothesis that x and v were independent. By the implicit function theorem this amounts (locally) to the hypothesis that ϕ_1 (partial derivative of the original ϕ) does not vanish. If ϕ_1 vanishes at a point (u_0, v_0), then from (1) it is clear that ϕ_2 does not vanish at (u_0, v_0), so assuming continuity, ϕ_2 does not vanish near (u_0, v_0). Hence locally we can solve for v in terms of x and u, and the argument proceeds as before with the roles of u and v interchanged. This leads to all other local solutions of (1). Take the other extreme. Suppose ϕ_1 vanishes everywhere, so that $x = \phi(v)$ is independent of u. Then (1) becomes $\phi' \psi_1 = -1$, and this equation can be integrated with respect to u since ϕ' depends only on v. We get

$$\phi' \psi = -u + k(v).$$

Thus

$$\text{(14)} \qquad \begin{aligned} x &= \phi(v) \\ y &= (-u + k(v))/\phi'(v) \end{aligned}$$

for any function ϕ having a non-vanishing derivative and for any function k. Equations (14) give a solution of (1) not included in the class previously obtained. These are the other desired solutions of (1).

Discussion and Justification. While much of the argument is valid for functions of class C^1, we shall first assume that only functions of class C^∞ are to be considered. Furthermore, we shall consider the problem only locally.

With these assumptions, the initial equation $x = \phi(u, v)$ can be solved for u in terms of x and v locally near any point where $\phi_1 \neq 0$, and the result expresses u as a C^∞ function of x and v. Once we have u as a C^∞ function of x and v, we can express y as a C^∞ function of x and v by substituting in $y = \psi(u, v)$. The derivation of (2) now proceeds without difficulty. This much of the argument requires only that the function ϕ be C^1.

In solving (2) we now insist that f, g, and h be of class C^∞. Say f is defined on the rectangular open set $I \times J$ where I and J are open intervals in R. Pick $a \in i$, $b \in J$, and define

$$y = \int_b^v f(x, \eta)d\eta + g(x)$$

$$u = \int_a^x f(\xi, v)d\xi + h(v).$$

These functions are C^∞ solutions of (2), and conversely every C^∞ solution of (2) with rectangular domain has this form. It is at this point that the restriction to C^∞ functions is important. It is not easy to characterize all solutions of (2) having a finite differentiability class, say C^1 or C^2.

REMARK. The expression $\phi_1\psi_2 - \phi_2\psi_1$ is the Jacobian of the transformation $x = \phi(u, v)$, $y = \psi(u, v)$, so the equation $\phi_1\psi_2 - \phi_2\psi_1 = 1$ indicates that this transformation is (locally) an area-preserving change of coordinates.

---

10. A particle moves under a central force inversely proportional to the kth power of the distance. If the particle describes a circle (the central force proceeding from a point *on the circumference of the circle*), find k.

Solution. Choose polar coordinates with pole at the center of force and initial ray a diameter of the circular orbit of the particle. The equation of the orbit is then

(1) $$r = A \cos \theta,$$

where A is the diameter of the circle.

The equations of motion are

(2) $$\frac{d^2r}{dt^2} - r\left(\frac{d\theta}{dt}\right)^2 = -\frac{1}{m}f$$

(3) $$r\frac{\partial^2\theta}{\partial t^2} + 2\frac{dr}{dt}\frac{d\theta}{dt} = 0,$$

where m is the mass of the particle and f is the magnitude of the central force. Since the sign is taken as negative, a positive f means an attractive force.

After multiplication by r, equation (3) can be integrated to give

$$r^2 \frac{d\theta}{dt} = h, \tag{4}$$

which asserts that the angular momentum of the particle is constant.

Differentiating (1) twice, using (4), we get

$$\frac{dr}{dt} = -A \sin\theta \frac{d\theta}{dt} = -\frac{Ah \sin\theta}{r^2},$$

$$\frac{d^2r}{dt^2} = -\frac{Ah\cos\theta}{r^2} \cdot \frac{d\theta}{dt} + \frac{2Ah\sin\theta}{r^3} \cdot \frac{dr}{dt}$$

$$= -\frac{h^2}{r^3} - \frac{2A^2h^2\sin^2\theta}{r^5}.$$

Then substituting in (2), we obtain

$$-\frac{f}{m} = -\frac{h^2}{r^3} - \frac{2A^2h^2\sin^2\theta}{r^5} - r\left(\frac{h}{r^2}\right)^2$$

$$= -\frac{2h^2}{r^5}(r^2 + A^2\sin^2\theta) = -\frac{2A^2h^2}{r^5}.$$

Thus $f = 2mA^2h^2r^{-5}$ and $k = 5$.

REMARK. See F. R. Moulton, *Introduction to Celestial Mechanics*, Macmillan, 1902, for a discussion of the general inverse power law of attraction.

---

11. Sketch the curve

$$y = \frac{x}{1 + x^6 \sin^2 x},$$

and show that

$$\int_0^\infty \frac{x\,dx}{1 + x^6 \sin^2 x}$$

exists.

Solution. Let

$$f(x) = \frac{x}{1 + x^6 \sin^2 x}.$$

Then f is an odd function, so its graph is symmetric with respect to the origin; hence we need consider only non-negative values of x.

Clearly

$$x \geq \frac{x}{1 + x^6 \sin^2 x} \geq \frac{x}{1 + x^6}.$$

At points $x = n\pi$, n an integer, the curve is tangent to the line $y = x$. At points $x = (n + \frac{1}{2})\pi$, it is tangent to the graph of $y = x/(1 + x^6)$.

So the graph of $f(x)$ oscillates between an upper curve $y = x$ and a lower curve $y = x/(1 + x^6)$.

For large values of x, except those very near an integral multiple of π, f is very small, i.e., near the lower curve. Indeed, the graph is characterized by tall, narrow spikes at integral multiples of π. (The spikes are much narrower than the drawing suggests. For example, the function f drops to less than .02 within .1 of 2π.)

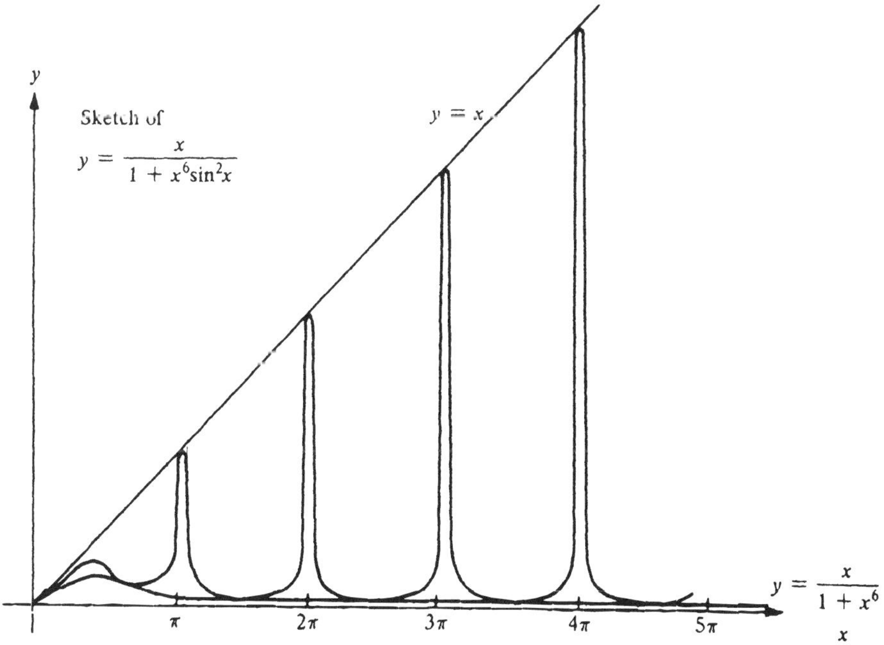

Since f is non-negative on the whole domain of integration, the question of the convergence of

$$\int_0^\infty f(x)dx$$

is equivalent to whether or not

$$\int_0^t f(x)dx$$

is bounded, and in turn to whether or not

$$\sum_{n=1}^{\infty} \int_{(n-1/2)\pi}^{(n+1/2)\pi} f(x)dx$$

is convergent.

Note that $|\sin t| \geq (2/\pi)\,|t|$ for $|t| \leq \pi/2$. Set $k_n = 2\pi^2(n - \frac{1}{2})^3$. Then, for any positive integer n, and for

$$\left(n - \frac{1}{2}\right)\pi \leq x \leq \left(n + \frac{1}{2}\right)\pi,$$

we have

$$\begin{aligned} 1 + x^6 \sin^2 x &= 1 + x^6 \sin^2 (x - n\pi) \\ &\geq 1 + \left(n - \frac{1}{2}\right)^6 \pi^6 \frac{4}{\pi^2} (x - n\pi)^2 \\ &= 1 + k_n{}^2 (x - n\pi)^2. \end{aligned}$$

So we get

$$f(x) \leq \frac{\left(n + \frac{1}{2}\right)\pi}{1 + k_n{}^2 (x - n\pi)^2}.$$

Therefore

$$\begin{aligned} \int_{(n-1/2)\pi}^{(n+1/2)\pi} f(x)dx &\leq \left(n + \frac{1}{2}\right)\pi \int_{-\pi/2}^{\pi/2} \frac{du}{1 + k_n{}^2 u^2} \\ &\leq \left(n + \frac{1}{2}\right)\pi \int_{-\infty}^{\infty} \frac{du}{1 + k_n{}^2 u^2} \end{aligned}$$

$$= \frac{\left(n + \frac{1}{2}\right) \pi^2}{k_n} \leq \frac{A}{n^2}$$

for some number A independent of n.

Since $\sum_{n=1}^{\infty} 1/n^2$ is convergent, we conclude that

$$\sum_{n=1}^{\infty} \int_{(n-1/2)\pi}^{(n+1/2)\pi} f(x)dx$$

is convergent.

Therefore $\int_0^\infty f(x)dx$ exists.

REMARK. The problem was first treated by G. H. Hardy, in *Messenger of Mathematics*, vol. 31, 1902, p. 177. He showed that

$$\int_0^\infty \frac{x^\mu}{1 + x^\nu \sin^2 x}\, dx$$

converges if and only if $\nu > 2\mu + 2$.

THE SIXTH WILLIAM LOWELL PUTNAM MATHEMATICAL COMPETITION

June 1, 1946

Morning Session

1. Suppose that the function $f(x) = ax^2 + bx + c$, where a, b, c are real constants, satisfies the condition $|f(x)| \le 1$ for $|x| \le 1$. Prove that $|f'(x)| \le 4$ for $|x| \le 1$.

First Solution. If $a \neq 0$ the graph of $y = ax^2 + bx + c$ is a parabola which can be assumed without loss of generality to open upward, i.e., $a > 0$. [We discuss the straight line case, $a = 0$, later.] By symmetry we may assume that b is non-negative. Then the vertex falls in the left half-plane and it is clear that $\max_{|x| \le 1} |f'(x)|$ occurs when $x = 1$, and this maximum value is $2a + b$. It remains to show that $2a + b \le 4$.

Now $f(1) = a + b + c \le 1$, and $f(0) = c \ge -1$. Thus $a + b \le 2$. Since a and b are both non-negative, $a \le 2$ and $2a + b \le 4$.

CASE $a = 0$. If $a = 0$, then

$$f'(x) = b = \frac{f(1) - f(-1)}{2},$$

so

$$|f'(x)| \le \frac{|f(1)| + |f(-1)|}{2} \le 1.$$

REMARK. The polynomial $f(x) = 2x^2 - 1$ satisfies the conditions of the problem and the absolute value of its derivative, $|4x|$, attains the bound 4 for $x = \pm 1$.

Second Solution. Since $f'(x) = 2ax + b$, a linear function, $|f'(x)|$ assumes its maximum on the closed interval $[-1, +1]$ at one of the two endpoints. Hence

$$\max_{|x| \le 1} |f'(x)| = |2a + b| \quad \text{or} \quad |2a - b|.$$

Now

$$2a + b = \frac{3}{2}(a + b + c) + \frac{1}{2}(a + c - b) - 2c$$

$$= \frac{3}{2}f(1) + \frac{1}{2}f(-1) - 2f(0).$$

So

$$|2a + b| \le \frac{3}{2}|f(1)| + \frac{1}{2}|f(-1)| + 2|f(0)| \le \frac{3}{2} + \frac{1}{2} + 2 = 4.$$

Also

$$2a - b = \frac{1}{2}f(1) + \frac{3}{2}f(-1) - 2f(0)$$

and

$$|2a - b| \le \frac{1}{2} + \frac{3}{2} + 2 = 4.$$

Hence max $|f'(x)| \le 4$.

Historical Note. The chemist Mendeleev raised the question as to the restrictions on $p_n'(x)$ for $-1 \le x \le 1$ when $|p_n(x)| \le 1$ on $-1 \le x \le 1$, where p_n is a polynomial of degree n.

A. A. Markoff answered this question in 1890 by proving that, if $|p_n(x)| \le 1$ on $-1 \le x \le 1$, then $|p_n'(x)| \le n^2$ on the same interval. The present problem is thus the special case $n = 2$. It is known that equality occurs if and only if, except for sign, $p_n(x) = \cos(n \arccos x)$, i.e., $p_n(x)$ is the polynomial such that $\cos n\theta = p_n(\cos\theta)$. For $n = 2$, $\cos 2\theta = 2\cos^2\theta - 1$, so $p_2(x) = \cos(2 \arccos x) = 2x^2 - 1$. The polynomials $p_n(x)$ are called Chebyshev polynomials. See John Todd, *A Survey of Numerical Analysis*, New York, 1962, pp. 138–139. The generalized version appears as problem 83, in Section 6, Pólya and Szegö, *Aufgaben und Lehrsatze aus der Analysis*, vol. 2, p. 91 and p. 287.

A slight variation on this problem was used as Problem A5 in the Twenty-ninth Competition held on December 7, 1968. That problem was phrased as follows: "Let V be the collection of all quadratic polynomials P with real coefficients such that $|P(x)| \le 1$ for all x on the closed interval $[0, 1]$. Determine $\sup[|P'(0)| : P \in V]$."

2. If $a(x)$, $b(x)$, $c(x)$, and $d(x)$ are polynomials in x, show that

$$\int_1^x a(x)c(x)\,dx \cdot \int_1^x b(x)d(x)\,dx - \int_1^x a(x)d(x)\,dx \cdot \int_1^x b(x)c(x)\,dx$$

is divisible by $(x-1)^4$.

First Solution. Since a, b, c, d are polynomials, the expression above is also a polynomial, say $F(x)$. For clarity we change the variable of integration to t, so that

$$F(x) = \int_1^x ac\,dt \cdot \int_1^x bd\,dt - \int_1^x ad\,dt \cdot \int_1^x bc\,dt.$$

It is obvious that $F(1) = 0$, whence $F(x)$ is divisible by $(x-1)$. Furthermore

$$F'(x) = ac\int_1^x bd\,dt + bd\int_1^x ac\,dt - ad\int_1^x bc\,dt - bc\int_1^x ad\,dt,$$

and $F'(1) = 0$. Also

$$F''(x) = (ac)'\int_1^x bd\,dt + (bd)'\int_1^x ac\,dt - (ad)'\int_1^x bc\,dt -$$
$$(bc)'\int_1^x ad\,dt + ac\,bd + bd\,ac - ad\,bc - bc\,ad,$$

and again $F''(1) = 0$. Finally

$$F'''(x) = (ac)''\int_1^x bd\,dt + (bd)''\int_1^x ac\,dt - (ad)''\int_1^x bc\,dt -$$
$$(bc)''\int_1^x ad\,dt + (ac)'\,bd + (bd)'\,ac - (ad)'\,bc - (bc)'\,ad.$$

The four terms not involving an integral are seen to be $[(ac)(bd)]' - [(ad)(bc)]' = 0$, and again $F'''(1) = 0$. Therefore $F(x)$ is divisible by $(x-1)^4$.

Second Solution. We prove a generalization. Suppose p, q, r, and s are functions of class C^3 such that $p(u) = q(u) = r(u) = s(u) = 0$ for some fixed u, and

$$\begin{vmatrix} p' & q' \\ r' & s' \end{vmatrix} = 0$$

identically. Then

$$f = \begin{vmatrix} p & q \\ r & s \end{vmatrix}$$

vanishes with its first three derivatives at u.

Evidently $f(u) = 0$, and we have

$$f' = \begin{vmatrix} p' & q' \\ r & s \end{vmatrix} + \begin{vmatrix} p & q \\ r' & s' \end{vmatrix}$$

$$f'' = \begin{vmatrix} p'' & q'' \\ r & s \end{vmatrix} + 2\begin{vmatrix} p' & q' \\ r' & s' \end{vmatrix} + \begin{vmatrix} p & q \\ r'' & s'' \end{vmatrix}$$

$$f''' = \begin{vmatrix} p''' & q''' \\ r & s \end{vmatrix} + 3\left\{\begin{vmatrix} p'' & q'' \\ r' & s' \end{vmatrix} + \begin{vmatrix} p' & q' \\ r'' & s'' \end{vmatrix}\right\} + \begin{vmatrix} p & q \\ r''' & s''' \end{vmatrix}.$$

The first and last determinants in each expression are zero at u. The middle determinant in f'' is identically zero by hypothesis, and the expression in braces is the derivative of the latter determinant, so it, too, is identically zero. Therefore, $f'(u) = f''(u) = f'''(u) = 0$.

The problem posed is the special case with $u = 1$,

$$p = \int_1^x ac, \; q = \int_1^x ad, \; r = \int_1^x bc, \; s = \int_1^x bd.$$

Third Solution. A solution using linear algebra can be given. The mapping

$$\phi\colon (a, b, c, d) \mapsto F$$

is a multilinear map $P^4 \to P$ where P is the space of polynomials. Since the set of polynomials divisible by $(x - 1)^4$ is a linear subspace of P, it is sufficient to verify the result as the given polynomials a, b, c, and d vary over a basis of P.

Take the basis $1, (x - 1), (x - 1)^2, \ldots$. Then if $a(x) = (x - 1)^p$, $b(x) = (x - 1)^q$, $c(x) = (x - 1)^r$, and $d(x) = (x - 1)^s$, we have $\phi(a, b, c, d) = F$, where

$$F(x) = (x - 1)^{p+q+r+s+2} \times \tag{1}$$

$$\left[\frac{1}{p + r + 1} \; \frac{1}{q + s + 1} - \frac{1}{p + s + 1} \cdot \frac{1}{q + r + 1}\right].$$

If $p + q + r + s \geq 2$, then clearly $F(x)$ is divisible by $(x - 1)^4$. If $p + q + r + s < 2$, either all the exponents are zero or all but one are. In each of these cases, the bracketed expression in (1) vanishes, so $F(x) = 0$ and is divisible by $(x - 1)^4$.

---

3. A projectile in flight is observed simultaneously from four radar stations which are situated at the corners of a square of side b. The distances of the projectile from the four stations, taken in order around the square, are found to be R_1, R_2, R_3, R_4. Show that

$$R_1^2 + R_3^2 = R_2^2 + R_4^2.$$

Show also that the height h of the projectile above the ground is given by

$$h^2 = -\frac{1}{2}b^2 + \frac{1}{4}(R_1^2 + R_2^2 + R_3^2 + R_4^2)$$

$$- \frac{1}{8b^2}(R_1^4 + R_2^4 + R_3^4 + R_4^4 - 2R_1^2R_3^2 - 2R_2^2R_4^2).$$

Solution. Choose the diagonals of the square as axes in the plane. Then the four vertices are $(\pm a, 0)$, $(0, \pm a)$, where $a^2 = b^2/2$, and we number them counterclockwise starting at $(a, 0)$. If the projectile is above the point (x, y), then

$$R_1^2 = (x - a)^2 + y^2 + h^2, \qquad R_2^2 = x^2 + (y - a)^2 + h^2,$$

$$R_3^2 = (x + a)^2 + y^2 + h^2, \qquad R_4^2 = x^2 + (y + a)^2 + h^2.$$

Hence, $R_1^2 + R_3^2 = R_2^2 + R_4^2$, as required, and

$$R_1^2 + R_2^2 + R_3^2 + R_4^2 - 4a^2 - 4h^2 = 4(x^2 + y^2).$$

Therefore,

$$R_1^4 + R_2^4 + R_3^4 + R_4^4 - 2R_1^2R_3^2 - 2R_2^2R_4^2$$

$$= (R_3^2 - R_1^2)^2 + (R_4^2 - R_2^2)^2 = 16a^2(x^2 + y^2)$$

$$= 4a^2(R_1^2 + R_2^2 + R_3^2 + R_4^2 - 4a^2 - 4h^2).$$

Hence,

$$16a^2h^2 = -16a^4 + 4a^2(R_1^2 + R_2^2 + R_3^2 + R_4^2)$$
$$- (R_1^4 + R_2^4 + R_3^4 + R_4^4 - 2R_1^2R_3^2 - 2R_2^2R_4^2).$$

Dividing through by $16a^2 = 8b^2$, we obtain the desired relation.

REMARK 1. To keep the algebra simple, it is essential to take advantage of the symmetry of the square. Another, almost equally good, choice of coordinates would make the vertices $(\pm c, \pm c)$, where $c = b/2$.

REMARK 2. A point above a plane is determined by its distances from any three non-collinear points in the plane. Hence h, x, and y are determined by any three of the R's. This explains why there must be a non-trivial relation connecting the four R's.

---

4. Let $g(x)$ be a function that has a continuous first derivative $g'(x)$ for all values of x. Suppose that the following conditions hold for every x: (i) $g(0) = 0$; (ii) $|g'(x)| \le |g(x)|$. Prove that $g(x)$ vanishes identically.

First Solution. We first convert the differential inequality into an integral inequality. Suppose $x \ge 0$. Using (i), (ii), and the continuity of g', we have

$$|g(x)| = \left| \int_0^x g'(t)dt \right| < \int_0^x |g'(t)|dt \le \int_0^x |g(t)|dt.$$

Thus

$$(1) \qquad |g(x)| \le \int_0^x |g(t)|dt, \quad \text{for } x \ge 0.$$

Let $a \ge 0$ be chosen arbitrarily. Since g is differentiable, it is continuous and therefore bounded on any finite interval. So there is a number K such that

$$|g(x)| \le K, \quad \text{for} \quad 0 \le x \le a.$$

If $0 \le t \le x \le a$, we have $|g(t)| \le K$, so (1) gives

$$|g(x)| \le \int_0^x Kdt = Kx, \quad \text{for} \quad 0 \le x \le a.$$

Now if $0 \le t \le x \le a$, we have $|g(t)| \le Kt$, and (1) gives

$$|g(x)| \le \int_0^x Kt\,dt = \frac{1}{2}Kx^2, \quad \text{for} \quad 0 \le x \le a.$$

Continuing in this way we find

$$(2) \qquad |g(x)| \le \frac{1}{n!}Kx^n, \quad \text{for} \quad 0 \le x \le a$$

and all positive integers n.

To prove (2) formally, we use mathematical induction. We have shown that (2) is true for $n = 1$. Suppose it is true for $n = p$. Then for $0 \le t \le x \le a$, we have $|g(t)| \le Kt^p/p!$. Using this in (1), we find

$$|g(x)| \le \int_0^x \frac{1}{p!}Kt^p dt = \frac{1}{(p+1)!}Kx^{p+1}, \quad \text{for} \quad 0 \le x \le a.$$

Thus (2) is true for $n = p + 1$. We conclude it is true for all n.

Setting $x = a$ in (2) and letting $n \to \infty$, we have

$$|g(a)| \le \lim_{n\to\infty} \frac{1}{n!}Ka^n = 0.$$

Therefore $g(a) = 0$. But a was arbitrary, so g vanishes on all of $[0, \infty)$.

Consider the function h defined by $h(x) = g(-x)$. Since h has a continuous derivative and satisfies (i) and (ii), h vanishes on $[0, \infty)$, by what we have already proved. Therefore g vanishes on $(-\infty, 0]$ as well.

By a slight variation of this reasoning, one can prove directly inequalities of the form

$$|g(x)| \le \frac{1}{n!}L|x|^n \quad \text{for} \quad -a \le x \le a,$$

and obtain the proof for positive and negative arguments simultaneously.

Second Solution. Suppose that $g(x)$ does not vanish identically. Then from the continuity of g it follows that there exist points a and b such that $g(a) = 0$, $g(b) \ne 0$, $|a - b| < 1$, and $|g(b)|$ is the maximum value of $|g(x)|$ on the closed interval with endpoints a and b. (For example, if g does not vanish on $[0, \infty)$, we can let a be the least non-negative half-integer such that g does not vanish on $[a, a + \frac{1}{2}]$, and let b be the point at which $|g(x)|$ attains its maximum for $a \le x \le a + \frac{1}{2}$.)

From the mean value theorem it follows that there is a number c between

a and b such that

$$|g'(c)| = \left|\frac{g(b) - g(a)}{b - a}\right| = \left|\frac{g(b)}{b - a}\right| > |g(b)|.$$

But our choice of b shows that $|g(b)| \geq |g(c)|$, hence $|g'(c)| > |g(c)|$, contrary to (ii). This contradiction proves that g vanishes identically.

This proof does not require the continuity of g'.

REMARK. This is a standard result that arises in proving the uniqueness of the solutions of differential equations.

---

5. Find the smallest volume bounded by the coordinate planes and by a tangent plane to the ellipsoid

$$\frac{x^2}{a^2} + \frac{y^2}{b^2} + \frac{z^2}{c^2} = 1.$$

Solution. The tangent plane to the given ellipsoid at the point (x_1, y_1, z_1) has the equation

$$\frac{xx_1}{a^2} + \frac{yy_1}{b^2} + \frac{zz_1}{c^2} = 1.$$

Its intercepts on the x, y, and z-axes, respectively, are

$$\frac{a^2}{x_1}, \frac{b^2}{y_1}, \frac{c^2}{z_1}.$$

The volume of the solid cut off by the tangent plane and the three coordinate planes is

$$V = \frac{1}{6}\left|\frac{a^2b^2c^2}{x_1y_1z_1}\right|. \tag{1}$$

(If $x_1y_1z_1 = 0$, then the four planes do not bound a finite region.) Hence

$$V^2 = \frac{1}{36}a^2b^2c^2\left(\frac{x_1^2}{a^2}\frac{y_1^2}{b^2}\frac{z_1^2}{c^2}\right)^{-1}. \tag{2}$$

But

$$\left(\frac{x_1^2}{a^2} \cdot \frac{y_1^2}{b^2} \cdot \frac{z_1^2}{c^2}\right)^{1/3} \leq \frac{1}{3}\left(\frac{x_1^2}{a^2} + \frac{y_1^2}{b^2} + \frac{z_1^2}{c^2}\right) = \frac{1}{3}$$

with equality if and only if

$$\frac{x_1^2}{a^2} = \frac{y_1^2}{b^2} = \frac{z_1^2}{c^2} = \frac{1}{3} \tag{3}$$

(the arithmetic-geometric mean inequality). Hence

$$V^2 \geq \frac{27}{36} a^2 b^2 c^2 \quad \text{and} \quad V \geq \frac{1}{2}\sqrt{3}\, abc$$

with equality if and only if (x_1, y_1, z_1) is one of the eight points for which (3) holds, namely

$$(\pm a/\sqrt{3}, \qquad \pm b/\sqrt{3}, \qquad \pm c/\sqrt{3}).$$

It is also possible, of course, to minimize V straightforwardly by maximizing the product $x_1 y_1 z_1$ under the constraint that (x_1, y_1, z_1) is a point of the ellipsoid.

---

6. A particle of unit mass moves on a straight line under the action of a force which is a function $f(v)$ of the velocity v of the particle, but the form of this function is not known. A motion is observed, and the distance x covered in time t is found to be connected with t by the formula $x = at + bt^2 + ct^3$, where a, b, c have numerical values determined by observation of the motion. Find the function $f(v)$ for the range of v covered by the experiment.

Solution. Newton's law of motion for a particle of unit mass takes the form

$$F = \text{force} = \frac{dv}{dt}.$$

Since we are given that

$$x = at + bt^2 + ct^3$$

it follows that

$$v = \frac{dx}{dt} = a + 2bt + 3ct^2,$$

$$\frac{dv}{dt} = 2b + 6ct.$$

We now express the force in terms of v:

$$\begin{aligned} F^2 &= 4b^2 + 24bct + 36c^2t^2 \\ &= 4b^2 + 12c(2bt + 3ct^2) \\ &= 4b^2 + 12c(v - a). \end{aligned}$$

Hence

$$F = f(v) = \pm\sqrt{4b^2 - 12ac + 12cv}.$$

The sign of the radical is taken to be the sign of $2b + 6ct$ which, if the hypotheses of the problem are satisfied, cannot change for the interval of time under consideration, since then v would take the same value twice but dv/dt would not.

Afternoon Session

1. Let K denote the circumference of a circular disc of radius one, and let k denote a circular arc that joins two points a, b on K and lies otherwise in the given circular disc. Suppose that k divides the circular disc into two parts of equal area. Prove that the length of k exceeds 2.

Solution. If a and b were diametrically opposite on K, there would exist no circular arc from a to b that bisects K. Hence we may choose coordinates such that K is the unit circle $x^2 + y^2 = 1$ and a and b have coordinates (c, d) and $(c, -d)$, respectively, where $c < 0$.

Now the arc k divides the circular disc into two parts of equal area, and hence it must intersect the positive x axis at a point e. If O is the origin, we get length $k > 2ae > 2aO = 2$.

REMARK. The requirement that k be a circular arc is not really important: Any curve k that bisects the disc has length at least 2 with equality

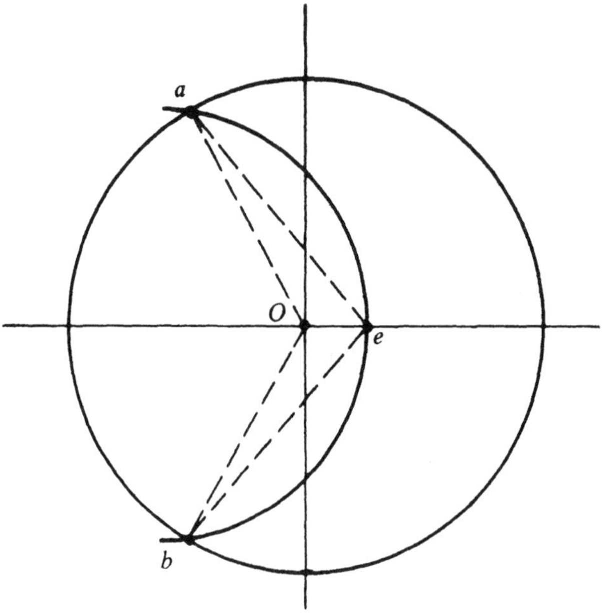

if and only if it is a diameter. If the endpoints a and b are diametrically opposite, the conclusion is immediate; otherwise, we can start as above. Then k must contain at least one point p of the open right half-plane, so its length is at least

$$ap + pb > aO + Ob = 2.$$

---

2. Let A, B be variable points on a parabola P, such that the tangents at A and B are perpendicular to each other. Show that the locus of the centroid of the triangle formed by A, B and the vertex of P is a parabola P_1. Apply the same process to P_1, obtaining a parabola P_2, and repeat the process, obtaining altogether the sequence of parabolas $P, P_1, P_2, \ldots, P_n$. If the equation of P is $y^2 = mx$, find the equation of P_n.

Solution. Since P is a parabola with equation $y^2 = mx$, any point of P has coordinates of the form (mt^2, mt) for some real t, and conversely, every such point is on P. The slope of the line tangent to P at (mt^2, mt) is $1/2t$.

Let A and B be the points (ms^2, ms) and (mt^2, mt), respectively. The tangents to P at A and B are perpendicular if and only if $(1/2s)(1/2t) = -1$, i.e., if and only if

$$st = -\frac{1}{4}. \tag{1}$$

The centroid of the points A, B, and the vertex (0, 0) of P is

$$\left(\frac{1}{3} m(s^2 + t^2),\ \ \frac{1}{3} m(s + t)\right) \tag{2}$$

and this centroid lies on a new parabola

$$y^2 = \frac{1}{3} m\left(x - \frac{m}{6}\right) \tag{3}$$

if the perpendicularity condition (1) is satisfied.

Conversely, any point (x, y) on the parabola (3) has the form (2), since the equations

$$\frac{1}{3} m(s + t) = y$$

$$st = -\frac{1}{4}$$

can always be solved to give real s and t. Indeed, s and t are zeros of $S^2 - (3y/m)S - \frac{1}{4}$, which has positive discriminant. Hence the locus in question is the entire parabola P_1 given by (3).

Now P_1 is obtained from P by changing the constant from m to $m/3$, and displacing the vertex to the right by $m/6$. Consequently, P_2 can be obtained from P_1 by changing the constant from $m/3$ to $(m/3)/3$ and displacing the vertex $(m/3)/6$ further to the right. The equation of P_2 is therefore

$$y^2 = \frac{1}{9} m\left(x - \frac{1}{6} m - \frac{1}{18} m\right).$$

Continuing this reasoning, we see that P_n has the equation

$$\begin{aligned} y^2 &= \frac{1}{3^n} m\left(x - \frac{1}{6} m - \frac{1}{6\cdot 3} m - \frac{1}{6\cdot 3^2} m - \cdots - \frac{1}{6\cdot 3^{n-1}} m\right) \\ &= \frac{1}{3^n} m\left(x - \frac{m}{4}\left(1 - \frac{1}{3^n}\right)\right). \end{aligned}$$

---

3. In a solid sphere of radius R the density ρ is a function of r, the distance from the center of the sphere. If the magnitude of the gravitational force of

attraction due to the sphere at any point inside the sphere is kr^2, where k is a constant, find ρ as a function of r. Find also the magnitude of the force of attraction at a point outside the sphere at a distance r from the center. (Assume that the magnitude of the force of attraction at a point P due to a thin spherical shell is zero if P is inside the shell, and is m/r^2 if P is outside the shell, m being the mass of the shell, and r the distance of P from the center.)

Solution. Let P be a point at distance r from the center. We regard the solid sphere as the union of many thin concentric spherical shells. The force of attraction at P is the sum of the forces of attraction due to the shells. All of these forces act in the same direction so we can simply add their magnitudes.

The shell at distance s from the center and thickness Δs has (approximately) volume $4\pi s^2 \Delta s$ and mass $4\pi\rho(s)s^2\Delta s$, so the force of attraction at P due to this shell is approximately

$$\frac{4\pi\rho(s)s^2\Delta s}{r^2},$$

if $r > s$. It is zero if $r < s$. The total force of attraction at P is therefore (exactly)

$$(1) \qquad F = \begin{cases} \displaystyle\int_0^r \frac{4\pi\rho(s)s^2\,ds}{r^2} & \text{if } r \le R, \\[2ex] \displaystyle\int_0^R \frac{4\pi\rho(s)s^2\,ds}{r^2} & \text{if } r > R. \end{cases}$$

Since we are given that $F = kr^2$ for $r \le R$, we have

$$kr^2 = \int_0^r \frac{4\pi\rho(s)s^2\,ds}{r^2} \qquad \text{for } r \le R,$$

and we must solve for ρ. Multiplying by r^2, and then differentiating with respect to r, we obtain

$$4kr^3 = 4\pi\rho(r)r^2.$$

Therefore

$$\rho(r) = \frac{k}{\pi} r$$

is the required formula for ρ.

Substituting this in (1), we have

$$F = \frac{4\pi}{r^2} \int_0^R \frac{ks}{\pi} s^2 ds = \frac{kR^4}{r^2} \quad \text{for } r > R.$$

REMARK. We could obtain the last formula directly because we were given that F is kR^2 at the surface and we know it falls off inversely with r^2 outside the sphere.

---

4. For each positive integer n, put

$$p_n = (1 + 1/n)^n, \; P_n = (1 + 1/n)^{n+1}, \; h_n = \frac{2p_n P_n}{p_n + P_n}.$$

Prove that $h_1 < h_2 < \cdots < h_n < \cdots$.

First Solution. We find $h_n = 2(n+1)^{n+1} n^{-n} (2n+1)^{-1}$. Consider the function g defined by

$$g(x) = \log 2 + (x+1) \log (x+1) - x \log x - \log (2x+1).$$

We have

$$g'(x) = \log (x+1) - \log x - \frac{2}{2x+1}$$

$$g''(x) = \frac{1}{x+1} - \frac{1}{x} + \frac{4}{(2x+1)^2} = -\frac{1}{x(x+1)(2x+1)^2} < 0$$

for $0 < x < \infty$. Hence g' decreases on $(0, \infty)$. Since

$$\lim_{x\to\infty} g'(x) = \lim_{x\to\infty} \log\left(\frac{x+1}{x}\right) - \lim_{x\to\infty} \frac{2}{2x+1} = 0,$$

it follows that g' is positive on $(0, \infty)$. Therefore g increases on $(0, \infty)$, so $h_n = \exp g(n)$ is strictly increasing for positive integers n.

Second Solution. We find

$$\frac{h_n}{h_{n-1}} = \left(1 - \frac{1}{n^2}\right)^n \cdot \frac{1 + \frac{1}{n}}{1 - \frac{1}{n}} \cdot \frac{1 - \frac{1}{2n}}{1 + \frac{1}{2n}}.$$

Since

$$\log(1-x) = -\sum_{k=1}^{\infty} \frac{x^k}{k}$$

and

$$\log \frac{1+x}{1-x} = 2\sum_{k=1}^{\infty} \frac{x^{2k-1}}{2k-1}$$

for $|x| < 1$, it follows that for $n > 1$,

$$\log \frac{h_n}{h_{n-1}} = -n\sum_{k=1}^{\infty} \frac{1}{kn^{2k}} + 2\sum_{k=1}^{\infty} \frac{1}{(2k-1)n^{2k-1}}$$

$$-2\sum_{k=1}^{\infty} \frac{1}{(2k-1)(2n)^{2k-1}}$$

$$= \sum_{k=1}^{\infty} \frac{1}{n^{2k-1}} \left(-\frac{1}{k} + \frac{2}{2k-1} - \frac{1}{(2k-1)2^{2k-2}} \right)$$

$$= \sum_{k=1}^{\infty} \frac{1}{n^{2k-1}} \frac{2^{2k-2}-k}{k(2k-1)2^{2k-2}} > 0.$$

Hence $h_n > h_{n-1}$ for $n > 1$, so $h_1 < h_2 < h_3 < \cdots$.

REMARK. Since h_n is the harmonic mean between p_n and P_n, and $\lim p_n = \lim P_n = e$, it follows that $\lim h_n = e$.

---

5. Show that the integer next above $(\sqrt{3}+1)^{2n}$ is divisible by 2^{n+1}.

Solution. The key to this and many similar problems is that the next integer above $(1+\sqrt{3})^{2n}$ is $(1+\sqrt{3})^{2n} + (1-\sqrt{3})^{2n}$. To establish this, note that for every positive integer n, there are integers A_n and B_n such that

$$(1+\sqrt{3})^{2n} = A_n + B_n\sqrt{3}$$

and

$$(1-\sqrt{3})^{2n} = A_n - B_n\sqrt{3}.$$

Thus $(1+\sqrt{3})^{2n}+(1-\sqrt{3})^{2n}=2A_n$ is certainly an integer. Since $|1-\sqrt{3}|<1$, we have $0<(1-\sqrt{3})^{2n}<1$. Hence $2A_n$ is indeed the next integer above $(1+\sqrt{3})^{2n}$.

The problem, therefore, is to show that $2A_n$ is divisible by 2^{n+1}, or that A_n is divisible by 2^n.

We claim that for all n, both A_n and B_n are divisible by 2^n. We prove this by induction. Since $A_1=4$ and $B_1=2$, it is true for $n=1$. Suppose it is true for $n=k$. Then we have

$$A_{k+1}+B_{k+1}\sqrt{3}=(1+\sqrt{3})^2(A_k+B_k\sqrt{3})$$

$$=(4A_k+6B_k)+(2A_k+4B_k)\sqrt{3},$$

so

$$A_{k+1}=2(2A_k+3B_k),$$

$$B_{k+1}=2(A_k+2B_k),$$

and it is clear from our inductive hypothesis that both of these are divisible by 2^{k+1}. This establishes our claim.

---

6. A particle moves on a circle with center O, starting from rest at a point P and coming to rest again at a point Q, without coming to rest at any intermediate point. Prove that the acceleration vector of the particle does not vanish at any point between P and Q and that, at some point R between P and Q, the acceleration vector points in along the radius RO.

Solution. Suppose the circle has radius r. If Cartesian coordinates are chosen with origin at the center of the circle, then the coordinates of the particle are $(r\cos\theta, r\sin\theta)$, where θ is a function of the time t. Differentiating this twice, we see that the acceleration vector is

$$r\frac{d\omega}{dt}(-\sin\theta,\cos\theta)+r\omega^2(-\cos\theta,-\sin\theta), \tag{1}$$

where $\omega=d\theta/dt$ is the angular velocity. Since $(-\sin\theta,\cos\theta)$ and $(-\cos\theta,-\sin\theta)$ are orthogonal unit vectors in the directions of the tangent and the inward normal, we see that the two terms of (1) are, respectively, the tangential and normal components of the acceleration. Since $\omega\neq0$ at any time during the motion, the normal component of acceleration, and hence the acceleration vector itself, is never zero. Since $\omega=0$ at the start and at the finish, by Rolle's theorem there is an intermediate time at which $d\omega/dt = 0$. At that time the acceleration vector points inward along the radius because $r\omega^2 > 0$.

REMARK. We have interpreted "coming to rest" to mean "having velocity zero." If "coming to rest" means "remaining stationary through some time interval," then the first statement is false, for it is certainly possible that ω and $d\omega/dt$ vanish simultaneously but not on an interval. The second statement is true, however, because the usual proof of Rolle's theorem shows that $d\omega/dt = 0$ at some point where $\omega \neq 0$, unless $\omega = 0$ identically, which is ruled out.

THE SEVENTH WILLIAM LOWELL PUTNAM MATHEMATICAL COMPETITION

May 24, 1947

Morning Session

1. If $\{a_n\}$ is a sequence of numbers such that for $n \geq 1$

$$(2 - a_n)a_{n+1} = 1,$$

prove that $\lim a_n$, as $n \to \infty$, exists and is equal to one.

First Solution. We begin by describing a graphical method of great utility in analyzing recursions of the form $a_{n+1} = f(a_n)$.

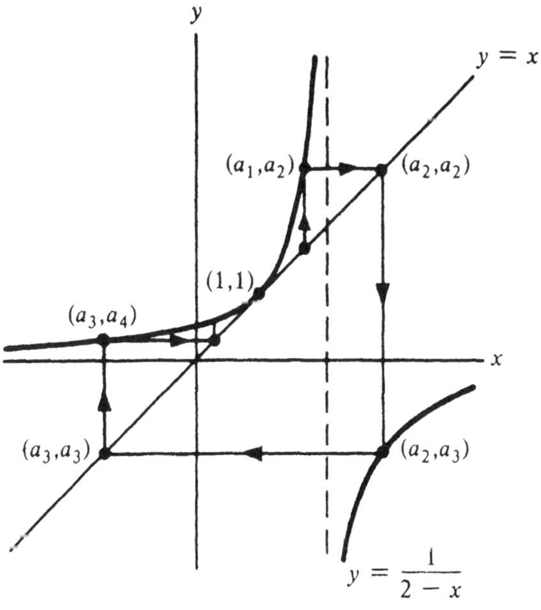

Draw the graph of f and the line $y = x$ on the same axes. (In this case $f(x) = 1/(2 - x)$.) Then start from the point $\langle a_1, a_1 \rangle$ on the line and move up or down to the point $\langle a_1, a_2 \rangle$ on the graph. Then move horizontally to the point $\langle a_2, a_2 \rangle$ on the line, then vertically to $\langle a_2, a_3 \rangle$, etc.

Connect the successive points to form a polygonal line. If the sequence $\{a_n\}$ is convergent with limit L, then the polygonal line must converge to $\langle L, L\rangle$, and if L is a point of continuity for f, then $f(L) = L$. Often it is possible to see at a glance how the sequence behaves. In this case, for example, it is clear that with any start (as long as we do not encounter the point 2 where f is undefined) the polygon must eventually reach the region below and to the left of $\langle 1, 1\rangle$, after which it must work its way up toward the point $\langle 1, 1\rangle$.

To make this precise, we prove first

$$x \le \frac{1}{2-x} \tag{1}$$

for $x < 2$, which follows immediately from

$$\frac{1}{2-x} - x = \frac{(1-x)^2}{2-x}.$$

Also, if $x \le 1$, then $1/(2-x) \le 1$.

Suppose $1 < a_1 < \frac{3}{2}$. We claim that for some n, $a_n \ge \frac{3}{2}$. For if not, then

$$1 \le a_1 \le a_2 \le \cdots \le \frac{3}{2}$$

by (1), so $\{a_n\}$ is convergent, say to L; then $1 < L \le \frac{3}{2}$ and $f(L) = L$; but there is no such L. If $a_n \ge \frac{3}{2}$, then $a_{n+2} \le 1$. (*Note*. We need not worry about the possibility that $a_k = 2$ for some k since we are given that the sequence does satisfy the recursion.) So in any event there is an index p such that $a_p \le 1$. Then we have

$$a_p \le a_{p+1} \le a_{p+2} \le \cdots \le 1.$$

Hence the sequence converges to a number M and $f(M) = M$. This implies $M = 1$.

Second Solution. Because of the special nature of f we can find a closed form for a_n. If $a_n = 1$ for one index n, then $a_n = 1$ for all n. Otherwise, let $b_n = 1/(1-a_n)$. Then

$$\begin{aligned} b_{n+1} &= \frac{1}{1-a_{n+1}} \\ &= \frac{1}{1-\dfrac{1}{2-a_n}} = \frac{2-a_n}{1-a_n} = b_n + 1. \end{aligned}$$

Hence $b_n = b_1 + n - 1$ and

$$a_n = 1 - \frac{1}{b_n} = \frac{b_1 + n - 2}{b_1 + n - 1}.$$

Therefore

$$\lim_{n \to \infty} a_n = 1,$$

as required.

REMARK. One might guess at the nature of the closed form for a_n after examining a possible sequence like

$$\frac{1}{4}, \frac{4}{7}, \frac{7}{10}, \frac{10}{13}, \ldots,$$

but there is a simple rationale for the trick we used. The function f is a fractional linear transformation with just one fixed point, at 1. We change the variable so that this fixed point is transformed to ∞; i.e., we set $b = 1/(a - 1)$ so $a = 1$ corresponds to $b = \infty$. In terms of the new variable, f is a translation; indeed, every fractional linear transformation which leaves ∞ fixed but has no finite fixed point is a translation.

---

2. A real valued continuous function satisfies for all real x and y the functional equation

$$f(\sqrt{x^2 + y^2}) = f(x)f(y).$$

Prove that

$$f(x) = [f(1)]^{x^2}.$$

Solution. A slight qualification in the statement of the problem is needed since the real valued continuous function $f(x) \equiv 0$ satisfies the functional equation for all real x and y, but does not satisfy the relation $f(0) = [f(1)]^0$, since 0^0 is undefined.

Assume then that for some y_0, $f(y_0) \neq 0$. Since

$$f(x)f(y_0) = f(\sqrt{x^2 + y_0^2}) = f(-x)f(y_0),$$

we have $f(x) = f(-x) = f(|x|)$ for all x. We now show by induction that for any positive integer n and any real number x, we have

$$f(\sqrt{n}\,x) = [f(x)]^n. \tag{1}$$

This is certainly true for $n = 1$, and assuming it true for $n = k$ we have

$$f(\sqrt{k+1}\,x) = f(\sqrt{k+1}\,|x|) = f(\sqrt{(\sqrt{k}\,x)^2 + x^2}) = f(\sqrt{k}\,x)f(x)$$

$$= [f(x)]^k f(x) = [f(x)]^{k+1}.$$

Therefore (1) is true for all positive integers n.

If p and q are non-zero integers, then

$$f(p) = f(|p|) = f(\sqrt{p^2} \cdot 1) = [f(1)]^{p^2}$$

and

$$f(|p|) = f\left(\sqrt{q^2}\left|\frac{p}{q}\right|\right) = \left[f\left(\left|\frac{p}{q}\right|\right)\right]^{q^2}.$$

From these two relations it follows that

$$\left[f\left(\frac{p}{q}\right)\right]^{q^2} = [f(1)]^{p^2}. \tag{2}$$

If $f(1) > 0$ then it follows that

$$\left[f\left(\frac{p}{q}\right)\right] = [f(1)]^{p^2/q^2};$$

that is, the required equation is valid for all rational values of x except, perhaps, $x = 0$. By continuity it follows for all values of x.

If $f(1) = 0$, then (2) implies that $f(p/q) = 0$ for all non-zero integers p and q, and thus $f(x) = 0$ for all rational x, hence for all real x.

Finally we show that $f(1) < 0$ is impossible. If p is even and q is odd, equation (2) implies that $f(p/q) > 0$. Hence $f(x) > 0$ for a dense set of x, and therefore $f(x) \geq 0$ for all x; in particular, $f(1) \geq 0$.

REMARK. If we consider the function g defined by

$$g(x) = \log f(\sqrt{x}) \qquad \text{for } x \geq 0, \tag{3}$$

then g satisfies the famous Cauchy functional equation

$$g(x + y) = g(x) + g(y)$$

whose only continuous solution is

$$g(x) = g(1)x$$

and it readily follows that $f(x) = f(1)^{x^2}$. This argument requires showing first that (3) defines a function; i.e., that $f(\sqrt{x}) > 0$ for $x \geq 0$, which is true unless f vanishes identically.

---

3. Given this figure and any two points Q_1, Q_2 in the plane not lying on any of the segments $s_1, s_2, \ldots, s_6$, show that there does not exist a polygonal line P joining Q_1 and Q_2 such that:

(i) P crosses each s_i, $i = 1, 2, \ldots, 6$, exactly once;

(ii) P does not intersect itself;

(iii) P does not pass through any vertex V_1, V_2, V_3, V.

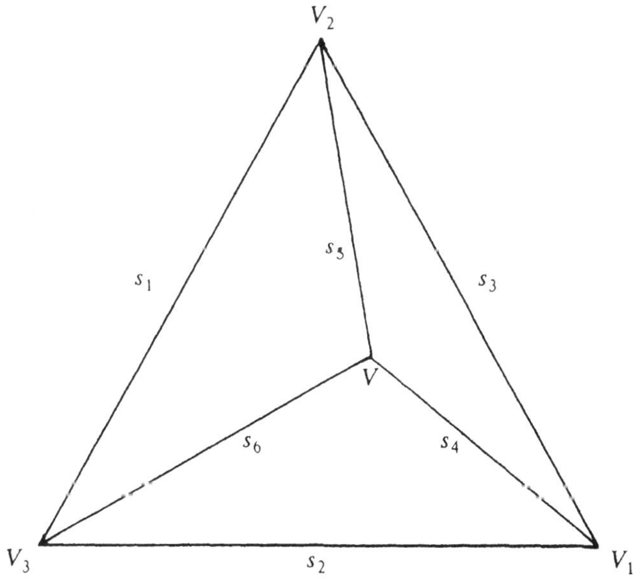

Solution. We begin with a lemma.

LEMMA. *If a polygonal line in the plane of a triangle passes through no vertex of the triangle, crosses each side exactly once, and has neither end-*

point on the boundary of the triangle, then one of its endpoints is interior to the triangle.

Suppose there exists a polygonal line P as described in the problem. P crosses each side of each of the triangles VV_1V_2, VV_2V_3, and VV_3V_1 exactly once, so it must have one end in the interior of each of these triangles. But the interiors of these triangles are disjoint and P has only two ends, so this is impossible.

Proof of the Lemma. Let T be the triangle, and suppose the polygonal line P with endpoints Q_1 and Q_2 satisfies the hypothesis.

If Q_1 is in the interior of T, there is nothing to prove, so we assume Q_1 is in the exterior of T. As we move along P from Q_1 toward Q_2 we encounter only exterior points until we cross the first side of T; then we encounter only interior points until we cross the second side of T. After that we encounter exterior points until we cross the third side of T and from there on all points, including the endpoint Q_2, are in the interior. Thus Q_2 is an interior point. ■

Discussion. The solution is easy enough and this is, no doubt, what the examining committee had in mind. However, the entire argument can be criticized because, for example, nowhere is it made clear what we mean by "P crosses a segment exactly once." The definition of this phrase must be carefully worded so as not to count the spurious crossing shown in Figure 1, but to allow for the crossing of Figure 2.

The restriction to non-self-intersecting polygonal lines is not actually necessary but was probably intended to obviate another technicality suggested by Figure 3.

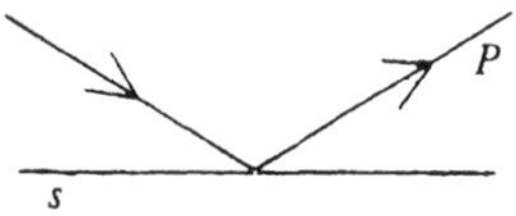

Fig. 1.

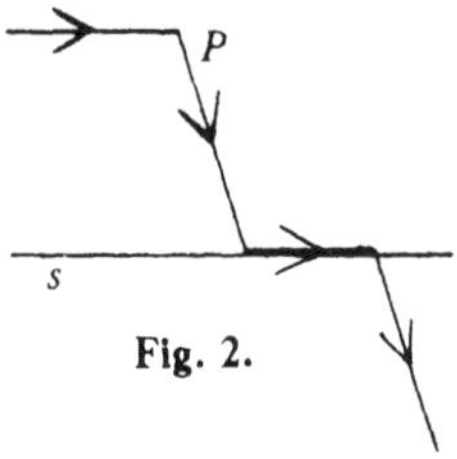

Fig. 2.

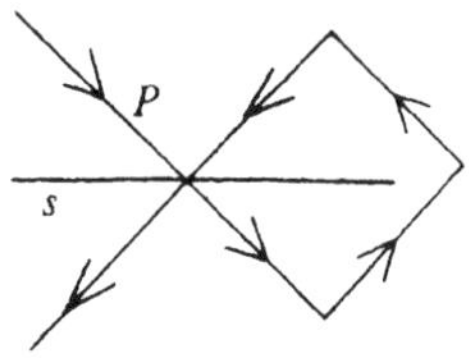

Fig. 3.

Such technicalities can all be avoided by restricting consideration to polygonal lines that are in general position with respect to the figure, that is, no vertex of P should lie on any of the segments s_i. With this restriction, the number of times P crosses a segment s is just the number of segments of P that meet s.

The problem is a simplification of the famous bridges-of-Königsberg problem that is discussed in many expository books on mathematics. See, for example, S. K. Stein, *Mathematics, the Man-Made Universe,* Freeman, San Francisco, 1963.

---

4. A coast artillery gun can fire at any angle of elevation between 0° and 90° in a fixed vertical plane. If air resistance is neglected and the muzzle velocity is constant ($= v_0$), determine the set H of points in the plane and above the horizontal which can be hit.

Solution. We take coordinates with origin at the gun, the y-axis vertical, and the x-axis horizontal in the direction of the fire. For a given angle α and the prescribed initial conditions the equations of motion

$$\frac{d^2x}{dt^2} = 0, \qquad \frac{d^2y}{dt^2} = -g$$

lead to

$$x = v_0 t \cos\alpha$$

$$y = v_0 t \sin\alpha - \frac{1}{2} gt^2.$$

Elimination of t gives

$$y = x \tan\alpha - \frac{g}{2v_0^2} x^2 \sec^2\alpha.$$

For a fixed positive x this can be written

$$y = \frac{v_0^2}{2g} - \frac{gx^2}{2v_0^2} - \frac{gx^2}{2v_0^2}\left(\tan\alpha - \frac{v_0^2}{gx}\right)^2,$$

whence it is clear that we can choose α so as to hit the point (x, y) if and only if

$$y \le \frac{v_0^2}{2g} - \frac{gx^2}{2v_0^2}. \tag{1}$$

To hit a point $(0, y)$, we fire straight up, i.e., take $\alpha = 90°$; then the parametric equation for y can be written

$$y = \frac{v_0^2}{2g} - \frac{v_0^2}{2g}\left(\frac{g}{v_0} t - 1\right)^2$$

and it is clear that we can reach $(0, y)$ if and only if $y \leq v_0^2/2g$. Therefore, the desired set H is defined by the inequalities (1) and $0 \leq x$, $0 < y$.

Historical Note. The parabola $y_{\max} = (v_0^2/2g) - (gx^2/2v_0^2)$ is sometimes called the "parabola of safety." An airplane staying outside of this parabola cannot be hit by the artillery gun.

See "Envelopes," by V. G. Boltyanskii (translated from the Russian by Robert B. Brown), vol. 12 in *Popular Lectures in Mathematics*, Pergamon Press, New York, 1964, where this problem is discussed.

REMARK. This problem is essentially the same as problem 6(ii) of the second competition. A different solution is given there.

---

5. a_1, b_1, c_1 are positive numbers whose sum is 1, and for $n = 1, 2, \ldots$ we define

$$a_{n+1} = a_n^2 + 2b_nc_n, \quad b_{n+1} = b_n^2 + 2c_na_n, \quad c_{n+1} = c_n^2 + 2a_nb_n.$$

Show that a_n, b_n, c_n approach limits as $n \to \infty$ and find these limits.

First Solution. First note that

$$a_{n+1} + b_{n+1} + c_{n+1} = (a_n + b_n + c_n)^2$$

so $a_k + b_k + c_k = 1$ for all k by induction. Also it is clear that the a_n's, b_n's, and c_n's are all positive.

Define $E_n = \max(a_n, b_n, c_n)$ and $F_n = \min(a_n, b_n, c_n)$. We will show

(1) $$F_1 \leq F_2 \leq F_3 \leq \cdots \leq F_n \leq F_{n+1} \leq \cdots$$
$$\leq E_{n+1} \leq E_n \leq \cdots \leq E_3 \leq E_2 \leq E_1$$

and also that

(2) $$\lim_{n\to\infty} (E_n - F_n) = 0.$$

It follows from (1) and (2) that E_n decreases weakly to some limit L and that F_n increases weakly to the same limit L. Since $F_n \leq a_n \leq E_n$, this implies $a_n \to L$. Similarly $b_n \to L$ and $c_n \to L$. Then since $a_n + b_n + c_n = 1$, we have $L = \frac{1}{3}$.

To prove (1), assume $a_n \geq b_n \geq c_n$ for some value of n, then

$$
\begin{aligned}
a_{n+1} &= a_n^2 + b_n c_n + b_n c_n \\
b_{n+1} &= a_n c_n + b_n^2 + a_n c_n \qquad (3) \\
c_{n+1} &= a_n b_n + a_n b_n + c_n^2.
\end{aligned}
$$

In all equations in (3), the right member is less than or equal to $a_n^2 + a_n b_n + a_n c_n = a_n$, and greater than or equal to $a_n c_n + b_n c_n + c_n^2 = c_n$. Hence $E_{n+1} \leq E_n$ and $F_{n+1} \geq F_n$, which proves (1).

To prove (2), we again assume $a_n \geq b_n \geq c_n$ for some n. Set

$$
\begin{aligned}
a_n - b_n &= \alpha \geq 0 \\
b_n - c_n &= \beta \geq 0 \\
a_n - c_n &= \delta = \alpha + \beta \geq 0.
\end{aligned}
$$

Then

$$
\begin{aligned}
|a_{n+1} - b_{n+1}| &= |a_n - b_n||a_n + b_n - 2c_n| = \alpha(\delta + \beta) = (\delta - \beta)(\delta + \beta) \leq \delta^2 \\
|a_{n+1} - c_{n+1}| &= |a_n - c_n||a_n + c_n - 2b_n| = \delta|\alpha - \beta| \leq \delta(\alpha + \beta) = \delta^2 \\
|c_{n+1} - b_{n+1}| &= |b_n - c_n||2a_n - b_n - c_n| = \beta(\alpha + \delta) \leq (\delta - \alpha)(\delta + \alpha) \leq \delta^2.
\end{aligned}
$$

This set of inequalities shows that

$$E_{n+1} - F_{n+1} \leq (E_n - F_n)^2$$

for all n. Therefore

$$E_{n+1} - F_{n+1} \leq (E_1 - F_1)^{2^n} \quad \text{for all } n.$$

Since we are given that $E_1 < 1$ and $F_1 > 0$, we have $E_1 - F_1 < 1$, and (2) follows. This conclues the proof.

REMARK. The proof shows that the result still holds if one of the numbers a_1, b_1, c_1 is zero and the other two are positive, since then $E_1 - F_1 < 1$.

Second Solution. Put

$$
\begin{aligned}
u_n &= a_n + b_n + c_n \\
v_n &= a_n + \omega b_n + \omega^2 c_n \\
w_n &= a_n + \omega^2 b_n + \omega c_n
\end{aligned}
$$

where ω is a complex cube root of unity. The recursion becomes

$$u_{n+1} = u_n{}^2$$
$$v_{n+1} = w_n{}^2$$
$$w_{n+1} = v_n{}^2.$$

Since $u_1 = 1$, it follows that $u_n = 1$ for all n. Moreover we see by induction that

$$v_{n+1} = v_1{}^{2^n} \text{ or } w_1{}^{2^n}$$
$$w_{n+1} = w_1{}^{2^n} \text{ or } v_1{}^{2^n}$$

depending on whether n is even or odd. Since v_1 and w_1 are convex combinations with positive coefficients of the three numbers 1, ω, ω^2, it follows that $|v_1| < 1$ and $|w_1| < 1$. (*Note.* For this conclusion it is sufficient that at least two of the numbers a_1, b_1, c_1 be positive.) Hence $v_n \to 0$ and $w_n \to 0$. Therefore

$$a_n = \frac{1}{3}(u_n + v_n + w_n) \to \frac{1}{3},$$

$$b_n = \frac{1}{3}(u_n + \omega^2 v_n + \omega w_n) \to \frac{1}{3},$$

$$c_n = \frac{1}{3}(u_n + \omega v_n + \omega^2 w_n) \to \frac{1}{3}.$$

REMARK. The general properties of the transformation

$$S: (a, b, c) \to (a^2 + 2bc,\ b^2 + 2ca,\ c^2 + 2ab)$$

are worth some attention. The points (a, b, c) for which $a + b + c = 1$ form a plane P, and those having also non-negative coordinates fill an equilateral triangular region T in P. S maps P into itself and T into itself, leaving the vertices of T fixed and carrying the rest of T into its interior. The function

$$v = a + b\omega + c\omega^2$$

maps P bijectively to the complex plane, and sends the circumcircle of T onto the unit circle. In terms of v, S takes the simple form $v \to \bar{v}^2$. From this form it is obvious that S is exactly two-to-one on all of P except at the origin (of v). Furthermore, when S is iterated, points interior to the unit circle move very rapidly toward the origin, which corresponds to the point $(\frac{1}{3}, \frac{1}{3}, \frac{1}{3})$ in P.

Third Solution. Define 3×3 matrices A_n by

$$A_n = \begin{pmatrix} a_n & b_n & c_n \\ c_n & a_n & b_n \\ b_n & c_n & a_n \end{pmatrix}.$$

The recursion now becomes

$$A_{n+1} = (A_n{}^T)^2$$

where the T stands for transpose. Hence

$$A_{n+1} = A_1{}^{2^n} \text{ or } (A_1{}^T)^{2^n}$$

according as n is even or odd.

A row-stochastic matrix is a square matrix in which all entries are non-negative and every row sums to one. It is proved in the theory of Markov processes that if M is a row-stochastic matrix with all entries positive, then $\lim_{k\to\infty} M^k$ exists and is a row-stochastic matrix with all rows identical. There is a similar result, of course, for column-stochastic matrices.

Now A_1 is doubly stochastic (i.e., both row- and column-stochastic) hence $\lim_k A_1{}^k$ exists and is a doubly stochastic matrix with all rows identical and all columns identical. Hence $\lim_{k\to\infty} A_1{}^k$ is the matrix B with all entries $\frac{1}{3}$. Of course,

$$(A_1{}^T)^k = (A_1{}^k)^T$$

has the same limit. Hence it follows that $A_n \to B$ and

$$a_n \to \frac{1}{3},\ b_n \to \frac{1}{3},\ c_n \to \frac{1}{3}.$$

REMARK. This solution is closely related to the second solution because the numbers u_n, v_n, and w_n are the eigenvalues of the matrix A_n.

---

6. A three-by-three matrix has determinant zero, and has the further property that the cofactor of any element is equal to the square of that element. (The cofactor of a_{ij} is $(-1)^{i+j}$ multiplied by the determinant obtained

by striking out the ith row and jth column.) Show that every element in the matrix is zero.

Solution. Let A be an $n \times n$ matrix ($n > 1$), and let B be the transpose of the matrix of its cofactors. Classically B is called the *adjoint* of A. Then

$$AB = BA = (\det A)I,$$

where I is the identity matrix. Furthermore, the adjoint of B is $(\det A)^{n-2}A$. (For the case $n = 3$, this can be verified quite directly.) Assuming that $n > 2$, this implies that if A is singular, then B has rank at most $n - 2$; that is, all $(n - 1) \times (n - 1)$ minors of B are zero.

In the present problem, let

$$A = \begin{pmatrix} a & b & c \\ d & e & f \\ g & h & i \end{pmatrix}.$$

Then the conditions imply that

$$B = \begin{pmatrix} a^2 & d^2 & g^2 \\ b^2 & e^2 & h^2 \\ c^2 & f^2 & i^2 \end{pmatrix}.$$

We shall show that at least one entry of B is zero. We know that all 2×2 minors of B are zero; in particular

$$a^2e^2 - b^2d^2 = 0$$
$$b^2f^2 - c^2e^2 = 0$$
$$c^2d^2 - a^2f^2 = 0.$$

These equations imply

$$(1) \qquad \begin{aligned} ae &= \pm bd \\ bf &= \pm ce \\ cd &= \pm af. \end{aligned}$$

If the minus sign is correct in all of these equations, then multiplying them all together, we find $abcdef = -abcdef$ and conclude that one of a, b, c, d, e, f is zero, so B has a zero entry. On the other hand, if the plus sign is correct in at least one of the equations (1), then a 2×2 minor of A is zero, and again B has a zero entry.

Any matrix of rank zero or one which has at least one entry zero has either a whole row or a whole column of zeros. If the column of the zero entry is not all zero, then this column spans the column space, and every other column is a multiple of it; so a whole row of zeros must appear. Thus we conclude that B has either a whole row or a whole column of zeros. Correspondingly, A has either a whole column or a whole row of zeros. Then the cofactors of all other entries of A are zeros, and this shows that all the rest of B is zeros. Hence $A = 0$, as required.

Afternoon Session

7. Let $f(x)$ be a function such that $f(1) = 1$ and for $x \geq 1$

$$f'(x) = \frac{1}{x^2 + f^2(x)}.$$

Prove that

$$\lim_{x \to \infty} f(x)$$

exists and is less than $1 + \pi/4$.

Solution. Since f' is everywhere positive, f is strictly increasing and therefore

$$f(t) > f(1) = 1 \qquad \text{for } t > 1.$$

Therefore

$$f'(t) = \frac{1}{t^2 + f^2(t)} < \frac{1}{t^2 + 1} \quad \text{for } t > 1. \tag{1}$$

So

$$f(x) = 1 + \int_1^x f'(t)dt$$

$$< 1 + \int_1^x \frac{1}{1 + t^2}\,dt < 1 + \int_1^\infty \frac{dt}{1 + t^2} = 1 + \pi/4.$$

Since f is increasing and bounded, $\lim_{x\to\infty} f(x)$ exists and is at most $1 + \pi/4$. Strict inequality also follows from (1) because

$$\lim_{x\to\infty} f(x) = 1 + \int_1^\infty f'(t)dt < 1 + \int_1^\infty \frac{1}{1+t^2}\,dt = 1 + \pi/4.$$

---

8. Let $f(x)$ be a differentiable function defined in the closed interval (0, 1) and such that

$$|f'(x)| \le M, \qquad 0 < x < 1.$$

Prove that

$$\left| \int_0^1 f(x)\,dx - \frac{1}{n}\sum_{k=1}^{n} f\left(\frac{k}{n}\right) \right| \le \frac{M}{n}.$$

Solution. Let

$$E_k = \int_{(k-1)/n}^{k/n} f(x)dx - \frac{1}{n} f\left(\frac{k}{n}\right)$$

for $k = 1, 2, \ldots, n$. Since f is differentiable, it is continuous, and therefore by the mean value theorem for integrals there exists a number η_k such that

$$\frac{k-1}{n} < \eta_k < \frac{k}{n}$$

and

$$\int_{(k-1)/n}^{k/n} f(x)dx = \frac{1}{n} f(\eta_k).$$

By the mean value theorem for derivatives there exists a number ξ_k such that $\eta_k < \xi_k < (k/n)$ and

$$f(\eta_k) - f\left(\frac{k}{n}\right) = \left(\eta_k - \frac{k}{n}\right) f'(\xi_k).$$

Then

$$|E_k| = \frac{1}{n}\left| f(\eta_k) - f\left(\frac{k}{n}\right)\right|$$

$$= \frac{1}{n}\left|\eta_k - \frac{k}{n}\right| \cdot |f'(\xi_k)| \leq \frac{1}{n^2}M.$$

Hence

$$\left|\int_0^1 f(x)dx - \frac{1}{n}\sum_{k=1}^{n} f\left(\frac{k}{n}\right)\right| = \left|\sum_{k=1}^{n} E_k\right| \leq \frac{M}{n}.$$

REMARK. The estimate can be improved by a factor of two. Let

$$F(x) = \int_0^x f(t)dt.$$

Expanding by Taylor's theorem about k/n, we obtain

$$F\left(\frac{k-1}{n}\right) = F\left(\frac{k}{n}\right) + \left(-\frac{1}{n}\right) F'\left(\frac{k}{n}\right) + \frac{1}{2}\left(-\frac{1}{n}\right)^2 F''(\theta_k),$$

where

$$\frac{k-1}{n} < \theta_k < \frac{k}{n}.$$

Since $F' = f$, this becomes

$$E_k = -\frac{1}{2n^2} f'(\theta_k),$$

and we have gained a factor of two.

---

9. Let x, y be Cartesian coordinates in the plane. I denotes the line segment $1 \leq x \leq 3, y = 1$. For every point P on I, let P^* denote that point that lies on the segment joining the origin to P and such that the distance PP^* is equal to $1/100$. As P describes I, the corresponding point P^* describes a certain curve C^*. Let $l(I)$, $l(C^*)$ be the lengths of I and C^* respectively. Which one of $l(I)$, $l(C^*)$ is greater? Prove your answer.

Solution. In polar coordinates the given line segment $1 \leq x \leq 3$, $y = 1$ lies on the line $r_1 = \csc\theta$, and the equation of the curve C^* is $r_2 = \csc\theta - h$ where $h = 1/100$. Then the respective arc lengths are given by

$$l(I) = \int_{\arctan 1/3}^{\arctan 1} \sqrt{r_1{}^2 + \left(\frac{dr_1}{d\theta}\right)^2}\, d\theta$$

$$l(C^*) = \int_{\arctan 1/3}^{\arctan 1} \sqrt{r_2{}^2 + \left(\frac{dr_2}{d\theta}\right)^2}\, d\theta.$$

But clearly $dr_1/d\theta = dr_2/d\theta$, and $r_2 < r_1$, so $l(C^*) < l(I)$.

---

10. Given $P(z) = z^2 + az + b$, a quadratic polynomial of the complex variable z with complex coefficients a, b. Suppose that $|P(z)| = 1$ for every z such that $|z| = 1$. Prove that $a = b = 0$.

First Solution. Let $P(z) = z^2 + az + b$. Let $a = p + iq$, $b = r + is$, where p, q, r, s are real. We are given that $|P(1)| = |P(-1)| = |P(i)| = |P(-i)| = 1$, and we find

$$\begin{aligned} |P(1)|^2 &= |1 + p + iq + r + is|^2 \\ &= 1 + p^2 + q^2 + r^2 + s^2 + 2p + 2r + 2pr + 2qs \end{aligned}$$

$$\begin{aligned} |P(-1)|^2 &= |1 - p - iq + r + is|^2 \\ &= 1 + p^2 + q^2 + r^2 + s^2 - 2p + 2r - 2pr - 2qs \end{aligned}$$

$$\begin{aligned} |P(i)|^2 &= |-1 + ip - q + r + is|^2 \\ &= 1 + p^2 + q^2 + r^2 + s^2 + 2q - 2r - 2qr + 2ps \end{aligned}$$

$$\begin{aligned} |P(-i)|^2 &= |-1 - ip + q + r + is|^2 \\ &= 1 + p^2 + q^2 + r^2 + s^2 - 2q - 2r + 2qr - 2ps. \end{aligned}$$

Adding these equations, we get

$$4 = 4 + 4(p^2 + q^2 + r^2 + s^2).$$

It follows that $p = q = r = s = 0$, so $a = b = 0$.

We can equally well use a different set of roots of one. If ξ is a primitive nth root of one, $n > 2$, then

$$n = \sum_{k=1}^{n} |P(\xi^k)|^2 = n(1 + |a|^2 + |b|^2),$$

and $a = b = 0$ follows. We can also replace the sum by an integral. For all real θ, we have

$$\begin{aligned} 1 = |P(e^{i\theta})|^2 &= (e^{2i\theta} + ae^{i\theta} + b)(e^{-2i\theta} + \bar{a}e^{-i\theta} + \bar{b}) \\ &= \bar{b}e^{2i\theta} + (\bar{a} + a\bar{b})e^{i\theta} + 1 + |a|^2 + |b|^2 \\ &\quad + (a + \bar{a}b)e^{-i\theta} + be^{-2i\theta}. \end{aligned}$$

If we integrate this over $[0, 2\pi]$, we get

$$2\pi = 2\pi(1 + |a|^2 + |b|^2),$$

and $a = b = 0$, as before.

Second Solution. If α, β, γ are complex numbers such that $|\alpha| = |\beta| = |\gamma| = 1$ and $\alpha + \beta + \gamma = 3$, then $\alpha = \beta = \gamma = 1$. Let $\omega = (-1 + i\sqrt{3})/2$, a cube root of one. Then $|P(1)| = |\omega P(\omega)| = |\omega^2 P(\omega^2)| = 1$ and $P(1) + \omega P(\omega) + \omega^2 P(\omega^2) = 3$, so

$$P(1) = \omega P(\omega) = \omega^2 P(\omega^2) = 1$$

and $P(1) = 1$, $P(\omega) = \omega^2$, $P(\omega^2) = \omega = \omega^4$. But there is only one polynomial of degree less than three that takes these values at these points, and it is $P(z) = z^2$.

This argument also works using higher roots of one or by integration.

Third Solution. Put $Q(z) = 1 + az + bz^2$. Then for $|z| = 1$, we have

$$|Q(z)| = |Q(z)z^{-2}| = |P(\bar{z})| = 1.$$

Since $Q(0) = 1$, we see that Q has the same absolute value at an interior point of the disc as its maximum absolute value on the boundary. By the maximum modulus principle, Q is constant; so $Q(z) = 1$ and $a = b = 0$.

Remark. All three of these proofs can be adapted to prove a more general result, namely, if

$$P(z) = z^n + a_1 z^{n-1} + a_2 z^{n-2} + \cdots + a_n$$

and $|P(z)| = 1$ for all z such that $|z| = 1$, then

$$a_1 = a_2 = \cdots = a_n = 0.$$

Blaschke studied functions f analytic on the open unit disc and continuous on the closed disc such that $|f(z)| = 1$ when $|z| = 1$. He showed that all such functions have the form

$$f(z) = \sigma \prod_{k=1}^{n} \frac{z - b_k}{1 - \bar{b}_k z},$$

where $b_1, b_2, \ldots, b_n$ satisfy $|b_k| < 1$ and σ is a constant of absolute value 1. See, for example, J. L. Walsh, *Interpolation and Approximation in the Complex Domain*, American Mathematical Society, Providence, R.I., 1935, pp. 281 ff.

---

11. a, b, c, d are distinct integers such that

$$(x - a)(x - b)(x - c)(x - d) - 4 = 0$$

has an integral root r. Show that $4r = a + b + c + d$.

Solution. Since r is a root,

$$(r - a)(r - b)(r - c)(r - d) = 4,$$

and since a, b, c, d are distinct integers, $r - a, r - b, r - c, r - d$ are distinct integers whose product is 4. But the only set of four distinct integers whose product is 4 is the set $\{1, -1, 2, -2\}$. Hence

$$(r - a) + (r - b) + (r - c) + (r - d) = 1 - 1 + 2 - 2 = 0,$$

so

$$4r = a + b + c + d.$$

---

12. C is a fixed point on OZ and U, V are variable points on OX, OY respectively, where OX, OY, OZ are mutually orthogonal lines. Find the locus of a point P such that PU, PV, PC are mutually orthogonal.

First Solution. Take OX, OY, OZ as coordinate axes and let $U = (u, 0, 0)$, $V = (0, v, 0)$, $C = (0, 0, c)$. Suppose $P(x, y, z)$ is a point of the locus. Then $(x - u, y, z)$, $(x, y - v, z)$, and $(x, y, z - c)$ must be perpendicular vectors. Therefore

$$
\begin{aligned}
&x^2 + y^2 + z^2 = xu + yv \\
(1) \qquad &x^2 + y^2 + z^2 = xu + zc \\
&x^2 + y^2 + z^2 = yv + zc.
\end{aligned}
$$

Adding the last two equations and subtracting the first, we get

$$
(2) \qquad x^2 + y^2 + z^2 = 2zc,
$$

and therefore P lies on the sphere $x^2 + y^2 + (z - c)^2 = c^2$, with center C and radius $|CO|$. Note that, if $c = 0$, i.e., $C = 0$, then P must be 0. But in that case PC is not a line, so there is no locus. We assume henceforth, therefore, that $c \neq 0$.

From (1) we also see that $xu = yv = zc$, so if $z \neq 0$ neither x nor y is 0. If $z = 0$, then $x = y = 0$ from (2).

Conversely, if $P = (x, y, z)$ is any point on the sphere (2) with $xy \neq 0$, then $z \neq 0$ and

$$
u = zc/x, \qquad v = zc/y
$$

gives a solution to (1), so the vectors PU, PV, and PC are pairwise orthogonal, while if $P = 0$, then for any non-zero choice of u and v, PU, PV, and PC are pairwise orthogonal. The locus therefore consists of the sphere less two great circles but including the point 0.

Second Solution. An elementary synthetic argument can be given based on the fact that the length of the median to the hypotenuse of a right triangle is half the length of the hypotenuse. Let M be the midpoint of UV. If P is on the locus, CP is perpendicular to the plane PUV and $\angle UPV$ is a right angle. By the result cited above, $PM = \frac{1}{2} UV = OM$, since PM is a median of $\triangle PUV$ and OM is a median of $\triangle OUV$. The right triangles CPM and COM have a common hypotenuse, CM, and a pair of congruent legs: hence these triangles are congruent. Thus $CP = CO$, and P is on the sphere centered at C with radius CO. PU and PV are tangent to this sphere.

Conversely, let P be any point on the sphere, and suppose the tangent plane at P intersects OX and OY respectively in U and V. Now $\overline{PM} = \overline{MO}$ since both lines are tangents to the sphere from the external point M. But $\overline{OM} = \frac{1}{2}\,\overline{UV}$, and thus $\overline{PM} = \frac{1}{2}\,\overline{UV}$. From the converse of the cited

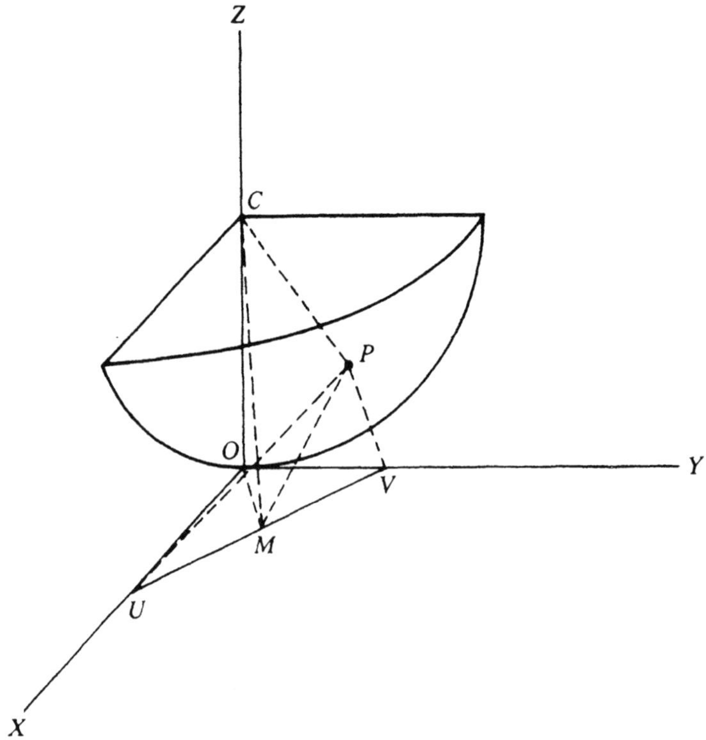

plane geometry theorem, $\angle UPV$ is a right angle. Also, $\angle CPV$ and $\angle CPU$ are clearly right angles since PU and PV are in the tangent plane to the sphere.

REMARK. If points at infinity on OX and OY are admitted, then all points of the sphere with center at C and radius $\overline{OC}$ can be considered as points of the locus.

THE EIGHTH WILLIAM LOWELL PUTNAM MATHEMATICAL COMPETITION

March 20, 1948

Morning Session

1. What is the maximum of $|z^3 - z + 2|$, where z is a complex number with $|z| = 1$?

Solution. Let $f(z) = z^3 - z + 2$. We may as well maximize $|f(z)|^2$. If $|z| = 1$, then $z = x + iy$, where $y^2 = 1 - x^2$ and $-1 \le x \le 1$, so

$$\begin{aligned}
|f(z)|^2 &= |(x + iy)^3 - (x + iy) + 2|^2 \\
&= |x^3 - 3x(1 - x^2) - x + 2 + iy(3x^2 - (1 - x^2) - 1)|^2 \\
&= (4x^3 - 4x + 2)^2 + (1 - x^2)(4x^2 - 2)^2 \\
&= 16x^3 - 4x^2 - 16x + 8 = L(x).
\end{aligned}$$

Hence we seek $\max_{1 \le x \le 1} L(x)$. This maximum must be attained either at a critical point or at an endpoint. The critical points, obtained by solving $L'(x) = 48x^2 - 8x - 16 = 0$, are $x = -\frac{1}{2}, \frac{2}{3}$. Since

$$L(-1) = 4, \qquad L\left(-\frac{1}{2}\right) = 13, \qquad L\left(\frac{2}{3}\right) = \frac{8}{27}, \qquad L(1) = 4,$$

the maximum value of L is 13, attained for $x = -\frac{1}{2}$. Hence the maximum value of $|f(z)|$ on the unit circle is $\sqrt{13}$, attained when $\operatorname{Re} z = -\frac{1}{2}$, i.e., when $z = (-1 \pm i\sqrt{3})/2$.

---

2. Two spheres in contact have a common tangent cone. These three surfaces divide the space into various parts, only one of which is bounded by all three surfaces; it is "ring-shaped." Being given the radii of the spheres, r and R, find the volume of the "ring-shaped" part. (The desired expression is a rational function of r and R.)

First Solution. Let α be the angle of the cone, and place the spheres as shown in the cross-section diagram, with centers at O_1 and O_2 and origin of the coordinate system at O, the point of contact.

The equation of the larger circle is

$$x^2 + y^2 + 2Rx = 0$$

and the equation of the smaller is

$$x^2 + y^2 - 2rx = 0.$$

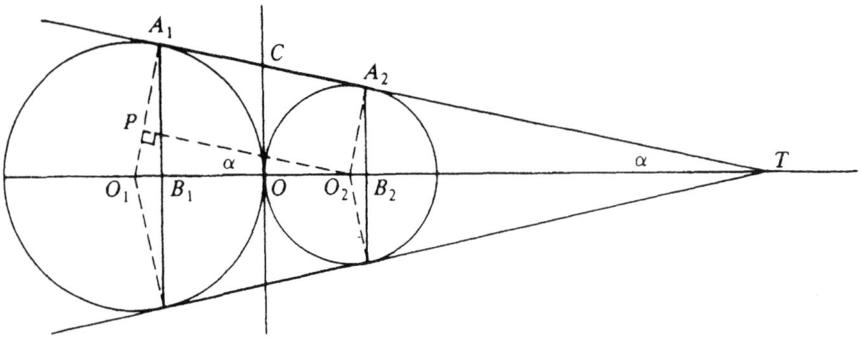

From the diagram,

$$\sin \alpha = \frac{O_1P}{O_1O_2} = \frac{R - r}{R + r},$$

so $\cos^2 \alpha = 4Rr/(R + r)^2$.

We compute V_1, the volume of the frustum of the cone obtained by revolving area $A_1A_2B_2B_1$ about the x-axis.

$$\text{Volume of larger cone} = \frac{1}{3}\pi A_1B_1{}^2 \cdot B_1T$$

$$= \frac{1}{3}\pi R^2 \cos^2 \alpha \cdot R \cot \alpha \cos \alpha$$

$$= \frac{16\pi}{3} R^3 \frac{r^2R^2}{(R - r)(R + r)^3}.$$

$$\text{Volume of smaller cone} = \frac{16\pi}{3} r^3 \frac{r^2R^2}{(R - r)(R + r)^3}.$$

Their difference is

$$V_1 = \frac{16\pi}{3} \cdot \frac{r^2R^2}{(R + r)^3} \cdot (R^2 + Rr + r^2).$$

Let V_2 and V_3 be, respectively, the volumes of the larger spherical segment and the smaller spherical segment enclosed in the frustum of the cone.

$$\begin{aligned} V_2 &= \pi \int_{R(\sin\alpha - 1)}^{0} (-x^2 - 2Rx)dx \\ &= \pi \left[-\frac{x^3}{3} - Rx^2 \right]_{R(\sin\alpha - 1)}^{0} \\ &= \frac{4\pi R^3 r^2}{3(R + r)^3} (3R + r), \end{aligned}$$

since $R(\sin\alpha - 1) = -2Rr/(R + r)$. Likewise

$$\begin{aligned} V_3 &= \pi \int_{0}^{r(1+\sin\alpha)} (-x^2 + 2rx)dx = \pi \left[-\frac{x^3}{3} + rx^2 \right]_{0}^{r(\sin\alpha + 1)} \\ &= \frac{4\pi}{3} \cdot \frac{R^2 r^3}{(R + r)^3} (R + 3r). \end{aligned}$$

The desired volume is given by

$$\begin{aligned} V_1 - V_2 - V_3 &= \frac{4\pi}{3} \frac{R^2r^2}{(R + r)^3} \\ &\quad \times [4(R^2 + Rr + r^2) - 3R^2 - Rr - 3r^2 - Rr] \\ &= \frac{4\pi}{3} \frac{R^2r^2}{R + r}. \end{aligned}$$

Second Solution. An elegant geometrical solution based on Cavalieri's principle can be given.

Consider the accompanying diagram labeled as shown and suppose the figure is rotated about the axis BT. The region Δ bounded by $\overline{AC}$, $\overline{CO}$ and the circular arc AO will sweep out a solid that is a frustum of a cone less a spherical segment. We shall prove that its volume is the same as that of the cone generated by the triangular region BOC.

Consider an arbitrary line perpendicular to BT between B and O crossing $\overline{AC}$ at P, $\widehat{AO}$ at Q, etc., as shown. When rotated the segment

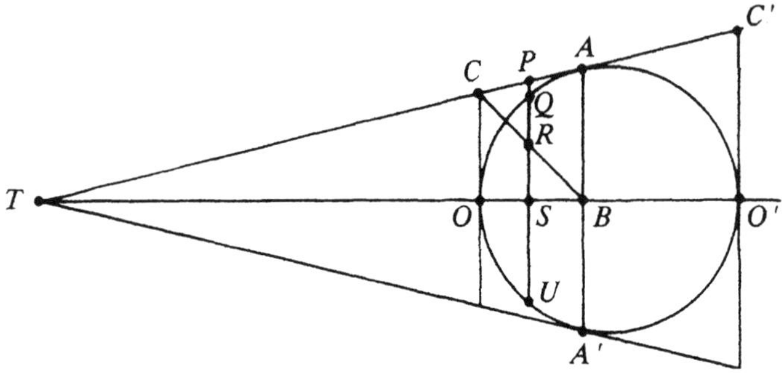

$\overline{PQ}$ sweeps out an annular region whose area is

$$\pi(PS^2 - QS^2) = \pi(PS + QS)(PS - QS) = \pi PU \cdot PQ = \pi(AP)^2.$$

By similarity

$$\frac{AP}{AC} = \frac{BS}{BO} = \frac{RS}{CO}.$$

Since the tangents AC and OC to the circle are equal, we have $AP = RS$. Thus the area of the annular region is $\pi(RS)^2$ which is the area of the circular region generated by $\overline{RS}$. By Cavalieri's principle the volume of the solid swept out by Δ is therefore the same as the volume of the cone swept out by the triangular region BOC.

This argument applies equally to the solid swept out by the curvilinear region Δ' bounded by $\overline{AC'}$, $\overline{C'O'}$ and the circular arc AO' and the cone swept out by the triangular region $BO'C'$.

Note that if $R = r$, this argument becomes a classical proof that the volume of the region between a hemisphere and the circumscribed cylinder is one-third the volume of the cylinder. Hence the volume of the hemisphere is two-thirds the volume of the cylinder.

Returning to the problem (see the first figure), the volume V required is the sum of the volumes of two cones with a common base generated by OC and altitudes OB_1 and OB_2. Hence $V = \frac{1}{3}\pi OC^2 \cdot B_1B_2$. Now $\angle O_1CO_2$ is a right angle since O_1C and O_2C bisect the supplementary angles A_1CO and OCA_2, so $OC^2 = O_1O \cdot OO_2 = Rr$. And $B_1B_2 = O_1O_2 \cos^2 \alpha = 4Rr/(R + r)$. Therefore

$$V = \frac{4\pi}{3} \cdot \frac{R^2r^2}{R + r}.$$

3. Let $\{a_n\}$ be a decreasing sequence of positive numbers with limit 0 such that

$$b_n = a_n - 2a_{n+1} + a_{n+2} \geq 0$$

for all n. Prove that

$$\sum_{n=1}^{\infty} nb_n = a_1.$$

Solution. Since the b's are the second differences of the a's, it is convenient to let $c_n = a_n - a_{n+1}$; then

$$c_n - c_{n+1} = a_n - 2a_{n+1} + a_{n+2} = b_n.$$

Since the a's decrease to zero, $c_n \geq 0$ for all n, and $c_n \to 0$.

For $k \geq m$, we have $\sum_{i=m}^{k} b_i = c_m - c_{k+1}$, and therefore

$$\sum_{i=m}^{\infty} b_i = c_m = a_m - a_{m+1}.$$

Similarly,

$$\sum_{m=1}^{k} (a_m - a_{m+1}) = a_1 - a_{k+1}, \quad \text{so} \sum_{m=1}^{\infty} (a_m - a_{m+1}) = a_1.$$

Thus

$$\sum_{m=1}^{\infty} \left(\sum_{i=m}^{\infty} b_i \right) = a_1.$$

The b's are non-negative, and when summing non-negative terms, rearrangement does not affect the value of the sum. For each index n, the term b_n appears exactly n times in the preceding double sum, once in each of the sums $\sum_{i=m}^{\infty} b_i$ for $m = 1, 2, \ldots, n$. Hence

$$\sum_{n=1}^{\infty} nb_n = \sum_{m=1}^{\infty} \left(\sum_{i=m}^{\infty} b_i \right) = a_1.$$

---

4. Let D be a plane region bounded by a circle of radius r. Let (x, y) be a point of D and consider a circle of radius δ and center at (x, y). Denote by

$l(x, y)$ the length of that arc of the circle which is outside D. Find

$$\lim_{\delta - 0} \frac{1}{\delta^2} \iint_D l(x, y)\, dx\, dy.$$

Solution. First convert the integral to polar coordinates, taking the origin at the center of the given circle. If a point (x, y) has polar coordinates (ρ, θ), then $l(x, y) = L(\rho)$, where

$$L(\rho) = 0 \qquad \text{if } 0 \le \rho \le r - \delta$$

and

$$L(\rho) = 2\delta\phi = 2\delta \arccos\left(\frac{r^2 - \rho^2 - \delta^2}{2\rho\delta}\right) \qquad \text{if } r - \delta \le \rho \le r,$$

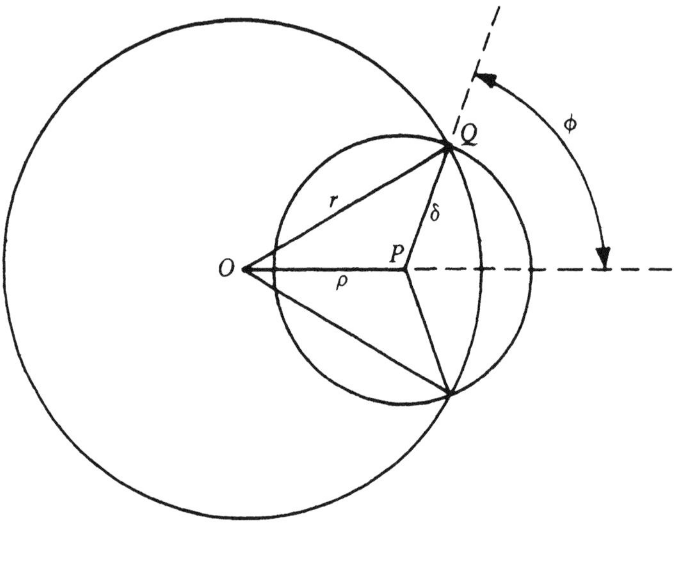

as we see by applying the law of cosines to triangle OPQ. Hence we must find

$$A = \lim_{\delta \to \infty} \frac{1}{\delta^2} \int_0^{2\pi} \int_0^r L(\rho)\rho\, d\rho\, d\theta$$

$$= \lim_{\delta \to 0} \frac{1}{\delta^2} \int_0^{2\pi} \int_{r-\delta}^{r} 2\delta\rho \arccos\left(\frac{r^2 - \rho^2 - \delta^2}{2\rho\delta}\right) d\rho d\theta$$

$$= \lim_{\delta \to 0} \frac{4\pi}{\delta} \int_{r-\delta}^{r} \rho \arccos\left(\frac{r^2 - \rho^2 - \delta^2}{2\rho\delta}\right) d\rho.$$

We make the substitution $\rho = r - \delta u$ and get

$$A = \lim_{\delta \to 0} 4\pi \int_0^1 (r - \delta u) \arccos\left(\frac{2ur - \delta(1 + u^2)}{2(r - \delta u)}\right) du.$$

Since the integrand is a continuous function of u and δ for $u \in [0, 1]$ and $\delta \in [0, \frac{1}{2}r]$, and the domain of integration is bounded, we can conclude that the limit A exists and

$$A = 4\pi \int_0^1 \lim_{\delta \to 0} (r - \delta u) \arccos\left(\frac{2ur - \delta(1 + u^2)}{2(r - \delta u)}\right) du$$

$$= 4\pi \int_0^1 r \arccos u \, du.$$

Integrating by parts, we obtain

$$A = 4\pi r[u \arccos u]_0^1 + 4\pi r \int_0^1 \frac{u\, du}{\sqrt{1 - u^2}}$$

$$= 0 + 4\pi r[-\sqrt{1 - u^2}]_0^1 = 4\pi r.$$

---

5. If $x_1, \ldots, x_n$ denote the nth roots of unity, evaluate

$$\Pi(x_i - x_j)^2 \qquad (i < j).$$

First Solution. Since

(1) $$t^n - 1 = (t - x_1)(t - x_2) \cdots (t - x_n)$$

we see that

(2) $$x_1 x_2 \cdots x_n = (-1)^{n-1}.$$

Differentiating (1) we have

$$nt^{n-1} = \sum_{i=1}^{n} (t - x_1) \cdots (t - x_{i-1})(t - x_{i+1}) \cdots (t - x_n).$$

Evaluating for $t = x_1, x_2, \ldots, x_n$, we obtain

$$\begin{array}{llllll}
nx_1^{n-1} = & & (x_1 - x_2) & (x_1 - x_3) \cdots & (x_1 - x_{n-1}) & (x_1 - x_n) \\
nx_2^{n-1} = & (x_2 - x_1) & \cdot & (x_2 - x_3) \cdots & (x_2 - x_{n-1}) & (x_2 - x_n) \\
nx_3^{n-1} = & (x_3 - x_1) & (x_3 - x_2) & \cdot \quad \cdots & (x_3 - x_{n-1}) & (x_3 - x_n) \\
\vdots & \vdots & \vdots & \vdots & \vdots & \vdots \\
nx_n^{n-1} = & (x_n - x_1) & (x_n - x_2) & (x_n - x_3) \cdots & (x_n - x_{n-1}) & \cdot
\end{array}$$

Multiplying these equations, we get

$$\begin{aligned} n^n(x_1x_2\cdots x_n)^{n-1} &= \prod_{i<j} [-(x_i - x_j)^2] \\ &= (-1)^{n(n-1)/2} \prod_{i<j} (x_i - x_j)^2. \end{aligned}$$

Using (2) we find

$$\begin{aligned} \prod_{i<j} (x_i - x_j)^2 &= n^n(-1)^{(n-1)^2}(-1)^{-n(n-1)/2} \\ &= (-1)^{(n-1)(n-2)/2} n^n. \end{aligned}$$

Second Solution. We number the roots so that $x_m = \exp(2\pi im/n)$. Then $x_n = 1$ and $x_sx_t = x_{s+t}$, where the subscripts are reckoned modulo n.

The factorization

$$t^n - 1 = (t - 1)(t^{n-1} + t^{n-2} + \cdots + 1)$$

gives

$$\prod_{k=1}^{n-1} (t - x_k) = t^{n-1} + t^{n-2} + \cdots + 1$$

and in particular

$$\prod_{k=1}^{n-1} (1 - x_k) = n.$$

For a moment let i be a fixed index. Then $|x_i - x_j| = |1 - x_{j-i}|$, and therefore

$$\prod_{j \neq i} |x_i - x_j| = \prod_{k=1}^{n-1} |1 - x_k| = n.$$

Multiplying these equations for $i = 1, 2, \ldots, n$, we obtain

$$\prod_{i=1}^{n} \prod_{j \neq i} |x_i - x_j| = n^n.$$

Since each linear factor appears twice in this double product,

$$\prod_{i<j} |x_i - x_j|^2 = n^n.$$

Therefore the required product is λn^n, where λ is a number of absolute value one. We must evaluate λ.

If $i < j$ and $i + j \neq n$, then $(x_i - x_j)^2$ and $(x_{n-j} - x_{n-i})^2$ are distinct factors in the product. Since they are conjugates of one another, their product is a positive real number. The factors of this type can thus be paired off so that each pair has a positive product. The remaining factors are $(x_i - x_{n-i})^2$ where $1 < i < n - i$. Since x_i and x_{n-i} are conjugate, each of these factors is a negative real number.

Suppose that n is odd. Then there are negative factors for $i = 1, 2, \ldots, (n-1)/2$, so $\lambda = (-1)^{(n-1)/2} = (-1)^{(n-1)(n-2)/2}$.

Suppose n is even. Then there are negative factors for $i = 1, 2, \ldots, (n-2)/2$, so $\lambda = (-1)^{(n-2)/2} = (-1)^{(n-1)(n-2)/2}$.

Thus, in either case

$$\prod_{i<j} (x_i - x_j)^2 = (-1)^{(n-1)(n-2)/2} n^n.$$

REMARK. The product of the squares of the differences of the zeros of a polynomial is known as the *discriminant* of the polynomial.

---

6. (i) A force acts on the element ds of a closed plane curve. The magnitude of this force is $r^{-1}\, ds$ where r is the radius of curvature at the point considered, and the direction of the force is perpendicular to the curve; it points to the convex side. Show that the system of such forces acting on all elements of the curve keep it in equilibrium.

First Solution. We interpret the problem a little more explicitly as follows: Let C be a closed plane curve of length L, represented in terms of its arc length s by $x = x(s)$ and $y = y(s)$, where $x(s)$ and $y(s)$ are periodic functions of class C^2 and period L. Let

$$\frac{dx}{ds}, \frac{dy}{ds}, \frac{d^2x}{ds^2}, \frac{d^2y}{ds^2}$$

be denoted by $\dot{x}$, $\dot{y}$, $\ddot{x}$, $\ddot{y}$, respectively. Now consider a force system F defined along the curve, where $F_x = (\dot{x}\ddot{y} - \dot{y}\ddot{x})\dot{y}$, $F_y = -(\dot{x}\ddot{y} - \dot{y}\ddot{x})\dot{x}$. Show that this force system is in equilibrium.

(For the curvature at (x, y) is given by $\dot{x}\ddot{y} - \dot{y}\ddot{x}$, and the signs for F_x and F_y are chosen to make F point to the convex side of C.)

The conditions for equilibrium are

$$(1) \qquad \int_0^L F_x ds = \int_0^L F_y ds = 0$$

and

$$(2) \qquad \int_0^L xF_y ds - \int_0^L yF_x ds = 0.$$

Here (1) means that the total force in any direction must vanish, and (2) means that the total moment of the force about the origin must vanish.

Keep in mind that $\dot{x}^2 + \dot{y}^2 = 1$ from which it follows that $\dot{x}\ddot{x} + \dot{y}\ddot{y} = 0$. Then

$$\int_0^L F_x ds = -\int_0^L (-\dot{x}\dot{y}\ddot{y} + \ddot{x}\dot{y}^2) ds = -\int_0^L (\ddot{x}\dot{x}^2 + \ddot{x}\dot{y}^2) ds$$

$$= -\int_0^L \ddot{x}(\dot{x}^2 + \dot{y}^2) ds = -\int_0^L \ddot{x} ds = -\dot{x}\Big|_0^L = 0.$$

Similarly $\int_0^L F_y ds = 0$ and thus condition (1) is met. Now

$$\int_0^L xF_y ds - \int_0^L yF_x ds$$

$$= -\int_0^L x\dot{x}(\dot{x}\ddot{y} - \dot{y}\ddot{x}) ds - \int_0^L y\dot{y}(\dot{x}\ddot{y} - \dot{y}\ddot{x}) ds$$

$$= -\int_0^L (x\ddot{y}\dot{x}^2 + x\ddot{y}\dot{y}^2 - y\ddot{x}\dot{x}^2 - y\ddot{x}\dot{y}^2) ds$$

$$= -\int_0^L [x\ddot{y}(\dot{x}^2 + \dot{y}^2) - y\ddot{x}(\dot{x}^2 + \dot{y}^2)]ds$$

$$= -\int_0^L (x\ddot{y} - y\ddot{x})ds = -(x\dot{y} - y\dot{x})\Big|_0^L = 0.$$

Thus the moment requirement (2) is met and the system F is in equilibrium.

Using vectors, this solution may be presented as follows:

Second Solution. Let $\rho = \rho(s)$ be the parametric vector equation of the given curve, where s is arc-length. We assume that ρ is a periodic function of class C^2 and period L, the length of the curve, so that ρ describes the curve once as s varies from 0 to L.

Then $d\rho/ds = \mathbf{t} = \mathbf{t}(s)$ is the unit tangent vector to the curve at $\rho(s)$, and

(1) $$\frac{d\mathbf{t}}{ds} = \kappa\mathbf{n}$$

where $\mathbf{n} = \mathbf{n}(s)$ is the unit normal vector at the point $\rho(s)$ directed toward the concave side of the curve and $\kappa = \kappa(s)$ is the curvature. The radius of curvature is $r = 1/\kappa$.

It is given that the force on an element ds of the curve has magnitude ds/r and direction $-\mathbf{n}$, i.e.,

(2) $$d\mathbf{F} = -\frac{\mathbf{n}ds}{r}$$

hence from (1),

$$d\mathbf{F} = -d\mathbf{t}.$$

The condition for equilibrium is that the total force should be zero and that the total moment of the force with respect to some point should be zero. Thus we must show

$$\oint d\mathbf{F} = 0 \quad \text{and} \quad \oint \rho \times d\mathbf{F} = 0.$$

Now

$$\oint d\mathbf{F} = -\oint d\mathbf{t} = -\mathbf{t}(s)\Big]_{s=0}^{s=L} = 0.$$

For the second requirement, note that

$$\frac{d}{ds}(\boldsymbol{\rho} \times \mathbf{t}) = \mathbf{t} \times \mathbf{t} + \boldsymbol{\rho} \times \frac{d\mathbf{t}}{ds} = \boldsymbol{\rho} \times \frac{d\mathbf{t}}{ds}.$$

Hence

$$\oint \boldsymbol{\rho} \times \mathbf{dF} = -\oint \boldsymbol{\rho} \times d\mathbf{t} = -\boldsymbol{\rho} \times \mathbf{t}\Big]_{s=0}^{s=L} = 0.$$

---

6. (ii) Show that

$$x + \frac{2}{3}x^3 + \frac{2}{3}\frac{4}{5}x^5 + \frac{2}{3}\frac{4}{5}\frac{6}{7}x^7 + \cdots = \frac{\arcsin x}{\sqrt{1 - x^2}}.$$

Solution. Let

$$f(x) = x + \frac{2}{3}x^3 + \frac{2}{3}\cdot\frac{4}{5}x^5 + \frac{2}{3}\cdot\frac{4}{5}\cdot\frac{6}{7}x^7 + \cdots.$$

Then

$$\begin{aligned} f'(x) &= 1 + x\left[2x + \frac{2}{3}\cdot 4x^3 + \frac{2}{3}\cdot\frac{4}{5}\cdot 6x^5 + \cdots\right] \\ &= 1 + x\frac{d}{dx}\left[x^2 + \frac{2}{3}x^4 + \frac{2}{3}\cdot\frac{4}{5}x^6 + \cdots\right] \\ &= 1 + x\frac{d}{dx}(xf(x)) = 1 + xf(x) + x^2 f'(x). \end{aligned}$$

Thus $f'(x)$ satisfies the differential equation

$$(1) \qquad (1 - x^2)f'(x) = 1 + xf(x)$$

and the initial condition

$$(2) \qquad f(0) = 0.$$

(We note that the series for f is convergent for $|x| < 1$ so that all formal manipulations are justified.)

Now (1) is a first order linear non-singular differential equation on the interval $(-1, 1)$, so it has a unique solution satisfying (2).

Since $(\arcsin x)/\sqrt{1-x^2}$ satisfies (1) and (2), it follows that

$$f(x) = \frac{\arcsin x}{\sqrt{1-x^2}}.$$

Afternoon Session

1. Let $f(x)$ be a cubic polynomial with roots x_1, x_2, and x_3. Assume that $f(2x)$ is divisible by $f'(x)$ and compute the ratios $x_1 : x_2 : x_3$.

Solution. Let $f(x) = x^3 + ax^2 + bx + c$. Since $f(2x)$ is divisible by $f'(x)$, we have

$$8x^3 + 4ax^2 + 2bx + c = (3x^2 + 2ax + b)(px + q)$$

for some p and q. Comparing coefficients we find

$$3p = 8, \qquad 2ap + 3q = 4a,$$
$$bp + 2aq = 2b, \qquad qb = c,$$

from which it follows that

$$p = 8/3, \qquad q = -4a/9, \qquad b = 4a^2/3, \qquad \text{and} \qquad c = -16a^3/27.$$

Hence

$$f(x) = x^3 + ax^2 + \frac{4}{3}a^2x - \frac{16}{27}a^3.$$

Now if $a = 0$, all the roots of $f(x) = 0$ are zero and their ratios are undefined; so we assume from now on that $a \neq 0$. Set $w = 3x/a$ and consider

$$F(w) = 27f(x) = 27f\left(\frac{aw}{3}\right) = a^3(w^3 + 3w^2 + 12w - 16)$$
$$= a^3(w-1)(w^2 + 4w + 16).$$

The roots of $F(w) = 0$ are 1, $-2 \pm 2\sqrt{3}i$. The roots of $f(x) = 0$ have the same ratios as the roots of $F(w) = 0$, so with suitable numbering we have

$$x_1 : x_2 : x_3 = 1 : (-2 + 2\sqrt{3}i) : (-2 - 2\sqrt{3}i).$$

---

2. "A penny in a corner." A circle moves so that it is continually in contact with all three coordinate planes of an ordinary rectangular system. Find the locus of the center of the circle.

Solution. As θ varies, the maximum and minimum values of $a \cos \theta + b \sin \theta$ are $\sqrt{a^2 + b^2}$ and $-\sqrt{a^2 + b^2}$. Hence, in order that zero should be an extreme value of the function

$$a \cos \theta + b \sin \theta + c,$$

it is necessary and sufficient that $c^2 = a^2 + b^2$.

Let the moving circle C have radius r and center $\mathbf{x} = (x_1, x_2, x_3)$. Let $\mathbf{u}$, $\mathbf{v}$, $\mathbf{w}$ be mutually orthogonal unit vectors such that $\mathbf{u}$ is perpendicular to the plane of C. Then C consists of points of the form

$$r \cos \theta \, \mathbf{v} + r \sin \theta \, \mathbf{w} + \mathbf{x}. \tag{1}$$

Since C is tangent to each of the coordinate planes, each of the functions

$$r v_i \cos \theta + r w_i \sin \theta + x_i \tag{2}$$

is zero for some value of θ, but otherwise takes only positive, or only negative, values. Hence

$$x_i^2 = r^2 v_i^2 + r^2 w_i^2 \tag{3}$$

by the result stated in the first paragraph. Therefore,

$$x_1^2 + x_2^2 + x_3^2 = r^2(v_1^2 + v_2^2 + v_3^2) + r^2(w_1^2 + w_2^2 + w_3^2) = 2r^2$$

because $\mathbf{v}$ and $\mathbf{w}$ are unit vectors. Thus $\mathbf{x}$ lies on the sphere S with center at the origin and radius $r\sqrt{2}$. Moreover, since $\mathbf{u}$, $\mathbf{v}$, $\mathbf{w}$ are mutually orthogonal unit vectors,

$$u_i^2 + v_i^2 + w_i^2 = 1 \quad \text{for } i = 1, 2, 3.$$

Hence (3) can be written

$$x_i^2 = r^2(1 - u_i^2), \tag{4}$$

and it follows that

$$x_i^2 \le r^2 \quad \text{for } i = 1, 2, 3. \tag{5}$$

Conversely, suppose $\mathbf{x}$ is any point of S that satisfies (5). Then we can solve (4) for u_1, u_2, u_3 and $\mathbf{u} = (u_1, u_2, u_3)$ will be a unit vector. We can then choose $\mathbf{v}$ and $\mathbf{w}$ so that $\mathbf{u}$, $\mathbf{v}$, $\mathbf{w}$ are mutually orthogonal unit vectors, and equation (3) will hold. The functions (2) will therefore have zero as an extreme value, so the circle given parametrically by (1) will be tangent to each of the coordinate planes. (In accord with the image of a "penny in a corner," we consider a circle lying wholly in a coordinate plane as tangent to that plane.)

We conclude that the required locus is that part of the sphere S within the cube $|x_i| \leq r$, $i = 1, 2, 3$. (If a circle lying in a coordinate plane is not regarded as tangent to that plane, then 12 points such as $(\pm r, \pm r, 0)$ must be deleted.)

REMARK. For most points $\mathbf{x}$ of the locus there will be eight choices of the unit vector $\mathbf{u}$ $(u_i = \pm\sqrt{1 - x_i^2/r^2})$ leading to four choices of a plane through $\mathbf{x}$ and thus to four circles of radius r with center at $\mathbf{x}$ and tangent to all three coordinate planes. Of these circles, one will be inscribed in the triangle defined in its plane by the three coordinate planes, and the other three circles will be escribed to the corresponding triangles.

---

3. If n is a positive integer, prove that

$$[\sqrt{n} + \sqrt{n+1}] = [\sqrt{4n+2}\,],$$

where $[x]$ denotes as usual the greatest integer not exceeding x.

Solution. Since $\sqrt{x}$ has negative second derivative for $x > 0$, its graph is concave downward and

$$\frac{\sqrt{x} + \sqrt{x+1}}{2} < \sqrt{x + \frac{1}{2}} \text{ for all } x \geq 0.$$

Thus $\sqrt{x} + \sqrt{x+1} < \sqrt{4x+2}$ for all $x \geq 0$, and hence $[\sqrt{x} + \sqrt{x+1}] \leq [\sqrt{4x+2}]$.

Suppose that for some positive integer n, $[\sqrt{n} + \sqrt{n+1}] \neq [\sqrt{4n+2}]$. Let $p = [\sqrt{4n+2}]$. Then

$$\sqrt{n} + \sqrt{n+1} < p \leq \sqrt{4n+2}.$$

Squaring, we get

$$2n + 1 + 2\sqrt{n(n+1)} < p^2 \leq 4n + 2.$$

Therefore $2\sqrt{n(n+1)} < p^2 - 2n - 1 \le 2n + 1$. Squaring again gives

$$4n(n+1) < (p^2 - 2n - 1)^2 \le 4n^2 + 4n + 1.$$

Since the outer numbers are consecutive integers, and since the middle number in the inequality is also an integer, we have $(p^2 - 2n - 1)^2 = (2n+1)^2$ and therefore $p^2 = 4n + 2$. But this last equation is impossible, for the square of an integer cannot be congruent to 2 modulo 4.

This contradiction proves that $[\sqrt{n} + \sqrt{n+1}] = [\sqrt{4n+2}]$ for every positive integer n.

---

4. Let $\min(x, y)$ denote the smaller of the numbers x and y. For what λ's does the equation

$$\int_0^1 \min(x,y)\, f(y)\, dy = \lambda f(x)$$

have continuous solutions which do not vanish identically in $(0, 1)$? What are these solutions?

Solution. The given equation can be written as

$$\lambda f(x) = \int_0^x y f(y) dy + x \int_x^1 f(y) dy \tag{1}$$

from which it is clear that, if $\lambda \neq 0$, f is differentiable and hence

$$\lambda f'(x) = xf(x) - xf(x) + \int_x^1 f(y) dy = \int_x^1 f(y) dy. \tag{2}$$

Thus f' is also differentiable and

$$\lambda f''(x) = -f(x). \tag{3}$$

If $\lambda = 0$, the same steps lead to the equation $0 = -f(x)$. Since we are only interested in functions not identically zero, we shall assume $\lambda \neq 0$ from now on.

The general solution of (3) is

$$f(x) = A \cos \mu x + B \sin \mu x,$$

where $\mu = \lambda^{-1/2}$ if $\lambda > 0$, or

$$f(x) = A \cosh \nu x + B \sinh \nu x,$$

where $\nu = (-\lambda)^{-1/2}$ if $\lambda < 0$.

It is evident from (1) that $\lim_{x \to 0} f(x) = 0$, so $A = 0$ whether λ is positive or negative. From (2) it follows that $\lim_{x \to 1} f'(x) = 0$, giving respectively

$$B\mu \cos \mu = 0$$

or

$$B\nu \cosh \nu = 0.$$

The last equation can hold only for $B = 0$ and hence $f = 0$; so we conclude that (1) has no non-zero solutions if $\lambda < 0$.

For $\lambda > 0$, a non-trivial solution is possible only if $\cos \mu = 0$, that is, μ is an odd multiple of $\pi/2$. In fact

$$f(x) = B \sin (2k + 1) \frac{\pi}{2} x \tag{4}$$

is readily checked to be a solution with

$$\lambda = \mu^{-2} = \frac{4}{(2k + 1)^2 \pi^2},$$

where k is any integer and B is arbitrary. Here we need only consider $k = 0, 1, 2, \ldots$, since negative values of k produce the same solutions.

REMARK. In the language of linear algebra, we have shown that the linear operator $T: C(0, 1) \to C(0, 1)$ defined by

$$(Tf)(x) = \int_0^1 \min (x, y) f(y) dy$$

has eigenvalues $4/[(2k + 1)^2 \pi^2]$ for $k = 0, 1, 2, \ldots$, and corresponding to each eigenvalue there is a one-dimensional family of eigenvectors given by (4).

---

5. The pairs of numbers (a, b) such that $|a + bt + t^2| \leq 1$ for $0 \leq t \leq 1$ fill a certain region in the (a, b)-plane. What is the area of this region?

First Solution. Consider

$$f(t) = a + bt + t^2.$$

This function has just one critical value, at $t = -b/2$ where its value is $a - b^2/4$. On the interval $[0, 1]$ f can have extreme values only at the endpoints and at the critical point if it should happen to fall in $[0, 1]$ (that is, if $b \in [-2, 0]$). Hence the extreme values of f are

$$f(0) = a \text{ and } f(1) = a + b + 1$$

if $b \notin [-2, 0]$, and they are in the set

$$\{a, a + b + 1, a - b^2/4\}$$

if $b \in [-2, 0]$.

Hence $|f(t)| \leq 1$ for all $t \in [0, 1]$ if and only if

$$b \notin [-2, 0] \text{ and } |a| \leq 1, |a + b + 1| \leq 1$$

or

$$b \in [-2, 0], \text{ and } |a| \leq 1, |a + b + 1| \leq 1, \left|a - \frac{b^2}{4}\right| \leq 1.$$

The region required is therefore as shown in the diagram where the arc from A to B is part of the parabola $a - b^2/4 = -1$.

The area of the required region can be obtained in several ways. The parallelogram $CEDF$ evidently has area 4. We must subtract the area of the piece AFB. Since $\overleftrightarrow{AF}$ and $\overleftrightarrow{FB}$ are tangents to the parabola, the area of AFB is $\frac{1}{3}$ that of the triangle AFB (Archimedes' rule). But triangle AFB has base AF of length 1 and altitude 1, so its area is $\frac{1}{2}$ and the curvilinear piece AFB has area $\frac{1}{6}$. Hence the area of $CEDBA$ is $4 - \frac{1}{6} = \frac{23}{6}$.

Second Solution. Let S_t be the strip in the (a, b)-plane bounded by the two parallel lines

$$a + bt + t^2 = 1 \tag{1}$$

$$a + bt + t^2 = -1. \tag{2}$$

We seek the area of the region Π defined by $\Pi = \bigcap_{0 \leq t \leq 1} S_t$. Now (1) and (2) are parallel tangents to the parabolas $b^2 = 4a - 4$ and $b^2 = 4a + 4$, and for the range $0 \leq t \leq 1$ these are tangents to the arcs $\widehat{AB}$ and $\widehat{A_1B_1}$ where $A(-1, 0)$, $B(0, -2)$, $A_1(1, 0)$, $B_1(2, -2)$ are points on the two parabolas.

Π is therefore the region bounded by the lines $a = 1$, $a = -1$, $a + b = 0$, $a + b = -2$ and the parabolic arc $\widehat{AB}$ of $b^2 = 4a + 4$. We now find the area as before.

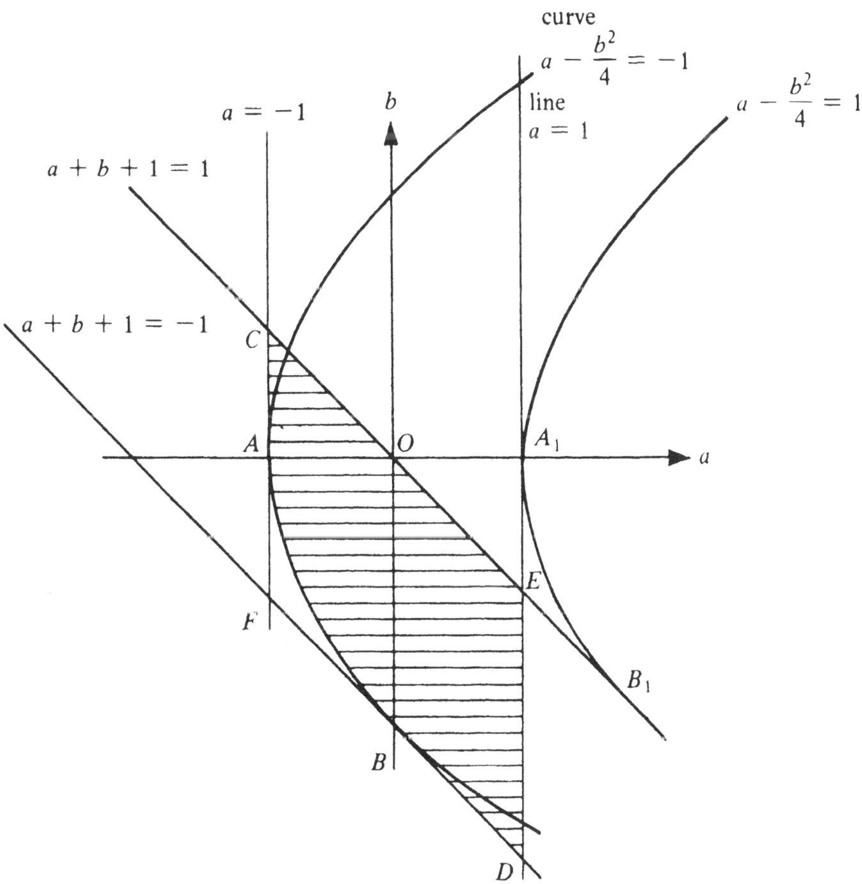

6. (i) Let V_1, V_2, V_3, and V denote four vertices of a cube. V_1, V_2, and V_3 are next neighbors of V, that is, the lines VV_1, VV_2, and VV_3 are edges of the cube. Project the cube orthogonally onto a plane (the z plane, the Gaussian plane) of which the points are marked with complex numbers. Let the projection of V fall in the origin and the projections of V_1, V_2, and V_3 in points marked with the complex numbers z_1, z_2, and z_3, respectively. Show that $z_1^2 + z_2^2 + z_3^2 = 0$.

Solution. Take Cartesian coordinates w, x, y in space so that the x-axis is the real axis and the y-axis is the imaginary axis in the given Gaussian plane. Then the projection maps each point (w, x, y) onto $(0, x, y)$.

Suppose (w_1, x_1, y_1), (w_2, x_2, y_2) and (w_3, x_3, y_3) are mutually orthogonal unit vectors in space. Then the matrix

$$\begin{pmatrix} w_1 & x_1 & y_1 \\ w_2 & x_2 & y_2 \\ w_3 & x_3 & y_3 \end{pmatrix}$$

is an orthogonal matrix, so its columns are also mutually orthogonal unit vectors, in particular

$$x_1^2 + x_2^2 + x_3^2 = 1$$
$$y_1^2 + y_2^2 + y_3^2 = 1$$
$$x_1y_1 + x_2y_2 + x_3y_3 = 0.$$

The orthogonal projections of the unit vectors into the Gaussian plane are the complex numbers $x_1 + iy_1$, $x_2 + iy_2$, and $x_3 + iy_3$, and

$$(x_1 + iy_1)^2 + (x_2 + iy_2)^2 + (x_3 + iy_3)^2 = x_1^2 + x_2^2 + x_3^2 - y_1^2$$
$$- y_2^2 - y_3^2 + 2i(x_1y_1 + x_2y_2 + x_3y_3) = 0.$$

Suppose the sides of the given cube are of length a. Since it is given that the vertex V projects onto the origin, it must be of the form $V = (b, 0, 0)$. There are mutually orthogonal unit vectors (w_j, x_j, y_j) such that

$$V_j = (b, 0, 0) + a(w_j, x_j, y_j) \quad \text{for } j = 1, 2, 3.$$

Then the projection of V_j into the Gaussian plane is

$$z_j = a(x_j + iy_j)$$

and

$$z_1^2 + z_2^2 + z_3^2 = a^2[(x_1 + iy_1)^2 + (x_2 + iy_2)^2 + (x_3 + iy_3)^2] = 0.$$

---

6. (ii) Let a_{ij} be a determinant in which each diagonal element exceeds in absolute value the sum of the absolute values of the other elements of its row, that is

$$|a_{ii}| > |a_{i1}| + |a_{i2}| + \cdots + |a_{i,i-1}| + |a_{i,i+1}| + \cdots + |a_{in}|.$$

Show that the determinant is not equal to zero. (Consider the corresponding system of linear homogeneous equations.)

Solution. The corresponding system of linear equations is

$$\sum_{j=1}^{n} a_{ij}x_j = 0, \qquad i = 1, 2, \ldots, n.$$

If the determinant of the matrix of the coefficients is zero, there exists a non-trivial solution, say $(\bar{x}_1, \bar{x}_2, \ldots, \bar{x}_n)$, of the system. Let m be an index for which $|\bar{x}_m|$ is largest, that is $|\bar{x}_m| \geq |\bar{x}_j|$ for $j = 1, 2, \ldots, n$. Clearly, $|\bar{x}_m| \neq 0$, since the solution is non-trivial. Consider the mth equation in the above system written in the form

$$-a_{mm}\bar{x}_m = \sum_{j \neq m} a_{mj}\bar{x}_j .$$

We have

$$|a_{mm}||\bar{x}_m| \leq \sum_{j \neq m} |a_{mj}||\bar{x}_j| \leq \left(\sum_{j \neq m} |a_{mj}|\right) |\bar{x}_m|$$

and therefore

$$|a_{mm}| \leq \sum_{j \neq m} |a_{mj}|$$

contrary to hypothesis. Hence the determinant cannot be zero.

REMARK. This is the same type of argument that was used by Gerŝgorin to obtain bounds on the eigenvalues of a matrix. See Marcus and Minc, *A Survey of Matrix Theory and Matrix Inequalities*. Allyn and Bacon, Boston, 1964, page 146.

THE NINTH WILLIAM LOWELL PUTNAM MATHEMATICAL COMPETITION

March 26, 1949

Morning Session

1. (i) Three straight lines pass through the three points $(0, -a, a)$, $(a, 0, -a)$, and $(-a, a, 0)$, parallel to the x-axis, y-axis, and z-axis, respectively; $a > 0$. A variable straight line moves so that it has one point in common with each of the three given straight lines. Find the equation of the surface described by the variable line.

Solution. Let L_1, L_2, L_3 be, respectively, the lines parallel to the x-axis through $(0, -a, a)$, parallel to the y-axis through $(a, 0, -a)$, and parallel to the z-axis through $(-a, a, 0)$. Let $\mathcal{S}$ be the required locus.

Let $P = (p, -a, a)$, $Q = (a, q, -a)$, $R = (-a, a, r)$ be three collinear points on L_1, L_2, L_3, respectively, and let $X = (x, y, z)$ be any point on the same line. Then the vectors PX, QX, and RX are proportional, that is, the matrix

$$(1) \qquad \begin{pmatrix} x - p & y + a & z - a \\ x - a & y - q & z + a \\ x + a & y - a & z - r \end{pmatrix}$$

has rank one. Thus, in particular

$$(2) \qquad \begin{aligned} (x - p)(y - a) &= (x + a)(y + a) \\ (x - p)(z + a) &= (x - a)(z - a). \end{aligned}$$

Therefore

$$(3) \qquad (x + a)(y + a)(z + a) = (x - p)(y - a)(z + a) = (x - a)(y - a)(z - a),$$

so

$$(4) \qquad (x + a)(y + a)(z + a) = (x - a)(y - a)(z - a),$$

which is equivalent to

$$xy + yz + zx + a^2 = 0. \tag{5}$$

This equation, then, is satisfied by every point (x, y, z) of $\mathcal{S}$.

To complete the discussion we must decide whether every point of the surface $\mathfrak{I}$ defined by (5), or equivalently by (4), is a point of the locus $\mathcal{S}$.

Let M_1, M_2, M_3, respectively, be the lines through $(0, a, -a)$ parallel to the x-axis, through $(-a, 0, a)$ parallel to the y-axis, and through $(a, -a, 0)$ parallel to the z-axis. From (5) it is clear that M_1, M_2, and M_3 all lie on the surface $\mathfrak{I}$. We shall prove that $\mathcal{S}$ is $\mathfrak{I}$ less M_1, M_2, and M_3.

Suppose Y is a point of M_1. Then there is no line through Y that meets L_1, L_2, and L_3. If such a line existed it would lie in the plane π_2 of Y and L_2 and in the plane π_3 of Y and L_3. These planes are different, since L_2 and L_3 are not coplanar, and they are not parallel since $Y \in \pi_2 \cap \pi_3$. Therefore $\pi_2 \cap \pi_3$ is a line, and this line happens to be M_1, which does not meet L_1. Hence $Y \notin \mathcal{S}$. Similarly, no point of M_2 or M_3 lies in $\mathcal{S}$. Hence $\mathcal{S} \subseteq \mathfrak{I} - (M_1 \cup M_2 \cup M_3)$.

We now show that M_1 is the only line parallel to L_1 that meets both L_2 and L_3. For such a line must be the intersection of the plane through L_2 parallel to L_1 and the plane through L_3 parallel to L_1. Similarly, M_2 is the only line parallel to L_2 that meets both L_1 and L_3, and M_3 is the only line parallel to L_3 that meets both L_1 and L_2.

Let Z be a point of L_1, but not on M_2 or M_3. Let N be the line of intersection of the planes determined by Z and L_2 and by Z and L_3. Since N is coplanar with L_2, it either meets L_2 or is parallel to L_2. Similarly, N either meets L_3 or is parallel to L_3. But N is not parallel to both L_2 and L_3 since these lines are skew. Hence either (1) N meets L_1 and L_2 and is parallel to L_3, or (2) N meets L_1 and L_3 and is parallel to L_2, or (3) N meets all three lines L_1, L_2 and L_3. As shown in the preceding paragraph possibilities (1) and (2) lead to the conclusions $N = M_3$ and $N = M_2$, respectively, and these are impossible since $Z \in N$ and $Z \notin M_3$, $Z \notin M_2$. So N meets all three lines L_1, L_2, and L_3; therefore $Z \in \mathcal{S}$. Similarly points of L_2 and L_3 not lying on M_1, M_2, or M_3 are in $\mathcal{S}$. This proves that $\mathfrak{I} - (M_1 \cup M_2 \cup M_3) \subseteq \mathcal{S}$. Combining the two inclusions, we have $\mathcal{S} = \mathfrak{I} - (M_1 \cup M_2 \cup M_3)$.

These arguments are most easily understood in the context of projective geometry. We have the following general results.

Given three mutually skew lines in projective 3-space, there is a unique quadric surface $\mathcal{Q}$ containing them. The rulings of $\mathcal{Q}$ (i.e., the lines contained in $\mathcal{Q}$) fall into two disjoint families $\mathcal{L}$ and $\mathfrak{M}$ such that (1) each member of $\mathcal{L}$ meets each member of $\mathfrak{M}$, and (2) through each point of $\mathcal{Q}$ there passes a unique member of $\mathcal{L}$ and a unique member of $\mathfrak{M}$.

In the present case, $\mathfrak{I}$ is the quadric surface $\mathcal{Q}$ determined by the skew lines L_1, L_2, and L_3, except for the points at infinity. Since these lines are mutually skew, they are in a single family, say $\mathfrak{L}$. Let p_1, p_2, and p_3 be the points at infinity on L_1, L_2, and L_3, respectively. Then M_1, M_2, and M_3 are the other rulings of $\mathcal{Q}$ through p_1, p_2, and p_3, respectively. These lines must be excluded from the locus $\mathcal{S}$ because they fail to intersect one of the L's at a finite point. Through any other point q of $\mathfrak{J}$, there is a ruling in the $\mathfrak{M}$-family and it meets the L's at finite points, so $q \in \mathcal{S}$.

---

1. (ii) Which planes cut the surface $xy + xz + yz = 0$ in (1) circles, (2) parabolas?

Solution. The given surface is a quadric cone containing the three coordinate axes. That it is a right circular cone can be seen as follows: The curve C of intersection of the given cone with the plane $x + y + z = 1$ is a circle, since

$$x^2 + y^2 + z^2 = (x + y + z)^2 - 2(xy + xz + yz) = 1 - 0 = 1$$

on C, and hence C is the intersection of the unit sphere and the plane $x + y + z = 1$. Thus the given surface is a right circular cone and its axis is the straight line $x = y = z$.

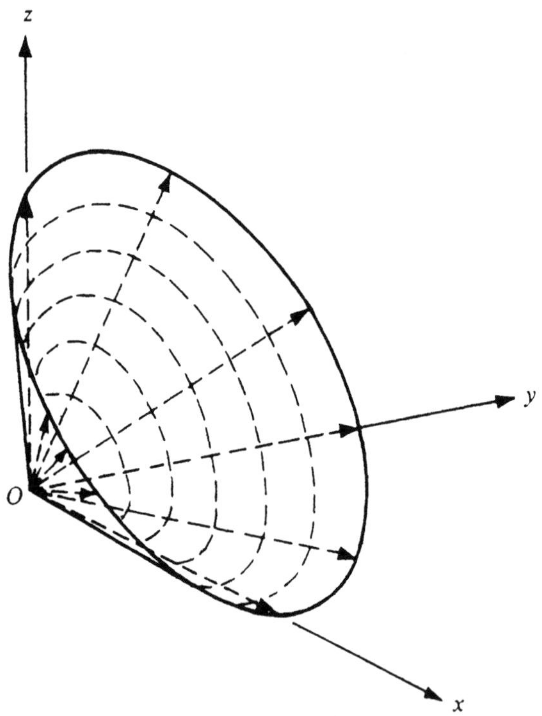

Now a plane cuts the cone in a circle if and only if the plane is perpendicular to the axis $x = y = z$ and does not pass through the origin. These planes have equations of the form $x + y + z = p, p \neq 0$.

A plane cuts the cone in a parabola if and only if it is parallel to, but does not contain, a generator, i.e., parallel, but not equal, to some plane tangent to the cone.

The plane tangent to the cone at (x_0, y_0, z_0) (not the origin) has the equation

$$(z_0 + y_0)x + (x_0 + z_0)y + (x_0 + y_0)z = 0.$$

Suppose the plane

$$ax + by + cz = d \tag{1}$$

cuts the cone in a parabola. Then $d \neq 0$ and $(a, b, c) \neq (0, 0, 0)$. Furthermore, there exists a point $(x_0, y_0, z_0) \neq (0, 0, 0)$ of the cone such that

$$(a, b, c) = \lambda(y_0 + z_0, x_0 + z_0, x_0 + y_0). \tag{2}$$

The three equations in (2) can be solved for x_0, y_0, z_0:

$$2\lambda(x_0, y_0, z_0) = (-a + b + c, a - b + c, a + b - c). \tag{3}$$

Then since (x_0, y_0, z_0) lies on the cone, we have

$$\begin{aligned}(-a + b + c)(a - b + c) + (-a + b + c)(a + b - c)\\ + (a - b + c)(a + b - c) = 4\lambda^2(x_0y_0 + y_0z_0 + z_0x_0) = 0.\end{aligned} \tag{4}$$

Simplifying this we see that a, b, c must satisfy

$$a^2 + b^2 + c^2 - 2ab - 2ac - 2bc = 0. \tag{5}$$

Conversely, suppose a, b, c are any three numbers not all zero satisfying (5), and $d \neq 0$. Take $\lambda = \frac{1}{2}$ and determine numbers x_0, y_0, z_0 by (3). They are not all zero, and (x_0, y_0, z_0) lies on the given cone by virtue of (4) and (5). Hence the plane (1) is parallel, but not equal, to the tangent plane at (x_0, y_0, z_0), so its intersection with the cone is a parabola.

Thus we have shown that the plane (1) cuts the cone in a parabola if and only if $(a, b, c) \neq (0, 0, 0)$, $d \neq 0$, and (5) holds.

---

2. We consider three vectors drawn from the same initial point O, of lengths a, b, and c, respectively. Let E be the parallelepiped with vertex O of

which the given vectors are the edges and H the parallelepiped with vertex O of which these vectors are the altitudes. Show that the product of the volumes of E and H equals $(abc)^2$, and generalize the theorem, with proof, to n dimensions.

Solution. We proceed at once to the general case. Let $\mathbf{v}_1, \mathbf{v}_2, \ldots, \mathbf{v}_n$ be vectors in an n-dimensional space. To say that these vectors span a parallelepiped P means that they are linearly independent and that

$$P = \{\Sigma \lambda_i \mathbf{v}_i : 0 \leq \lambda_i \leq 1\}.$$

The volume of the parallelepiped P is $|\det A|$ where A is the $n \times n$ matrix whose rows are the components of $\mathbf{v}_1, \mathbf{v}_2, \ldots, \mathbf{v}_n$ in some Cartesian coordinate system.

Since the $\mathbf{v}$'s are linearly independent there are linear functionals $f_1, \ldots, f_n$ such that

$$f_j(\mathbf{v}_i) = \delta_{ij} \|\mathbf{v}_i\|^2$$

for all i, j, where δ_{ij} is the Kronecker delta ($\delta_{ij} = 0$ if $i \neq j$, $\delta_{ii} = 1$). Since every linear functional is realizable with an inner product there are vectors $\mathbf{w}_1, \ldots, \mathbf{w}_n$ such that

$$(1) \qquad (\mathbf{v}_i, \mathbf{w}_j) = \delta_{ij} \|\mathbf{v}_i\|^2$$

for all i, j. Now the $\mathbf{w}$'s are linearly independent, for if $\Sigma \alpha_j \mathbf{w}_j = 0$, then

$$(\mathbf{v}_i, \Sigma \alpha_j \mathbf{w}_j) = \alpha_i \|\mathbf{v}_i\|^2 = 0, \qquad i = 1, 2, \ldots, n,$$

whence $\alpha_1 = \alpha_2 = \cdots = \alpha_n = 0$.

Therefore the vectors $\mathbf{w}_1, \mathbf{w}_2, \ldots, \mathbf{w}_n$ span a parallelepiped Q. The vector $\mathbf{v}_i$ is the altitude of Q on the face spanned by all the $\mathbf{w}$'s except $\mathbf{w}_i$, since it is perpendicular to that face, and the projection of $\mathbf{w}_i$ in the direction of $\mathbf{v}_i$ has length $\|\mathbf{v}_i\|$ by (1).

Let B be the $n \times n$ matrix whose rows are $\mathbf{w}_1, \mathbf{w}_2, \ldots, \mathbf{w}_n$. Then $\operatorname{vol} Q = |\det B| = |\det B^T|$, where B^T is the transpose of B.

Now equation (1) shows that AB^T is the diagonal matrix

$$\operatorname{diag}(\|\mathbf{v}_1\|^2, \|\mathbf{v}_2\|^2, \ldots, \|\mathbf{v}_n\|^2).$$

Hence

$$(\operatorname{vol} P)(\operatorname{vol} Q) = |\det A| \cdot |\det B^T| = |\det AB^T|$$
$$= \|\mathbf{v}_1\|^2 \cdot \|\mathbf{v}_2\|^2 \cdot \cdots \cdot \|\mathbf{v}_n\|^2.$$

In the three-dimensional case, the problem calls the parallelepipeds E and H instead of P and Q, and, since $\|\mathbf{v}_1\| = a$, $\|\mathbf{v}_2\| = b$, and $\|\mathbf{v}_3\| = c$, the formula above is the required result for the three-dimensional case.

---

3. Assume that the complex numbers $a_1, a_2, \ldots, a_n, \ldots$ are all different from zero, and that $|a_r - a_s| > 1$ for $r \neq s$. Show that the series

$$\sum_{n=1}^{\infty} \frac{1}{a_n{}^3}$$

converges.

Solution. Let $S_k = \{n : k < |a_n| \leq k + 1\}$ for $k = 0, 1, 2, \ldots$. The discs $|z - a_n| \leq \frac{1}{2}$ are all disjoint by hypothesis, and for $n \in S_k$ these discs all lie in the annulus

$$\left\{z : k - \frac{1}{2} \leq |z| \leq k + \frac{3}{2}\right\},$$

(a disc if $k = 0$). Let the cardinality of the set S_k be denoted by $|S_k|$. Then adding areas gives

$$|S_k| \frac{\pi}{4} \leq \pi\left[\left(k + \frac{3}{2}\right)^2 - \left(k - \frac{1}{2}\right)^2\right] = 2\pi(2k + 1)$$

so that $|S_k| \leq 8(2k + 1)$ for $k > 0$. A separate calculation shows that $|S_0| \leq 9$.

Then

$$\sum_{n \in S_k} \frac{1}{|a_n|^3} \leq \frac{|S_k|}{k^3} \leq \frac{8(2k + 1)}{k^3} \leq \frac{24}{k^2}$$

for $k \geq 1$ because $2k + 1 \leq 3k$. Since S_0 is finite,

$$\sum_{n \in S_0} \frac{1}{|a_n|^3}$$

is finite.

Hence we have

$$\sum_{n=1}^{\infty} \frac{1}{|a_n|^3} = \sum_{k=0}^{\infty} \sum_{n \in S_k} \frac{1}{|a_n|^3} \leq \sum_{n \in S_0} \frac{1}{|a_n|^3} + \sum_{k=1}^{\infty} \frac{24}{k^2} < \infty.$$

The rearrangement of the sum in the first step is permissible since the terms are all positive. Thus the original series converges absolutely.

---

4. Given that P is a point inside a tetrahedron with vertices at A, B, C, and D, such that the sum of the distances $PA + PB + PC + PD$ is a minimum, show that the two angles $\angle APB$ and $\angle CPD$ are equal and are bisected by the same straight line. What other pairs of angles must be equal?

First Solution. Consider the ellipsoid of revolution $\mathcal{E}_1$, with foci A and B, which passes through the given point P. $\mathcal{E}_1$ is the locus of all points X such that $|AX| + |XB| = |AP| + |PB|$ and points Y interior to $\mathcal{E}_1$ satisfy $|AY| + |YB| < |AP| + |PB|$. Let $\mathcal{E}_2$ be the ellipsoid of revolution, with foci C and D, which passes through P. Since P minimizes the sum of the distances $|PA| + |PB| + |PC| + |PD|$, the ellipsoids $\mathcal{E}_1$ and $\mathcal{E}_2$ can have no interior point in common. Thus they are tangent and have a common normal line l at P, which intersects the segments AB and CD. By the reflection property of the ellipse, l bisects the angles APB and CPD.

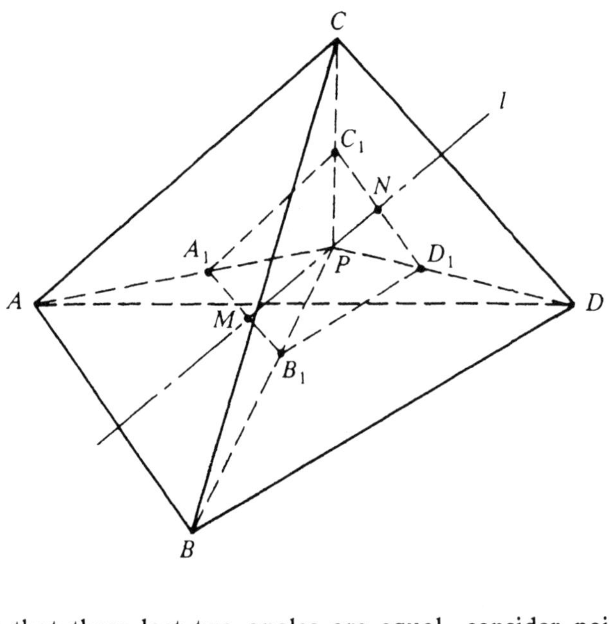

To show that these last two angles are equal, consider points A_1, B_1, C_1, and D_1 on the rays PA, PB, PC, and PD, respectively, such that $|PA_1| = |PB_1| = |PC_1| = |PD_1| = 1$, for example. Line l meets segment A_1B_1 at its midpoint M and meets segment C_1D_1 at its midpoint N. Thus P is on the line joining the midpoints of two opposite edges of the new tetrahedron $A_1B_1C_1D_1$. A similar argument shows that P is on the line joining the midpoints of A_1C_1 and B_1D_1, say Q and R. But M, N, R, Q are the vertices of a parallelogram, and P is the intersection of its diagonals. Hence P bisects MN, and it follows that triangles A_1PB_1 and C_1PD_1 are congruent. Hence $\angle APB = \angle CPD$. Likewise, it can be shown that $\angle APC = \angle BPD$ and $\angle APD = \angle BPC$.

Second Solution. In Euclidean 3-space (or even in Euclidean n-space), the function $f_A(X) = |AX|$, the distance between the fixed point A and the variable point X, is differentiable at all points except A. The gradient $(\nabla f_A)(X)$ is the unit vector in the direction from A to X. [This is geometrically obvious, since at any point the distance from A to that point increases most rapidly in the direction away from A. It is a unit vector, since the distance from A increases at the same rate as the distance from X.]

Choose a coordinate system with P at the origin. We are given that the function

$$g = f_A + f_B + f_C + f_D$$

has its minimum at P and that P is none of the points $A, B, C,$ or D. Since g is differentiable at P, we have

$$(\nabla g)(P) = 0,$$

that is,

$$(\nabla f_A)(P) + (\nabla f_B)(P) + (\nabla f_C)(P) + (\nabla f_D)(P) = 0. \tag{1}$$

We have already noted that $(\nabla f_A)(P)$ is the unit vector from P to A. Hence, with a sign change, (1) becomes

$$\mathbf{a} + \mathbf{b} + \mathbf{c} + \mathbf{d} = 0, \tag{2}$$

where $\mathbf{a}$, $\mathbf{b}$, $\mathbf{c}$, and $\mathbf{d}$ are the unit vectors from P to $A, B, C,$ and D, respectively. Since P is inside the given tetrahedron, no two of these unit vectors are collinear. The sum of two non-collinear unit vectors has the direction of the bisector of the angle formed by them. Hence, when (2) is rewritten as

$$\mathbf{a} + \mathbf{b} = -(\mathbf{c} + \mathbf{d}), \tag{3}$$

we see that the bisector of $\angle APB$ is opposite to that of $\angle CPD$; that is, these angles are bisected by the same straight line. It also follows from (3) that

$$(\mathbf{a} + \mathbf{b}, \mathbf{a} + \mathbf{b}) = (\mathbf{c} + \mathbf{d}, \mathbf{c} + \mathbf{d}),$$

which reduces (because **a**, **b**, **c**, and **d** are unit vectors) to

$$(\mathbf{a}, \mathbf{b}) = (\mathbf{c}, \mathbf{d}),$$

which tells us that the angle between **a** and **b** is the same as the angle between **c** and **d**, i.e.,

$$\angle APB = \angle CPD.$$

Similarly, $\angle APC = \angle BPD$ and $\angle APD = \angle BPC$.

---

5. How many roots of the equation $z^6 + 6z + 10 = 0$ lie in each quadrant of the complex plane?

First Solution. Let $P(z) = z^6 + 6z + 10$. The minimum value of $P(z)$ for real z is $P(-1) = 5$. Hence the equation has no real roots. There can be no purely imaginary roots since $\operatorname{Im}(P(iy)) = 6y \neq 0$ unless $y = 0$, and $P(0) \neq 0$.

The roots sum to zero so they do not all lie in the right half-plane or all in the left half-plane. Since P has real coefficients, the roots occur in conjugate pairs, so the number in the first quadrant is the same as the number in the fourth, and the number in the second is the same as the number in the third. There are, therefore, just two possibilities:

(1) There is one root in each of the first and fourth quadrants, and there are two roots in each of the second and third; or

(2) there are two roots in each of the first and fourth quadrants and there is one root in each of the second and third.

We can decide between these two possibilities through the argument principle of complex variable theory. This states:

If a function f is analytic at each point of a simple closed curve and its interior and does not vanish at any point of the curve, then the number of zeros of f in the interior of the curve (counting multiplicities) is $1/2\pi$ times the variation of $\arg f(z)$ as z describes the curve once in the positive direction.

In this problem take $f = P$ and the simple closed curve formed by the real axis from O to R, the arc of the circle $|z| = R$ from R to iR, and the imaginary axis from iR to O, where R is a large positive number.

The argument variation of P along $[O, R]$ is zero since P remains real and positive. If R is large, then the argument variation of $P(z)$ along the circular arc is approximately the same as that of z^6 along that same arc, since $|(P(z)/z^6) - 1|$ is small for all z on this arc. Therefore the argument variation of $P(z)$ along the circular arc is about $6 \cdot \pi/2 = 3\pi$. Finally, as z goes from iR to O along the imaginary axis, $P(z)$ goes back to the value

10, keeping always in the upper half-plane; hence the argument variation along this part of the path is about $-\pi$.

z-plane:
Simple closed curve

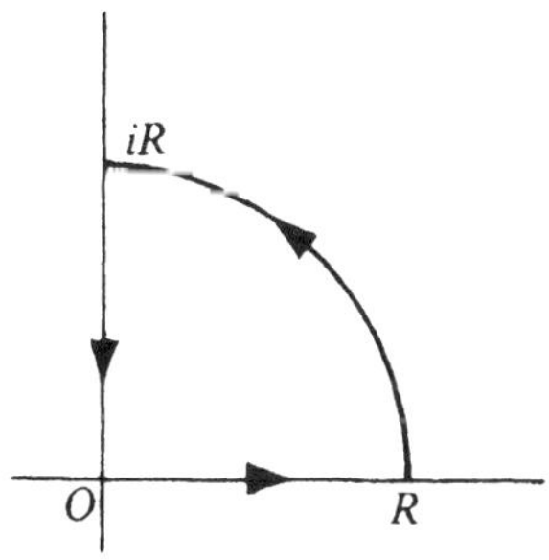

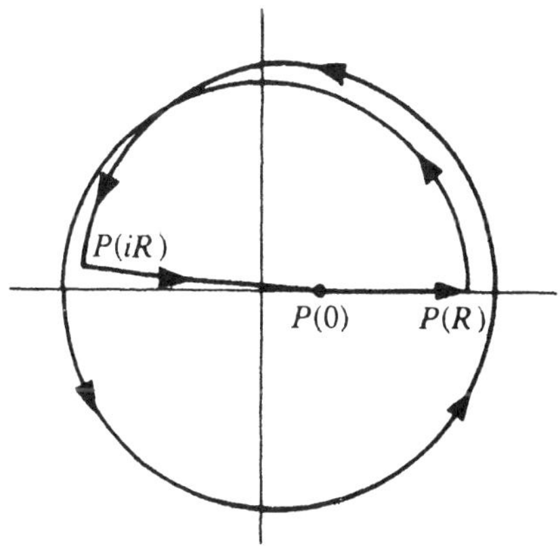

But the total argument variation, seen to be about $(3\pi - \pi) = 2\pi$, must be an integral multiple of 2π, so it must be exactly 2π (assuming that R is large enough). Hence the total number of zeros enclosed by the path is one for any large R. Therefore the number of zeros in the first quadrant is one. Hence the distribution of zeros by quadrants is given by (1).

Second Solution. Another way to decide between possibilities (1) and (2) is to study carefully a root in the first quadrant.

Suppose $z_1 = re^{i\theta}$, $0 < \theta < \pi/2$, is a root of $z^6 + 6z + 10 = 0$ in the first quadrant. Then $6z_1 + 10$ also lies in the first quadrant, and $\arg(6z_1 + 10) < \arg z_1 = \theta$. Since $z_1{}^6 = r^6e^{6i\theta} = -(6z_1 + 10)$, $z_1{}^6$ lies in the third quadrant. Since also $0 < 6\theta < 3\pi$, we see that $\pi < 6\theta < \frac{3}{2}\pi$, and hence $6\theta - \pi$ is that argument of $-z_1{}^6$ which lies between 0 and $\pi/2$. Therefore $6\theta - \pi = \arg(6z_1 + 10) < \theta$, giving $\theta < \pi/5$. Thus any root in the first quadrant has argument in $(0, \pi/5)$.

Suppose that z_1 and z_2 are different roots in the first quadrant. Then

$$(3) \qquad 0 = \frac{P(z_1) - P(z_2)}{z_1 - z_2}$$

$$= z_1{}^5 + z_1{}^4z_2 + z_1{}^3z_2{}^2 + z_1{}^2z_2{}^3 + z_1z_2{}^4 + z_2{}^5 + 6.$$

Since both z_1 and z_2 have arguments in $(0, \pi/5)$ every term in the right member of (3) has argument in $[0, \pi)$, so the sum of these terms cannot be zero, a contradiction. Therefore there cannot be two distinct roots of the equation in the first quadrant, and we conclude that (1) holds.

Third Solution. The zeros of $z^6 + 6z + t$ are continuous functions of the real parameter t. For no positive t does this polynomial have a purely imaginary root. Hence for all positive t the number N of roots in the right half-plane remains fixed. As we saw above, N is even for large positive t, so N is even for all positive t. For $t = 0$, there are three roots in the left half-plane, two in the right half-plane, and one at zero. For small positive t, N must be either two or three by continuity. So $N = 2$ for small, and hence all, positive t. Therefore (1) holds.

---

6. Prove that for every real or complex x

$$\prod_{k=1}^{\infty} \frac{1 + 2\cos\frac{2x}{3^k}}{3} = \frac{\sin x}{x}.$$

Solution. The identity

$$\sin 3\theta - \sin\theta = 2\sin\frac{1}{2}(3\theta - \theta)\cos\frac{1}{2}(3\theta + \theta)$$

leads to the identity

$$\sin 3\theta = \sin\theta\,(1 + 2\cos 2\theta).$$

Hence for any x,

$$\begin{aligned}\sin x &= \sin(x/3)(1 + 2\cos(2x/3)) \\ &= \sin(x/9)(1 + 2\cos(2x/9))(1 + 2\cos(2x/3))\end{aligned}$$

and after n iterations we have

(1) $$\sin x = \sin(x/3^n)\prod_{k=1}^{n}(1 + 2\cos(2x/3^k)).$$

For $x = 0$, the right member of the required relation is not defined. However, if we interpret $(\sin 0)/0$ as 1 ($= \lim_{x\to 0}\sin x/x$), then the required equation is correct since every factor in the infinite product is 1.

If $x \neq 0$, we divide (1) by x and rearrange to get

$$\frac{\sin x}{x} = \frac{\sin(x/3^n)}{x/3^n}\prod_{k=1}^{n}\left(\frac{1 + 2\cos(2x/3^k)}{3}\right).$$

Now

$$\lim_{n\to\infty} \frac{\sin (x/3^n)}{x/3^n} = 1,$$

and it follows that

$$\lim_{n\to\infty} \prod_{k=1}^{n} \frac{1 + 2 \cos (2x/3^k)}{3} = \frac{\sin x}{x}.$$

REMARK. Because of the special behavior of 0 in multiplication, the convergence of infinite products is not usually decided simply by the convergence of the sequence of partial products. Rather, it is required that only finitely many factors be zero and that the partial products of the others have a non-zero limit. (See, for example, Ahlfors, *Complex Analysis*, 2nd ed., McGraw-Hill, New York, 1966, p. 189.) The infinite product above converges in this restricted sense because for any fixed x,

$$1 + 2 \cos \frac{2x}{3^k}$$

does not vanish for large k and the above proof shows that

$$\lim_{n\to\infty} \prod_{k=t+1}^{n} \left(1 + 2 \cos\frac{2x}{3^k}\right) = \frac{\sin (x/3^t)}{x/3^t} \neq 0$$

for large t.

Afternoon Session

1. Each rational number p/q (p, q relatively prime positive integers) of the open interval (0, 1) is covered by a closed interval of length $1/2q^2$, whose center is at p/q. Prove that $\sqrt{2}/2$ is not covered by any of the above closed intervals.

Solution. The problem may be restated as follows:

Show that

$$\left|\frac{\sqrt{2}}{2} - \frac{p}{q}\right| \leq \frac{1}{4q^2} \tag{1}$$

is impossible if p and q are integers, $0 < p < q$.

Hence suppose p and q are integers such that $0 < p < q$. Then

$$\frac{\sqrt{2}}{2} + \frac{p}{q} < 2.$$

Therefore, if (1) holds, we have

$$\left|\frac{1}{2} - \frac{p^2}{q^2}\right| = \left|\frac{\sqrt{2}}{2} - \frac{p}{q}\right|\left|\frac{\sqrt{2}}{2} + \frac{p}{q}\right| < \frac{1}{2q^2},$$

whence $|q^2 - 2p^2| < 1$. But $q^2 - 2p^2$ is an integer, so this implies that $q^2 - 2p^2 = 0$. But this is impossible since $\sqrt{2}$ is irrational. Clearly (1) is impossible for integers p, q such that $p \geq q > 0$.

REMARKS. The hypothesis that p and q be relatively prime is unimportant.

Hurwitz (1891) proved the following theorem concerning the approximation of irrational numbers by rational numbers.

For any irrational number α there are infinitely many pairs of integers p, q such that

$$\left|\alpha - \frac{p}{q}\right| < \frac{1}{\sqrt{5}\,q^2}.$$

On the other hand, the inequality

$$\left|\frac{1 + \sqrt{5}}{2} - \frac{p}{q}\right| < \frac{1}{kq^2}$$

has only finitely many solutions if $k > \sqrt{5}$, so the constant appearing in Hurwitz' theorem is best possible. See Hardy and Wright, *An Introduction to the Theory of Numbers*, Oxford, 1938, p. 163.

---

2. (i) Prove that

$$\sum_{n=2}^{\infty} \frac{\cos(\log \log n)}{\log n}$$

diverges.

Solution. A convergent series cannot have blocks of terms whose sum is arbitrarily large. We shall show that the given series has such blocks.

For a positive integer k consider the set N_k of integers n such that

$$2\pi k - \frac{1}{3}\pi \le \log \log n \le 2\pi k$$

and let

$$T_k = \sum_{N_k} \frac{\cos(\log \log n)}{\log n}. \tag{1}$$

We shall prove that $T_k \to \infty$ as $k \to \infty$. Now $N_k = \{n: \exp\exp(2\pi k - \frac{1}{3}\pi) \le n \le \exp\exp(2\pi k)\}$ and therefore $|N_k|$, the number of elements in N_k, satisfies

$$|N_k| \ge \exp\exp 2\pi k - \exp(\alpha \exp 2\pi k) - 1$$

where $\alpha = \exp(-\frac{1}{3}\pi)$.

Each term in the sum (1) is at least

$$\frac{\cos\left(-\frac{1}{3}\pi\right)}{\exp 2\pi k} = \frac{1}{2\exp 2\pi k}$$

so

$$T_k \ge \frac{1}{2x_k}(\exp(x_k) - \exp(\alpha x_k) - 1)$$

where $x_k = \exp 2\pi k$.

Now

$$\lim_{x\to\infty} \frac{1}{x}(\exp x - \exp \alpha x) = \lim_{x\to\infty}(\exp x - \alpha \exp \alpha x) = \infty$$

by L'Hospital's rule, using the fact that $\alpha < 1$. Since $x_k \to \infty$ as $k \to \infty$, it follows that $T_k \to \infty$ as $k \to \infty$, and this proves that the given series diverges.

2. (ii) Assume that $p > 0$, $a > 0$, and $ac - b^2 > 0$, and show that

$$\int_{-\infty}^{\infty}\int_{-\infty}^{\infty} \frac{dx\, dy}{(p + ax^2 + 2bxy + cy^2)^2} = \pi p^{-1}(ac - b^2)^{-1/2}.$$

Solution. By a rotation of axes,

$$(1) \qquad \begin{aligned} x &= u\cos\theta + v\sin\theta \\ y &= -u\sin\theta + v\cos\theta, \end{aligned}$$

with proper choice of the parameter θ, the quadratic form $ax^2 + 2bxy + cy^2$ becomes

$$(2) \qquad Au^2 + Cv^2.$$

The discriminant $ac - b^2$ is an invariant under (1), so

$$AC = ac - b^2.$$

Since the original form is positive definite, we also have $A > 0$ and $C > 0$. The transformation is area-preserving, so the required integral in the problem becomes

$$\int_{-\infty}^{\infty}\int_{-\infty}^{\infty} \frac{du\, dv}{(p + Au^2 + Cv^2)^2}.$$

Under the substitutions

$$u = \sqrt{\frac{p}{A}}\, s, \qquad v = \sqrt{\frac{p}{C}}\, t$$

this becomes

$$(3) \qquad \frac{1}{p\sqrt{AC}} \int_{-\infty}^{\infty}\int_{-\infty}^{\infty} \frac{ds\, dt}{(1 + s^2 + t^2)^2}.$$

We need only show that the value of the integral in (3) is π. Passing to polar coordinates

$$\begin{aligned} s &= r\cos\theta \\ t &= r\sin\theta, \end{aligned}$$

we get

$$\int_0^{2\pi}\int_0^{\infty}\frac{r\,dr\,d\theta}{(1+r^2)^2} = 2\pi\int_0^{\infty}\frac{r\,dr}{(1+r^2)^2}$$

$$= 2\pi\left[\frac{-1}{2(1+r^2)}\right]_0^{\infty} = \pi.$$

---

3. Let K be a closed plane curve such that the distance between any two points of K is always less than 1. Show that K lies inside a circle of radius $1/\sqrt{3}$.

First Solution. We shall prove a more general result: If K is a closed bounded set in the plane such that the distance between any two points of K is less than 1, then K lies inside a circle of radius $1/\sqrt{3}$.

This result is trivial if K has less than two points; so we assume that K has at least two points. Then there is a closed circular disk D of smallest radius r containing K. [This can be proved by a compactness argument; the details are given in a lemma later.] It is sufficient to prove that $r < 1/\sqrt{3}$, since K then lies inside the circle of radius $1/\sqrt{3}$ concentric with D.

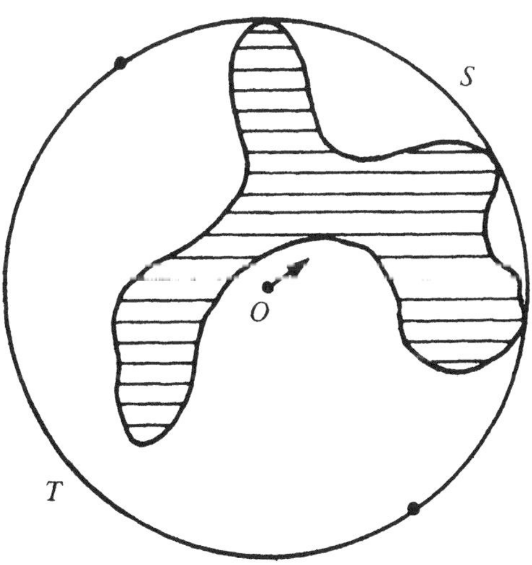

Striped area is K

Let O be the center of D and let C be its bounding circle. We shall show first that $C \cap K$ does not lie on an open semicircle of C. Suppose, on the contrary, that it lies on the open semicircle S. Let T be the complementary closed semicircle. Then K and T are disjoint compact sets, and hence they are at positive distance, say δ, from one another. Now if D is moved $\frac{1}{2}\delta$ in the direction from O toward the mid-point of S, K will lie entirely in the interior of D. But then D can be replaced by a concentric disk of smaller radius that still contains K, which is impossible.

Let P and Q be points of $C \cap K$ that are as far apart as possible. (The existence of such points follows from the compactness of $C \cap K$.) If P and Q are diametrically opposite on C, then $2r = |PQ| < 1$, so

$$r < \frac{1}{2} < \frac{1}{\sqrt{3}},$$

as required. If P and Q are not opposite, then $C \cap K$ does not lie on minor arc PQ, as we have seen above, so we can choose a third point, $R \in K$ on major arc PQ.

Now for any three points P, Q, R on a circle with center O we must have one of the four equations

$$\angle POQ + \angle QOR = \angle POR \tag{1}$$

$$\angle QOP + \angle POR = \angle QOR \tag{2}$$

$$\angle POR + \angle ROQ = \angle POQ \tag{3}$$

$$\angle POQ + \angle QOR + \angle ROP = 2\pi. \tag{4}$$

In the present case, (1) and (2) are eliminated by the choice of P and Q, and (3) is impossible because then R would be on minor arc PQ. Hence (4) holds, and we conclude that $\angle POQ$, the largest of the three angles, is at least $2\pi/3$. Hence we have

$$|PQ| = 2r \sin\left(\frac{1}{2} \angle POQ\right) \geq 2r \sin \pi/3 = r\sqrt{3}.$$

Then since $|PQ| < 1$, we have $r < 1/\sqrt{3}$, as required.

LEMMA. *If K is a bounded set in the plane (closed or not) containing at least two points, then there is a closed circular disk of smallest radius containing K.*

Proof. Since K is bounded, there is some closed circular disk containing K, and since K contains at least two points, say P and Q, all such disks have radius at least $\frac{1}{2}|PQ|$.

Let r be the greatest lower bound of the radii of all such disks. Let D_1, D_2, D_3, ... be a sequence of closed circular disks containing K and with centers at O_1, O_2, O_3, ... respectively, so chosen that $r_n \to r$.

Now $\{O_n\}$ is a bounded sequence in the plane, so by the Bolzano-Weierstrass theorem, it has a convergent subsequence. We may as well assume that $\{O_n\}$ itself converges, say to O.

Let D be the closed disk of radius r about O. We claim $K \subseteq D$. Let Z be any point of K. For all n, Z is a point of D_n, so $|O_nZ| \le r_n$. Hence $|OZ| = \lim |O_nZ| \le \lim r_n = r$. Thus $z \in D$. Hence $K \subseteq D$, as claimed. It is clear that there is no closed disk of smaller radius that contains K. ■

Second Solution. We shall apply Helly's theorem. In the plane, Helly's theorem asserts that if each three members of a family of bounded closed convex sets have a point in common, then there is a point common to all members of the family.

LEMMA. *If, P, Q, R are three points in a plane and $|PQ| \le d$, $|PR| \le d$, $|QR| \le d$, then the closed disks of radius $d/\sqrt{3}$ about P, Q, and R have a point in common.*

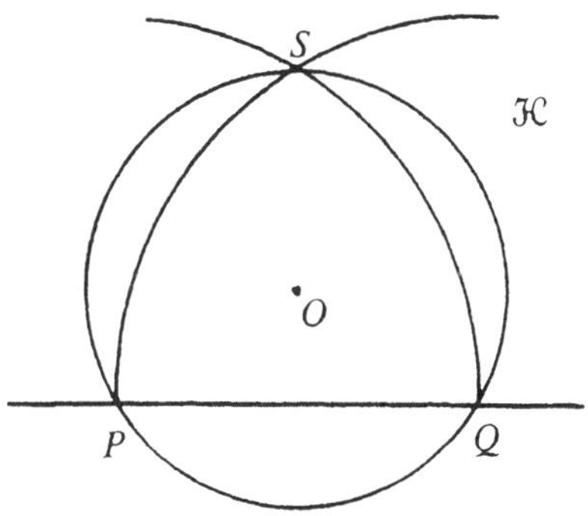

Proof. We may suppose that $|PQ| \ge |PR|, |QR|$. Let $\mathcal{H}$ be the closed half-plane with the edge PQ in which R lies. The circles of radius $|PQ|$ about P and Q intersect in $\mathcal{H}$, say at S. Let O be the center of the equilateral triangle PQS. Now R lies in the closed region bounded by PQ, $\widehat{QS}$, $\widehat{SP}$, so R is on or inside the circumcircle of triangle PQS, which has radius $|PQ|/\sqrt{3}$. Hence, $|OR| \le |OP| = |OQ| = |PQ|/\sqrt{3} \le d/\sqrt{3}$. Therefore, the closed disks of radius $d/\sqrt{3}$ have the point O in common. ■

We return to the problem. Let K be any closed set in the plane, any two points of which are at distance less than 1. We assume K has at least two points to avoid triviality. Let

$$d = \sup\{|PQ| : P, Q \in K\}.$$

Because K is compact, this supremum is attained, so $d < 1$. Consider the family of all closed disks of radius $d/\sqrt{3}$ having center a point of K. According to the lemma, any three of these disks have a point in common, and therefore by Helly's theorem, they all do. If O lies in all these disks, then the closed disk of radius $d/\sqrt{3}$ about O contains K, and K is inside the circle of radius $1/\sqrt{3}$ about O.

REMARKS. The result of the problem is known in the literature as the theorem of Jung.

The hypothesis that K be closed is essential. For suppose K is an equilateral triangle of side 1 with two vertices removed. Then the distance between any two points of K is less than 1, but K does not lie *inside* any circle of radius $1/\sqrt{3}$.

In Euclidean n-dimensional space, Helly's theorem asserts that if each $(n + 1)$ members of a family of bounded closed convex sets have a point in common, then there is a point common to all members of the family. The argument of the second solution extends easily to prove that if K is a closed set in n-space, any two points of which are at distance less than 1, then K is inside a hypersphere of radius

$$\sqrt{\frac{n}{2(n+1)}}.$$

This is the least radius with this property, as can be seen by taking K to be a regular n-simplex of edge $1 - \epsilon$.

For Helly's theorem, Jung's theorem, and a general introduction to the theory of convex sets, see Yaglom and Boltyanskii, *Convex Figures*, Holt, Rinehart & Winston, New York, 1961.

---

4. Show that the coefficients $a_1, a_2, a_3, \ldots$ in the expansion $\frac{1}{4}[1 + x - (1 - 6x + x^2)^{1/2}] = a_1x + a_2x^2 + a_3x^3 + \cdots$ are positive integers.

Solution. Let

$$y = y(x) = \frac{1}{4}[1 + x - (1 - 6x + x^2)^{1/2}].$$

By the general binomial theorem $(1 - 6x + x^2)^{1/2}$, and hence y, can be expanded in a power series convergent for values of x such that $|x^2| + |6x| < 1$. Since $y(0) = 0$, the series has the form

(1) $$y(x) = a_1x + a_2x^2 + a_3x^3 + \cdots.$$

Now

(2) $$2y^2 - (1 + x)y + x = 0,$$

so if we substitute the power series for y in (2) we have

$$2(a_1x + a_2x^2 + a_3x^3 + \cdots)^2 = (1 + x)(a_1x + a_2x^2 + a_3x^3 + \cdots) - x.$$

Comparing coefficients of x, we see that $a_1 = 1$, and for $n > 1$,

$$2(a_1a_{n-1} + a_2a_{n-2} + \cdots + a_{n-1}a_1) = a_n + a_{n-1}.$$

Hence $a_2 = 1$ and

$$a_n = 3a_{n-1} + 2\sum_{i=2}^{n-2} a_ia_{n-i} \quad \text{for } n > 2.$$

Therefore, if $a_1, a_2, \ldots, a_{n-1}$ are positive integers, a_n is also a positive integer. Since a_1 and a_2 are positive integers, all the coefficients $\{a_i\}$ are positive integers.

---

5. Let $a_1, a_2, \ldots, a_n, \ldots$ be an arbitrary sequence of positive numbers. Show that

$$\limsup_{n\to\infty}\left(\frac{a_1 + a_{n+1}}{a_n}\right)^n \geq e.$$

Solution. We shall show that there are infinitely many integers n for which

$$\frac{a_1 + a_{n+1}}{a_n} > 1 + 1/n.$$

Our proof is indirect. Suppose it is false. Then for some integer k and for all $n \geq k$

(1) $$\frac{a_1 + a_{n+1}}{a_n} \leq \frac{n+1}{n}$$

whence

$$\frac{a_n}{n} \geq \frac{a_1}{n+1} + \frac{a_{n+1}}{n+1}.$$

Therefore

$$\frac{a_k}{k} \geq \frac{a_1}{k+1} + \frac{a_{k+1}}{k+1} \geq \frac{a_1}{k+1} + \frac{a_1}{k+2} + \frac{a_{k+2}}{k+2}$$

$$\geq \frac{a_1}{k+1} + \frac{a_1}{k+2} + \frac{a_1}{k+3} + \frac{a_{k+3}}{k+3}.$$

By induction

$$\frac{a_k}{k} \geq a_1\left(\sum_{i=k+1}^{p} \frac{1}{i}\right) + \frac{a_p}{p}$$

for any $p \geq k$. Then

$$\sum_{i=k+1}^{p} \frac{1}{i} \leq \frac{a_k}{ka_1} \quad \text{for } p \geq k.$$

But the harmonic series diverges, so the sums $\Sigma_{i=k+1}^{p} 1/i$ are unbounded. Thus relation (1) cannot hold for all $n \geq k$, and hence there must be infinitely many integers n for which

$$\frac{a_1 + a_{n+1}}{a_n} > 1 + \frac{1}{n}.$$

Then

$$\limsup_{n\to\infty} \left(\frac{a_1 + a_{n+1}}{a_n}\right)^n \geq \lim_{n\to\infty}\left(1 + \frac{1}{n}\right)^n = e.$$

REMARK. The lower bound e cannot be improved because there is a sequence such that

$$\lim \left(\frac{a_1 + a_{n+1}}{a_n}\right)^n = e.$$

Such a sequence is given by

$$a_1 = 1, \qquad a_n = n \log n \qquad \text{for } n \geq 1.$$

For this sequence

$$\frac{a_1 + a_{n+1}}{a_n} = 1 + \frac{b_n}{n}$$

where

$$b_n = \frac{1}{\log n}\left(1 + n \log\left(\frac{n+1}{n}\right) + \log(n+1)\right).$$

Then

$$\left(\frac{a_1 + a_{n+1}}{a_n}\right)^n = \left(1 + \frac{b_n}{n}\right)^n \to e$$

since $b_n \to 1$ as $n \to \infty$.

---

6. Let C be a closed convex curve with a continuously turning tangent and let O be a point inside C. With each point P on C we associate the point $T(P)$ on C which is defined as follows: Draw the tangent to C at P and from O drop the perpendicular to that tangent. $T(P)$ is then the point at which this perpendicular intersects the curve C.

Starting now with a point P_0 on C, we define points P_n by the formula $P_n = T(P_{n-1})$, $n \geq 1$. Prove that the points P_n approach a limit, and characterize those points which can be limits of sequences P_n. (You may consider the facts that T is a continuous transformation and that a convex curve lies on one side of each of its tangents as not requiring proofs.)

Solution. The phrase "with a continuously turning tangent" means that C is a curve of class C^1.

Let $g(P)$ be the distance from O to a point P; this defines a differentiable function on the whole plane except for the point O. Since C does not pass through O, the restriction of g to C is differentiable. It has a critical point at P if and only if the gradient of g at P is normal to C at P, that is the line $\overleftrightarrow{OP}$ is perpendicular to the tangent to C at P. This is precisely the condition that $T(P) = P$.

Since T is a continuous map of C into itself, the set F of fixed points of T is a closed set, and $C - F$, an open set relative to C, falls into components each of which is an open arc bounded by two members of F. The function g is strictly monotonic on any such arc since it has no critical points there.

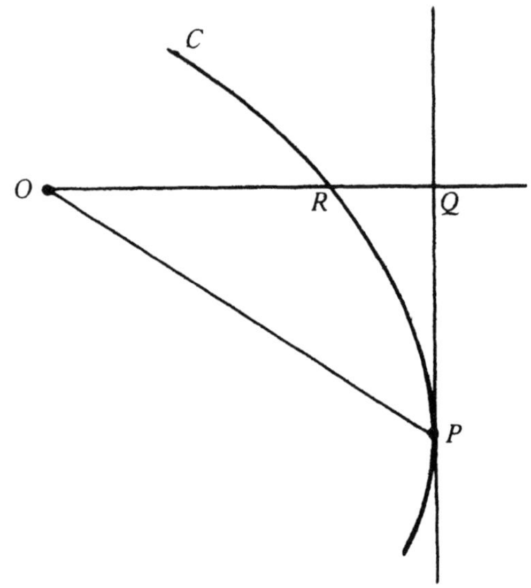

Suppose P is not a fixed point of T. Consider the diagram. $\overleftrightarrow{PQ}$ is tangent to the curve C at P, and $\overleftrightarrow{OQ}$ is perpendicular to $\overleftrightarrow{PQ}$. Since $T(P) \neq P$, $P \neq Q$, so $|OP| > |OQ|$.

Since the origin O and all of C lie on the same side of the tangent $\overleftrightarrow{PQ}$, and $R = T(P)$ is on the ray $\overrightarrow{OQ}$, we have $|OQ| \geq |OR|$. Hence

$$g(T(P)) < g(P).$$

Moreover, there can be no fixed point S of T on the part of C lying in the interior of $\angle POQ$. For such a point S would lie in the closed triangular region POQ, and the perpendicular l to OS at S would contain a point of the interior of segment OP, and this point would lie in the interior of C. But if $S = T(S)$, then l would be tangent to C at S and could not contain a point interior to C. Hence we conclude that the open arc $\widehat{PR}$ of C lies in C–F and therefore either $R = T(P)$ is in the same component of C–F as P or $T(P)$ is one endpoint of that component and $T(P) \in F$. [Note that C might contain the whole segment PQ, in which case $T(P) = Q$ and $T(Q) = Q$.]

Now suppose P_0 is given and $P_n = T(P_{n-1})$ for $n \geq 1$. If $P_0 \in F$, then clearly $P_0 = P_1 = P_2 = \cdots$ and the sequence converges to P_0. If $P_0 \notin F$, let A and B be the endpoints of the component D of C–F containing P_0, and choose the notation so that

$$g(A) < g(P_0) < g(B).$$

(This is possible since g is strictly monotonic on D.) We have seen that

$g(P_1) < g(P_0)$ and either $P_1 \in D$ or P_1 is an endpoint of D, in which case $P_1 = A$. Repeating this argument, it follows that either the points $P_0, P_1, P_2, \ldots$ are all in D or eventually $P_k = A$ for some k, and then $P_n = A$ for all $n \geq k$. In the latter case, we clearly have $P_n \to A$. In the former case, $P_0, P_1, P_2, \ldots$ is a monotonic sequence in D since $g(P_0) > g(P_1) > g(P_2) > \cdots$, so it converges to some point X in $\overline{D}$.

Now $T(X) = T(\lim_{n\to\infty} P_n) = \lim_{n\to\infty} T(P_n) = \lim_{n\to\infty} P_{n+1} = X$. Thus X is a fixed point of T and $X \in \overline{D}$. So $X = A$ or $X = B$. But $g(X) = \lim g(P_n) \leq g(P_0) < g(B)$, so $X = A$. Thus we have shown that, in any case, $P_n \to A$, i.e., if $P_0 \in C-F$, then P_n converges to that endpoint of the component of $C - F$ containing P_0 which is closer to O. And $\lim P_n$ is a fixed point of T.

It is clear that any fixed point Y of T is the limit of such a sequence: Take $P_0 = Y$.

THE TENTH WILLIAM LOWELL PUTNAM MATHEMATICAL COMPETITION

March 25, 1950

Morning Session

1. For what values of the ratio a/b is the limaçon $r = a - b \cos \theta$ a convex curve? $(a > b > 0)$

Solution. The graph of $r = f(\theta)$ in polar coordinates is a simple closed curve surrounding the origin if f is periodic with period 2π and everywhere positive. This is the case in the present problem, since by hypothesis $a > b > 0$. Such a curve is nonsingularly parametrized by θ if $r^2 + (r')^2 > 0$, again true in the present problem. The curvature is given by

$$\frac{r^2 + 2(r')^2 - rr''}{[r^2 + (r')^2]^{3/2}} = \kappa$$

(whenever f is of class C^2).

The curve is convex if and only if the curvature is everywhere non-negative, i.e., if and only if

$$r^2 + 2(r')^2 - rr'' \geq 0.$$

For $r = a - b \cos \theta$, we have $r' = b \sin \theta$, $r'' = b \cos \theta$ and

$$r^2 + 2(r')^2 - rr'' = a^2 + 2b^2 - 3ab \cos \theta.$$

This last expression is always non-negative if and only if

$$a^2 + 2b^2 - 3ab \geq 0.$$

(Since a and b are positive, the least value occurs for $\theta = 0$.) This is equivalent to

$$(a - 2b)(a - b) \geq 0,$$

and since $a - b > 0$ by hypothesis, to

$$a \geq 2b.$$

Thus the limaçon is convex if and only if $a \geq 2b$.

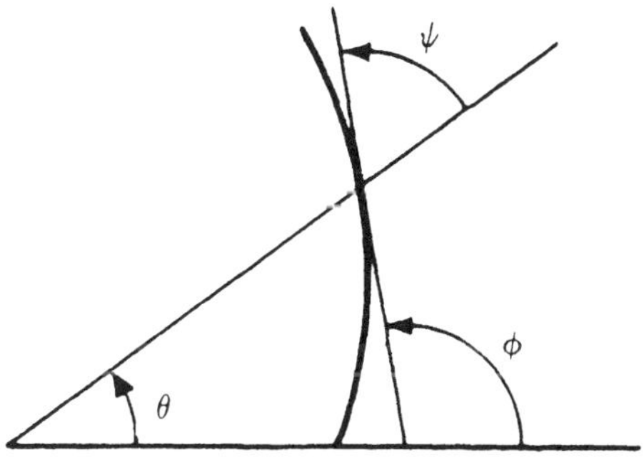

The formula for the curvature used above is easily derived. If ϕ is the direction angle of the tangent vector, then $\phi = \theta + \psi$, where ψ is given by $\tan \psi = r/(dr/d\theta)$. Then by definition the curvature is

$$\frac{d\phi}{ds} = \frac{d\phi}{d\theta}\bigg/\frac{ds}{d\theta} = (r^2 + (r')^2)^{-1/2} \frac{d}{d\theta}\left(\theta + \arctan \frac{r}{r'}\right)$$

$$= \frac{r^2 + 2(r')^2 - rr''}{(r^2 + (r')^2)^{3/2}}.$$

Alternatively, the curvature can be computed from the formula

$$\kappa = \frac{x'y'' - x''y'}{(x'^2 + y'^2)^{3/2}},$$

where in the present case

$$x = r\cos\theta = a\cos\theta - b\cos^2\theta$$

$$y = r\sin\theta = a\sin\theta - b\cos\theta\sin\theta.$$

---

2. Answer both (i) and (ii). Test for convergence the series

(i) $$\frac{1}{\log(2!)} + \frac{1}{\log(3!)} + \frac{1}{\log(4!)} + \cdots + \frac{1}{\log(n!)} + \cdots$$

(ii) $$\frac{1}{3} + \frac{1}{3\sqrt{3}} + \frac{1}{3\sqrt{3}\sqrt[3]{3}} + \cdots + \frac{1}{3\sqrt{3}\sqrt[3]{3}\cdots\sqrt[n]{3}} + \cdots.$$

Solution. For $n \geq 2$, we have $n^n > n!$, hence $n \log n > \log(n!)$ and

$$\frac{1}{\log(n!)} > \frac{1}{n \log n}.$$

Series (i) therefore dominates the series

$$\sum_{n=2}^{\infty} \frac{1}{n \log n}. \tag{1}$$

Since

$$\int_2^x \frac{dt}{t \log t} = \log\log x - \log\log 2,$$

the improper integral $\int_2^\infty dt/(t \log t)$ diverges, and hence, by the integral test, so does (1). Therefore series (i) is divergent.

The denominator of the nth term of series (ii) is $3^{1+(1/2)+\cdots+(1/n)}$, and $1 + \frac{1}{2} + \cdots + (1/n) \sim \log n$. Hence the nth term of (ii) is about

$$\frac{1}{3^{\log n}} = \frac{1}{n^{\log 3}}.$$

Now $\Sigma\, n^{-p}$ converges if $p > 1$ and $\log 3 > 1$, so series (ii) converges.

We shall give the details of this argument. Since

$$\sum_{k=1}^{n} \frac{1}{k} > \sum_{k=1}^{n} \int_k^{k+1} \frac{dt}{t} = \int_1^{n+1} \frac{dt}{t} = \log(n+1) > \log n,$$

we have

$$3^{1+(1/2)+\cdots+(1/n)} > 3^{\log n} = n^{\log 3}.$$

Hence series (ii) is dominated by $\Sigma\, n^{-\log 3}$. Since the latter converges, so does (ii).

---

3. The sequence $x_0, x_1, x_2, \ldots$ is defined by the conditions

$$x_0 = a,\; x_1 = b,\; x_{n+1} = \frac{x_{n-1} + (2n-1)x_n}{2n} \quad \text{for } n \geq 1,$$

where a and b are given numbers. Express $\lim_{n\to\infty} x_n$ concisely in terms of a and b.

Solution. The recursion can be rearranged as

$$x_{n+1} - x_n = -\frac{1}{2n}(x_n - x_{n-1})$$

whence it follows that

$$x_{n+1} - x_n = \left(-\frac{1}{2}\right)^n \frac{1}{n!}(x_1 - x_0) = \left(-\frac{1}{2}\right)^n \frac{1}{n!}(b - a).$$

But

$$x_{n+1} = x_0 + \sum_{i=0}^{n}(x_{i+1} - x_i) = a + (b - a)\sum_{i=0}^{n}\left(-\frac{1}{2}\right)^i \frac{1}{i!}$$

The sum on the right is the partial sum of the power series for $e^{-1/2}$. Therefore

$$\lim_{n\to\infty} x_n = a + (b - a)e^{-1/2}.$$

Remark. There is a close analogy between linear difference equations and linear differential equations. To bring out this analogy we write the given recursion relation in terms of the difference operator Δ (defined by $\Delta x_n = x_{n+1} - x_n$, whence $\Delta^2 x_n = x_{n+2} - 2x_{n+1} + x_n$). We find

$$(2n + 2)\Delta^2 x_n + (2n + 3)\Delta x_n = 0. \tag{1}$$

Because it contains no term in x_n, (1) becomes a first-order difference equation for $v_n = \Delta x_n$,

$$(2n + 2)\Delta v_n + (2n + 3)v_n = 0.$$

We can solve this directly since

$$\begin{aligned} v_{n+1} &= \frac{-1}{2(n+1)} v_n \\ &= \frac{-1}{2(n+1)} \cdot \frac{-1}{2n} \cdot v_{n-1} = \cdots \\ &= \left(\frac{-1}{2}\right)^{n+1} \frac{1}{(n+1)!} v_0. \end{aligned}$$

In terms of the original variables $\{x_n\}$ this becomes

$$\Delta x_n = \left(-\frac{1}{2}\right)^n \frac{1}{n!} \Delta x_0$$

whence

$$x_n = x_0 + \sum_{i=0}^{n-1} \Delta x_i$$

$$= x_0 + \Delta x_0 \sum_{i=0}^{n-1} \left(-\frac{1}{2}\right)^i \frac{1}{i!}$$

and

$$\lim x_n = x_0 + \Delta x_0 \sum_{i=0}^{\infty} \left(-\frac{1}{2}\right)^i \frac{1}{i!}.$$

The force of the analogy is quite striking if we now solve the following problem:

A function $y = y(t)$ is defined on $[0, \infty)$ by the differential equation

$$(2t + 2)y'' + (2t + 3)y' = 0$$

(suggested by (1)) and the initial conditions $y(0) = a$, $y'(0) = b - a$. Find $\lim_{t\to\infty} y(t)$.

---

4. (i) In a right prism with triangular base, given the sum of the areas of three mutually adjacent faces (that is, of two lateral faces and one base), show that these faces are of equal area and perpendicular to each other when the volume attains its maximum.

Solution. Let the base triangles have sides a and b with included angle θ, and let the right prism have altitude c. If L denotes the given sum of the three face areas, then $L = ac + bc + \frac{1}{2}ab \sin\theta$, and the volume is $V = \frac{1}{2}abc \sin\theta$. Let $X = ac$, $Y = bc$, $Z = \frac{1}{2}ab\sin\theta$ be the areas of the three faces. Then $V^2 = \frac{1}{2}XYZ\sin\theta$ where X, Y, Z are three positive numbers whose sum is L. Now the arithmetic mean-geometric mean inequality for positive numbers yields $(XYZ)^{1/3} \le (X + Y + Z)/3$, and hence

$$V^2 \le \frac{1}{2}\left(\frac{L}{3}\right)^3 \sin\theta, \quad \text{and} \quad V \le \frac{1}{\sqrt{2}}\left(\frac{L}{3}\right)^{3/2}$$

with equality occurring only if $X = Y = Z = \frac{1}{3}L$ and $\sin\theta = 1$. But if all these conditions are satisfied, the three faces all have equal area and are mutually perpendicular.

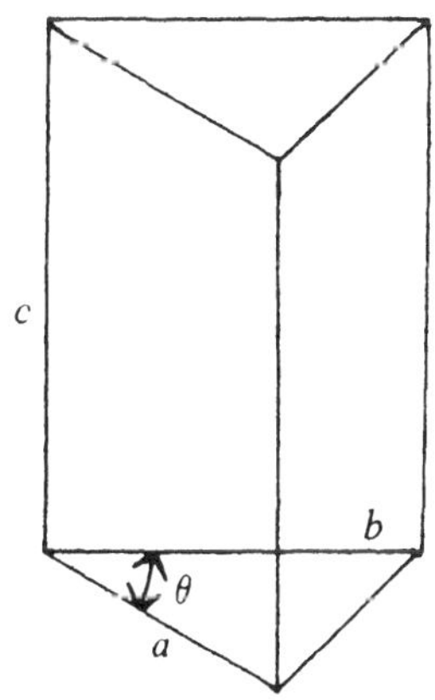

It remains to be shown that these conditions can be satisfied by a right prism. For this we need only take $a = b = 2c = \sqrt{2L/3}$ and $\theta = \pi/2$.

REMARK. The Lagrange multiplier method leads quickly to the critical point, but the proof that it is a maximum seems to require an argument similar to the above.

---

4. (ii) Show that

$$\frac{\dfrac{x}{1} + \dfrac{x^3}{1\cdot 3} + \dfrac{x^5}{1\cdot 3\cdot 5} + \dfrac{x^7}{1\cdot 3\cdot 5\cdot 7} + \cdots}{1 + \dfrac{x^2}{2} + \dfrac{x^4}{2\cdot 4} + \dfrac{x^6}{2\cdot 4\cdot 6} + \cdots} = \int_0^x e^{-t^2/2}\,dt.$$

Solution. The series appearing in the numerator converges for all x by the ratio test. Let its sum be $f(x)$. Since the denominator is immediately seen to be

$$\sum_{n=0}^{\infty} \frac{1}{n!}\left(\frac{x^2}{2}\right)^n = e^{x^2/2},$$

the problem is equivalent to proving that

$$f(x) = e^{x^2/2} \int_0^x e^{-t^2/2}\,dt. \tag{1}$$

Differentiating the series for f term by term, we find

$$f'(x) = 1 + xf(x). \tag{2}$$

This first-order linear differential equation, together with the initial condition $f(0) = 0$ determines the function f, so we can establish (1) either by differentiating the right member of (1) and noting that it satisfies (2) and the initial condition, or by solving (2) by the usual method. (The corresponding homogeneous equation is $g'(x) - xg(x) = 0$, with general solution $g(x) = ce^{x^2/2}$. Then $e^{-x^2/2}$ is an integrating factor for (2):

$$\frac{d}{dx}\left(f(x)e^{-x^2/2}\right) = e^{-x^2/2}$$

and (1) follows immediately.)

REMARK. This is essentially Problem 4623, *American Mathematical Monthly*, vol. 63 (1956), page 260.

---

5. A function $D(n)$ of the positive integral variable n is defined by the following properties: $D(1) = 0$, $D(p) = 1$ if p is a prime, $D(uv) = uD(v) + vD(u)$ for any two positive integers u and v. Answer all three parts below.

(i) Show that these properties are compatible and determine uniquely $D(n)$. (Derive a formula for $D(n)/n$, assuming that $n = p_1^{\alpha_1} p_2^{\alpha_2} \cdots p_k^{\alpha_k}$ where $p_1, p_2, \ldots, p_k$ are different primes.)

(ii) For what values of n is $D(n) = n$?

(iii) Define $D^2(n) = D[D(n)]$, etc., and find the limit of $D^m(63)$ as m tends to ∞.

Solution. (i) Suppose there is a function D with the required properties. We have

$$\frac{D(uv)}{uv} = \frac{D(u)}{u} + \frac{D(v)}{v} \tag{1}$$

and by induction

$$\frac{D(u_1u_2\cdots u_k)}{u_1u_2\cdots u_k} = \sum_i \frac{D(u_i)}{u_i}.$$

Hence

$$\frac{D(p^\alpha)}{p^\alpha} = \alpha\frac{D(p)}{p} = \frac{\alpha}{p}$$

if p is a prime, and for any integer n with prime factorization $p_1^{\alpha_1}p_2^{\alpha_2}\cdots p_k^{\alpha_k}$ we have

$$(2) \qquad \frac{D(n)}{n} = \sum_i \frac{\alpha_i}{p_i}.$$

This equation shows that there is at most one function with the given properties.

On the other hand, since every integer $n > 1$ has a factorization into primes that is unique apart from order, and since the order in which the primes are numbered does not affect the sum in (2), we can define $D(1) = 0$ and use (2) to define $D(n)$ for $n > 1$. So defined, $D(p) = 1$ for p a prime, and D clearly satisfies (1), which is equivalent to $D(uv) = uD(v) + vD(u)$. Thus there is a unique function with the prescribed properties.

We note for future reference that $D(n) > 0$ for $n > 1$.

(ii) The equation $D(n) = n$ is equivalent to

$$(3) \qquad \frac{\alpha_1}{p_1} + \frac{\alpha_2}{p_2} + \cdots + \frac{\alpha_k}{p_k} = 1,$$

where $n = p_1^{\alpha_1}p_2^{\alpha_2}\cdots p_k^{\alpha_k}$ is the prime factorization of n. If (3) is multiplied through by $p_1p_2\cdots p_{k-1}$, we see that $p_1p_2\cdots p_{k-1}\alpha_k/p_k$ is an integer. Since the p's are all different, we conclude that p_k divides α_k. So α_k/p_k is an integer, and it is clear from (3) that $k = 1$ and $\alpha_k = p_k$. Thus any solution of $D(n) = n$ has the form $n = p^p$ where p is prime. Conversely, any such n is a solution.

(iii)

$$\begin{aligned} D(63) &= 51, \\ D^2(63) &= D(51) = 20, \\ D^3(63) &= D(20) = 24, \\ D^4(63) &= D(24) = 44, \\ D^5(63) &= D(44) = 48. \end{aligned}$$

It appears that $D^m(63)$ has started to increase with m.

Suppose $n = 4k$ where $k > 1$. Then $D(n) = D(4)k + 4D(k) = 4(k + D(k)) > 4k = n$. Thus, if $n > 4$ and n is divisible by 4, then $D(n) > n$ and $D(n)$ is divisible by 4. This implies that the sequence

$$D(n),\ D^2(n),\ D^3(n),\ \ldots,\ D^m(n),\ \ldots$$

is strictly increasing. Since D takes integral values, $D^m(n) \to \infty$ as $m \to \infty$ whenever $n = 4k$, $k > 1$.

Applying this result to the case above, we see that $D^m(63) = D^{m-2}(20) \to \infty$ as $m \to \infty$.

---

6. Each coefficient a_n of the power series

$$a_0 + a_1x + a_2x^2 + a_3x^3 + \cdots = f(x)$$

has either the value 1 or the value 0. Prove the easier of the two assertions:

(i) If $f(0.5)$ is a rational number, $f(x)$ is a rational function.

(ii) If $f(0.5)$ is not a rational number, $f(x)$ is not a rational function.

Solution. (i) Suppose $f(0.5) = f(1/2)$ is a rational number. To prove that f is a rational function, note that

$$f\left(\frac{1}{2}\right) = a_0 + \frac{a_1}{2} + \frac{a_2}{2^2} + \cdots,$$

so $a_0 \cdot a_1a_2a_3 \cdots$ can be regarded as a binary expansion of $f(1/2)$. It is well known that the binary (or decimal) expansion of a number is eventually periodic if and only if the number is rational.

(A dyadic rational number has two binary expansions, both eventually periodic; for example, $\frac{3}{8} = 0.011000\ldots = 0.010111111\ldots$. This ambiguity is immaterial in the argument below.)

Thus if $f(\frac{1}{2})$ is a rational number, then its binary expansion must be eventually periodic, that is, there exist integers N and k such that $a_{k+n} = a_n$ for all $n \geq N$. Then

$$\begin{aligned} f(x) &= a_0 + a_1x + \cdots + a_Nx^N \\ &\qquad + x^{N+1}[a_{N+1} + a_{N+2}x + \cdots a_{N+k}x^{k-1}][1 + x^k + x^{2k} + \cdots] \\ &= a_0 + a_1x + \cdots a_Nx^N + \frac{x^{N+1}}{1 - x^k}[a_{N+1} + a_{N+2}x + \cdots a_{N+k}x^{k-1}]. \end{aligned}$$

This shows that f is a rational function and proves assertion (i).

Note that all formal manipulations are justified for $|x| < 1$ because the power series for $f(x)$ converges for $|x| < 1$ in view of the boundedness of the coefficients.

(ii) We prove (ii) in the contrapositive form: If f is a rational function whose power series at zero exists and has every coefficient either 0 or 1, then $f(\frac{1}{2})$ is a rational number.

Suppose

$$f(x) = \frac{b_0 + b_1 x + \cdots b_m x^m}{c_0 + c_1 x + \cdots c_k x^k}. \tag{1}$$

We may assume that any powers of x dividing both numerator and denominator have been cancelled. Then $c_0 \neq 0$ since f is analytic at 0. Using the given series for f it follows that

$$(c_0 + c_1 x + \cdots + c_k x^k)(a_0 + a_1 x + a_2 x^2 + \cdots)$$
$$= b_0 + b_1 x + \cdots + b_m x^m.$$

Comparing the coefficients of x^{k+n} where $k + n > m$ we find

$$c_0 a_{n+k} + c_1 a_{n+k-1} + \cdots + c_k a_n = 0$$

and, since $c_0 \neq 0$, we can solve for a_{n+k}

$$a_{n+k} = -\frac{1}{c_0}(c_1 a_{n+k-1} + \cdots + c_k a_n). \tag{2}$$

This is a linear recursion relation that expresses a_{n+k} in terms of the preceding k coefficients $a_{n+k-1}, \ldots, a_n$.

Suppose that, for some integers r and s with $r + k > m$ and $s > 0$, we have

$$a_{r+s} = a_r$$

$$a_{r+s+1} = a_{r+1}$$

$$\cdots$$

$$a_{r+s+k-1} = a_{r+k-1}.$$

It then follows from (2) that $a_{r+s+k} = a_{r+k}$ and by induction that $a_{r+s+t} = a_{r+t}$ for all positive t.

Now since we know that every a is either 0 or 1, there can be at most 2^k

distinct k-tuples of consecutive coefficients, and hence there must be cases in which the k-tuple

$$\langle a_r, a_{r+1}, \ldots, a_{r+k-1}\rangle$$

is the same as the k-tuple

$$\langle a_{r+s}, a_{r+s+1}, \ldots, a_{r+s+k-1}\rangle$$

for $r > m - k$ and $s > 0$; indeed, there must be an example where $m - k < r < r + s \leq m - k + 1 + 2^k$. Once such a repetition occurs, the coefficients are periodic with period s from that point on as we have seen above. Thus the a's are eventually periodic. Then

$$f\left(\frac{1}{2}\right) = \Sigma\, a_i \frac{1}{2^i}$$

has an eventually periodic binary expansion, so it is a rational number. This proves assertion (ii).

REMARK. A result much stronger than (ii) is true. If f is a rational function regular at 0, whose power series at 0 has all rational coefficients, then f is the quotient of two polynomials with rational coefficients. Hence $f(r)$ is rational for any rational number r for which $f(r)$ is defined.

Proof. With k as above, let $\mathbf{a}_n = \langle a_{n+k}, a_{n+k-1}, \ldots, a_n\rangle \in \mathbf{Q}^{k+1}$. Then (2) shows that the vectors $\{\mathbf{a}_n : n > m - k\}$ do not span $\mathbf{R}^{k+1}$. Hence they do not span $\mathbf{Q}^{k+1}$ either, and there exists a nonzero vector

$$\mathbf{c}' = \langle c_0', c_1', \ldots, c_k'\rangle \in \mathbf{Q}^{k+1}$$

such that $\mathbf{c}' \cdot \mathbf{a}_n = 0$ for $n > m - k$. Then

$$(c_0' + c_1'x + \cdots + c_k'x^k)f(x) = b_0' + b_1'x + \cdots + b_m'x^m,$$

and it is clear that $b_0', b_1', \ldots, b_m' \in \mathbf{Q}$. Thus we can replace (1) by a representation of f as a quotient of polynomials with rational coefficients.

Afternoon Session

1. In each of n houses on a straight street are one or more boys. At what point should all the boys meet so that the sum of the distances that they walk is as small as possible?

First Solution. Suppose the linear coordinate of the ith boy's house is x_i. We can number the boys so that $x_1 \le x_2 \le \cdots \le x_n$. Let y be the coordinate of the best meeting point. (There is a best meeting point because the total distance walked to a meeting at z is a continuous function of z that tends to infinity as $z \to \pm\infty$.)

Suppose that r boys live to the right of y and l boys to the left of y. If y' is a point to the right of y but not beyond the next house and the boys congregated at y' instead of y, then r boys would walk $y' - y$ less and $n - r$ would walk $y' - y$ farther. If $n < 2r$, this would make the total distance walked less by $(2r - n)(y' - y)$, contrary to the choice of y. Hence $n \ge 2r$. Similarly, $n \ge 2l$.

Suppose n is odd, say, $n = 2k - 1$. We cannot have $y < x_k$ because then at least $k + 1$ boys would live to the right of y, that is $r \ge k + 1$, contradicting the previous paragraph. Similarly, we cannot have $y > x_k$. So $y = x_k$.

Now suppose n is even, $n = 2k$. By the same reasoning, we cannot have $y < x_k$ or $y > x_{k+1}$; therefore, $x_k \le y \le x_{k+1}$. Moreover, the total distance walked will be the same for any choice of y in this interval, as shown in the second paragraph.

Summarizing, if n is odd, the boys should meet at the home of the middle boy; if n is even, they should meet at any point between (or at) the homes of the two middle boys. If the two middle boys happen to live in the same house, the interval degenerates to a point and the meeting place is uniquely determined.

Second Solution. Number the boys as above. Wherever they meet, the first and nth boys together must walk at least $x_n - x_1$; the second and $(n-1)$st boys together must walk at least $x_{n-1} - x_2$; etc.

If n is even, $n = 2k$, the boys must walk altogether at least

$$(x_n - x_1) + \cdots + (x_{k+1} - x_k)$$

with equality if and only if the meeting place y is in each of the intervals $[x_1, x_n], \ldots, [x_k, x_{k+1}]$. Since these intervals are nested, this is equivalent to $y \in [x_k, x_{k+1}]$.

If n is odd, $n = 2k - 1$, the pairing above leaves the kth boy unpaired; but he must walk at least 0, so the total distance walked is at least

$$(x_n - x_1) + \cdots + (x_{k+1} - x_{k-1}) + 0$$

with equality if and only if $y = x_k$.

This second proof was adapted from A. R. Kokan, "The Minimum Property of the Mean Deviation," *Mathematical Gazette,* 59 (1975), 111.

2. Two obvious approximations to the length of the perimeter of the ellipse with semi-axes a and b are $\pi(a+b)$ and $2\pi(ab)^{1/2}$. Which one comes nearer the truth when the ratio b/a is very close to 1?

Solution. Let the ellipse be taken in the parametric form $x = a \cos t$, $y = b \sin t$. Then the length L is given by

$$L = \int_0^{2\pi} \sqrt{\left(\frac{dx}{dt}\right)^2 + \left(\frac{dy}{dt}\right)^2}\, dt = \int_0^{2\pi} \sqrt{a^2 \sin^2 t + b^2 \cos^2 t}\, dt,$$

a well-known elliptic integral. Since we are asked to consider L when b/a is nearly 1 we put $b = (1+\lambda)a$ and consider

$$L(\lambda) = a \int_0^{2\pi} \sqrt{1 + (2\lambda + \lambda^2)\cos^2 t}\, dt.$$

This is evidently an analytic function of λ, and we calculate its power series to terms of degree 2

$$L(\lambda) = a \int_0^{2\pi} \left(1 + \frac{1}{2}(2\lambda + \lambda^2)\cos^2 t - \frac{1}{8}(2\lambda + \lambda^2)^2 \cos^4 t + \cdots\right) dt$$

$$= 2\pi a \left[1 + \frac{1}{4}(2\lambda + \lambda^2) - \frac{3}{64}(2\lambda + \lambda^2)^2 + \cdots\right]$$

$$= 2\pi a \left[1 + \frac{1}{2}\lambda + \frac{1}{16}\lambda^2 + \cdots\right].$$

The first expression was obtained using the binomial expansion of $(1+z)^{1/2}$. All the terms omitted are of degree at least three in λ.

The formal manipulation is justified because the series in question converges absolutely for small values of λ (in fact if $|2\lambda| + \lambda^2 < 1$). (We could also find the power series by differentiating under the integral sign.)

The proposed approximations to the perimeter are

$$\pi(a+b) = 2\pi a\left(1 + \frac{1}{2}\lambda\right)$$

and

$$2\pi\sqrt{ab} = 2\pi a\sqrt{1+\lambda} = 2\pi a\left(1 + \frac{1}{2}\lambda - \frac{1}{8}\lambda^2 \cdots\right).$$

Since the three functions have the same constant and first degree terms their differences are controlled by the second degree terms for small λ. We have

$$L(\lambda) > 2\pi a\left(1 + \frac{1}{2}\lambda\right) > 2\pi a\left(1 + \frac{1}{2}\lambda - \frac{1}{8}\lambda^2 + \cdots\right)$$

for small λ. Thus $\pi(a + b)$ is a better approximation to the length of an ellipse than $2\pi\sqrt{ab}$; in fact it is roughly three times better since

$$L(\lambda) - 2\pi a\left(1 + \frac{1}{2}\lambda\right) \sim \frac{1}{16}\lambda^2 \text{ and } L(\lambda) - 2\pi a\sqrt{1 + \lambda} \sim \frac{3}{16}\lambda^2.$$

Continuation. It is clear that L is a differentiable function of λ for all values of $\lambda > -1$. If we calculate the second derivative we find

$$L''(\lambda) = a\int_0^{2\pi} \frac{\sin^2 t\cos^2 t}{(1 + (2\lambda + \lambda^2)\cos^2 t)^{3/2}}\,dt,$$

which is evidently positive for all $\lambda > -1$. Hence by Taylor's Theorem

$$L(\lambda) = L(0) + \lambda L'(0) + \frac{1}{2}\lambda^2 L''(\xi) >$$

$$L(0) + \lambda L'(0) = 2\pi a\left(1 + \frac{1}{2}\lambda\right)$$

for all $\lambda \neq 0$. Thus the length of a (non-circular) ellipse always exceeds $\pi(a + b)$. By the arithmetic-geometric mean inequality we also have $\pi(a + b) > 2\pi\sqrt{ab}$, so $\pi(a + b)$ is always a better approximation to the perimeter of an ellipse than $2\pi\sqrt{ab}$.

The inequality $L > \pi(a + b)$ can be demonstrated directly as follows: First note that

$$\sqrt{\mu A + (1 - \mu)B} + \sqrt{(1 - \mu)A + \mu B} \geq \sqrt{A} + \sqrt{B}$$

whenever A and B are positive and $0 \leq \mu \leq 1$, with strict inequality unless $A = B$ or $\mu = 0$ or 1. This is easy to prove by successive squaring and canceling.

Replacing t by $t + \pi/2$ in the original integral for L we find that

$$L = \int_0^{2\pi} \sqrt{a^2\cos^2 t + b^2\sin^2 t}\,dt.$$

Then

$$2L = \int_0^{2\pi} \left(\sqrt{a^2\cos^2 t + b^2\sin^2 t} + \sqrt{a^2\sin^2 t + b^2\cos^2 t}\right) dt$$

$$> \int_0^{2\pi} (a + b)dt = 2\pi(a + b),$$

using the inequality just stated with $A = a^2$, $B = b^2$, $\mu = \cos^2 t$. See M. S. Klamkin, "Elementary Approximations to the Area of N-dimensional Ellipsoids," *American Mathematical Monthly*, vol. 78 (1971), pages 280–283.

---

3. In the Gregorian calendar:

(i) years not divisible by 4 are common years;
(ii) years divisible by 4 but not by 100 are leap years;
(iii) years divisible by 100 but not by 400 are common years;
(iv) years divisible by 400 are leap years;
(v) a leap year contains 366 days; a common year 365 days.

Prove that the probability that Christmas falls on a Wednesday is not 1/7.

Solution. According to the rules given, any 400 consecutive Gregorian years will involve 303 ordinary years and 97 leap years making a total of $400 \cdot 365 + 97$ days. Since this number is divisible by 7 ($400 \equiv 1$, $365 \equiv 1$, $97 \equiv -1 \pmod 7$), there is an integral number of weeks in 400 years (in fact, 20,871 weeks) and therefore the day of the week on which Christmas (or any other calendar date) falls repeats in cycles of 400. If there are N years in such a period on which Christmas falls on Wednesday, then the probability that Christmas falls on Wednesday is $N/400$. But $N/400 \neq 1/7$ for any integer N.

REMARK. The following table gives the number of years in each 400-year cycle on which Christmas falls on each day of the week.

Sun.	Mon.	Tues.	Wed.	Thurs.	Fri.	Sat.
58	56	58	57	57	58	56

The superstitious may also be interested to know that the thirteenth of the month is more likely to fall on Friday than on any other day of the week. See: *American Mathematical Monthly*, vol. 40 (1933), page 607; also

Emanuel Parzen, *Modern Probability Theory and Its Applications.* New York, 1960, page 26 ff.

---

4. The cross-section of a right cylinder is an ellipse, with semi-axes a and b, where $a > b$. The cylinder is very long, made of very light homogeneous material. The cylinder rests on the horizontal ground which it touches along the straight line joining the lower endpoints of the minor axes of its several cross-sections. Along the upper endpoints of these minor axes lies a very heavy homogeneous wire, straight and just as long as the cylinder. The wire and the cylinder are rigidly connected. We neglect the weight of the cylinder, the breadth of the wire, and the friction of the ground.

The system described is in equilibrium, because of its symmetry. This equilibrium seems to be stable when the ratio b/a is very small, but unstable when this ratio comes close to 1. Examine this assertion and find the value of the ratio b/a which separates the cases of stable and unstable equilibrium.

First Solution. Because the cylinder is long, the only displacements we need to consider are the rolling motions of the cylinder. Hence we may confine our attention to a plane perpendicular to the axis of the cylinder, and the problem becomes effectively two-dimensional. It is equivalent to the following. An ellipse, whose equation we take to be

$$\frac{x^2}{a^2} + \frac{y^2}{b^2} = 1,$$

is restricted to the plane but is free to roll along the line $y = -b$. The weight of the ellipse is negligible but there is a heavy particle attached to the ellipse at $P = (0, b)$. The force of gravity acts parallel to the y-axis in the negative direction. We are to determine, in terms of a/b, the condition for stability. Suppose the ellipse rolls a little way from the starting position. If this motion raises P it will be opposed by the force of gravity and the ellipse will be in stable equilibrium when resting on Q. On the other hand, if a slight motion lowers P the equilibrium at Q will be unstable.

Thus, to decide whether the equilibrium is stable or not, we consider the distance $|PZ|$ from P to a variable point Z on the ellipse. If this function has a strict local minimum at Q then the equilibrium is stable. On the other hand, if $|PZ|$ has a strict local maximum for $Z = Q$, the equilibrium is unstable. (For a sharper form of this criterion, see the solution to Problem A.M. 7 of the Fourth Competition.) We can equally well consider the function $|PZ|^2$.

Suppose Z is the point ($a \sin t$, $-b$ cost) (which is on the ellipse). Then $t = 0$ corresponds to $Z = Q$, and $|PZ|^2 = a^2 \sin^2 t + b^2(1 + \cos t)^2$. If we

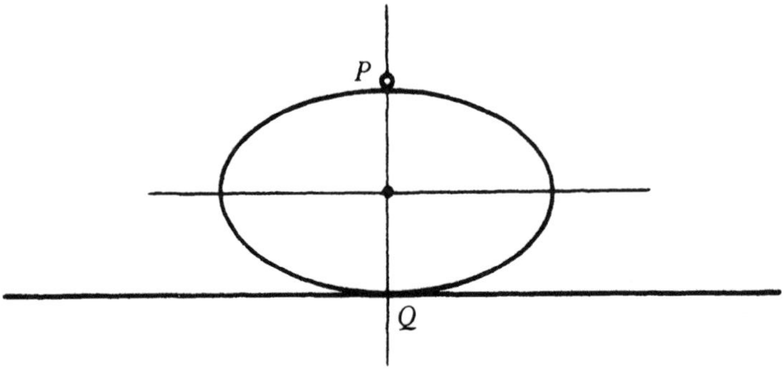

put $a = rb$ this becomes

$$b^2[(r^2 - 1)\sin^2 t + 2 + 2\cos t].$$

The first derivative (with respect to t) is

$$b^2[(r^2 - 1)\sin 2t - 2\sin t],$$

which vanishes for $t = 0$; and the second derivative is

$$b^2[2(r^2 - 1)\cos 2t - 2\cos t]$$

which is $2b^2(r^2 - 2)$ for $t = 0$. Hence we have a strict local maximum at $t = 0$ (and therefore instability) if $r^2 < 2$ and a strict local minimum (and therefore stability) if $r^2 > 2$. The critical value of b/a $(= 1/r)$ is therefore $\frac{1}{2}\sqrt{2}$.

Continuation. If $r^2 = 2$, the critical case, our function becomes

$$b^2[\sin^2 t + 2 + 2\cos t] = b^2[4 - (1 - \cos t)^2],$$

which evidently has a strict local maximum for $t = 0$, so the critical case is unstable.

Another way to calculate whether P rises or falls as the ellipse rolls is to compute its actual height when the point of contact is at $Z = (a\sin t, -b\cos t)$. This is the distance from P to the tangent to the ellipse at Z. The equation of this tangent is

$$(x - a\sin t)\, b\sin t = (y + b\cos t)\, a\cos t,$$

and the distance from P to this line is

$$\frac{(b + b\cos t)\, a\cos t - (-a\sin t)\, b\sin t}{\sqrt{a^2\cos^2 t + b^2\sin^2 t}}$$

$$= a\frac{1 + \cos t}{\sqrt{r^2 \cos^2 t + \sin^2 t}}.$$

The solution now proceeds in the same way, but the algebra is more complicated.

Second Solution. Suppose the ellipse rolls so that the point of contact with the ground is at Z. The support of the ground acts upward along the

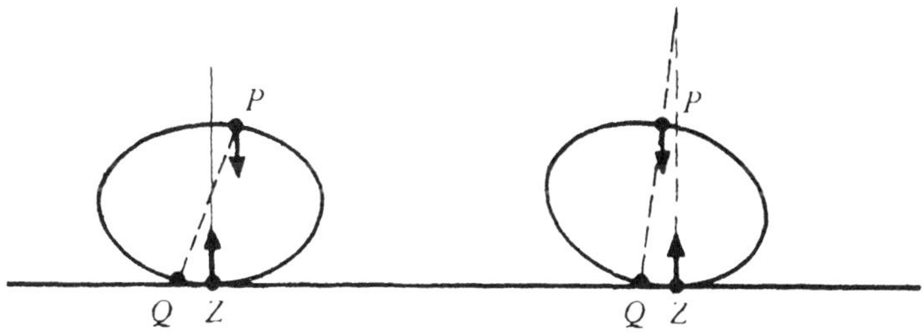

normal to the ellipse at Z. If this normal crosses the segment PQ as in the left-hand figure, the force of gravity acting downward through P together with the support will cause the body to roll more. On the other hand, if the normal crosses the line PQ above P, as in the right-hand figure, the force of gravity and the support will act to reverse the rolling. Hence we obtain the following:

Criterion. The equilibrium is stable if the normal to the ellipse at Z cuts PQ above P for all Z sufficiently near Q (other than Q itself), and it is unstable if the normal cuts PQ between P and Q for Z sufficiently near Q.

If Z is $(a \sin t, -b \cos t)$, the equation of the normal is

$$a \cos t\,(x - a \sin t) + b \sin t\,(y + b \cos t) = 0.$$

This line cuts the y axis (i.e., the line PQ) at

$$\left(0, \frac{a^2 - b^2}{b} \cos t\right),$$

which is between P and Q for small $t(\neq 0)$ if

$$\frac{a^2 - b^2}{b} \leq b$$

and is above P for small t if

$$\frac{a^2 - b^2}{b} > b.$$

Hence we have stability if $a^2 > 2b^2$ and instability if $a^2 \leq 2b^2$.

Since the limiting position of the intersection of the normal at Z with the normal at Q is the center of curvature for the ellipse at Q, the above argument gives the following result.

Suppose a (two-dimensional) convex body rests on a horizontal line so that the center of mass is directly over the point of support; the body will then be in equilibrium. If the center of mass (in this case, P) is above the center of curvature at the point of support (in this case, Q) the equilibrium is unstable; while if the center of mass is below the center of curvature, the equilibrium is stable.

This result, which is true for an arbitrary smooth convex body, gives no conclusion in the critical case in which the center of mass and the center of curvature coincide.

---

5. (i) Given that the sequence whose nth term is $(s_n + 2s_{n+1})$ converges, show that the sequence $\{s_n\}$ converges also.

Solution. Suppose $\lim (s_n + 2s_{n+1}) = 3L$. Then

$$\lim [(s_n - L) + 2(s_{n+1} - L)] = 0.$$

Put $t_n = s_n - L$; then $\lim (t_n + 2t_{n+1}) = 0$. We shall prove that $\lim t_n = 0$. This will prove that $\lim s_n = L$, and establish the convergence of the sequence.

Given $\epsilon > 0$, choose k so that

$$|t_n + 2t_{n+1}| < \epsilon \quad \text{for all } n \geq k.$$

By induction on p we find that

$$t_k - (-2)^p t_{k+p} = \sum_{i=0}^{p-1} (-2)^i (t_{k+i} + 2t_{k+i+1}).$$

Therefore

$$|t_k - (-2)^p t_{k+p}| \leq \sum_{i=0}^{p-1} 2^i |t_{k+i} + 2t_{k+i+1}| < 2^p \epsilon,$$

provided $p \geq 1$. Dividing by 2^p, we get

$$\left| t_{k+p} - \left(-\frac{1}{2}\right)^p t_k \right| < \epsilon$$

so

$$|t_{k+p}| < \epsilon + \frac{1}{2^p}|t_k|.$$

Hence

$$\limsup_{p\to\infty} |t_{k+p}| \leq \epsilon.$$

Therefore

$$\limsup_{n\to\infty} |t_n| \leq \epsilon.$$

Since ϵ was arbitrary, this implies $\lim t_n = 0$. As we remarked before, this proves that the sequence $\{s_n\}$ converges to L.

REMARK. The argument generalizes to prove that if $s_n - \lambda s_{n+1} \to A$, where $|\lambda| > 1$, then $s_n \to A/(1 - \lambda)$.

---

5. (ii) A plane varies so that it includes a cone of constant volume equal to $\pi a^3/3$ with the surface the equation of which in rectangular coordinates is $2xy = z^2$. Find the equation of the envelope of the various positions of this plane.

State the result so that it applies to a general cone (that is, conic surface) of the second order.

Solution. Make the change of coordinates

$$x = \frac{1}{2}\sqrt{2}\,(u + v)$$

$$y = \frac{1}{2}\sqrt{2}\,(u - v).$$

Then the uv-axes are orthogonal and rotated by $\pi/4$ radians from the xy-axes. The equation of the given surface in the new coordinates is

$$z^2 + v^2 = u^2,$$

which is a right circular cone with axis the u-axis.

Next we find the volume of the conical region cut off from the solid cone by a plane. Because of rotational symmetry we need only consider planes of the form $u = mv + b$. In order that the plane should cut off a bounded region it is necessary that $|m| < 1$. The region cut off is then a cone with an elliptical base. We will find the altitude of that cone and the area of its base.

The altitude is the distance from the origin to the plane, namely

$$\frac{|b|}{\sqrt{1 + m^2}}.$$

The area of the base is $\sqrt{1 + m^2}$ times the area of the ellipse obtained by projecting it orthogonally onto the vz-plane. To find the equation of the projected ellipse we eliminate u between the equations of the cone and the plane, getting

$$z^2 + v^2 = (mv + b)^2.$$

Collecting the v terms and completing the square we get

$$z^2 + (1 - m^2)\left(v - \frac{mb}{1 - m^2}\right)^2 = b^2\left(\frac{1}{1 - m^2}\right).$$

[Note that this is indeed an ellipse because $|m| < 1$]. The area of this ellipse is

$$A = \pi b^2 \frac{1}{(1 - m^2)^{3/2}}.$$

The volume of the conical region is therefore

$$\frac{1}{3}\text{ base} \times \text{altitude} = \frac{1}{3}\left(\sqrt{1 + m^2}\,\pi b^2 \frac{1}{(1 - m^2)^{3/2}}\right)\left(\frac{|b|}{\sqrt{1 + m^2}}\right)$$

$$= \frac{1}{3}\pi |b|^3 \frac{1}{(1 - m^2)^{3/2}}.$$

The problem restricts consideration to those planes which cut off a volume $\frac{1}{3}\pi a^3$, that is, to planes for which

$$|b| = a\sqrt{1 - m^2},$$

where m is the tangent of the angle between the plane and the zv-plane.

We are to find the envelope E of all such planes. Clearly E must share the rotational symmetry of the entire configuration, so it suffices to find the intersection I of E with the uv-plane.

Consider the plane P tangent to E at a point of I. Reflection in the uv-plane preserves P, since it preserves E and fixes each point of I. Therefore P is perpendicular to the uv-plane, so it has an equation of the form $u = mv + b$ where, as we have seen, $|b| = a\sqrt{1-m^2}$. The line l in which P meets the uv-plane is clearly tangent to I. Hence I is the envelope of all lines l in the uv-plane having equations of the form

$$u = mv \pm a\sqrt{1-m^2}. \tag{1}$$

The envelope problem is thus reduced to two dimensions.

To find the envelope, we first take the positive sign in (1), and eliminate m between the equation

$$u = mv + a\sqrt{1-m^2} \tag{2}$$

and the equation obtained by differentiating (2) with respect to m, namely

$$0 = v - \frac{am}{\sqrt{1-m^2}}. \tag{3}$$

From (3) we find

$$m^2 = \frac{v^2}{v^2+a^2}, \quad \text{so} \quad 1 - m^2 = \frac{a^2}{v^2+a^2}.$$

Equation (2) now becomes

$$u = \sqrt{1-m^2}\left(\frac{m}{\sqrt{1-m^2}}v + a\right) = \sqrt{1-m^2}\left(\frac{v^2+a^2}{a}\right),$$

giving finally

$$u^2 = v^2 + a^2 \tag{4}$$

as the equation of the envelope I. The same equation results if we start with the negative sign.

[Since the algebraic work could introduce extraneous points not in the locus, we should check to see that the given family of lines is exactly the family of tangent lines to the hyperbola (4). This examination reveals that the tangents to the upper branch of the hyperbola involve the plus sign in (1), while the tangents to the lower branch involve the minus sign. Thus the lower branch is indeed extraneous to the problem of finding the envelope of the family of lines (2).]

The three-dimensional envelope E is obtained by revolving I around the

u axis; hence its equation is

$$u^2 = v^2 + z^2 + a^2.$$

Transforming back to the original coordinates its equation is

$$2xy = z^2 + a^2.$$

Generalization. Any non-degenerate quadratic cone is affinely equivalent to a right circular cone, and since ratios of volumes are preserved under an affine transformation, the calculations above lead to the conclusion that if all planes of a family cut off solids of the same fixed volume from a quadratic cone, the planes must be tangent to a hyperboloid which is asymptotic to the cone.

REMARK. We have seen that all planes tangent to a hyperboloid of two sheets cut off the same volume from the asymptotic cone. This is analogous to the fact that all lines tangent to a hyperbola form with the asymptotes triangles of the same area. A similar statement is true in higher dimensions.

Consider H the "hyperboloid of two sheets" in $\mathbf{R}^n$ given by

$$x_1^2 + x_2^2 + \cdots + x_{n-1}^2 = x_n^2 - a^2.$$

Its asymptotic cone C has equation

$$x_1^2 + x_2^2 + \cdots + x_{n-1}^2 = x_n^2.$$

If P is any hyperplane tangent to H, then P and C divide $\mathbf{R}^n$ into five regions (seven, if $n = 2$) of which only one is bounded. The n-dimensional volume of this bounded region B_P is the same for all choices of P.

To see this, consider the group G of all linear transformations of $\mathbf{R}^n$ that preserve the quadratic form $x_1^2 + x_2^2 + \cdots + x_{n-1}^2 - x_n^2$. G clearly preserves H and C, and any hyperplane tangent to H is mapped by any element of G to a hyperplane tangent to H. Since G acts transitively on H, it acts transitively on the set of hyperplanes tangent to H, and hence on the set of regions B_P. Since the elements of G all have determinant ± 1, they are all volume-preserving, and thus all the regions B_P have the same volume.

As in the case of three dimensions, all "hyperboloids of two sheets" (i.e., non-degenerate, disconnected quadric surfaces) in $\mathbf{R}^n$ are equivalent to H under a linear transformation.

6. Consider the closed plane curves C_i and C_o, their respective lengths $|C_i|$ and $|C_o|$, the closed surfaces S_i and S_o, and their respective areas $|S_i|$ and $|S_o|$. Assume that C_i lies inside C_o and S_i inside S_o. (Subscript i stands for "inner," o for "outer.") Prove the correct assertions among the following four, and disprove the others.

(i) If C_i is convex, $|C_i| \le |C_o|$.
(ii) If S_i is convex, $|S_i| \le |S_o|$.
(iii) If C_o is the smallest convex curve containing C_i, then $|C_o| \le |C_i|$.
(iv) If S_o is the smallest convex surface containing S_i, then $|S_o| \le |S_i|$.

You may assume that C_i and C_o are polygons and S_i and S_o polyhedra. (Why?)

Solution. Statements (i), (ii), and (iii) are true, while (iv) is false. As suggested in the problem, we shall assume that the sets involved are polyhedral and later discuss the role of this assumption.

(i) Suppose C_i is a closed convex polygon inside of the closed polygon C_o. We shall prove that $|C_i| \le |C_o|$.

The meanings of the words "inside" and "outside" require some clarification. We shall need only the fact that every infinite ray emanating from a point of C_i meets C_o.

Let the vertices of C_i in order be $A_0, A_1, \ldots, A_{n-1}, A_n = A_0$. On the segment $A_{q-1}A_q$ construct a semi-infinite rectangular strip S_q outside of C_i, including the open segment $A_{q-1}A_q$ but not the infinite edges. Since C_i is convex, these strips are disjoint.

Consider the orthogonal projection of $C_o \cap S_q$ into $A_{q-1}A_q$. This projection is surjective because if $X \in A_{q-1}A_q$ is not in the range then the ray in S_q from X perpendicular to $A_{q-1}A_q$ would not meet C_o.

Now orthogonal projection never increases the length of a polygon, so

$$|C_o \cap S_q| \ge |A_{q-1}A_q|.$$

Therefore

$$|C_o| \ge \sum_q |C_o \cap S_q| \ge \sum_q |A_{q-1}A_q| = |C_i|.$$

(ii) The three-dimensional analog of (i) (indeed, the analog in any dimension) is also true and a strictly analogous proof applies. On the qth face F_q of the convex polyhedron S_i erect a semi-infinite rectangular prism P_q outside of S_i, including the open qth face but no point of the infinite faces. Then these prisms are disjoint. Projecting $S_o \cap P_q$ orthogonally into F_q does not increase its area, and the projection must cover the interior of

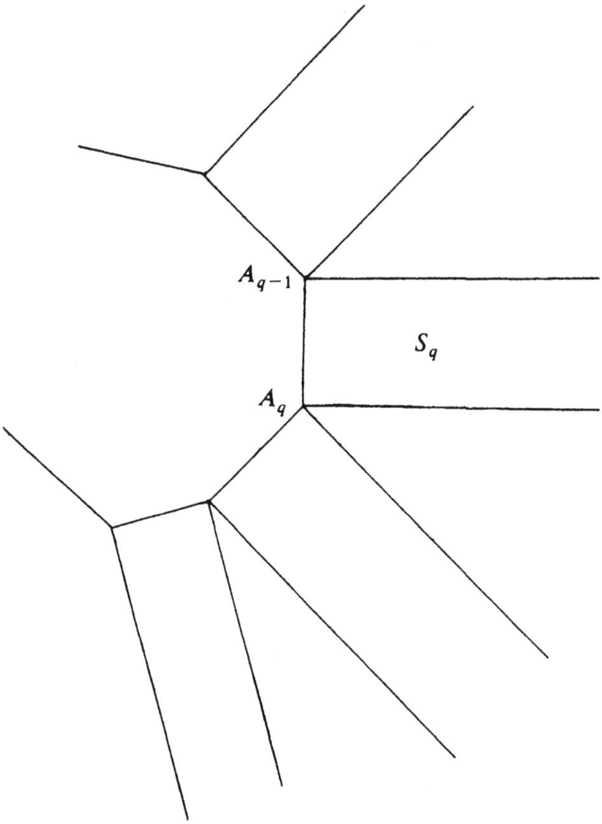

F_q. Therefore we have, as in case (i),

$$|S_o| \geq \sum_q |S_o \cap P_q| \geq \sum_q |F_q| = |S_i|$$

where the absolute signs refer to areas.

(iii) Let C_o be a closed convex polygon with vertices, in order, $A_0, A_1, A_2, \ldots, A_n = A_0$. Let C be any closed (possibly self-intersecting) polygon whose vertices include all the A_k. We assert that

$$|C| \geq |C_o| \tag{1}$$

with equality if and only if $C = C_o$.

Assuming this inequality for the moment, suppose C_i is a polygonal closed curve in the plane and let C_o be the boundary of the convex hull of C_i. Then C_o is polygonal and its vertices are among those of C_i. Hence the assertion above applies, and

$$|C_i| \geq |C_o|.$$

Now we prove inequality (1). Since the A's are the vertices of a convex polygon, no three of them are collinear. If C has vertices in addition to A_0, $A_1, \ldots, A_n$, we can replace it by a strictly shorter polygon from which these additional vertices have been eliminated. Hence we assume that C can be described as $B_0B_1 \ldots B_n$ where $B_n = B_0$ and the B's are a permutation of the A's.

If the B's are the A's in the same cyclic order or in reverse order, then $C = C_o$. If not, we shall show how to reorder the B's to obtain a strictly shorter closed polygon C'. Since there are only a finite number of possible polygons having the A's as vertices, there must be a shortest, and it follows that C_o is the shortest one, as claimed.

Suppose then that two consecutive B's, which we take to be B_0 and B_1 by cyclically renumbering the B's if necessary, are not consecutive vertices of C_o. Then the line $\overleftrightarrow{B_0B_1}$ is not a line of support for C_o, and there must be vertices on both sides of $\overleftrightarrow{B_0B_1}$. Hence there must be an integer k, $2 \leq k \leq n - 2$ such that B_k is on one side of $\overleftrightarrow{B_0B_1}$ while B_{k+1} is on the other. Now the four points B_0, B_1, B_k, B_{k+1} are the vertices of a convex quadrilateral Q, and since $\overleftrightarrow{B_0B_1}$ separates B_k and B_{k+1}, the diagonals of Q are B_0B_1 and B_kB_{k+1}. Hence (as proved below)

(2) $$|B_0B_1| + |B_kB_{k+1}| > |B_0B_k| + |B_1B_{k+1}|.$$

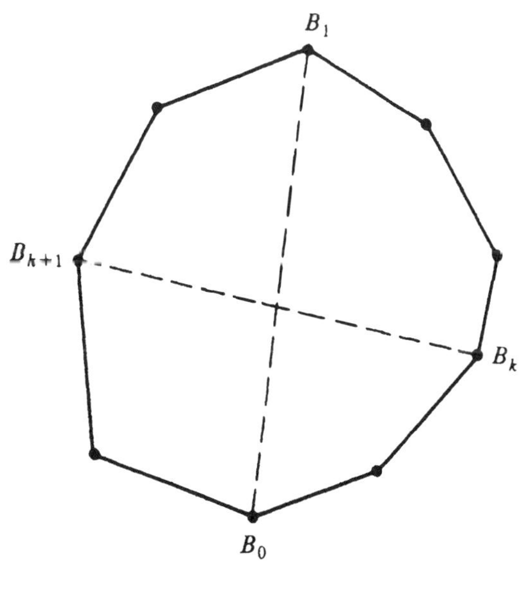

Consider now the polygonal closed curve C' described by the vertices $B_0B_kB_{k-1} \ldots B_2B_1B_{k+1}B_{k+2} \ldots B_n$. We have

$$|C'| = |C| - |B_0B_1| - |B_kB_{k+1}| + |B_0B_k| + |B_1B_{k+1}| < |C|.$$

Thus C' is a closed polygon having the same vertices which is strictly shorter than C. This proves inequality (1).

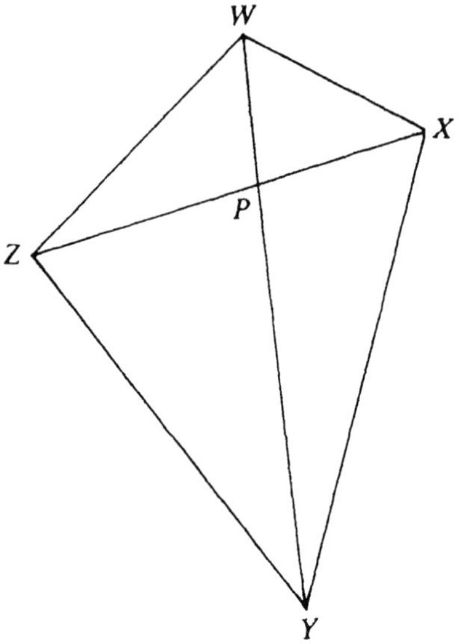

To prove (2), let $WXYZ$ be a convex quadrilateral. The diagonals intersect at P, and we have

$$|WY| + |XZ| = |WP| + |PZ| + |YP| + |PX| > |WZ| + |XY|.$$

(iv) This assertion is false. Suppose $ABCD$ is a regular tetrahedron in space with center O. Let S be the union of the four segments OA, OB, OC, OD. The smallest convex set containing S is clearly the tetrahedron $ABCD$, but the surface area of S is zero. To be sure, S is not a surface at all, but we can "fatten" it a bit to produce a surface S_ϵ with a small area and with the same convex hull as S. Then we shall have $|S_0| = |ABCD| > |S_\epsilon|$.

To be explicit, let A', B', C', D' be chosen so that O is on each of segments AA', BB', CC', DD' with $|OA'| = |OB'| = |OC'| = |OD'| = \epsilon$ where ϵ is a small positive number. Let S_ϵ be the surface of the polyhedral solid

$$A'B'C'D' \cup AB'C'D' \cup A'BC'D' \cup A'B'CD' \cup A'B'C'D.$$

By choosing ϵ small enough we can make $|S_\epsilon|$ as small as we please.

DISCUSSION. Assertions (i) and (iii) can easily be extended to non-polygonal curves. This is because the length of a non-polygonal curve can be obtained as the limit of the lengths of suitable approximating polygons. In fact, the length of a curve is usually defined as the limit of the lengths of approximating polygons.

Suppose, for example, C_i is an arbitrary closed convex curve in the plane and C_o is an arbitrary curve surrounding C_i. Suppose $|C_i| > |C_o|$. Then we could replace C_i by a closed convex polygon C_i' inside C_i but only a trifle shorter, so that $|C_i'| > |C_o|$. C_o would still surround C_i'. Then we could replace C_o by a polygonal approximation C_o', still surrounding C_i', so that $|C_i'| > |C_o'|$. But this contradicts assertion (a). We conclude that $|C_i| \leq |C_o|$. Thus (i) holds without the restriction to polygons.

To extend assertion (ii) is more difficult. The theory of area is complicated by the fact that one cannot define the area of an arbitrary surface simply as the limit of the areas of inscribed polyhedra. (See, for example, R. Courant, *Differential and Integral Calculus*, vol. 2, 2nd English ed., Interscience Publishers, New York, 1947, pp. 268 ff. and 341-342.) This difficulty is easily overcome for smooth surfaces by requiring that each face of an approximating polyhedron be nearly parallel to that portion of the given surface it is supposed to approximate. There are several plausible ways to define the area of a "rough" surface, but unfortunately they do not all agree. Assertion (ii) is true, and the proof given above remains valid, if the convex surface S_i is polyhedral, and the notion of area employed to evaluate $|S_o|$ is such that an orthogonal projection into a plane never increases area. Fortunately, this property is shared by the best-known definitions of area. (For a thorough discussion of area and more generally of k-dimensional measure for subsets of n-space, see H. Federer, *Geometric Measure Theory*, Springer, New York, 1969.)

Now allow S_i to be the surface of a convex set, not necessarily a polyhedron. If $|S_i| > |S_o|$, then we could replace S_i by a convex polyhedral surface S_i' lying inside S_i with $|S_i'| > |S_o|$; but this would contradict the last paragraph.

The choice of S_i' relies on the following fact. The surface area of a convex body is the supremum of the areas of inscribed polyhedral convex bodies. To see this, note that assertion (ii) implies that the area of any convex polyhedron containing a convex body K exceeds the area of any convex polyhedron inside K. Moreover, it is not hard to prove that there are inner and outer convex polyhedra whose areas differ by as little as we please. Hence the supremum of the areas of inner convex polyhedra coincides with the infimum of the areas of outer convex polyhedra. We may appropriately define the area of K to be this common value, or we may prove that this value is forced upon us either by some general definition of area or by some axiom concerning area. In his famous treatise on the cylinder and sphere, Archimedes begins with a fundamental assumption which amounts to the statement that, if one convex body is inside another, the area of the inner one is less than that of the outer one. In view of the remarks above, this assumption determines the area of K to be the supremum of the areas of convex polyhedra inside K.

Alternate Solution to (i). It is possible to give a relatively short and simple proof of (i) which, however, becomes more difficult in the higher-dimensional cases.

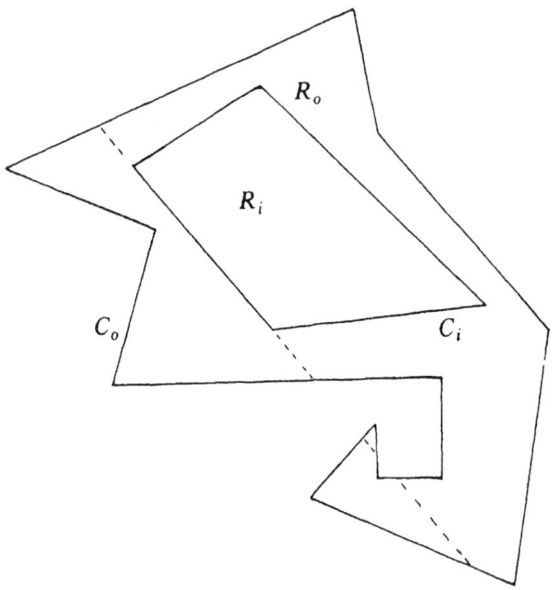

Suppose R_o is a polygonal region in the plane and R_i is a convex polygonal region inside R_o. We wish to prove that the boundary C_o of R_o is at least as long as the boundary C_i of R_i (actually longer unless $C_o = C_i$). By extending some side of C_i we cut off a piece (or pieces) of R_o to obtain a new region R_o' containing R_i. The perimeter of R_o' is evidently smaller than that of R_o unless $R_o' = R_o$. After a finite number of such cuts (extending the sides of R_i in turn) we reduce the perimeter of R_o to that of R_i. It follows immediately that $|C_o| > |C_i|$ unless $C_o = C_i$.

REMARK. Part (iii) of the problem is the subject of a paper by Jonathan Schaer, "An 'Obvious' but Useful Theorem About Closed Curves," *Mathematics Magazine*, vol. 45 (1972), pages 154–155.

THE ELEVENTH WILLIAM LOWELL PUTNAM MATHEMATICAL COMPETITION
March 31, 1951

Morning Session

1. Show that the determinant:

$$\begin{vmatrix} 0 & a & b & c \\ -a & 0 & d & e \\ -b & -d & 0 & f \\ -c & -e & -f & 0 \end{vmatrix}$$

is non-negative, if its elements a, b, c, etc., are real.

Solution. Assume $f \neq 0$. Divide the third row and the third column by f, and write $\bar{b} = b/f$, $\bar{d} = d/f$. Then, if Δ is the determinant,

$$\frac{1}{f^2}\Delta = \begin{vmatrix} 0 & a & \bar{b} & c \\ -a & 0 & \bar{d} & e \\ -\bar{b} & -\bar{d} & 0 & 1 \\ -c & -e & -1 & 0 \end{vmatrix}.$$

Subtract c times the third row from the first row and e times the third row from the second row.

$$\frac{1}{f^2}\Delta = \begin{vmatrix} \bar{b}c & a + \bar{d}c & \bar{b} & 0 \\ -a + \bar{b}e & \bar{d}e & \bar{d} & 0 \\ -\bar{b} & -\bar{d} & 0 & 1 \\ -c & -e & -1 & 0 \end{vmatrix}.$$

Make similar transformations of the columns.

$$\frac{1}{f^2}\Delta = \begin{vmatrix} 0 & a + \bar{d}c - \bar{b}e & \bar{b} & 0 \\ -a + \bar{b}e - \bar{d}c & 0 & \bar{d} & 0 \\ -\bar{b} & -\bar{d} & 0 & 1 \\ 0 & 0 & -1 & 0 \end{vmatrix}$$

$$= \begin{vmatrix} 0 & a + c\bar{d} - \bar{b}e \\ -a + \bar{b}e - c\bar{d} & 0 \end{vmatrix}$$

$$= (a + c\bar{d} - \bar{b}e)^2.$$

Finally, $\Delta = (af + cd - be)^2$, and therefore Δ is non-negative for any real choice of a, b, c, d, e, f.

Note that if $f = 0$, the matrix has more zeros, and an elementary application of Laplace's expansion of a determinant gives $\Delta = (cd - be)^2$. (Alternatively, this can also be seen by letting $f \to 0$ in the above formula $\Delta = (af + cd - be)^2$.) More generally, since Δ is obviously a polynomial in the matrix entries, the assumption $f \neq 0$ is really no loss of generality in evaluating Δ. In the language of algebra, the computation was really made in the field $Q(a, b, c, d, e, f)$ where a, b, c, d, e, and f are indeterminates. In this field, $f \neq 0$.

REMARK. This proof was chosen because it generalizes easily to prove the following fact by induction: The determinant of a skew-symmetric matrix of even order is the square of a polynomial in the matrix entries. (We show below that the determinant of a skew-symmetric matrix of odd order is zero.) In fact the determinant of the $2n \times 2n$ skew-symmetric matrix (a_{ij}) is the square of

$$\frac{1}{2 \cdot 4 \cdots 2n} \sum_{\sigma} (-1)^{\sigma} a_{\sigma(1),\sigma(2)}\, a_{\sigma(3),\sigma(4)} \cdots a_{\sigma(2n-1),\sigma(2n)},$$

where $(-1)^{\sigma}$ denotes the sign of the permutation σ and the sum is taken over all members σ of the symmetric group. This polynomial was called the Pfaffian by Cayley because Jacobi had used it in his work on Pfaff's problem. Because the a's are skew-symmetric, the Pfaffian can be written as the sum of $1 \cdot 3 \cdot 5 \cdots (2n - 1)$ products without denominators. See Thomas Muir, *The Theory of Determinants in the Historical Order of Development*, Vol. 1, Dover, New York, pages 401–406.

The determinant of a skew-symmetric matrix S of odd order n is always zero. We have

$$(-1)^n \det S = \det(-S) = \det S^T = \det S;$$

so, for n odd, $\det S = 0$.

2. In the plane, what is the locus of points the sum of the squares of whose distances from n fixed points is a constant? What restrictions, stated in geometric terms, must be put on the constant so that the locus is non-null?

Solution. Let the given points be $\{(x_i, y_i)\}$, $i = 1, 2, \ldots, n$. Let a point on the locus be (x, y). Then the required condition is

$$\Sigma[(x - x_i)^2 + (y - y_i)^2] = C$$

where C is a constant. This can be rewritten as $nx^2 - 2(\Sigma x_i)x + \Sigma x_i^2 + ny^2 - 2(\Sigma y_i)y + \Sigma y_i^2 = C$ and by completing the squares in the form

$$\left(x - \frac{\Sigma x_i}{n}\right)^2 + \left(y - \frac{\Sigma y_i}{n}\right)^2 = \frac{C}{n} + \left(\frac{\Sigma x_i}{n}\right)^2 + \left(\frac{\Sigma y_i}{n}\right)^2 - \frac{\Sigma x_i^2}{n} - \frac{\Sigma y_i^2}{n}.$$

The locus is therefore empty if the right member is negative; a single point (namely the centroid $((1/n)\Sigma x_i, (1/n)\Sigma y_i)$ of the given points) if the right member is zero; and a circle with center at the centroid if the right member is positive. If coordinates are chosen with the origin at the centroid of the given points, the condition for a real non-null locus then becomes

$$C \geq \Sigma(x_i^2 + y_i^2) = \Sigma r_i^2$$

where r_i is the distance of the ith point from the new center. That is, C must be at least as large as the sum of the squares of the distances from the centroid. This last formulation is a special instance of the general fact that a planar mass has its minimal moment of inertia about an axis perpendicular to the plane through the centroid of the mass.

---

3. Find the sum to infinity of the series:

$$1 - \frac{1}{4} + \frac{1}{7} - \frac{1}{10} + \cdots + \frac{(-1)^{n+1}}{3n - 2} + \cdots.$$

Solution. We first show that

$$1 - \frac{1}{4} + \frac{1}{7} - \frac{1}{10} + \cdots + \frac{(-1)^{n+1}}{3n - 2} + \cdots = \int_0^1 \frac{dt}{1 + t^3}.$$

Note that

$$\frac{1}{1 + t^3} = \sum_{n=1}^{k} (-1)^{n+1} t^{3n-3} + \frac{(-1)^k t^{3k}}{1 + t^3}$$

for any $t \neq -1$.

Integrate from 0 to 1 to obtain

$$\int_0^1 \frac{dt}{1+t^3} - \sum_{n=1}^{k} (-1)^{n+1} \int_0^1 t^{3n-3}\,dt = (-1)^k \int_0^1 \frac{t^{3k}\,dt}{1+t^3}.$$

Hence

$$\left| \int_0^1 \frac{dt}{1+t^3} - \sum_{n=1}^{k} \frac{(-1)^{n+1}}{3n-2} \right| = \left| \int_0^1 \frac{t^{3k}\,dt}{1+t^3} \right| \le \int_0^1 t^{3k}\,dt = \frac{1}{3k+1}.$$

Letting $k \to \infty$, we get

$$\sum_{n=1}^{\infty} \frac{(-1)^{n+1}}{3n-2} = \int_0^1 \frac{dt}{1+t^3}.$$

This integral can be evaluated by partial fractions:

$$\begin{aligned} \int_0^1 \frac{dt}{1+t^3} &= \frac{1}{3} \int_0^1 \left[\frac{1}{1+t} + \frac{2-t}{1-t+t^2} \right] dt \\ &= \frac{1}{3} \left[\log(1+t) - \frac{1}{2} \log(1-t+t^2) + \sqrt{3} \arctan \frac{2t-1}{\sqrt{3}} \right]_0^1 \\ &= \frac{1}{3} \left[\log 2 + \sqrt{3} \left(\frac{\pi}{6} + \frac{\pi}{6} \right) \right] = \frac{1}{3} \left(\log 2 + \frac{\pi}{\sqrt{3}} \right). \end{aligned}$$

REMARK. By the same method it can be shown that

$$\frac{1}{a} - \frac{1}{a+b} + \frac{1}{a+2b} - \frac{1}{a+3b} + \cdots = \int_0^1 \frac{t^{a-1}\,dt}{1+t^b}$$

whenever $a, b > 0$, a result due to Gauss.

---

4. Trace the curve whose equation is:

$$y^4 - x^4 - 96y^2 + 100x^2 = 0.$$

Solution. On completing squares we obtain

$$(x^2 - 50)^2 - (y^2 - 48)^2 = 14^2.$$

Letting $X = x^2$, $Y = y^2$, we find

$$(X - 50)^2 - (Y - 48)^2 = 14^2$$

and the graph of this equation in the XY-plane is readily identified as a rectangular hyperbola with center at (50, 48) and asymptotes

$$\begin{cases} X + Y = 98 \\ X - Y - 2. \end{cases}$$

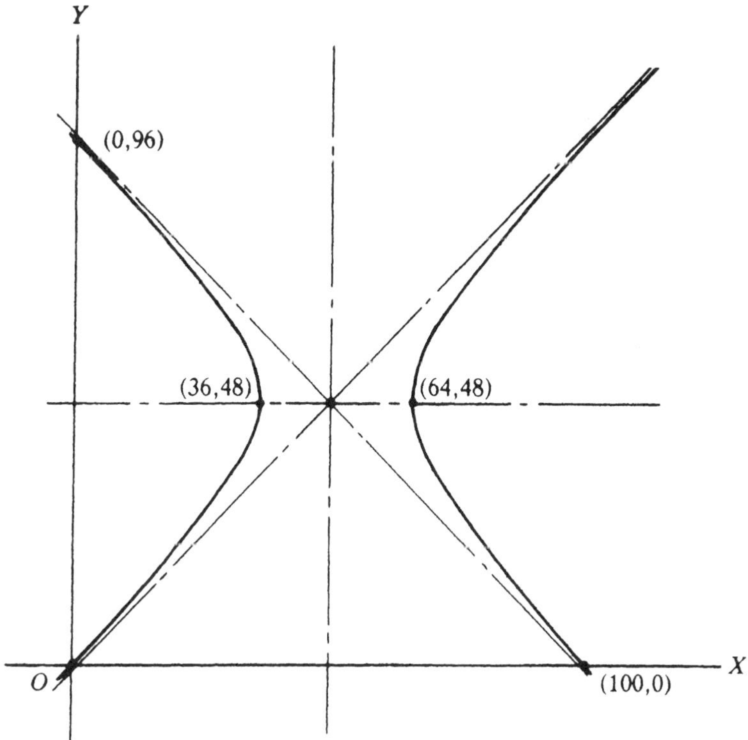

$(X - 50)^2 - (Y - 48)^2 = 14^2$
Auxiliary Graph

In the present situation interest is confined to the first quadrant of the XY-plane. For each point (X, Y) in the first quadrant there are four points of the required locus L, namely $(\pm\sqrt{X}, \pm\sqrt{Y})$. Because the transformation $(x, y) \mapsto (X, Y)$ mapping each quadrant of the xy-plane onto the first quadrant of the XY-plane is differentiable in both directions, the smooth arcs of the auxiliary graph correspond to smooth arcs of L.

The three points (0, 0), (0, 96), and (100, 0) of the auxiliary graph lying on the boundary of the first quadrant correspond to five points (0, 0), $(0, \pm\sqrt{96})$, $(\pm 10, 0)$ of L which require special consideration.

The function

$$f(x, y) = y^4 - x^4 - 96y^2 + 100x^2$$

has gradient

$$\nabla f = (-4x^3 + 200x,\ 4y^3 - 192y).$$

Since ∇f does not vanish at the points $(0, \pm\sqrt{96})$, $(\pm 10, 0)$, the curve L is smooth at those points. The first component of ∇f vanishes at $(0, \pm\sqrt{96})$, so L has horizontal tangents at these two points. The second component of ∇f vanishes at $(\pm 10, 0)$, so L has vertical tangents at these points. At the origin, both components of ∇f vanish, so we consider the quadratic terms

$$100x^2 - 96y^2$$

of f at this point. Since these constitute a non-degenerate quadratic form which vanishes along the lines $y = \pm(100/96)x$, the curve L crosses itself at the origin, the two branches being tangent to these two lines.

Examination shows that the first component of ∇f does not vanish at any other points of L, so L has no more horizontal tangents, but the second component of ∇f vanishes at eight more points of L, namely $(\pm 6, \pm\sqrt{48})$, $(\pm 8, \pm\sqrt{48})$, so L has vertical tangents at these points, corresponding to the vertical tangents on the auxiliary graph.

Since the line $X - Y = 2$ is an asymptote to the auxiliary hyperbola, $X - Y \to 2$ along the unbounded arc of this hyperbola in the first quadrant. Correspondingly,

$$x - y = \sqrt{X} - \sqrt{Y} = \frac{X - Y}{\sqrt{X} + \sqrt{Y}} \to 0$$

along the unbounded arc of L in the first quadrant. Hence L is asymptotic to the line $y = x$ in the first quadrant. By symmetry L is asymptotic to the same line in the third quadrant and to $y = -x$ in the second and fourth quadrants.

Since there are no points of the auxiliary locus in the strip $36 < X < 64$, there are no points of L in the strips $6 < x < 8$ and $-8 < x < -6$.

The graph of $f(x, y) = 0$ is therefore as shown.

For general methods of graphing higher-degree plane curves, see C. A. Stewart, *Advanced Calculus*, 3rd ed., Methuen, London, 1951, or J. H.

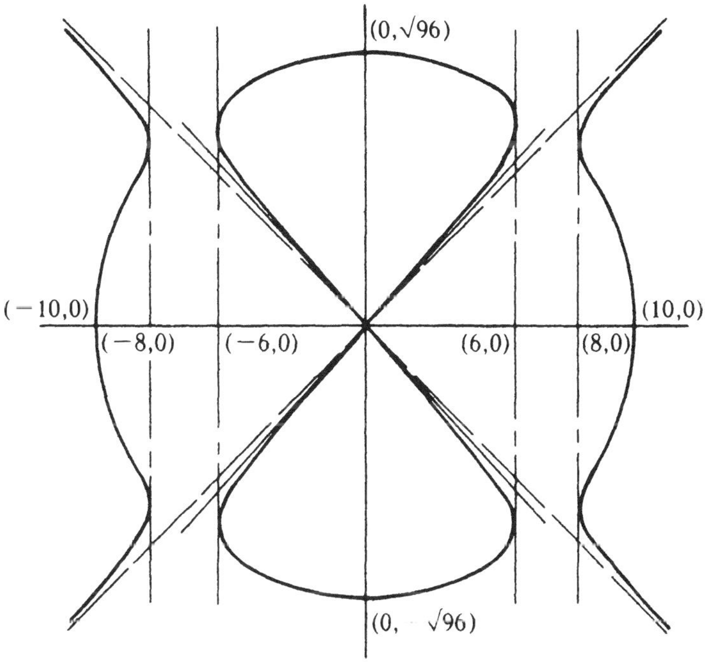

Graph of
$y^4 - x^4 + 96y^2 + 100x^2 = 0.$

Cadwell, *Topics in Recreational Mathematics,* Cambridge University Press, New York, 1966.

---

5. Consider in the plane the network of points having integral coordinates. For lines having rational slope show that:

(i) the line passes through no points of the network or through infinitely many;

(ii) there exists for each line a positive number d having the property that no point of the network, except such as may be on the line, is closer to the line than the distance d.

Solution. (i) If L is a line with rational slope its equation may be written as $ax + by + c = 0$ where a and b are integers not both zero. Suppose (x_1, y_1) is a point on the line and that x_1 and y_1 are both integers. Then points of the form $(x_1 + kb, y_1 - ka)$, $k = 0, \pm 1, \pm 2, \ldots$ all lie on the line, since

$$a(x_1 + kb) + b(y_1 - ka) + c = ax_1 + by_1 + c = 0.$$

Thus, if there is one point with integer coordinates on a line with rational slope, there are infinitely many.

(ii) The distance from the point (p, q) to the line L with equation $ax + by + c = 0$ is

$$d = \frac{|ap + bq + c|}{\sqrt{a^2 + b^2}}.$$

Since L has rational slope we can choose a and b to be integers. Then $ap + bq$ takes on only integer values. Therefore, if c is an integer, d is either 0 or at least $1/\sqrt{a^2 + b^2}$. If c is not an integer, d is at least $e/\sqrt{a^2 + b^2}$, where e is the distance from c to the nearest integer.

REMARK. If we choose the equation of L so that a and b are relatively prime integers, then $ap + bq$ takes all integer values and the least distance to L from a point with integral coordinates not on L is exactly $1/\sqrt{a^2 + b^2}$ if c is an integer and $e/\sqrt{a^2 + b^2}$ if it is not.

Properties of the integral lattice were discussed by Martin Gardner, "Mathematical Games," *Scientific American,* May 1965, pages 120–122, and also by Fritz Herzog and B. M. Stewart, "Patterns of Visible and Non-Visible Lattice Points," *American Mathematical Monthly,* vol. 78 (1971), pages 487–493. Further references to the extensive literature can be found in these sources.

---

6. Determine the position of a normal chord of a parabola such that it cuts off of the parabola a segment of minimum area.

Solution. Choose coordinates so that the equation of the parabola is $4ay = x^2$, $a > 0$.

The chord connecting the point $P(2as, as^2)$ to the point $Q(2at, at^2)$ has the equation

$$y = \frac{1}{2}(t + s)x - ast \tag{1}$$

and the tangent line at $(2at, at^2)$ has slope t. Hence the line (1) will be normal to the parabola at Q if and only if

$$\frac{1}{2}t(t + s) = -1$$

which may be written as

$$s = -\frac{2}{t} - t.$$

We see, therefore, that s and t have opposite signs. Take $s < 0$ and $t > 0$. Then the area cut off by the chord is

$$\int_{2as}^{2at} \left[\frac{1}{2}(t + s)x - ast - \frac{1}{4a}x^2 \right] dx = \frac{1}{3}a^2(t - s)^3.$$

This area will be minimal when $t - s$ is minimal. But

$$t - s = 2t + \frac{2}{t} = 2\left(\sqrt{t} - \frac{1}{\sqrt{t}}\right)^2 + 4 \geq 4.$$

Equality is attained only when $\sqrt{t} = 1$ and hence $t = 1$.

Thus, of all normals to the parabola at points to the right of the axis the normal at $(2a, a)$ cuts off the least area. The area cut off is $64a^2/3$.

By symmetry, the normal at $(-2a, a)$ cuts off the least area among normals at points to the left of the axis.

The critical normals can be characterized as those which meet the axis at an angle of 45°.

---

7. Show that if the series $a_1 + a_2 + a_3 + \cdots + a_n + \cdots$ converges, then the series $a_1 + a_2/2 + a_3/3 + \cdots + a_n/n + \cdots$ converges also.

Solution. This is a special case of a result sometimes called Abel's summation theorem: If the partial sums of the series $\Sigma\, a_n$ are bounded and the sequence $\{b_n\}$ decreases to zero, then $\Sigma\, a_n b_n$ is convergent.

This theorem can be proved as follows. Let $s_k = \Sigma_{i=1}^{k} a_i$ and let M be a bound for $\{|s_k|\}$. Then

$$\begin{aligned} a_1b_1 + a_2b_2 + \cdots + a_nb_n & \\ &= s_1b_1 + (s_2 - s_1)b_2 + \cdots + (s_n - s_{n-1})b_n \\ &= s_1(b_1 - b_2) + s_2(b_2 - b_3) + \cdots + s_{n-1}(b_{n-1} - b_n) + s_n b_n, \end{aligned}$$

that is,

$$\sum_{i=1}^{n} a_ib_i = \sum_{i=1}^{n-1} s_i(b_i - b_{i+1}) + s_n b_n .$$

Now the series $\sum_{i=1}^{\infty} s_i(b_i - b_{i+1})$ is absolutely convergent, for

$$|s_i(b_i - b_{i+1})| \le M(b_i - b_{i+1})$$

and $\sum_{i=1}^{\infty} (b_i - b_{i+1})$ converges to b_1.

Moreover, $\lim_{n\to\infty} s_n b_n = 0$ because $\{s_n\}$ is bounded. Therefore,

$$\lim_{n\to\infty} \sum_{i=1}^{n} a_i b_i = \sum_{i=1}^{\infty} s_i(b_i - b_{i+1}).$$

Thus the theorem is proved.

For the particular problem, take $b_n = 1/n$ and the result is immediate.

Afternoon Session

1. Find the condition that the functions $M(x, y)$ and $N(x, y)$ must satisfy in order that the differential equation $Mdx + Ndy = 0$ shall have an integrating factor of the form $f(xy)$. You may assume that M and N have continuous partial derivatives of all orders.

Solution. If there is such an integrating factor, say $f(xy)$, then

$$f(xy)M(x, y)dx + f(xy)N(x, y)dy = 0 \tag{1}$$

must be a closed (i.e., locally exact) differential form. For this it is necessary that

$$\frac{\partial}{\partial y}[f(xy)M(x, y)] = \frac{\partial}{\partial x}[f(xy)N(x, y)], \tag{2}$$

that is,

$$xf'M + f\frac{\partial M}{\partial y} = yf'N + f\frac{\partial N}{\partial x} \tag{3}$$

whence

$$\frac{f'}{f} = \frac{1}{xM - yN}\left(\frac{\partial N}{\partial x} - \frac{\partial M}{\partial y}\right) \tag{4}$$

assuming $f(xy) \neq 0$ and $xM - yN \neq 0$.

Now the left-hand side of (4) is a function of (xy), hence the right-hand side must be also. Thus, assuming $xM - yN \neq 0$, a necessary condition for the existence of the desired integrating factor is that there exist a function R of one variable such that

$$\frac{1}{xM - yN}\left(\frac{\partial N}{\partial x} - \frac{\partial M}{\partial y}\right) = R(xy).$$

Conversely, if such a function R exists, then an integrating factor is given by

$$f(xy) = \exp \int^{xy} R(t)dt$$

since this function will certainly satisfy (2).

Justification. The theory of integrating factors is usually considered as a local theory; that is, one seeks to convert a differential form into one that is only locally exact and for this the cross derivative condition is both necessary and sufficient. Hence (2') is both necessary and sufficient to find a local integrating factor.

The condition $f \neq 0$ involved in (4) was not investigated. Suppose $xM - yN$ does not vanish at a point (x_0, y_0). By continuity we confine ourselves to a convex open neighborhood U of (x_0, y_0) on which $xM - yN$ never vanishes. If $f(xy)$ is an integrating factor defined on U and $f(x_0y_0) = 0$, then we may regard (3), restricted to some straight line through (x_0, y_0) as a non-singular homogeneous linear differential equation satisfied by f. But the solution of such an equation, if zero at one point, is zero everywhere. Thus if $f(x_0y_0) = 0$, f is everywhere zero. But, by definition, an integrating factor is not identically zero. Hence the condition stated is indeed necessary and sufficient for the existence of a local integrating factor near a point at which $xM - yN$ does not vanish. The situation near a point at which $xM - yN$ vanishes seems to be complicated.

A differential condition can be found that tells whether or not a smooth function $L(x, y)$ can be expressed in the form $R(xy)$. It is evidently necessary that

$$x\frac{\partial L}{\partial x} = y\frac{\partial L}{\partial y} \tag{5}$$

since if $L(x, y) = R(xy)$, then $\partial L/\partial x = yR'(xy)$ and $\partial L/\partial y = xR'(xy)$.

Conversely (5) is sufficient locally at any point except the origin. To prove this note that if (5) holds, then

$$\frac{d}{dt}L\left(t, \frac{a}{t}\right) = 0$$

for any fixed a, and therefore L is constant along any connected set on which xy is constant. It is clear that this implies that L can be written as $R(xy)$ locally, except at the origin.

Our previous condition becomes

$$(6)\quad [xM - yN]\left[x\frac{\partial^2 N}{\partial x^2} - x\frac{\partial^2 M}{\partial x\,\partial y} - y\frac{\partial^2 N}{\partial x\,\partial y} + y\frac{\partial^2 M}{\partial y^2}\right]$$

$$= \left[\frac{\partial N}{\partial x} - \frac{\partial M}{\partial y}\right]\left[xM + yN + x^2\frac{\partial M}{\partial x} + y^2\frac{\partial N}{\partial y} - xy\left(\frac{\partial N}{\partial x} + \frac{\partial M}{\partial y}\right)\right]$$

and we can assert that (6) is a necessary and sufficient condition for the local existence of an integrating factor of the form $f(xy)$ near a point (x_0, y_0) at which $xM - yN$ does not vanish.

REMARK. Condition (5) is not sufficient that L be expressible as $R(xy)$ in the neighborhood of the origin, for we can take $L(x, y) = |x|xy^2$, which is a C^1-function that satisfies (5) everywhere, but which cannot be written as $R(xy)$ in any neighborhood of the origin.

---

2. Two functions of x are differentiable and not identically zero. Find an example of two such functions having the property that the derivative of their quotient is the quotient of their derivatives.

Solution. We must find functions f and g such that

$$(1)\qquad \frac{f'g - fg'}{g^2} = \frac{f'}{g'},$$

which is essentially equivalent to

$$(2)\qquad g(g' - g)f' - g'^2 f = 0.$$

Now, if we choose any interval I and any function g such that neither g, nor $g' - g$, nor g' vanishes on I, then (2) is a non-singular first-order linear differential equation for f on I with the general solution

$$(3)\qquad f(x) = C\exp\int^x \frac{g'^2}{g(g' - g)}\,dt.$$

With any choice of $C \neq 0$, the functions f and g will satisfy (1) and not vanish identically.

To be completely explicit, try $I = \mathbf{R}$ and $g(x) = e^{\lambda x}$. Then (2) becomes

$$(\lambda - 1)f' - \lambda^2 f = 0$$

and we can take

$$f(x) = \exp\left(\frac{\lambda^2}{\lambda - 1}\right)x$$

if $\lambda \neq 1$. A convenient choice is $\lambda = 2$. Then $f(x) = e^{4x}$, $g(x) = e^{2x}$, $f(x)/g(x) = e^{2x}$ and

$$(f/g)' = 2e^{2x} = f'/g'.$$

REMARK. Formula (3), with any choice of g such that $g(g' - g)g'$ does not vanish on I, gives the "general solution" of the problem. It does not, however, give all solutions because the requirement that $g' - g$ not vanish on I is too restrictive. Thus

$$f(x) = x \exp \int_0^x \frac{(t+2)^2}{1-t^3}\,dt$$

$$g(x) = 1 + x + x^2$$

is a solution of the problem on $(-\frac{1}{2}, 1)$, but $g' - g$ vanishes at 0.

---

3. Show that if x is positive, then

$$\log_e (1 + 1/x) > 1/(1 + x).$$

First Solution.

$$\log\left(1 + \frac{1}{x}\right) = \int_x^{1+x} \frac{dt}{t} > \int_x^{1+x} \frac{dt}{1+x} = \frac{1}{1+x}.$$

Second Solution. In the well-known inequality

$$\log(1 + y) < y \qquad \text{for } -1 < y < 0$$

put $y = -1/(1 + x)$. We obtain

$$\log\left(\frac{x}{1+x}\right) < -\frac{1}{1+x}$$

and, on changing signs,

$$\log\left(1+\frac{1}{x}\right) > \frac{1}{1+x}.$$

---

4. Investigate, in any way which yields significant results, the existence, in the plane, of the configuration consisting of an ellipse simultaneously tangent to four distinct concentric circles.

First Solution. We shall show that from any point P near the center of a (non-circular) ellipse four distinct normals to the ellipse can be drawn. We shall then show that if P is on neither axis of the ellipse the lengths of these normals are all different. Hence with any such point P as center, four distinct concentric circles can be drawn tangent to the given ellipse. Thus the configuration called for certainly does exist.

Suppose the ellipse has center O, major axis AC of length $2a$, and minor axis BD of length $2b$, where $a > b$.

Suppose P is any point such that $|OP| < \frac{1}{2}(a-b)$. Then the triangle law shows that

$$\begin{aligned}
|PA| &\geq |OA| - |OP| > \tfrac{1}{2}(a+b)\\
|PB| &\leq |OB| + |OP| < \tfrac{1}{2}(a+b)\\
|PC| &> \tfrac{1}{2}(a+b)\\
|PD| &< \tfrac{1}{2}(a+b).
\end{aligned} \tag{1}$$

Hence, as X varies along the ellipse, $|PX|$ will have a maximum at some point A' along the arc DAB, a minimum at some point B' along the arc ABC, a maximum at some point C' along the arc BCD, and a minimum at some point D' along the arc CDA. None of these extrema are taken at the endpoints of the arcs cited because of the inequalities (1), hence the segments PA', PB' PC', PD' are all normal to the ellipse (because for example, the circle with center P through A' lies on or outside the ellipse near A').

Now assume also that P is not on the major axis. For definiteness say that P lies above $\overleftrightarrow{AC}$. Let E be the point obtained by reflecting D' in $\overleftrightarrow{AC}$. Then E is on the ellipse; and $|PE| < |PD'|$, because $\overleftrightarrow{AC}$ is the perpendicular bisector of $D'E$ and P is on the same side of $\overleftrightarrow{AC}$ as E. Now $|PE| \geq |PB'|$ because $|PB'|$ is the least value of $|PX|$ as X varies along the elliptical arc ABC which contains E. Therefore $|PB'| < |PD'|$.

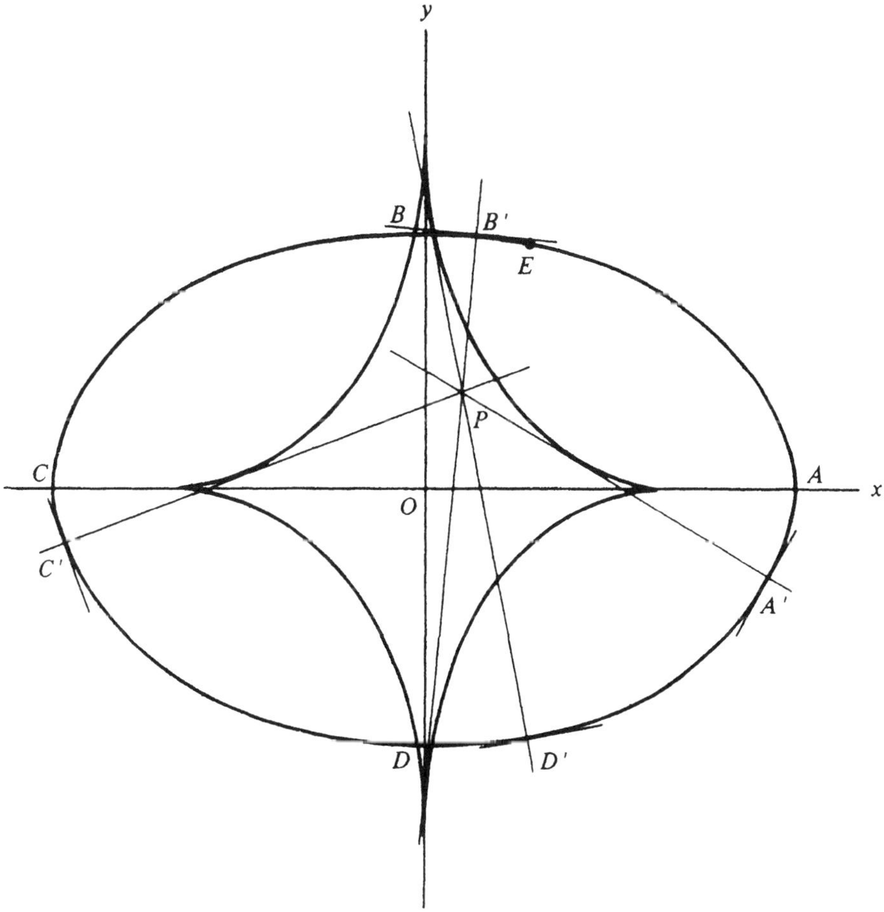

A similar argument shows that $|PA'| \neq |PC'|$ if P is not on the minor axis.

It follows from the inequalities (1) and the choice of A', B', C', D' that $|PA'| > \frac{1}{2}(a + b)$, $|PB'| < \frac{1}{2}(a + b)$, $|PC'| > \frac{1}{2}(a + b)$, and $|PD'| < \frac{1}{2}(a + b)$. Hence $|PA'|$, $|PB'|$, $|PC'|$, $|PD'|$ are all different. This completes the proof that the configuration exists.

The four critical points of the function $|PX|$ can easily be located analytically. If the ellipse has equation

$$\frac{x^2}{a^2} + \frac{y^2}{b^2} = 1$$

and P has coordinates (h, k), then

$$|PX|^2 = (x - h)^2 + (y - k)^2.$$

Taking λ as a Lagrange multiplier we consider

$$(x-h)^2+(y-k)^2-\lambda\left(\frac{x^2}{a^2}-\frac{y^2}{b^2}-1\right).$$

Setting the partial derivatives equal to zero, we obtain

$$(x-h)=\lambda\frac{x}{a^2}$$

$$(y-k)=\lambda\frac{y}{b^2}.$$

Eliminating λ, we see that the four critical points are the intersection of the ellipse with the (possibly degenerate) hyperbola

$$(a^2-b^2)xy=a^2hy-b^2kx.$$

Since there are four intersections for $h=k=0$, there must be four intersections when h and k are near zero.

The situation is easy to visualize in terms of the evolute of the ellipse. From a point P within the evolute, four tangents can be drawn to the evolute, and these will be normal to the ellipse.

Second Solution. We shall now sketch a proof that, given any four distinct concentric circles, there is an ellipse tangent to all four.

Let C_1, C_2, C_3, C_4 be four such circles with center O and respective radii $r_1<r_2<r_3<r_4$. Fix a point A on C_1 (using rotational symmetry we can insist that our ellipse be tangent to C_1 at A). Let l be the tangent to C_1 at A and let Q_1 and Q_2 be the quarter-planes bounded by l and the ray AO. Let l_1 be the ray of l that bounds Q_1. From the point $B=l_1\cap C_3$, draw the tangent BD to C_2 in Q_1, and from any point X of l_1 outside C_3, draw the tangent XY to C_2 in Q_1.

There is a one-parameter family of conics tangent to l at A and to $\overleftrightarrow{XY}$ at Y. All of these are tangent to both C_1 and C_2. Degenerate members of this family include the double line connecting A to Y and the union of the lines $\overleftrightarrow{XY}$ and l. Continuity considerations show that there is a member $K(X)$ of this family that touches C_3 at a point $P(X)$ of $Q_1\cap C_3$.

As $X\to B$, $K(X)$ degenerates to $l\cup\overleftrightarrow{BD}$; hence for X near B, $K(X)$ is a hyperbola (with one branch in the lower left of the diagram). As $X\to\infty$, $K(X)$ approaches an ellipse symmetric about $\overleftrightarrow{OA}$ and hence tangent to C_3 a second time in Q_2; this ellipse does not meet C_4 at all. Assuming

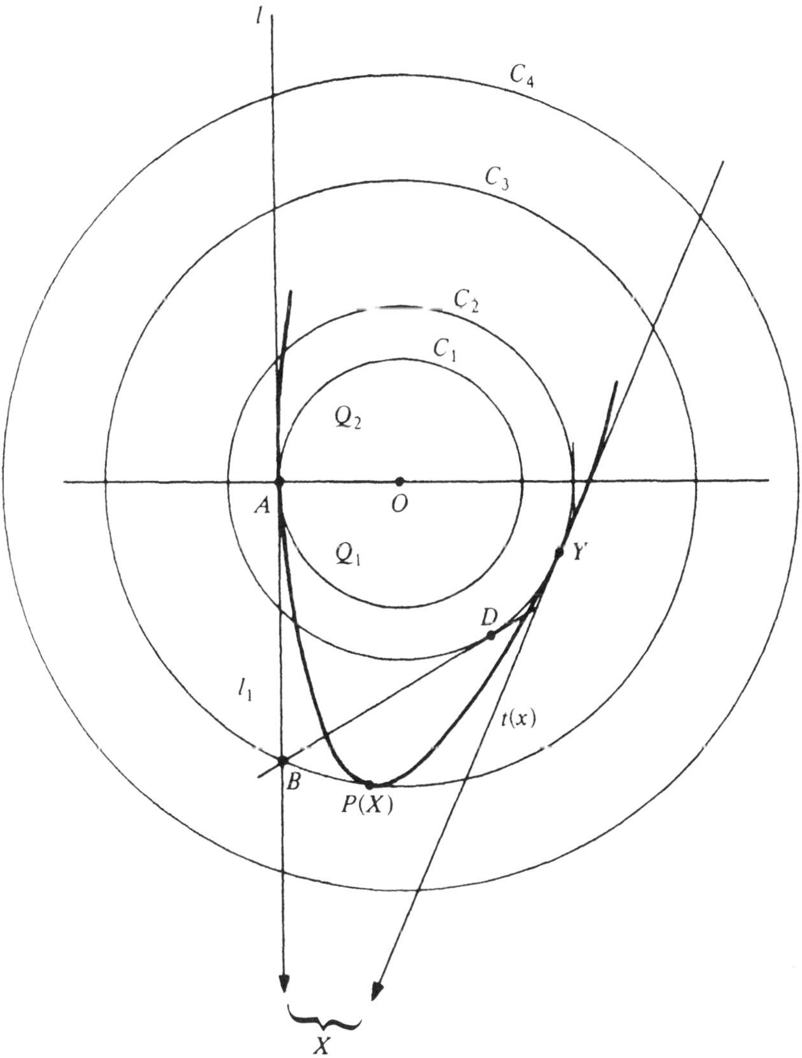

$K(X)$ depends continuously on X, there must be a value of X such that $K(X)$ meets C_4 at a double point in Q_2; i.e., $K(X)$ is tangent to C_4. Evidently this conic has no points outside C_4 so it is an ellipse. So we have found an ellipse tangent to all four given circles.

Another solution (still with A as point of tangency with C_1) is found by reflection in $\overleftrightarrow{OA}$, but this is not essentially different. However, by choosing Q_1 and Q_2 on the other side of l, an entirely new solution is found.

Remark. The configuration of four normals to an ellipse from a point P evoked much interest in the nineteenth century; for example, it was shown by Joachimstal that the feet of three such normals and the reflection through P of the fourth foot are concyclic. See J. Casey, *A Treatise on the Analytical Geometry of the Point Line and Conic Sections*, Dublin, 1893, pages 218–219.

---

5. A plane through the center of a torus is tangent to the torus. Prove that the intersection of the plane and the torus consists of two circles.

Solution. We can choose a coordinate system so that the origin is at the center of the torus and the z-axis is the axis of rotational symmetry. Then the equation of the torus is

$$[\sqrt{(x^2 + y^2)} - a]^2 + z^2 = b^2$$

where $a > b > 0$. Let Π be a plane tangent to the torus at a point Q. Rotate the coordinate system about the z-axis so that Q lies in the xz-plane. Now Π intersects the xz-plane in a line $z = \lambda x$. We can suppose, without loss of generality, that $\lambda > 0$.

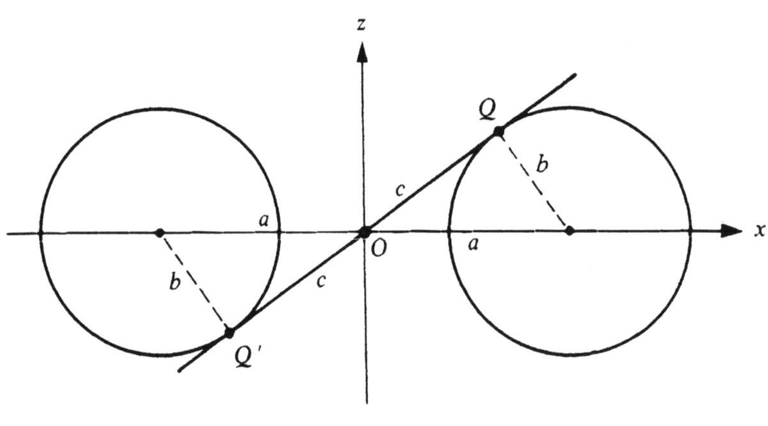

The line through Q normal to the xz-plane is tangent to the torus and therefore lies in Π. Hence the equation of Π is $z = \lambda x$.

It is clear from symmetry that Π is also tangent to the torus at a second point Q'. Let c denote the distance from O to Q. Then $c^2 + b^2 = a^2$ and $\lambda = b/c$.

The intersection set I of Π and the torus certainly contains the following six points: four points of the form $(0, \pm a \pm b, 0)$ on the y axis, and the two points Q and Q'. We note that the set I is also symmetric in the y-axis. Hence if I consists of two circles, they must be the circles of radius a and centers $(0, \pm b, 0)$ lying in the plane Π. (See figure.)

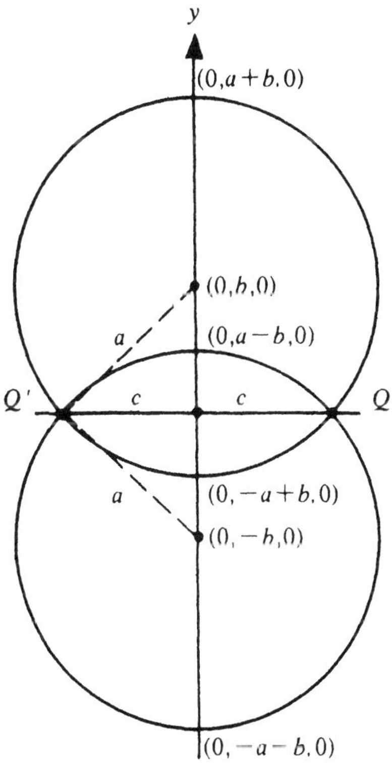

Orthogonal View of the plane Π.

Parametric equations for the first of these circles are

(1)
$$y = b + a\sin\theta,\quad x = c\cos\theta,\quad z = b\cos\theta.$$

To check this, note that these values satisfy both $x^2 + (y - b)^2 + z^2 = a^2$ and $z = \lambda x$, so (1) describes a closed curve that lies on the sphere of radius a about $(0, b, 0)$ and on the plane $z = \lambda x$. (To get the second circle just reverse the sign of b.)

Next we show that these circles lie on the torus. For any point on (1) we have

$$\begin{aligned} x^2 + y^2 &= c^2\cos^2\theta + b^2 + 2ab\sin\theta + a^2\sin^2\theta \\ &= (a^2 - b^2)\cos^2\theta + b^2 + 2ab\sin\theta + a^2\sin^2\theta \\ &= a^2 + 2ab\sin\theta + b^2\sin^2\theta = (a + b\sin\theta)^2. \end{aligned}$$

Since $a > b$, $a + b \sin\theta > 0$; so $\sqrt{x^2 + y^2} = a + b \sin\theta$. Hence

$$[\sqrt{x^2 + y^2} - a]^2 + z^2 = b^2 \sin^2\theta + b^2 \cos^2\theta = b^2,$$

which is the defining equation of the torus.

Thus the first circle lies on the torus. That the second circle does also follows by similar algebra, or by noting that the two circles are interchanged by a reflection in the xz-plane.

REMARK. This solution is not quite complete, because it only shows that I contains the two circles; conceivably there are further points in the intersection set I, although it is intuitively evident that there are not.

A complete solution can be obtained by purely algebraic manipulations. If the radical is eliminated from the original equation for the torus, we obtain the equivalent equation

$$(x^2 + y^2 + z^2 + a^2 - b^2)^2 = 4a^2x^2 + 4a^2y^2.$$

This can be rewritten as

$$(x^2 + y^2 + z^2 + b^2 - a^2)^2 - 4b^2y^2 = 4b^2x^2 - 4c^2z^2$$

and then as

$$(x^2 + (y - b)^2 + z^2 - a^2)(x^2 + (y + b)^2 + z^2 - a^2) = 4(bx - cz)(bx + cz).$$

From this equation it appears that any point on both the torus and the plane $bx - cz = 0$ is also on one of the two spheres

$$x^2 + (y - b)^2 + z^2 = a^2 \quad \text{and} \quad x^2 + (y + b)^2 + z^2 = a^2.$$

Conversely, any point on the plane and either sphere lies also on the torus. Thus the intersection set I is exactly the union of the two circles.

REMARK. This problem is discussed in H. S. M. Coxeter, *Introduction to Geometry*, Wiley, New York, 1961, pages 132–133. References to other papers may be found in Coxeter's book.

---

6. Assuming that all the roots of the cubic equation $x^3 + ax^2 + bx + c =$

0 are real, show that the difference between the greatest and the least roots is not less than $(a^2 - 3b)^{1/2}$ or greater than $2(a^2 - 3b)^{1/2}/3^{1/2}$.

Solution. Let the roots be ordered so that $r_1 \le r_2 \le r_3$ and let $r_2 = r_1 + u$ and $r_3 = r_2 + v$ where u and v are non-negative. Then

$$a = -(r_1 + r_2 + r_3)$$
$$b = r_1r_2 + r_2r_3 + r_3r_1;$$

and

$$\begin{aligned} a^2 - 3b &= \tfrac{1}{2}\,[(r_1 - r_2)^2 + (r_2 - r_3)^2 + (r_3 - r_1)^2] \\ &= \tfrac{1}{2}\,[u^2 + v^2 + (u + v)^2] = (u + v)^2 - uv. \end{aligned}$$

Since u and v are non-negative, this gives

$$a^2 - 3b \le (u + v)^2.$$

Since the difference between the greatest and the least roots is $u + v$, this difference is at least $(a^2 - 3b)^{1/2}$.

On the other hand,

$$u^2 + v^2 = \tfrac{1}{2}\,[(u + v)^2 + (u - v)^2] \ge \tfrac{1}{2}\,(u + v)^2,$$

so

$$a^2 - 3b \ge \tfrac{3}{4}\,(u + v)^2$$

and the difference $u + v$ is at most $(2/\sqrt{3})\,(a^2 - 3b)^{1/2}$.

Remark. The proof shows that the given lower bound for $u + v$ is attained precisely when two of the roots are equal, and the given upper bound precisely when $u = v$, that is, when the roots are in arithmetic progression.

---

7. Find the volume of the four-dimensional hypersphere $x^2 + y^2 + z^2 + t^2 = r^2$, and also the hypervolume of its interior $x^2 + y^2 + z^2 + t^2 < r^2$.

Solution. Let $V_4(r)$ be the hypervolume of the interior of the hypersphere, i.e., the four-dimensional ball of radius r. If the hypervolume is "sliced" perpendicular to the x-axis, one gets

$$V_4(r) = \int_{-r}^{r} V_3(\sqrt{r^2 - x^2})\, dx$$

where

$$V_3(\rho) = (4/3)\pi\rho^3$$

is the ordinary volume of the three-dimensional ball of radius ρ. Hence

$$V_4(r) = (4/3)\pi \int_{-r}^{r} (r^2 - x^2)^{3/2}\, dx.$$

Making the substitution $x = r \sin \theta$

$$V_4(r) = \frac{4\pi r^4}{3} \int_{-\pi/2}^{\pi/2} \cos^4\theta \, d\theta$$

$$= \frac{\pi r^4}{3} \int_{-\pi/2}^{\pi/2} (\tfrac{3}{2} + 2 \cos 2\theta + \tfrac{1}{2} \cos 4\theta)\, d\theta,$$

where we have used the double-angle formula $2 \cos^2\theta = 1 + \cos 2\theta$ twice to simplify the integrand. Hence

$$V_4(r) = \frac{\pi r^4}{3} \cdot \tfrac{3}{2}\pi = \frac{\pi^2 r^4}{2}$$

is the hypervolume of the four-dimensional ball.

Let $A_4(r)$ be the three-dimensional volume of the hyperspherical surface of radius r. We can relate A_4 to V_4 as follows. If we compute $V_4(r)$ by considering concentric spherical shells we see that

$$V_4(r) = \int_0^r A_4(\rho)\, d\rho.$$

Hence by the fundamental theorem of calculus

$$\frac{d}{dr} V_4(r) = A_4(r),$$

so that

$$A_4(r) = 2\pi^2 r^3.$$

Extension. This method shows that the n-dimensional volume of an n-dimensional ball of radius r is given by

$$V_n(r) = \gamma_n r^n$$

where the γ's can be calculated recursively by the formula

$$\gamma_n = \gamma_{n-1} \int_{-\pi/2}^{\pi/2} \cos^n\theta \, d\theta.$$

The values of these integrals are obtainable from the reduction formula

$$\int_{-\pi/2}^{\pi/2} \cos^n\theta \, d\theta = \frac{n-1}{n} \int_{-\pi/2}^{\pi/2} \cos^{n-2} d\theta \qquad \text{for } n \geq 2,$$

which leads to the result

$$\gamma_{2n} = \frac{\pi^n}{n!}, \qquad \gamma_{2n+1} = \frac{2(2\pi)^n}{1 \cdot 3 \cdot 5 \cdots (2n+1)}.$$

Using the Γ-function we can express both of these equations by

$$\gamma_m = \frac{\pi^{m/2}}{\Gamma\left(\frac{m}{2} + 1\right)}.$$

The $(n - 1)$-dimensional volume of the surface of the n-ball is given by

$$A_n(r) = \frac{d}{dr}(V_n(r)) = n\gamma_n r^{n-1} = \frac{2\pi^{n/2} r^{n-1}}{\Gamma\left(\frac{n}{2}\right)}.$$

See H. P. Evans, *American Mathematical Monthly*, vol. 54 (1947), pages 592–594.

THE TWELFTH WILLIAM LOWELL PUTNAM MATHEMATICAL COMPETITION

March 22, 1952

Morning Session

1. Let

$$f(x) = \sum_{i=0}^{i=n} a_i x^{n-i}$$

be a polynomial of degree n with integral coefficients. If a_0, a_n, and $f(1)$ are odd, prove that $f(x) = 0$ has no rational roots.

Solution. Suppose $f(p/q) = 0$ where p and q are integers having no common factor. Then

$$(1) \qquad q^n f\left(\frac{p}{q}\right) = a_0 p^n + q a_1 p^{n-1} + \cdots + q^{n-1} a_{n-1} p + q^n a_n = 0.$$

It follows that q divides a_0 and p divides a_n. Therefore p and q are both odd. Hence

$$\begin{aligned} a_0 p^n + q a_1 p^{n-1} + \cdots + q^{n-1} a_{n-1} p + q^n a_n &\equiv a_0 + a_1 + \cdots + a_{n-1} + a_n \\ &= f(1) \equiv 1 \pmod 2, \end{aligned}$$

contradicting (1).

---

2. Show that the equation

$$(9 - x^2)\left(\frac{dy}{dx}\right)^2 = (9 - y^2)$$

characterizes a family of conics touching the four sides of a fixed square.

Solution. Equation (1) defines a two-valued direction field in the open square

$$S: -3 < x < 3, \qquad -3 < y < 3$$

and in the four quadrants

$$\begin{aligned} Q_1&: \ 3 < x, & 3 < y \\ Q_2&: \ x < -3, & 3 < y \\ Q_3&: \ x < -3, & y < -3 \\ Q_4&: \ 3 < x, & y < -3. \end{aligned}$$

Let U be the union of these five open regions. The direction field can be extended continuously to the boundary of U except at the points $(\pm 3, \pm 3)$ where the boundary lines intersect. This leaves four open semi-infinite strips of width 6 where no direction field is defined at all. We shall discuss the solutions of the differential equation first on the open region U.

On U the two-valued direction field can be resolved into two ordinary direction fields given by the differential equations

$$\frac{dy}{dx} = +\sqrt{\frac{9-y^2}{9-x^2}} \tag{2}$$

and

$$\frac{dy}{dx} = -\sqrt{\frac{9-y^2}{9-x^2}}\,. \tag{3}$$

The right members of both of these equations are continuously differentiable (on U), so there is a unique maximal solution for each equation through each point of U. These solutions can be found explicitly since the variables are separable. On S, equation (2) becomes

$$\frac{dy}{\sqrt{9-y^2}} = \frac{dx}{\sqrt{9-x^2}}$$

whence

$$\arcsin\frac{y}{3} = \arcsin\frac{x}{3} + C. \tag{4}$$

Here C must be between $-\pi$ and π because arcsin takes values between $-\pi/2$ and $\pi/2$. Similarly, equation (3) leads to

(5) $$\arcsin \frac{y}{3} = -\arcsin \frac{x}{3} + D$$

on S with $-\pi < D < \pi$.

On the quadrants (2) becomes

$$\frac{dy}{\sqrt{y^2 - 9}} = \frac{dx}{\sqrt{x^2 - 9}}$$

giving

(6) $$\operatorname{arccosh} \tfrac{1}{3}|y| = \operatorname{arccosh} \tfrac{1}{3}|x| + E,$$

where E may have any value. Equation (3) leads to

(7) $$\operatorname{arccosh} \tfrac{1}{3}|y| = -\operatorname{arccosh} \tfrac{1}{3}|x| + F.$$

If we take the sine of both sides of (4) and use the addition formula we get

$$\frac{y}{3} = \frac{x}{3} \cos C + \sqrt{1 - \frac{x^2}{9}} \sin C.$$

Clearing the radical, this reduces to

(8) $$x^2 + y^2 - 2xy \cos C = 9 \sin^2 C.$$

Similar transformations applied to (5), (6), and (7) give

(9) $$x^2 + y^2 + 2xy \cos D = 9 \sin^2 D,$$

(10) $$x^2 + y^2 - 2|xy| \cosh E = -9 \sinh^2 E,$$

and

(11) $$x^2 + y^2 + 2|xy| \cosh F = -9 \sinh^2 F.$$

Since $|xy| = +xy$ or $-xy$ throughout any single quadrant Q_i, these equations show that the solution curves for all solutions of (2) and (3) on the open domain U are parts of curves in the one parameter family

(12) $$x^2 + y^2 - 2\lambda xy = 9(1 - \lambda^2).$$

It follows from our work that there are at least two curves of the family

(12) passing through each point of U. There are, in fact, exactly two because, given (x, y), there will be exactly two values of λ satisfying (12) if and only if the discriminant of (12), regarded as a polynomial in λ, is positive. This condition simplifies to $(x^2 - 9)(y^2 - 9) > 0$, which is valid if and only if $(x, y) \in U$.

Except for the values $\lambda = \pm 1$ the curves (12) are conics tangent to the lines $x = \pm 3$ and $y = \pm 3$. The latter fact is easily seen; for example, if we put $x = 3$ in (12) and solve for y, we get a double root $y = 3\lambda$. For $\lambda = \pm 1$, the conics degenerate into straight lines, $(x - y)^2 = 0$ and $(x + y)^2 = 0$. Equation (12) remains the same if we interchange x and y or $-x$ and y; hence, the conics are all symmetric in the lines $y = x$ and $y = -x$. For $|\lambda| < 1$, the discriminant of (12) is negative, so the curves are ellipses (a circle for $\lambda = 0$) lying in $\bar{S}$. For $|\lambda| > 1$, the discriminant is positive and the curves are hyperbolas, each branch lying in one of the closed quadrants $\overline{Q_i}$. (See Figure 1.)

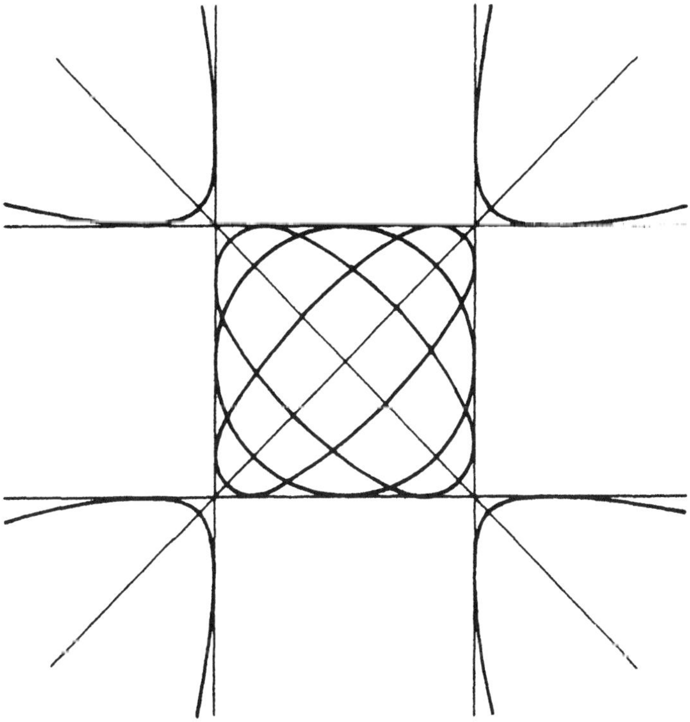

Fig. 1. Some curves in the family (12).

The family (12) is, of course, the family of conics referred to in the problem. It is worth noting that we can derive the differential equation (1) directly from (12). Differentiating (12) implicitly we find

$$\frac{dy}{dx} = -\frac{x - \lambda y}{y - \lambda x}.$$

Therefore,

$$\left(\frac{dy}{dx}\right)^2 = \frac{(x^2 + y^2 - 2\lambda xy) + (\lambda^2 - 1)y^2}{(x^2 + y^2 - 2\lambda xy) + (\lambda^2 - 1)x^2}$$

$$= \frac{9(1 - \lambda^2) + (\lambda^2 - 1)y^2}{9(1 - \lambda^2) + (\lambda^2 - 1)x^2} = \frac{9 - y^2}{9 - x^2},$$

which is essentially equivalent to (1).

Now we turn our attention to solutions of (1) on the closed region $\overline{U}$. Because the right members of (2) and (3) do not represent functions differentiable on $\overline{U}$, the uniqueness of solutions may and does fail at points of the boundary of U. For example, $y = 3$ and $y = \sqrt{9 - x^2}$ are both solutions of (1) passing through $(0, 3)$. We can piece together the old solutions to (1) on U with various parts of the lines $y = \pm 3$ to obtain a great variety of solutions to (1). Thus,

$$\begin{aligned} y &= \sqrt{9 - x^2} && \text{for } -3 < x < 0 \\ &= 3 && \text{for } 0 \le x \le 4\tfrac{1}{2} \\ &= \tfrac{3}{2}x - \tfrac{1}{2}\sqrt{5(x^2 - 9)} && \text{for } 4\tfrac{1}{2} < x, \end{aligned}$$

defines a solution to (1) whose graph includes a quadrant of a circle in S and part of a hyperbola (corresponding to $\lambda = 3/2$) in Q_1. (See Figure 2.)

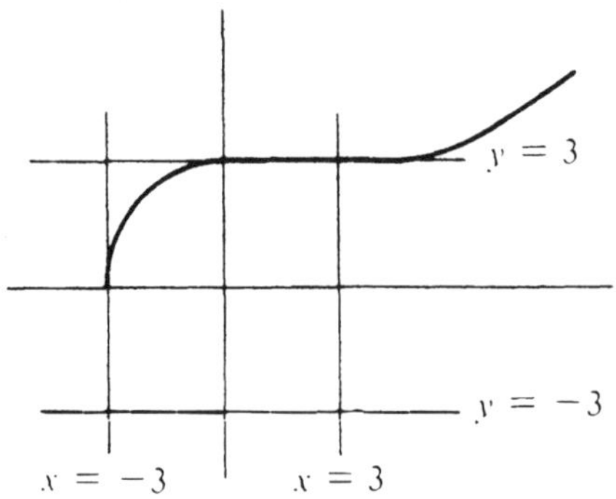

Fig. 2.

If we interpret (1) as a differential equation for curves in the plane (including curves which may not be the graphs of functions), then the whole of any of the conics defined by (12) is a solution curve. In this interpretation we can also splice segments of the vertical lines $x = \pm 3$ into solutions obtaining curves that wander around in $\overline{S}$ in a complicated fashion, as shown in Figure 3.

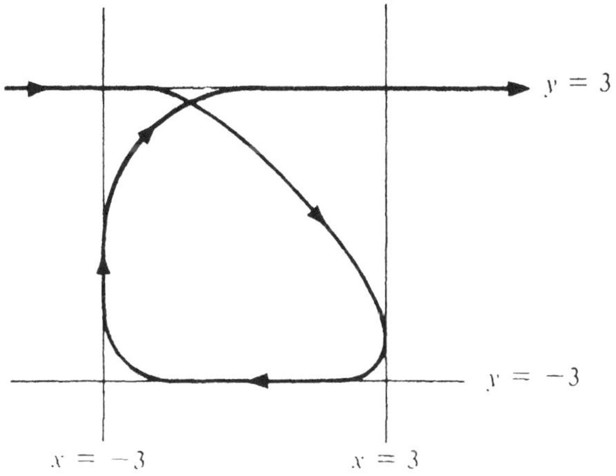

Fig. 3.

---

3. Develop necessary and sufficient conditions which ensure that r_1, r_2, r_3 and r_1^2, r_2^2, r_3^2 are simultaneously roots of the equation $x^3 + ax^2 + bx + c = 0$.

Solution. It seems clear that r_1, r_2, r_3 are intended to be all the roots of the equation in the algebraic sense, i.e.,

$$x^3 + ax^2 + bx + c = (x - r_1)(x - r_2)(x - r_3).$$

However, it is not so clear that r_1^2, r_2^2, r_3^2 must also be *all* the roots or merely among the roots. For example, $x(x - 1)(x + 1) = 0$ has the roots $r_1 = 0$, $r_2 = 1$, $r_3 = -1$, and $r_1^2 = 0$, $r_2^2 = 1$, $r_3^2 = 1$ are among the roots but are not *all* the roots. We shall find all polynomials for each interpretation.

Interpretation 1. r_1^2, r_2^2, r_3^2 are all of the roots.

(1) $$(x - r_1)(x - r_2)(x - r_3) = x^3 + ax^2 + bx + c$$
$$= (x - r_1^2)(x - r_2^2)(x - r_3^2).$$

Using symmetric functions of the roots, we get

$$r_1^2 + r_2^2 + r_3^2 = (r_1 + r_2 + r_3)^2 - 2(r_1r_2 + r_2r_3 + r_3r_1)$$
$$= a^2 - 2b \text{ from the left equation of (1),}$$
$$= -a \text{ from the right equation of (1).}$$

Also

$$r_1^2r_2^2 + r_2^2r_3^2 + r_3^2r_1^2 = (r_1r_2 + r_2r_3 + r_3r_1)^2 - 2r_1r_2r_3(r_1 + r_2 + r_3)$$
$$= b^2 - 2ac, \text{ and also } = b.$$

Again

$$r_1^2r_2^2r_3^2 = c^2, \text{ and also } = -c.$$

Thus, we have the three equations

$$c^2 = -c$$
$$b^2 - 2ac = b,$$
$$a^2 - 2b = -a.$$

The first relation has only two possible solutions, $c = 0$ and $c = -1$. It is quite easy to find the solution triplets for $c = 0$.

These are

$$c = 0, \quad b = 0, \quad a = 0$$
$$c = 0, \quad b = 0, \quad a = -1$$
$$c = 0, \quad b = 1, \quad a = 1$$
$$c = 0, \quad b = 1, \quad a = -2.$$

If $c = -1$, then $a^2 + a = 2b$ and $b^2 - b = -2a$. Substituting the second of these equations into the first we get

$$b^4 - 2b^3 - b^2 - 6b = b(b - 3)(b^2 + b + 2) = 0.$$

This gives four solution triplets

$$c = -1, \quad b = 0, \quad a = 0$$
$$c = -1, \quad b = 3, \quad a = -3$$
$$c = -1, \quad b = \frac{-1 + i\sqrt{7}}{2}, \quad a = \frac{1 + i\sqrt{7}}{2}$$
$$c = -1, \quad b = \frac{-1 - i\sqrt{7}}{2}, \quad a = \frac{1 - i\sqrt{7}}{2}.$$

These eight cases yield eight explicit polynomials

$$f_1(x) = x^3$$

$$f_2(x) = x^3 - x^2 = x^2(x - 1)$$

$$f_3(x) = x^3 + x^2 + x = x(x^2 + x + 1)$$

$$f_4(x) = x^3 - 2x^2 + x = x(x - 1)^2$$

$$f_5(x) = x^3 - 1 = (x - 1)(x^2 + x + 1)$$

$$f_6(x) = x^3 - 3x^2 + 3x - 1 = (x - 1)^3$$

$$f_7(x) = x^3 + \left(\frac{1 + i\sqrt{7}}{2}\right) x^2 + \left(\frac{-1 + i\sqrt{7}}{2}\right) x - 1$$

$$f_8(x) = x^3 + \left(\frac{1 - i\sqrt{7}}{2}\right) x^2 + \left(\frac{-1 - i\sqrt{7}}{2}\right) x - 1.$$

Second Solution for Interpretation 1. There are essentially three different ways that the sequences r_1, r_2, r_3 and r_1^2, r_2^2, r_3^2 can be identified. That is, by renumbering the roots we can arrange that one of the following is true:

(i) $r_1^2 = r_1, \quad r_2^2 = r_2, \quad r_3^2 = r_3,$

(ii) $r_1^2 = r_1, \quad r_2^2 = r_3, \quad r_3^2 = r_2,$

(iii) $r_1^2 = r_2, \quad r_2^2 = r_3, \quad r_3^2 = r_1.$

Relations (i) yield $r_i = 0$ or 1 for $i = 1, 2, 3$, and hence correspond to the four polynomials x^3, $x^2(x - 1)$, $x(x - 1)^2$, $(x - 1)^3$ and hence to f_1, f_2, f_4, f_6 already found in the first method of solution.

Relations (ii) yield $r_1 = 0$ or 1, and $r_2^4 = r_2$. If $r_2 = 0$ or 1, then $r_3 = r_2^2 = r_2$ and the resulting polynomials have been included under (a). However, there are two new solutions $r_2 = \omega$ and $r_2 = \omega^2$ where ω is a complex cube root of unity. These cases yield two new polynomials, $x(x^2 + x + 1)$ and $(x - 1)(x^2 + x + 1)$, previously called f_3 and f_5.

Relations (iii) yield $r_1^8 = r_1$. This can be written in the form $r_1(r_1 - 1) \cdot (r_1^6 + r_1^5 + r_1^4 + r_1^3 + r_1^2 + r_1 + 1) = 0$. The trivial roots $r_1 = 0$ and $r_1 = 1$ lead to cases already considered. Let $\alpha = \exp(2\pi i/7)$, a seventh root of unity. Then we can have $r_1 = \alpha, \alpha^2, \alpha^3, \alpha^4, \alpha^5, \alpha^6$. These six cases lead to two polynomials, one having the roots α, α^2, α^4, and the other having the roots α^3, α^5, α^6. These polynomials must be f_7 and f_8.

It is easy to check that $\eta = \alpha + \alpha^2 + \alpha^4$ satisfies $\eta^2 + \eta + 2 = 0$; hence

$$\eta = \frac{-1 \pm i\sqrt{7}}{2}.$$

From the definitions of α and η, it follows easily that the imaginary part of η is positive, so that

$$\eta = \frac{-1 + i\sqrt{7}}{2}.$$

Also, if $\overline{\eta} = \alpha^3 + \alpha^5 + \alpha^6$, then $\overline{\eta}$ is also a root of $\eta^2 + \eta + 2 = 0$, and the imaginary part of $\overline{\eta}$ is negative, so

$$\overline{\eta} = \frac{-1 - i\sqrt{7}}{2}.$$

Thus $r_1 = \alpha$ leads to the polynomial

$$x^3 - (\alpha + \alpha^2 + \alpha^4)x^2 + (\alpha^3 + \alpha^5 + \alpha^6)x - 1,$$

or

$$x^3 - \eta x^2 + \overline{\eta}x - 1,$$

which is f_8.

Examination of the other possible choices $r_1 = \alpha^2, \alpha^3, \alpha^4, \alpha^5, \alpha^6$, gives f_8 for $r_1 = \alpha^2, \alpha^4$, while f_7 is obtained for $r_1 = \alpha^3, \alpha^5, \alpha^6$.

Interpretation 2. $r_1{}^2, r_2{}^2, r_3{}^2$ are among the roots.

In addition to the solutions already found under Interpretation 1, the following additional cases arise under Interpretation 2.

$$\begin{array}{lll} \text{(iv)} & r_1{}^2 = r_2{}^2 = r_1, & r_3{}^2 = r_3 \\ \text{(v)} & r_1{}^2 = r_2{}^2 = r_1, & r_3{}^2 = r_2 \\ \text{(vi)} & r_1{}^2 = r_2{}^2 = r_3, & r_3{}^2 = r_1 \\ \text{(vii)} & r_1{}^2 = r_2{}^2 = r_3{}^2 = r_1. & \end{array}$$

These cases may give additional polynomials.

Relations (iv) yield new polynomials only for $r_1 = 1$, $r_2 = -1$ and $r_3 = 0$ or 1. These new polynomials are $x(x^2 - 1)$ and $(x^2 - 1)(x - 1)$.

Relations (v) yield new polynomials for $r_1 = 1$, $r_2 = -1$, and $r_3 = \pm i$, namely $(x^2 - 1)(x - i)$ and $(x^2 - 1)(x + i)$.

Relations (vi) require that $r_3^4 = r_3$. The roots $r_3 = 0, 1$ give previously obtained polynomials. The roots $r_3 = \omega$, $r_1 = \omega^2$, $r_2 = \pm\omega^2$ and $r_3 = \omega^2$, $r_1 = \omega$, $r_2 = \pm\omega$, where ω is a complex cube root of unity, produce four new polynomials.

Relations (vii) yield one new case, $r_1 = 1$, $r_2 = r_3 = -1$ with corresponding polynomial $(x + 1)^2(x - 1)$.

The new polynomials obtained under Interpretation 2 are therefore seen to be

$$f_9 = x(x^2 - 1)$$

$$f_{10} = (x^2 - 1)(x - 1) = (x + 1)(x - 1)^2$$

$$f_{11} = (x^2 - 1)(x - i)$$

$$f_{12} = (x^2 - 1)(x + i)$$

$$f_{13} = (x - \omega^2)(x - \omega^2)(x - \omega) = (x^2 + x + 1)(x - \omega^2)$$

$$f_{14} = (x - \omega^2)(x + \omega^2)(x - \omega) = (x^2 + x + 1)(x + \omega^2)$$

$$f_{15} = (x - \omega)(x - \omega)(x - \omega^2) = (x^2 + x + 1)(x - \omega)$$

$$f_{16} = (x - \omega)(x + \omega)(x - \omega^2) = (x^2 + x + 1)(x + \omega)$$

$$f_{17} = (x + 1)^2(x - 1).$$

---

4. The flag of the United Nations consists of a polar map of the world, with the North Pole as center, extending approximately to 45° South Latitude. The parallels of latitude are concentric circles with radii proportional to their co-latitudes. Australia is near the periphery of the map and is intersected by the parallel of latitude 30° S. In the very close vicinity of this parallel how much are East and West distances exaggerated as compared to North and South distances?

Solution. Let a point on the earth with co-latitude α and longitude θ be represented by a point on the flag map with polar coordinates $k\alpha$ and θ where k is a constant of proportionality.

Let R be the radius of the earth. A meridian on the earth has length πR and on the (complete) flag map is represented by a segment of length πk. Hence a short distance d in the North-South direction is represented on the flag map by the distance $d_1 = kd/R$.

On the earth the small circle of latitude 30°S ($=$ co-latitude $2\pi/3$) is represented on the flag map by a circle of radius $2\pi k/3$ and of circumference $4\pi^2k/3$. On the earth that circle has length $2\pi R \sin(2\pi/3) = \pi R\sqrt{3}$.

Hence a short distance d in the East-West direction is represented on the flag map by a distance

$$d\,\frac{\frac{4\pi^2 k}{3}}{\pi R\sqrt{3}} = \frac{4\pi kd}{3\sqrt{3}R} = \frac{4\pi}{3\sqrt{3}}\,d_1.$$

Therefore distances in the East-West direction at latitude 30°S are magnified relative to distances in the North-South direction by the factor $4\pi/3\sqrt{3} \sim 2.42$.

---

5. Let a_j $(j = 1, 2, \ldots, n)$ be entirely arbitrary numbers except that no one is equal to unity. Prove

$$a_1 + \sum_{i=2}^{n} a_i \prod_{j=1}^{i-1} (1 - a_j) = 1 - \prod_{j=1}^{n} (1 - a_j).$$

Solution. The given statement is true for $n = 1$ (interpreting the empty sum as 0) and for $n = 2$. Suppose it is true for $n = k$, i.e.,

$$a_1 + \sum_{i=2}^{k} a_i \prod_{j=1}^{i-1} (1 - a_j) = 1 - \prod_{i=1}^{k} (1 - a_i).$$

Then

$$\begin{aligned}
a_1 + \sum_{i=2}^{k+1} a_i \prod_{j=1}^{i-1} (1 - a_j) &= a_1 + \sum_{i=2}^{k} a_i \prod_{j=1}^{i-1} (1 - a_j) + a_{k+1} \prod_{j=1}^{k} (1 - a_j) \\
&= 1 - \prod_{i=1}^{k} (1 - a_i) + a_{k+1} \prod_{j=1}^{k} (1 - a_j) \\
&= 1 - \left[\prod_{i=1}^{k} (1 - a_i)\right] (1 - a_{k+1}) \\
&= 1 - \prod_{i=1}^{k+1} (1 - a_i).
\end{aligned}$$

Thus the statement is true for $n = k + 1$. It follows by induction that it is true for all positive integers n.

REMARK. It is not necessary to require that none of the a's be unity.

---

6. A man has a rectangular block of wood m by n by r inches (m, n, and r are integers). He paints the entire surface of the block, cuts the block into inch cubes, and notices that exactly half the cubes are completely unpainted. Prove that the number of essentially different blocks with this property is finite. (Do *not* attempt to enumerate them.)

Solution. The unpainted cubes originally form a rectangular block of size $(m - 2) \times (n - 2) \times (r - 2)$. Hence the condition of the problem can be expressed

$$mnr = 2(m - 2)(n - 2)(r - 2),$$

which can be rewritten as

$$\frac{1}{2} = \frac{m-2}{m} \cdot \frac{n-2}{n} \cdot \frac{r-2}{r}.$$

Assume $m \leq n \leq r$. Then

$$\left(\frac{m-2}{m}\right)^3 \leq \frac{1}{2} < \frac{m-2}{m}.$$

Hence

$$\frac{1}{2} < \frac{m-2}{m} \leq \left(\frac{1}{2}\right)^{1/3}.$$

This gives $4 < m < 10$. Thus there are only a finite number of possibilities for the smallest integer m.

With m fixed, the equation becomes

$$\frac{m}{2(m-2)} = \frac{n-2}{n} \cdot \frac{r-2}{r}.$$

and the same reasoning as above shows that

$$\left(\frac{n-2}{n}\right)^2 \leq \frac{m}{2(m-2)} < \frac{n-2}{n},$$

whence

$$\text{(1)} \qquad \frac{m}{2(m-2)} < \frac{n-2}{n} < \left(\frac{m}{2(m-2)}\right)^{1/2} \le \left(\frac{5}{6}\right)^{1/2} < 1.$$

Thus for a fixed m, there are only a finite number of possibilities for n. Evidently, for fixed m and n there can be at most one integer r which satisfies the equation. Hence altogether there are only a finite number of cases. Note that the relaxation of the ordering assumption $m \le n \le r$ can at most multiply the number of solution triplets by six.

REMARK. It is clear that the same kind of reasoning will show that a Diophantine equation like

$$\alpha = \frac{m-2}{m} \cdot \frac{n-2}{n} \cdot \frac{r-2}{r} \cdot \frac{s-2}{s} \cdot \frac{t-2}{t}$$

has only a finite number of solutions in positive integers.

FURTHER REMARK. Disregarding the instructions of the problem, it is an easy matter to enumerate all solutions. For each value of m, the range of possible values for n is given in the following table. For the larger values of m, the inequality $n \ge m$ is more restrictive then the first inequality in (1).

m	inclusive bounds for n	
5	13	22
6	9	14
7	7	12
8	8	10
9	9	10

This makes 27 pairs m, n; of these, twenty have integral solutions for r.
A complete list of the ordered non-decreasing triples is

(5, 13, 132), (5, 14, 72), (5, 15, 52), (5, 16, 42),
(5, 17, 36), (5, 18, 32), (5, 20, 27), (5, 22, 24),
(6, 9, 56), (6, 10, 32), (6, 11, 24), (6, 12, 20),
(6, 14, 16), (7, 7, 100), (7, 8, 30), (7, 9, 20),
(7, 10, 16), (8, 8, 18), (8, 9, 14), (8, 10, 12).

7. Directed lines are drawn from the center of a circle, making angles of 0, ± 1, ± 2, ± 3, ... (measured in radians from a prime direction). If these lines meet the circle in points $P_0, P_1, P_{-1}, P_2, P_{-2}, \ldots$, show that there is no interval on the circumference of the circle which does not contain some $P_{\pm i}$. (You may assume that π is irrational.)

Solution. The various points P_0, $P_{\pm 1}$, $P_{\pm 2}$, ... must all be distinct, else π would be a rational number. (For $P_i = P_j$ implies that $j - i$ is an integral multiple of 2π.)

Let I be a given interval on the circumference, say of length ϵ. Choose N so that $\epsilon > 2\pi/N$. The points $P_0, P_1, P_2, \ldots, P_{N-1}$ are all different and divide the circle into N arcs, of which one must have length at most $2\pi/N$. Hence there are indices m and $m + k$ $(k > 0)$ such that P_m and P_{m+k} bound an arc of length $\delta < \epsilon$. If q is the greatest integer in $2\pi/\delta$, then

$$(1) \qquad P_m, P_{m+k}, P_{m+2k}, \ldots, P_{m+qk}$$

divide the circle into $q + 1$ arcs each of length δ or less. Since I is longer than any of these arcs, it contains one of the points listed in (1).

REMARK. We needed only the points $P_0, P_1, P_2, \ldots$. Clearly, the same argument will apply if P_n is at $n\alpha\pi$ radians from P_0, where α is any irrational number.

Afternoon Session

1. A mathematical moron is given two sides and the included angle of a triangle and attempts to use the Law of Cosines: $a^2 = b^2 + c^2 - 2bc \cos A$, to find the third side a. He uses logarithms as follows. He finds $\log b$ and doubles it; adds to that the double of $\log c$; subtracts the sum of the logarithms of 2, b, c, and $\cos A$; divides the result by 2; and takes the antilogarithm. Although his method may be open to suspicion, his computation is accurate. What are the necessary and sufficient conditions on the triangle that this method should yield the correct result?

Solution. The given sequence of operations will produce the correct answer for side a if and only if

$$\frac{b^2c^2}{2bc \cos A} = b^2 + c^2 - 2bc \cos A.$$

This yields

$$(b - 2c \cos A)\left(b - \frac{c}{2 \cos A}\right) = 0.$$

Hence either $b = 2c \cos A$ or $c = 2b \cos A$.

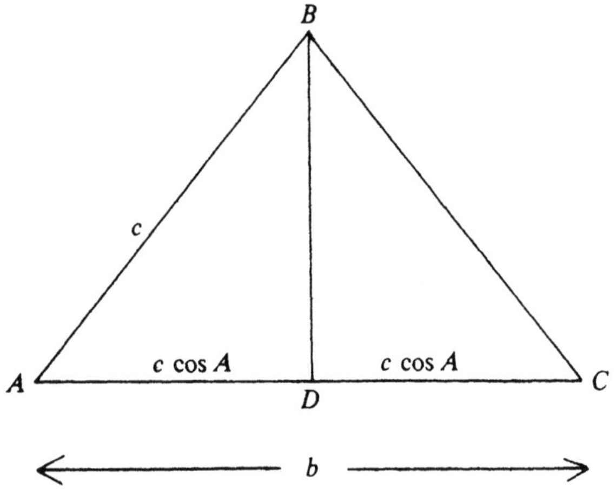

The condition $b = 2c \cos A$ implies that D, the foot of the altitude from B, is the midpoint of $\overline{AC}$, whence $\angle A = \angle C$.

Conversely, if $\angle A = \angle C$, then $b = 2c \cos A$.

Similarly, the condition $c = 2b \cos A$ is equivalent to $\angle A = \angle B$.

Hence, in order for the moron's procedure to lead to the correct answer it is necessary and sufficient that the triangle ABC be isosceles with one of the base angles at A.

---

2. Find the surface generated by the solutions of

$$\frac{dx}{yz} = \frac{dy}{zx} = \frac{dz}{xy},$$

which intersects the circle $y^2 + z^2 = 1, x = 0$.

Solution. The given differential equations mean that at the point (x, y, z) we assign a line element having the direction numbers yz, zx, and xy, and we seek curves having at each point the assigned line element as tangent. At points of the coordinate axes, however, no line element is assigned, since 0, 0, 0 is not a set of direction numbers. Furthermore, the direction field cannot be extended to the coordinate axes by continuity, since the line elements assigned on a coordinate plane are always perpendicular to that plane. On the rest of space, however, the equations assign an evidently smooth family of line elements, and so through each point of this domain passes a unique maximal solution curve. We shall determine these curves.

Multiply the given equations by $2xyz$ to get

$$2x\,dx = 2y\,dy = 2z\,dz.$$

Then integration yields

$$x^2 = y^2 + c_1 = z^2 + c_2. \tag{1}$$

In general, these equations represent the intersection of two hyperbolic cylinders which falls into four connected smooth curves, each of which is a maximal solution curve of the differential system. For example, with the choice $c_1 = -\alpha^2$, $c_2 = -\beta^2$, $\alpha\beta \neq 0$, the surfaces (1) pass through the point $(0, \alpha, \beta)$ and have the form

$$y^2 = x^2 + \alpha^2, \qquad z^2 = x^2 + \beta^2. \tag{2}$$

Their four curves of intersection are given by

$$y = \pm\sqrt{x^2 + \alpha^2}, \qquad z = \pm\sqrt{x^2 + \beta^2}. \tag{3}$$

One of these curves passes through $(0, \alpha, \beta)$. If $\alpha = 0$ but $\beta \neq 0$, two of these curves pass through each of the points $(0, 0, \pm\beta)$. Since these points are not in the domain of the original differential system, in this case each of the four curves (3) breaks into two maximal solution curves after removing these points.

Now we determine the surface formed by the solution curves which intersect the circle C: $y^2 + z^2 = 1$, $x = 0$. These solutions are evidently given by (2) with $\alpha^2 + \beta^2 = 1$, $\alpha\beta \neq 0$, and therefore they lie on the surface given by

$$y^2 + z^2 = 2x^2 + 1. \tag{4}$$

This is the equation of a hyperboloid of one sheet rotationally symmetric about the x-axis. Since four points of C must be excluded from consideration, we expect that only part of this hyperboloid is generated by the solution curves through C. Indeed, if (x, y, z) lies on (2) with α^2 and β^2 positive, then

$$y^2 > x^2, \qquad z^2 > x^2. \tag{5}$$

Conversely, if (x, y, z) satisfies (4) and (5), then positive numbers α and β satisfying (2) and $\alpha^2 + \beta^2 = 1$ can be chosen. With the appropriate choice of signs, the point (x, y, z) will lie on the curve (3), which meets the circle C at one of the points $(0, \pm\alpha, \pm\beta)$. Thus the required surface is given by the equation (4) and the inequalities (5).

The four quadrants of C each generate a portion of the required surface and these portions are bounded by the exceptional curves

$$y = \pm x,\ z = \pm\sqrt{x^2+1} \quad \text{and}\ y = \pm\sqrt{x^2+1},\ z = \pm x.$$

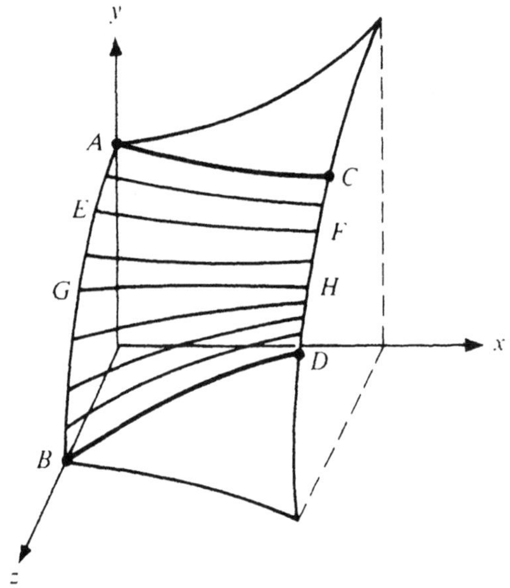

Hyperboloid: $y^2 + z^2 = 2x^2 + 1$

$A(0,1,0)$ and $B(0,0,1)$ are excluded points. Curves AC and BD are exceptional solution curves. Curves EF, GH, etc., are ordinary solution curves.

The solution curves through the illustrated quadrant of C generate that portion of the hyperboloid that lies between the exceptional solution curves.

---

3. Develop necessary and sufficient conditions that the equation

$$\begin{vmatrix} 0 & a_1 - x & a_2 - x \\ -a_1 - x & 0 & a_3 - x \\ -a_2 - x & -a_3 - x & 0 \end{vmatrix} = 0 \qquad (a_i \neq 0)$$

shall have a multiple root.

Solution. The given determinant is

$$-2x^3 + 2(a_1a_2 + a_2a_3 - a_1a_3)x = 0.$$

The necessary and sufficient condition for a multiple root is

$$a_1a_2 + a_2a_3 - a_1a_3 = 0.$$

If $a_1a_2a_3 \neq 0$, this condition can be expressed in the form

$$\frac{1}{a_1} + \frac{1}{a_3} = \frac{1}{a_2}.$$

COMMENT. We might consider a slightly more general problem. Let A be a 3×3 skew-symmetric matrix and S a 3×3 symmetric matrix. Let $f(x) = \det(A - xS)$. Then

$$\begin{aligned} f(-x) &= \det(A + xS) = \det(A^T + xS^T) \\ &= \det(-A + xS) = -\det(A - xS) = -f(x). \end{aligned}$$

So f must be an odd function, $f(x) = \alpha x^3 + \beta x$. A multiple root exists if and only if $\beta = 0$ and $\alpha \neq 0$.

---

4. A homogeneous solid body is made by joining a base of a circular cylinder of height h and radius r, and the base of a hemisphere of radius r. This body is placed with the hemispherical end on a horizontal table, with the axis of the cylinder in a vertical position, and then slightly oscillated. It is intuitively evident that if r is large as compared to h, the equilibrium will be stable; but if r is small as compared to h, the equilibrium will be unstable. What is the critical value of the ratio r/h which enables the body to rest in neutral equilibrium in any position?

Solution. Since the body is homogeneous, the center of gravity and the centroid are the same. Assume a coordinate system such that $z = 0$ is the equation of the plane where the hemisphere is joined to the cylinder (when the body is in a vertical position); the equation of the table is $z = -r$.

By cylindrical symmetry, it suffices to consider a section perpendicular to the plane of the table through the contact point. If the solid is slightly tilted, the normal force N at the point of contact will act through the point O.

If the centroid of the solid is above O, then the couple formed by N and the weight W will produce a toppling rotation. If the centroid of the solid is below O then the couple will produce a restoring rotation. Hence the problem amounts to determining the ratio of r to h which will give a centroid at O.

By symmetry we need only consider the z-coordinate of the centroid.

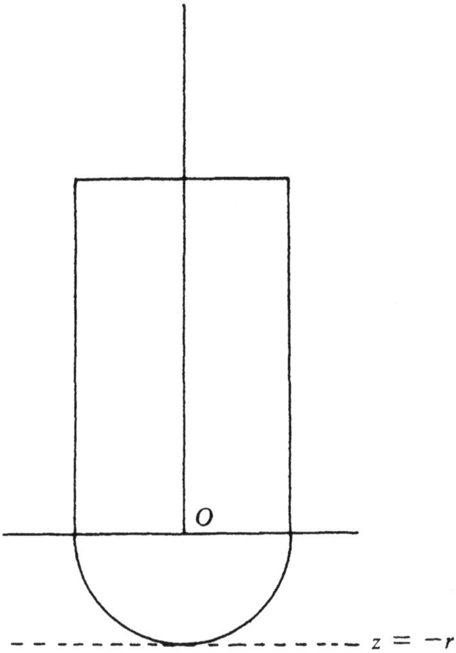

The z-coordinate of the centroid of the cylinder is $h/2$, and the volume is $\pi r^2 h$. For the hemisphere the volume is $(2/3)\pi r^3$, and the z-coordinate of the centroid is given by

$$\bar{z} = \frac{\int_{-r}^{0} z\pi(r^2 - z^2)\,dz}{(2/3)\pi r^3} = -(3/8)r.$$

The centroid of the whole body will be at the origin if and only if

$$(\pi r^2 h)\frac{h}{2} - \left(\frac{2}{3}\pi r^3\right)\left(\frac{3r}{8}\right) = 0,$$

that is $r^2 = 2h^2$, whence $r/h = \sqrt{2}$ is the critical value. For this shape, the center of gravity will be over the point of support whenever the body rests on a point of the hemispherical surface, so there will be neutral equilibrium.

For problems involving stable equilibrium see also Problem A.M. 7b of the Fourth Competition and Problem P.M. 4 of the Tenth Competition.

5. If the terms of a sequence, a_n, are monotonic, and if $\sum_1^\infty a_n$ converges, show that $\sum_1^\infty n(a_n - a_{n+1})$ converges.

First Solution. Since $\Sigma\, a_n$ converges, $\lim_{n\to\infty} a_n = 0$. Since the sequence is monotonic we have either

$$a_1 \geq a_2 \geq a_3 \geq \cdots \geq a_n \geq \cdots \geq 0,$$

or

$$a_1 \leq a_2 \leq a_3 \leq \cdots \leq a_n \leq \cdots \leq 0.$$

In the second case we can change the sign of each term and thus without loss of generality we need only consider the case where the a's are non-negative and decrease to zero.

Let

$$S_k = \sum_{n=1}^{k} n(a_n - a_{n+1}).$$

Then

$$S_k = a_1 + a_2(2 - 1) + \cdots + a_k(k - (k - 1)) - ka_{k+1}$$

$$= \sum_{n=1}^{k} a_n - ka_{k+1}.$$

Now since $\sum_1^\infty a_n$ converges, the Cauchy criterion implies $\lim_{n\to\infty}(a_n + a_{n+1} + \cdots + a_{2n}) = 0$. But $a_n + a_{n+1} + \cdots + a_{2n} \geq na_{2n}$, and hence $\lim_{n\to\infty} 2na_{2n} = 0$; from this it follows that $\lim_{n\to\infty} na_{n+1} = 0$.

In the expression

$$S_k = \sum_{n=1}^{k} a_n - ka_{k+1}$$

both $\lim_{k\to\infty} \sum_{n=1}^{k} a_n$ and $\lim\,(ka_{k+1})$ exist, so

$$\lim_{k\to\infty} S_k = \lim_{k\to\infty} \sum_1^k a_n - \lim_{k\to\infty}(ka_{k+1}) = \lim_{k\to\infty} \sum_1^k a_n.$$

This establishes the desired convergence.

REMARK. Our proof shows that if Σa_n converges, where $\{a_n\}$ is a monotone sequence, then $\lim_{n\to\infty} na_n = 0$. This result is known as Abel's theorem (or lemma).

Second Solution. We may assume, as we have seen above, that the sequence $\{a_k\}$ decreases to zero. Then for each k

$$a_k = \sum_{n=k}^{\infty} (a_n - a_{n+1}).$$

So

$$\sum_{k=1}^{\infty} a_k = \sum_{k=1}^{\infty} \sum_{n=k}^{\infty} (a_n - a_{n+1}) = \sum_{n=1}^{\infty} \sum_{k=1}^{n} (a_n - a_{n+1})$$

$$= \sum_{n=1}^{\infty} n(a_n - a_{n+1}).$$

Reversing the order of summation is justified because all terms are non-negative.

---

6. Prove the necessary and sufficient condition that a triangle inscribed in an ellipse shall have maximum area is that its centroid coincide with the center of the ellipse.

Solution. Let the equation of the ellipse be

$$\frac{x^2}{a^2} + \frac{y^2}{b^2} = 1,$$

and make the affine transformation

$$u = x/a$$
$$v = y/b.$$

Then the ellipse in the (x, y) plane becomes a circle in the (u, v) plane with equation

$$u^2 + v^2 = 1.$$

Under an affine transformation, triangles inscribed in the ellipse go into triangles inscribed in the circle; midpoints of line segments go into midpoints; centroids of triangles go into centroids; the center of the ellipse goes into the center of the circle; and all areas are multiplied by a fixed constant. Hence we need only prove that:

(*) A necessary and sufficient condition that a triangle inscribed in a circle shall have maximum area is that its centroid coincide with the center of the circle.

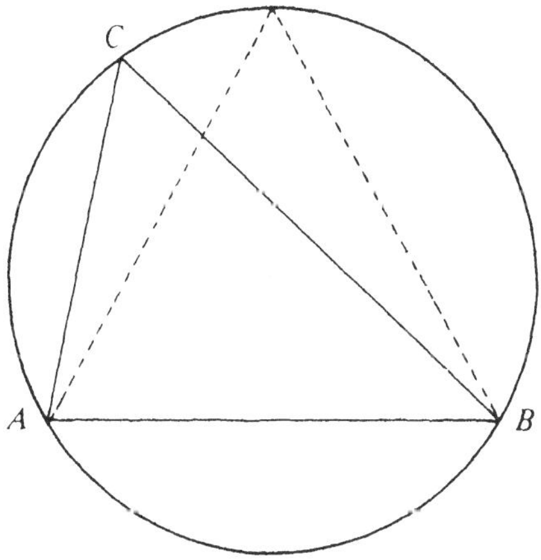

Now a triangle inscribed in a circle has a maximum area if and only if it is equilateral.

Indeed, it is clear from the figure that if we fix one side AB of an inscribed triangle ABC, the area is maximized when the altitude on AB is maximized and this occurs when $AC = BC$. Thus if there is a triangle of maximum area, it must be equilateral.

The existence of triangles of maximum area follows from the fact that the circle is compact, and the area is a continuous function of the vertices.

It is obvious that the centroid of an equilateral triangle coincides with its circumcenter. Conversely, if the centroid of a triangle coincides with its circumcenter, then the triangle is equilateral. This can be proved as follows. If the centroid is the same point as the circumcenter, then each median of the triangle has two points in common with the perpendicular bisector of the corresponding side, namely the midpoint of the side and the centroid. These points cannot coincide since the centroid must be an interior point of the triangle. Therefore each median is perpendicular to the corresponding side and the triangle is equilateral.

---

7. Given any real number N_0, if $N_{j+1} = \cos N_j$, prove that $\lim_{j\to\infty} N_j$ exists and is independent of N_0.

Solution. The graphical method for recursions (explained on p. 223) makes it clear that for any choice of N_0, the sequence converges to the unique root of the equation $x = \cos x$. We formalize this analysis.

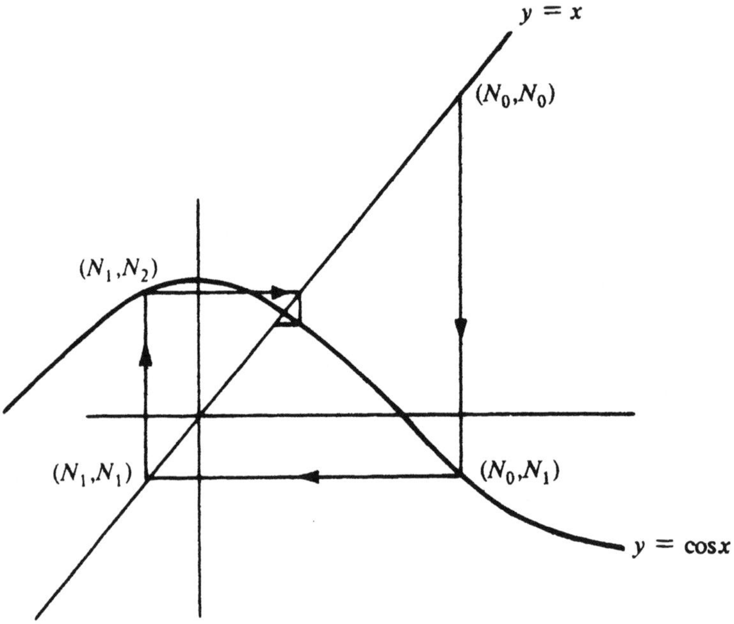

Since $f(x) = x - \cos x$ defines a strictly increasing function (its derivative is non-negative with only isolated zeros), $f(0) < 0$, and $f(1) > 0$, there is a unique number ξ in $(0, 1)$ such that $f(\xi) = 0$; i.e., $\cos \xi = \xi$.

By the mean value theorem, for any x there is an η between x and ξ such that

$$\cos x - \cos \xi = -(\sin \eta)(x - \xi).$$

If $x \in [0, 1]$, then $\eta \in (0, 1)$ and $|\sin \eta| < \sin 1$; hence,

$$(1) \qquad |\cos x - \cos \xi| \leq (\sin 1)|x - \xi|.$$

Now $N_1 = \cos N_0 \in [-1, 1]$, $N_2 = \cos N_1 \in [0, 1]$, and $N_j \in [0, 1]$ for $j \geq 2$ by induction. Hence (1) yields

$$|N_{j+1} - \xi| = |\cos N_j - \cos \xi| \leq (\sin 1)|N_j - \xi|$$

for $j \geq 2$, and therefore

$$|N_j - \xi| \leq (\sin 1)^{j-2}|N_2 - \xi|$$

for all $j \geq 2$. Since $\sin 1 < 1$, it follows that $|N_j - \xi| \to 0$ and therefore

$$\lim_{j\to\infty} N_j = \xi.$$

REMARKS. Using the recursion and a hand calculator, it is easy to find ξ to as many decimals as the capacity of the machine. In fact, $\xi \sim 0.739085$.

The problem is an example of a very general theorem about recursions. Suppose f is any function of class C^1, $f(\xi) = \xi$, and $|f'(\xi)| < 1$. Then the recursion

$$x_{n+1} = f(x_n) \tag{2}$$

produces a sequence $\{x_n\}$ that converges to ξ, provided x_0, or some later member of the sequence, is sufficiently close to ξ. Precisely, if I is an interval symmetric about ξ on which $|f'| \leq \alpha < 1$, and $x_k \in I$, then

$$|x_n - \xi| \leq \alpha^{n-k}|x_k - \xi|,$$

for $n > k$, so $x_n \to \xi$. In this situation, ξ is said to be an attractive fixed point of f.

There is a vast literature on the subject of calculating roots of equations by iteration. See, for example, Ortega and Rheinboldt, *Iterative Solutions of Non-linear Equations in Several Variables*, Academic Press, New York, 1970, pages 120–125.

THE THIRTEENTH WILLIAM LOWELL PUTNAM MATHEMATICAL COMPETITION

March 23, 1953

Morning Session

1. Prove that, for every positive integer n,

$$\sqrt{1} + \sqrt{2} + \cdots + \sqrt{n}$$

is more than $\frac{2}{3} n \sqrt{n}$ and less than

$$\frac{4n + 3}{6} \sqrt{n}.$$

Solution. For k a positive integer and $k - 1 \leq x < k$, we have $\sqrt{k} > \sqrt{x}$. Therefore $\sqrt{k} > \int_{k-1}^{k} \sqrt{x}\, dx$. Adding, we get

$$\sqrt{1} + \sqrt{2} + \cdots + \sqrt{n} > \int_0^n \sqrt{x}\, dx = \frac{2}{3} n^{3/2}.$$

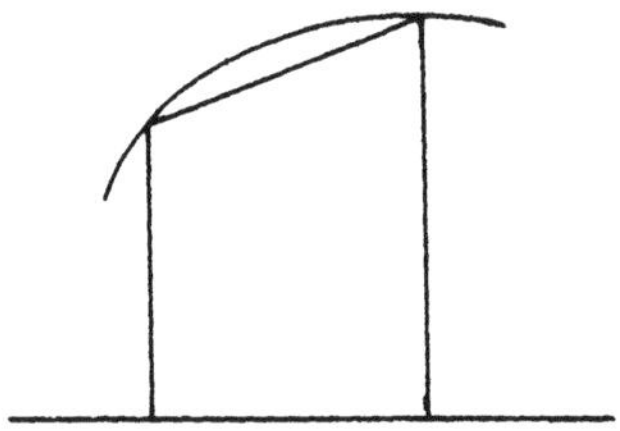

Since the graph of $\sqrt{x}$ is concave downward, it is clear from the diagram that

$$\frac{1}{2}(\sqrt{k-1} + \sqrt{k}) < \int_{k-1}^{k} \sqrt{x}\, dx \tag{1}$$

(i.e., the trapezoidal approximation to the integral is too small). Hence

$$\sqrt{1} + \sqrt{2} + \cdots + \sqrt{n} = \frac{1}{2}(\sqrt{0} + \sqrt{1}) + \frac{1}{2}(\sqrt{1} + \sqrt{2})$$

$$+ \cdots + \frac{1}{2}(\sqrt{n-1} + \sqrt{n}) + \frac{1}{2}\sqrt{n}$$

$$< \int_0^n \sqrt{x}\,dx + \frac{1}{2}\sqrt{n} = \frac{4n+3}{6}\sqrt{n}.$$

The inequality (1) can be proved analytically, of course. We need the standard theorem for estimating the error in the trapezoidal rule for approximating an integral. It is:

Let f be continuous on the closed interval $[a, b]$ and assume f'' exists on the open interval (a, b). Then there exists a point η in (a, b) such that

$$\int_a^b f(x)dx = \frac{1}{2}(b-a)[f(a) + f(b)] - \frac{1}{12}f''(\eta)(b-a)^3.$$

For the case at hand, take $f(x) = \sqrt{x}$, $a = k - 1$, $b = k$. Since f'' is negative on $(0, \infty)$, inequality (1) follows.

---

2. Six points are in general position in space (no three in a line, no four in a plane). The fifteen line segments joining them in pairs are drawn and then painted, some segments red, some blue. Prove that some triangle has all its sides the same color.

Solution. Let P be any one of the six points. Five of the line segments end at P, and of these at least three, say PQ, PR, and PS, must have the same color, say blue. Then, if any one of the segments QR, RS, SQ is blue we will have a blue triangle, and if not QRS will be a red triangle. Thus in any event at least one triangle has all its sides the same color.

REMARKS. This problem later appeared as Problem E 1321, *American Mathematical Monthly*, vol. 66 (1959), pages 141–142. A. W. Goodman, "On Sets of Acquaintances and Strangers at Any Party," *American Mathematical Monthly*, vol. 66 (1959), pages 778–783, shows that there must always be at least two monochromatic triangles. G. J. Simmons, in "The Game of Sim," *Journal of Recreational Mathematics*, vol. 2 (1969), page 66, proposed the following game for two players.

Players alternately color the segments red and blue, the object being to avoid making a triangle in one's own color. Since a monochromatic triangle must eventually be formed, this game cannot end in a draw. It has been shown that the second player can force a win. See Rounds and Yau, "A Winning Strategy for Sim," *Journal of Recreational Math.*, vol. 7

(1974), pages 193–202; or Mead, Rosa, and Huang, "The Game of Sim: A Winning Strategy for the Second Player," *Mathematics Magazine*, vol. 47 (1974), pages 243–247.

Generalizations to more points (and hence more line segments) and/or more colors have attracted considerable attention. For example, if the segments connecting 18 points are colored with two colors, there must be a monochromatic tetrahedron, but for 17 points there need not be such a tetrahedron. See A. M. Gleason and R. E. Greenwood, "Combinatorial Relations and Chromatic Graphs," *Canadian Journal of Mathematics*, vol. 7 (1955), pages 1–7. A very general existence theorem along these lines was proved by F. P. Ramsey, "On a Problem in Formal Logic," *Proceedings of the London Mathematical Society*, Ser. 2, vol. 30 (1930), pages 264–286; hence the least numbers which guarantee the existence of certain configurations are frequently called "Ramsey numbers."

See also J. G. Kalbfleisch, "Upper Bounds for Some Ramsey Numbers," *Journal of Combinatorial Theory*, vol. 2 (1967), pages 35–42; Keith Walker, "Dichromatic Graphs and Ramsey Numbers," *Journal of Combinatorial Theory*, vol. 5 (1968), pages 238–243; and J. Spencer, "Ramsey's Theorem —A New Lower Bound," *Journal of Combinatorial Theory*, vol. 18 (1975), pages 108–115.

---

3. If x_1, x_2, x_3 are real numbers and the sum of any two is greater than the third, show that

$$\frac{2}{3}\sum_{i=1}^{3} x_i \sum_{i=1}^{3} x_i^2 > \sum_{i=1}^{3} x_i^3 + x_1x_2x_3.$$

First Solution. Note that the required inequality is equivalent to

$$2(\Sigma x_i)(\Sigma x_i^2) > 3\,\Sigma x_i^3 + 3x_1x_2x_3. \tag{1}$$

Put

$$2a = x_2 + x_3 - x_1, \qquad 2b = x_3 + x_1 - x_2, \qquad 2c = x_1 + x_2 - x_3.$$

Then

$$x_1 = b + c, \qquad x_2 = c + a, \qquad x_3 = a + b$$

and

$$\sum x_i = 2 \sum a$$
$$\sum x_i^2 = 2 \sum a^2 + 2 \sum ab$$
$$\sum x_i^3 = 2 \sum a^3 + 3 \sum a^2b$$
$$x_1x_2x_3 = \sum a^2b + 2abc$$

where $\sum a^2b$ stands for

$$a^2b + a^2c + b^2a + b^2c + c^2a + c^2b,$$

with similar interpretations for the other summations. The left side of (1) is

$$8(\sum a)(\sum a^2 + \sum ab) = 8(\sum a^3 + 2 \sum a^2b + 3abc) \tag{2}$$

and the right side of (1) is

$$6 \sum a^3 + 12 \sum a^2b + 6abc \tag{3}$$

and it is obvious that (2) exceeds (3) since a, b, and c are all positive numbers.

Second Solution. We first note that x_1, x_2, and x_3 are all positive since, for example, $2x_1 = (x_1 + x_2 - x_3) + (x_1 + x_3 - x_2)$. We also know that

$$\begin{aligned} 0 &< (x_1 + x_2 - x_3)(x_2 + x_3 - x_1)(x_3 + x_1 - x_2) \\ &= \sum x_i^2x_j - \sum x_i^3 - 2x_1x_2x_3. \end{aligned} \tag{4}$$

Since the left member of (1) is

$$2(\sum x_i^2x_j + \sum x_i^3)$$

the required inequality is equivalent to

$$2 \sum x_i^2x_j > \sum x_i^3 + 3x_1x_2x_3. \tag{5}$$

But (5) follows immediately from (4) and

$$\sum x_i^2x_j > x_1x_2x_3$$

which is obvious since all terms on the left are positive and if for example, $x_1 \le x_2 \le x_3$, then $x_1x_2x_3 \le x_2^2x_3$. [In fact the much stronger inequality $\frac{1}{6} \sum x_i^2x_j \ge x_1x_2x_3$ is valid for positive x_1, x_2, x_3, since this asserts that the arithmetic mean of the six numbers $x_i^2x_j$ exceeds their geometric mean.]

4. From the identity

$$\int_0^{\pi/2} \log \sin 2x \, dx = \int_0^{\pi/2} \log \sin x \, dx + \int_0^{\pi/2} \log \cos x \, dx + \int_0^{\pi/2} \log 2 \, dx,$$

deduce the value of

$$\int_0^{\pi/2} \log \sin x \, dx.$$

Solution. Let

$$I = \int_0^{\pi/2} \log \sin x \, dx.$$

[This is, of course, an improper integral which we assume for the moment to exist.] Making the substitutions $x = \pi/2 - u$ and $x = \pi - v$ we see that

$$I = \int_0^{\pi/2} \log \cos u \, du = \int_{\pi/2}^{\pi} \log \sin v \, dv.$$

Therefore

$$\int_0^{\pi} \log \sin v \, dv = \int_0^{\pi/2} \log \sin v \, dv + \int_{\pi/2}^{\pi} \log \sin v \, dv = 2I.$$

Making the substitution $v = 2w$ we have also

$$\int_0^{\pi} \log \sin v \, dv = 2 \int_0^{\pi/2} \log \sin 2w \, dw.$$

Then

$$I = \int_0^{\pi/2} \log \sin 2w \, dw = \int_0^{\pi/2} (\log 2) dw + \int_0^{\pi/2} \log \sin w \, dw + \int_0^{\pi/2} \log \cos w \, dw$$

$$= \frac{\pi}{2} \log 2 + 2I.$$

Hence

$$I = -\frac{\pi}{2}\log 2.$$

To justify these manipulations, note that

$$\frac{2}{\pi}x < \sin x \leq 1 \quad \text{for} \quad 0 < x \leq \frac{\pi}{2}$$

and hence

$$\log x - \log\frac{\pi}{2} < \log \sin x \leq 0.$$

Now

$$\int_{\epsilon}^{1} \log x \, dx = -1 - \epsilon \log \epsilon + \epsilon.$$

So

$$\lim_{\epsilon \to 0} \int_{\epsilon}^{1} \log x \, dx = -1.$$

It follows that the improper integral

$$\int_{0}^{\pi/2} \log \sin x \, dx = \lim_{\epsilon \to 0} \int_{\epsilon}^{\pi/2} \log \sin x \, dx$$

exists. The several substitutions employed above can now be justified routinely.

---

5. Let P be a point from which three distinct normals can be drawn to a parabola. Show that the sum of the angles which these three normals make with the axis exceeds by a multiple of π the angle which the line joining P to the focus makes with the axis.

Solution. Let the equation of the parabola in parametric form be $x = at^2$, $y = 2at$. The normal at the point $(at^2, 2at)$ has slope $-t$ and equation $tx + y - at^3 - 2at = 0$.

Now if the normals at $Q_i(at_i{}^2, 2at_i)$, $i = 1, 2, 3$, pass through $P(h, k)$ then the t_i must be roots of the cubic equation $at^3 + (2a - h)t - k = 0$ and hence $t_1 + t_2 + t_3 = 0$, $t_1t_2 + t_2t_3 + t_1t_3 = (2a - h)/a$ and $t_1t_2t_3 = k/a$.

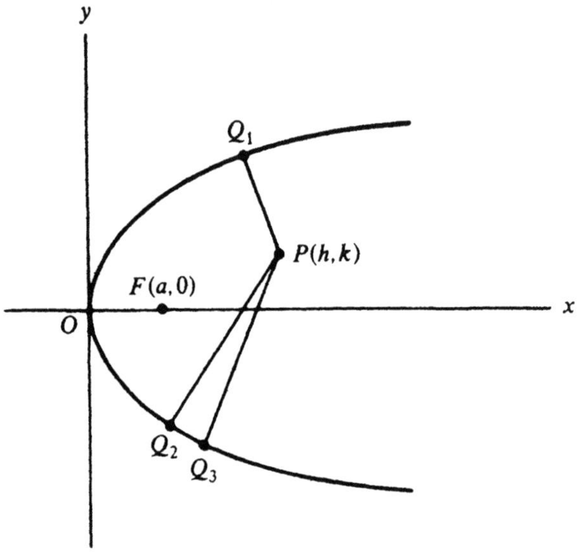

Let the slope angles of Q_iP be α_i and the slope angle of PF be β. Then $\alpha_1 + \alpha_2 + \alpha_3 = \beta + n\pi$ for some integer n if and only if $\tan(\alpha_1 + \alpha_2 + \alpha_3) = \tan\beta$. But

$$\tan(\alpha_1 + \alpha_2 + \alpha_3) = \frac{\tan\alpha_1 + \tan\alpha_2 + \tan\alpha_3 - \tan\alpha_1\tan\alpha_2\tan\alpha_3}{1 - \tan\alpha_1\tan\alpha_2 - \tan\alpha_2\tan\alpha_3 - \tan\alpha_1\tan\alpha_3}$$

$$= \frac{-(t_1 + t_2 + t_3) + t_1t_2t_3}{1 - (t_1t_2 + t_2t_3 + t_1t_3)} = \frac{\dfrac{k}{a}}{1 - \dfrac{2a - h}{a}}$$

$$= \frac{k}{h - a} = \tan\beta.$$

---

6. Show that the sequence

$$\sqrt{7},\ \sqrt{7 - \sqrt{7}},\ \sqrt{7 - \sqrt{7 + \sqrt{7}}},\ \sqrt{7 - \sqrt{7 + \sqrt{7 - \sqrt{7}}}},\ \ldots$$

converges, and evaluate the limit.

Solution. Let $x_0 = \sqrt{7}$, $x_1 = \sqrt{7 - \sqrt{7}}$, $x_2 = \sqrt{7 - \sqrt{7 + \sqrt{7}}}$, $\ldots$. The later terms of the intended sequence are given by the recursion

$$x_{n+2} = \sqrt{7 - \sqrt{7 + x_n}} \quad \text{for} \quad n \geq 0.$$

We have therefore two interlocked recursions of the form $a_{n+1} = f(a_n)$, where

$$f(x) = \sqrt{7 - \sqrt{7 + x}}.$$

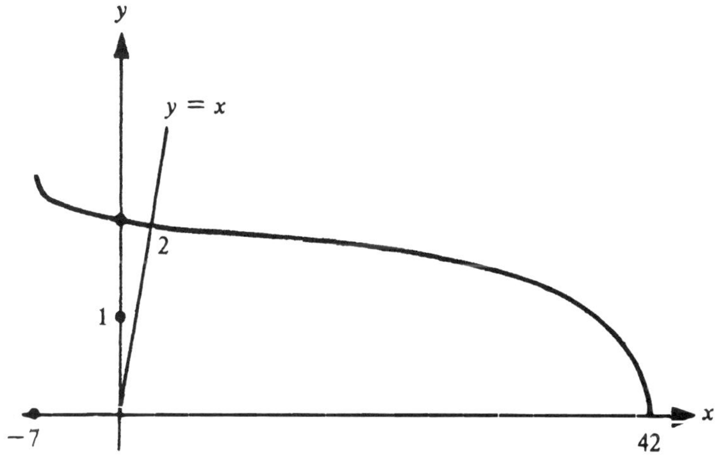

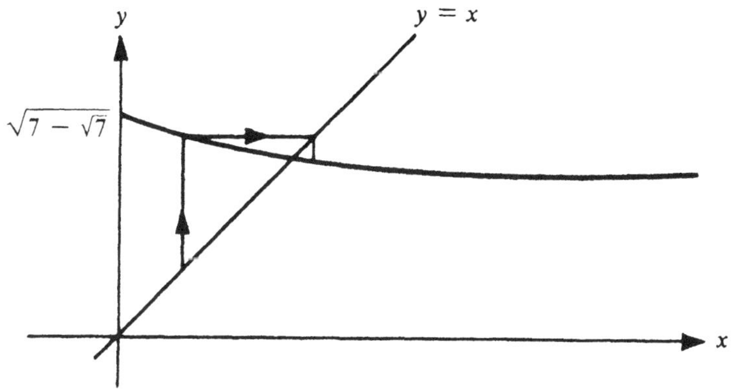

The diagram (explained on p. 223) indicates that for any start in the domain of f the recursion converges to 2, which is the unique fixed point of f. We shall prove this analytically.

By the mean value theorem

$$|f(x) - 2| = |f(x) - f(2)| = |f'(\xi)||x - 2|$$

for some number ξ between 2 and x. If $0 \le x \le 7$, then surely $0 \le \xi \le 7$ and

$$(1)\quad |f'(\xi)| = \left| \frac{-1}{4\sqrt{7 - \sqrt{7+\xi}}\sqrt{7+\xi}} \right| \le \frac{1}{4\sqrt{7 - \sqrt{14}}\sqrt{7}} = \alpha < 1.$$

Hence

$$|x_{n+2} - 2| = |f(x_n) - 2| \leq \alpha |x_n - 2|$$

if $0 \leq x_n \leq 7$. Since x_0 and x_1 are both in this interval it follows that $0 \leq x_n \leq 7$ for all n and that

$$|x_{2k} - 2| \leq \alpha^{2k} |x_0 - 2|$$

and

$$|x_{2k+1} - 2| \leq \alpha^{2k} |x_1 - 2|$$

for all positive integers k. Since $\alpha < 1$, these inequalities imply $x_n \to 2$ as $n \to \infty$.

REMARK. The inequality (1) together with the mean value theorem shows that f satisfies

$$|f(x) - f(y)| \leq \alpha |x - y| \quad \text{for} \quad 0 \leq x, y \leq 7.$$

Hence f is a contraction on the interval $[0, 7]$ and the whole problem can be viewed as an instance of the Banach contraction fixed point theorem.

References. Ortega and Rheinboldt, *Iterative Solution of Nonlinear Equations in Several Variables*, Academic Press, New York, 1970, pages 120-125; J. Dieudonné, *Foundations of Modern Analysis*, Academic Press, New York, 1969; and A. N. Kolmogorov and S. V. Fomin, *Elements of the Theory of Functions and Functional Analysis*, Graylock Press, Rochester, N.Y., 1957.

---

7. Assuming that the roots of $x^3 + px^2 + qx + r = 0$ are all real and positive, find the relation between p, q, and r which is a necessary and sufficient condition that the roots may be the cosines of the angles of a triangle.

Solution. For any triangle ABC, we have

$$a = b \cos C + c \cos B$$
$$b = c \cos A + a \cos C$$
$$c = a \cos B + b \cos A.$$

Regarding these as three homogeneous linear equations for a, b, and c

having a non-trivial solution we see that

$$\det \begin{vmatrix} 1 & -\cos C & -\cos B \\ -\cos C & 1 & -\cos A \\ -\cos B & -\cos A & 1 \end{vmatrix} = 0,$$

that is

(1) $$\cos^2 A + \cos^2 B + \cos^2 C + 2\cos A \cos B \cos C = 1.$$

If the roots of the equation

(2) $$x^3 + px^2 + qx + r = 0$$

are $\cos A$, $\cos B$, $\cos C$, then

$$\begin{aligned} -p &= \cos A + \cos B + \cos C \\ q &= \cos A \cos B + \cos B \cos C + \cos C \cos A \\ -r &= \cos A \cos B \cos C \end{aligned}$$

and (1) becomes

(3) $$p^2 - 2q - 2r = 1,$$

which is thus a necessary condition.

Now suppose that (3) holds and that the roots, say x_1, x_2, x_3, of (2) are all real and positive. Then

(4) $$x_1^2 + x_2^2 + x_3^2 + 2x_1x_2x_3 = 1.$$

From this it is clear that each root lies between 0 and 1; hence there are unique acute angles A, B, and C such that $x_1 = \cos A$, $x_2 = \cos B$, $x_3 = \cos C$.

To prove that these are the angles of a triangle, it is sufficient to show that $A + B + C = \pi$. Substituting in (4), we get

$$\cos^2 C + 2\cos A \cos B \cos C = 1 - \cos^2 A - \cos^2 B.$$

Completing the square on the left, we obtain

$$(\cos C + \cos A \cos B)^2 = \sin^2 A \sin^2 B.$$

Since the angles are all acute, taking the positive square root gives

$$\cos C + \cos A \cos B = \sin A \sin B$$

and therefore

$$\begin{aligned}\cos C &= -(\cos A \cos B - \sin A \sin B)\\ &= -\cos (A + B) = \cos (\pi - A - B).\end{aligned}$$

Since both C and $\pi - A - B$ are in $(0, \pi)$, we have $C = \pi - A - B$, as required.

Thus, if the roots of (2) are all real and positive, then (3) is necessary and sufficient that the roots be the cosines of the angles of some triangle.

Afternoon Session

1. Is the infinite series

$$\sum_{n=1}^{\infty} \frac{1}{n^{(n+1)/n}}$$

convergent? Prove your statement.

Solution. For every positive integer n, $n < 2^n$. Hence $n^{1/n} < 2$, so

$$\frac{1}{n^{(n+1)/n}} > \frac{1}{2n}.$$

Since $\sum_{n=1}^{\infty} \frac{1}{2n}$ diverges, so does $\sum_{n=1}^{\infty} \frac{1}{n^{(n+1)/n}}$.

---

2. Let $a_0, a_1, \ldots, a_n$ be real numbers and let $f(x) = a_0 + a_1 x + \cdots + a_n x^n$. Suppose that, for every integer i, $f(i)$ is an integer. Prove that $n! \cdot a_k$ is an integer for each k.

Solution. For any function g defined on $\mathbf{R}$, let Δg be the first difference defined by

$$\Delta g(x) = g(x + 1) - g(x)$$

and let $\Delta^2 g = \Delta(\Delta g)$, $\Delta^3 g = \Delta(\Delta^2 g)$, etc. Then if f is a polynomial of degree $\leq n$,

$$f(x) = f(0) + (\Delta f(0))x + \frac{1}{2!}(\Delta^2 f(0))x(x-1) + \frac{1}{3!}(\Delta^3 f(0))x(x-1)(x-2) + \cdots + \frac{1}{n!}(\Delta^n f(0))x(x-1)\cdots(x-n+1). \tag{1}$$

(This is Taylor's series for the calculus of finite differences. To prove the validity of (1), let $g(x)$ represent the right-hand member of (1). Then evidently g is a polynomial of degree at most n. Moreover, $\Delta^i g(0) = \Delta^i f(0)$ for $i = 0, 1, \ldots, n$, and it follows by induction that $g(i) = f(i)$ for $i = 0, 1, \ldots, n$. But two polynomials of degree at most n that agree at $n + 1$ points are identical, so $f = g$.)

Since the given polynomial f takes integer values for integer arguments, it is obvious that $\Delta^i f(0)$ is an integer for $i = 0, 1, 2, \ldots$. Then from (1) it is obvious that $n!f(x)$ is a polynomial in x with integer coefficients. But the coefficients in the representation of a function by a polynomial are unique (if such a representation exists), so $n!a_k$ is an integer for $k = 0, 1, 2, \ldots, n$.

REMARK. If $f(x) = x^n$, the numbers

$$[\Delta^k x^n]_{x=0}, \qquad k = 0, 1, \ldots, n$$

are called the differences of zero, and the numbers

$$\frac{1}{k!}[\Delta^k x^n]_{x=0}, \qquad k = 0, 1, \ldots, n$$

are the Stirling numbers of the second kind. They have many applications in the calculus of finite differences and in combinatorial analysis.

See Francis B. Hildebrand, *Finite Difference Equations and Simulations*, Prentice-Hall, 1968, Englewood Cliffs, N.J., page 117; and Marshall Hall Jr., *Combinatorial Theory*, Blaisdell, Waltham, Mass., 1967, pages 26-27.

---

3. Solve the equations

$$\frac{dy}{dx} = z(y+z)^n \qquad \frac{dz}{dx} = y(y+z)^n,$$

given the initial conditions $y = 1$ and $z = 0$ when $x = 0$.

Solution. Let $u = y + z$. Adding the two given equations we get

$$\frac{du}{dx} = u^{n+1} \tag{1}$$

where $u = 1$ when $x = 0$.

We temporarily assume $n \neq 0$. Then we can solve (1) by separating the variables. We have

$$-n\frac{du}{u^{n+1}} = -n\,dx$$

and therefore

$$\frac{1}{u^n} = a - nx.$$

Using the initial condition we see that $a = 1$ and

$$u^n = \frac{1}{1 - nx} \tag{2}$$

for $-\infty < x < 1/n$ if $n > 0$ and for $1/n < x < \infty$ if $n < 0$.

Now let $v = y - z$ and subtract the original equations,

$$\frac{dv}{dx} = -vu^n = \frac{-v}{1 - nx} \tag{3}$$

where $v = 1$ when $x = 0$. Again the variables separate:

$$n\frac{dv}{v} = -n\frac{dx}{1 - nx};$$

whence

$$n \ln v = b + \ln(1 - nx);$$

and, using the initial condition,

$$v^n = 1 - nx.$$

Therefore,

$$
\begin{aligned}
y &= \frac{1}{2}(u + v) = \frac{1}{2}[(1 - nx)^{-1/n} + (1 - nx)^{1/n}] \\
z &= \frac{1}{2}(u - v) = \frac{1}{2}[(1 - nx)^{-1/n} - (1 - nx)^{1/n}]
\end{aligned}
\tag{4}
$$

for $-\infty < x < 1/n$ if $n > 0$ and for $1/n < x < \infty$ if $n < 0$.

The case $n = 0$ can be solved by the same method. In this case $du/dx = u$, $dv/dx = -v$, so $u = e^x$, $v = e^{-x}$, and

$$
\begin{aligned}
y &= \frac{1}{2}(e^x + e^{-x}) = \cosh x \\
z &= \frac{1}{2}(e^x - e^{-x}) = \sinh x,
\end{aligned}
\tag{5}
$$

for all x.

If n is a non-negative integer, the original equations are real analytic on the whole of $\mathbf{R}^3$. For other values of n, however, the equations exhibit some pathology along the plane $y + z = 0$; hence some discussion of the uniqueness of the solution is necessary. Let S be the region where $y + z > 0$. The differential equations are real analytic on S and therefore a solution curve cannot "split" in S. The solution (4) remains in S and is unbounded as $x \to 1/n$. It follows that it is the unique maximal solution of the given differential equation satisfying the initial conditions. Note that if $-1 < n < 0$, then (2) does not give the maximal solution of (1), because the solution can be continued to the left by setting $u = 0$ for $x \leq 1/n$.

REMARK. As $n \to 0$, the solution given by (4) approaches that given by (5), since for example $(1 - nx)^{-1/n} \to e^x$ as $n \to 0$ for each fixed x. This is in accord with the general theory since the original equations depend smoothly on the parameter n in the region S.

---

4. Determine the equation of a surface in three dimensional cartesian space which has the following properties: (a) it passes through the point (1, 1, 1); and (b) if the tangent plane be drawn at any point P, and A, B and C are the intersections of this plane with the x, y and z axes respectively, then P is the orthocenter (intersection of the altitudes) of the triangle ABC.

Solution. Consider any plane Π that crosses the coordinate axes at three distinct points A, B, and C, and let P be the orthocenter of the triangle ABC. Treat the points as vectors from the origin, use "(,)" to denote the inner product of two vectors, and recall that $(A, B) = (B, C) = (C, A) = 0$ since the axes are orthogonal. Then the orthocenter property becomes

$$(P - A, B - C) = (P - B, C - A) = (P - C, A - B) = 0.$$

Using the linearity properties of the inner product, we get

$$(P, B - C) = (P, C - A) = (P, A - B) = 0.$$

Thus P, as a vector, is orthogonal to the plane Π. This means that P is the foot of the perpendicular on Π from the origin O.

The condition on the surface is, then, that every normal passes through the origin. Obviously the sphere

$$x^2 + y^2 + z^2 = 3 \tag{1}$$

has this property and passes through the required point (1, 1, 1).

A surface that satisfies the original conditions, however, cannot contain a point of a coordinate plane, since it is impossible (by the above analysis) that such a point be the orthocenter of a triangle whose vertices are on the coordinate axes. We shall show that the largest connected surface satisfying the conditions is that portion of the sphere (1) satisfying $x > 0$, $y > 0$, $z > 0$. [We consider only connected surfaces, since without this restriction we can adjoin portions of other spheres at will.] The condition on the surface is that the differential form

$$xdx + ydy + zdz = \frac{1}{2} d(x^2 + y^2 + z^2)$$

vanish on the surface. We conclude that $x^2 + y^2 + z^2$ is constant.

Since the surface is required to contain (1, 1, 1) it must be a part of the sphere (1). Since, as we have seen, it can contain no point of the coordinate planes, it must lie entirely in the first octant.

---

5. Show that the roots of (1): $x^4 + ax^3 + bx^2 + cx + d = 0$, if suitably numbered, satisfy the relation $r_1/r_2 = r_3/r_4$, provided $a^2d = c^2 \neq 0$.

Solution. We shall show that, for any quartic (1) with roots r_1, r_2, r_3, r_4, we have

$$(r_1r_2 - r_3r_4)(r_1r_3 - r_4r_2)(r_1r_4 - r_2r_3) = a^2d - c^2. \tag{2}$$

Multiplying out the left member, we obtain

$$\sum r_1{}^3r_2r_3r_4 - \sum r_1{}^2r_2{}^2r_3{}^2,$$

where in each case the sum is over the distinct terms obtained by permuting the subscripts. On the other hand, since

$$a = -\sum r_1, \qquad c = -\sum r_1r_2r_3, \qquad \text{and } d = r_1r_2r_3r_4,$$

we have

$$\begin{aligned} a^2d - c^2 &= (\sum r_1{}^2 + 2\sum r_1r_2)r_1r_2r_3r_4 - (\sum r_1{}^2r_2{}^2r_3{}^2 + 2\sum r_1{}^2r_2{}^2r_3r_4) \\ &= \sum r_1{}^3r_2r_3r_4 - \sum r_1{}^2r_2{}^2r_3{}^2. \end{aligned}$$

Thus (2) is established.

Given that $a^2d = c^2$, we know that

$$(r_1r_2 - r_3r_4)(r_1r_3 - r_4r_2)(r_1r_4 - r_2r_3) = 0,$$

so one of the factors must vanish. By renumbering the roots we can arrange that

$$r_1r_4 - r_2r_3 = 0.$$

Since $d \neq 0$, none of the roots vanish, so we can divide by r_2r_4 to obtain

$$\frac{r_1}{r_2} = \frac{r_3}{r_4}$$

as required.

---

6. P and Q are any points inside a circle (C) with center C, such that $CP = CQ$. Determine the location of a point Z on (C) such that $PZ + QZ$ shall be a minimum.

First Solution. Let A and B be the points where the perpendicular bisector of PQ meets (C). Since $|CP| = |CQ|$, C lies on AB. We choose the labels so that $|AP| \leq |BP|$.

Consider the function

$$X \mapsto |PX| + |XQ|$$

as X varies along (C). It has critical points at A and B by symmetry. Suppose Z is any other critical point. The ellipse (E) with foci at P and Q passing through Z must be tangent to (C) at Z. The normal to (E) at Z is therefore $\overleftrightarrow{CZ}$. If P, Q, and Z are collinear, then $\overleftrightarrow{PQ}$ is normal to (E) at Z, so C is on $\overleftrightarrow{PQ}$. Assume P, Q, and Z are not collinear. Let (D) be the circle through P, Q, and Z. By the reflection property of the ellipse, $\overleftrightarrow{CZ}$ bisects $\angle PZQ$. The bisector of $\angle PZQ$ meets (D) at a point which bisects arc PQ, that is at a point of $\overleftrightarrow{AB}$. Since Z is not on $\overleftrightarrow{AB}$, $\overleftrightarrow{CZ}$ and $\overleftrightarrow{AB}$ are distinct and meet at just one point, which must be C. Thus P, Q, Z, and C are concyclic.

This shows that we can locate any critical points other than A and B by the following construction. Draw the circle (or line) containing P, Q, and C. If this circle fails to meet (C) or meets it only at the point A, there are no other critical points. If, however, this circle (or line) meets (C) at two points Z and Z', these points are both critical, for the arguments of the preceding paragraph reverse easily to show that the ellipse with foci P and Q passing through Z is tangent to (C) at Z and Z'.

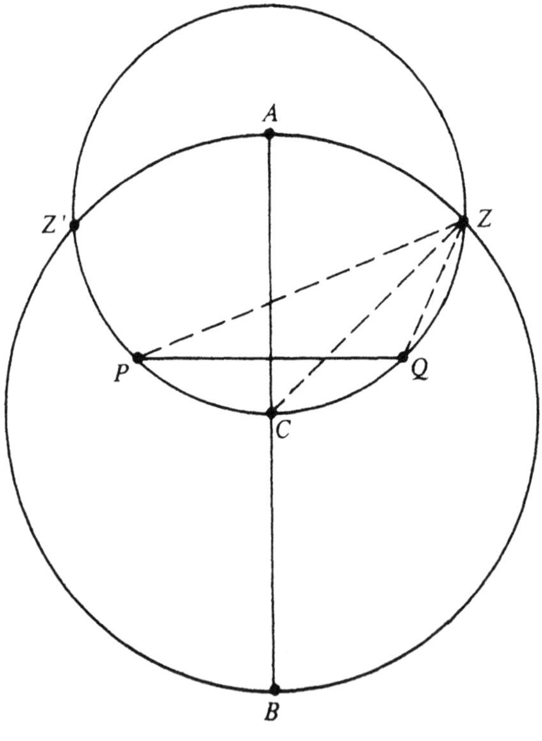

It remains to decide the nature of these new critical points. We make the following observation: If an ellipse and a circle are tangent at two distinct points, then one of the two curves lies completely inside the other except for the points of contact. In the non-collinear case, the ellipse (E) certainly has at least one point inside (C), namely, the reflection in $\overleftrightarrow{PQ}$ of Z; therefore the circle is outside the ellipse and we have

$$|PB| + |BQ| > |PA| + |AQ| > |PZ| + |ZQ|.$$

In the collinear case, $|PB| + |BQ| = |PA| + |AQ| > 2|AC| = |PZ| + |ZQ|$. Since our function must be monotone between critical points, Z is, in either case, a minimum. By symmetry, Z' is also a minimum.

Summarizing, if the circle (or possibly line) containing P, Q, and C meets (C) in two points, these are the minimum points. If not, the minimum is attained at the nearer point where the perpendicular bisector of PQ meets (C).

The quoted result on circles and ellipses is easily proved analytically as follows. Let $x^2 + y^2 - a^2 = 0$ be the equation of the circle and let $L = bx + cy - d = 0$ be the equation of the line joining the two points of contact. Any conic tangent to the circle at these two points has an equation of the form

$$x^2 + y^2 - a^2 + \lambda L^2 = 0.$$

If this conic contains one point inside the circle, then at that point $x^2 + y^2 - a^2 < 0$, $L^2 > 0$, so λ must be positive. Similarly, if the conic contains a point outside the circle, $\lambda < 0$. It follows that the whole conic (except the points of contact) lies either inside or outside the circle. [Note: we can also argue as follows: The curves do not cross at a new point because that would be a fifth point of intersection (counting multiplicity); nor can they cross at one of the points of contact, because that would then be a point of multiplicity three. However, this argument is harder to make precise, because we have to know about counting multiplicity of contact.]

Using a modern version of Ptolemy's theorem (proved below), we can easily prove that when the circle (or line) determined by P, C, and Q meets (C), the intersection points Z and Z' are the minimum points. Because $|CP| = |CQ| < |CZ| = |CZ'|$, the points P and Q separate C from Z and Z'. Consider any point X on (C). According to Ptolemy's theorem

$$|PX| \cdot |CQ| + |QX| \cdot |CP| \geq |CX| \cdot |PQ|$$

with equality if and only if P, Q, C, and X are concyclic or collinear with

P and Q separating C and X; that is, if and only if $X = Z$ or Z'. Since $|CP| = |CQ|$ and $|CX| = r$, the radius of (C), we have

$$|PX| + |QX| \geq r \cdot |PQ|/|CP|$$

with equality if and only if $X = Z$ or Z'.

REMARK. This problem, without the requirement that $PC = QC$, is known as Alhazen's problem. Alhazen was an Arabic mathematician (ca. 965–1039), who posed the problem in the context of optics. (See Dörrie, *100 Great Problems of Elementary Mathematics*, Dover, New York, 1965.)

The theorem referred to above is as follows:

THEOREM. *Suppose A, B, C, and D are four points in the plane. Then*

$$|AB| \cdot |CD| + |AD| \cdot |BC| \geq |AC| \cdot |BD| \tag{1}$$

with equality if and only if A, B, C, and D are concyclic or collinear with A and C separating B and D.

Proof. We take a single complex coordinate in the plane with A as origin and regard the plane as part of the Riemann sphere S. Then there is one ideal point at infinity which is counted as lying on every line and the so augmented lines are regarded as ideal circles.

Let the coordinates of B, C, D be b, c, d, respectively. Then

$$\left|\frac{1}{b} - \frac{1}{c}\right| + \left|\frac{1}{c} - \frac{1}{d}\right| \geq \left|\frac{1}{b} - \frac{1}{d}\right|.$$

Multiply through by $|b| \cdot |c| \cdot |d|$ to get

$$|d| \cdot |c - b| + |b| \cdot |d - c| \geq |c| \cdot |d - b|,$$

which is (1). Equality holds if and only if $1/c$ is on the segment connecting $1/b$ and $1/d$, that is, if and only if ∞, $1/b$, $1/c$, $1/d$ are on an ideal circle with ∞ and $1/c$ separating $1/b$ and $1/d$.

The inverse transformation $z \mapsto 1/z$ is everywhere defined on S and carries the set of circles (ordinary or ideal) into itself, preserving the cyclic order of points on every circle. Hence the equality condition becomes 0, b, c, d, that is, A, B, C, D, are on a circle (ordinary or ideal) with A and C separating B and D.

The original theorem of Ptolemy asserts only the equality in (1) whenever A, B, C, D are concyclic in that order. It appears in the first book of Ptolemy's great work *The Almagest.* For an English translation of this second-century scientific masterpiece see "Great Books of the Western World," Vol. 16: *Ptolemy, Copernicus, Kepler,* Encyclopaedia Britannica, Chicago, 1952.

Second Solution. The method of inversion makes the previous solution extremely neat.

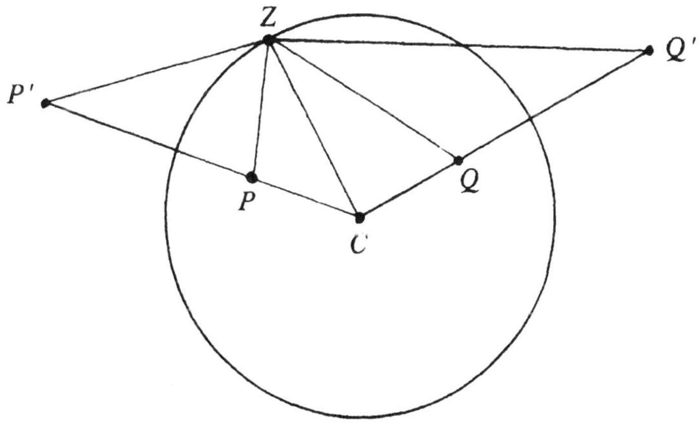

Let the circle (C) have radius r, and let $\lambda = |PC|/r = |QC|/r$. Choose P' on $\overleftrightarrow{CP}$ and Q' on $\overleftrightarrow{CQ}$ so that $|CP'| \cdot |CP| = |CQ'| \cdot |CQ| = r^2$ (i.e., invert the points P and Q in the circle.) Let Z be any point of (C). Since $\Delta CZP' \sim \Delta CPZ$ and $\Delta CZQ' \sim \Delta CQZ$, we have

$$\lambda = \frac{|PC|}{|ZC|} = \frac{|PZ|}{|ZP'|} = \frac{|QZ|}{|ZQ'|} = \frac{|PZ| + |QZ|}{|ZP'| + |ZQ'|}.$$

It is therefore clear that the choice of Z on (C) that minimizes $|PZ| + |QZ|$ is the same as that which minimizes $|ZP'| + |ZQ'|$. But the solution of the latter problem is obvious: If $\overleftrightarrow{P'Q'}$ meets (C), then the minimum is achieved at either of the two (conceivably just one) points of intersection. If $\overleftrightarrow{P'Q'}$ does not meet (C), the minimum is achieved at the point A of (C) nearest to the line $\overleftrightarrow{P'Q'}$. (For if ℓ is the line tangent to (C) at A, then $|AP'| + |AQ'| < |ZP'| + |ZQ'|$ for all other points Z of ℓ, *a fortiori* for all other points Z of (C).)

The line $P'Q'$ is the inverse of the circle (D) constructed in the first solution. For more on the method of inversion see, for example, H. S. M. Coxeter, *Introduction to Geometry,* John Wiley and Sons, 1961.

7. Let w be an irrational number with $0 < w < 1$. Prove that w has a unique convergent expansion of the form

$$w = \frac{1}{p_0} - \frac{1}{p_0p_1} + \frac{1}{p_0p_1p_2} - \frac{1}{p_0p_1p_2p_3} + \cdots,$$

where $p_0, p_1, p_2, \ldots$ are integers and $1 \le p_0 < p_1 < p_2 < \cdots$. If $w = \frac{1}{2}\sqrt{2}$, find p_0, p_1, p_2.

Solution. Before starting the construction we make an observation.

(1) If α is an irrational number such that $0 < \alpha < 1$, there is a unique integer p such that

$$\frac{1}{p+1} < \alpha < \frac{1}{p};$$

moreover, p is positive, $1 - p\alpha$ is irrational, and $0 < 1 - p\alpha < 1/(p+1)$.

Now define a sequence $w_0, w_1, w_2, \ldots$ of irrational numbers between 0 and 1 and a sequence of positive integers by induction as follows:

Let $w_0 = w$ and let p_0 be the integer such that

$$\frac{1}{p_0 + 1} < w_0 < \frac{1}{p_0}.$$

Let $w_1 = 1 - p_0w_0$ (which is irrational and between 0 and 1, by (1)) and let p_1 be the integer such that

$$\frac{1}{p_1 + 1} < w_1 < \frac{1}{p_1}.$$

After w_{k-1} and p_{k-1} have been defined so that

(2) $$\frac{1}{p_{k-1} + 1} < w_{k-1} < \frac{1}{p_{k-1}},$$

let $w_k = 1 - p_{k-1}w_{k-1}$, which is irrational and between 0 and 1, and let p_k be the integer such that

$$\frac{1}{p_k + 1} < w_k < \frac{1}{p_k}.$$

From (1) and (2) it follows that

$$w_k < \frac{1}{p_{k-1} + 1}$$

so $p_k \geq p_{k-1} + 1$. Hence the p's increase strictly. Now

$$w_k = \frac{1}{p_k} - \frac{w_{k+1}}{p_k} \quad \text{for} \quad k = 0, 1, 2, \ldots,$$

so

$$w = w_0 = \frac{1}{p_0} - \frac{w_1}{p_0}$$

$$= \frac{1}{p_0} - \frac{1}{p_0 p_1} + \frac{w_2}{p_0 p_1}$$

(3) $$= \sum_{n=0}^{k} \frac{(-1)^n}{p_0 p_1 \cdots p_n} + (-1)^{k+1} \frac{w_{k+1}}{p_0 p_1 \cdots p_k}$$

for all k. The p's increase strictly, so $p_0 p_1 \cdots p_k \geq (k+1)!$ Furthermore, the w's are bounded, so the last term of (3) approaches 0 as $k \to \infty$. Hence we have

$$w = \sum_{n=0}^{\infty} \frac{(-1)^n}{p_0 p_1 \cdots p_n},$$

as required.

Next we show uniqueness. Suppose

(4) $$w = \sum_{n=0}^{\infty} \frac{(-1)^n}{q_0 q_1 \cdots q_n}$$

where $q_0, q_1, q_2, \ldots$ is a strictly increasing sequence of positive integers. We shall prove that, if the preceding construction is carried out, $p_n = q_n$ for all n.

Since the series appearing in (4) is alternating with terms strictly decreasing in absolute value, we have

$$\frac{1}{q_0} > w_0 > \frac{1}{q_0} - \frac{1}{q_0 q_1} \geq \frac{1}{q_0} - \frac{1}{q_0(q_0+1)} = \frac{1}{q_0+1}.$$

Therefore $p_0 = q_0$, and

$$w_1 = 1 - p_0 w_0 = \sum_{n=1}^{\infty} \frac{(-1)^{n-1}}{q_1 q_2 \cdots q_n}.$$

Repeating this argument inductively we find $p_{k-1} = q_{k-1}$ and

$$w_k = \sum_{n=k}^{\infty} \frac{(-1)^{n-k}}{q_k q_{k+1} \cdots q_n}$$

for $k = 1, 2, 3, \ldots$. This proves the uniqueness of the expansion.

If $w_0 = \frac{1}{2}\sqrt{2}$, then $p_0 = 1$.

$$w_1 = 1 - \frac{1}{2}\sqrt{2} = \frac{1}{2+\sqrt{2}}, \quad \text{so} \quad p_1 = [2 + \sqrt{2}] = 3,$$

$$w_2 = 1 - 3\left(1 - \frac{1}{2}\sqrt{2}\right) = \frac{1}{4+3\sqrt{2}}, \quad \text{so} \quad p_2 = [4 + 3\sqrt{2}] = 8.$$

REMARK. If we start with a rational number w in the interval (0, 1), we are led to a terminating expansion of the form (4). The expansion is not unique, however; for example,

$$\frac{2}{3} = \frac{1}{1} - \frac{1}{1\cdot 3} = \frac{1}{1} - \frac{1}{1\cdot 2} + \frac{1}{1\cdot 2\cdot 3}.$$

The series (4) will converge for any strictly increasing sequence of positive integers $\{q_i\}$, and it is easy to see that the sum will always be irrational. Hence

$$w_0 \leftrightarrow \{p_0, p_1, p_2, \ldots\}$$

is an explicit bijective correspondence between the irrational numbers in (0, 1) and the strictly increasing sequences of positive integers.

THE FOURTEENTH WILLIAM LOWELL PUTNAM MATHEMATICAL COMPETITION

March 6, 1954

Morning Session

1. Let n be an odd integer greater than 1. Let A be an n by n symmetric matrix such that each row and each column of A consists of some permutation of the integers $1, \ldots, n$. Show that each one of the integers $1, \ldots, n$ must appear in the main diagonal of A.

Solution. Each integer of the given set must appear exactly n times in the matrix A. The off-diagonal appearances occur in pairs because of the symmetry of A. Since n is odd, therefore, each integer must appear at least once on the main diagonal.

REMARK. It follows that each integer appears exactly once on the main diagonal.

---

2. Consider any five points P_1, P_2, P_3, P_4, P_5 in the interior of a square S of side-length 1. Denote by d_{ij} the distance between the points P_i and P_j. Prove that at least one of the distances d_{ij} is less than $\sqrt{2}/2$. Can $\sqrt{2}/2$ be replaced by a smaller number in this statement?

Solution. Let the square be divided into four small squares, as indicated in the sketch, each of side-length $\frac{1}{2}$. Considering each of the smaller squares as closed sets, two of the five points must fall in the same small square, and these two points are at a distance less than $\frac{1}{2}\sqrt{2}$ from each other. For a formal proof of this, note that with axes parallel to the sides of the square, both coordinate differences are less than $\frac{1}{2}$.

Given $\epsilon > 0$, choose the center and four points on the diagonals, one within ϵ of each corner. This gives five points in the interior of the square such that the minimum distance is more than $\frac{1}{2}\sqrt{2} - \epsilon$. Hence no number smaller than $\frac{1}{2}\sqrt{2}$ will do.

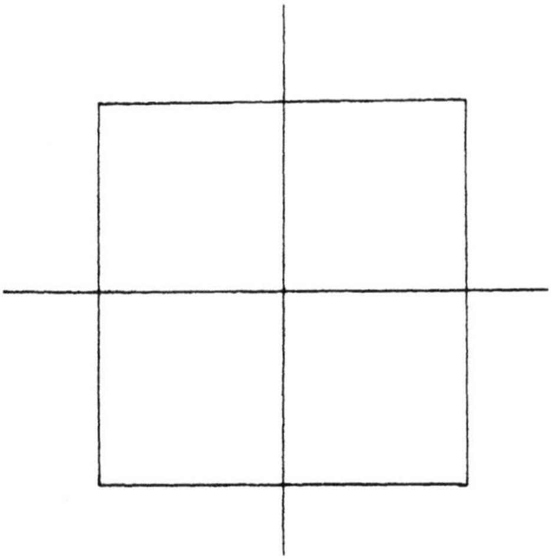

---

3. Prove that if the family of integral curves of the differential equation

$$\frac{dy}{dx} + p(x)y = q(x) \qquad p(x) \cdot q(x) \neq 0$$

is cut by the line $x = k$, the tangents at the points of intersection are concurrent.

Solution. The equation of the line tangent to the smooth curve $y = f(x)$ at the point (k, m) is

$$y - m = f'(k)(x - k).$$

If f is a solution of the given differential equation this becomes

$$y - m = [q(k) - mp(k)](x - k).$$

For any value of m, this line passes through the point

$$\left(k + \frac{1}{p(k)}, \frac{q(k)}{p(k)}\right).$$

---

4. A uniform rod of length $2k$ and weight w rests with the end A against a smooth vertical wall, while to the lower end B is fastened a string BC of

length $2b$ coming from a point C in the wall directly above A. If the system is in equilibrium, determine the angle ABC.

Solution. Obviously the system will be in equilibrium if the rod and the string are both vertical, i.e., if B is on the wall and $\angle ABC = 0$.

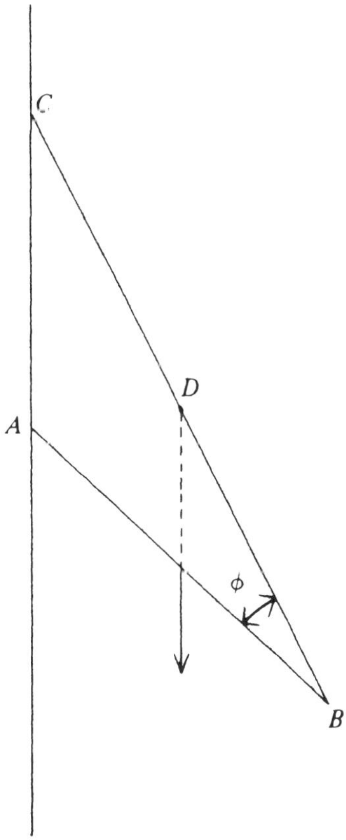

Assume that the rod is in equilibrium in some non-vertical position, as pictured. Let D be the midpoint of BC. The force of tension in the string and the force of gravity on the rod both act through the point D (since the latter force acts vertically through the midpoint of the rod). Hence for equilibrium the reaction of the wall must also act through D. Since the wall is smooth this force is perpendicular to the wall. Hence we have AD perpendicular to AC and

$$|AC|^2 + |AD|^2 = |CD|^2 = b^2.$$

By the law of cosines

$$|AC|^2 = 4k^2 + 4b^2 - 8bk \cos \phi$$

$$|AD|^2 = 4k^2 + b^2 - 4bk \cos \phi.$$

Hence

$$\cos \phi = \frac{2k^2 + b^2}{3bk} = \frac{2}{3}\frac{k}{b} + \frac{1}{3}\frac{b}{k}.$$

Thus

$$\angle ABC = \arccos\left(\frac{2}{3}\frac{k}{b} + \frac{1}{3}\frac{b}{k}\right). \tag{1}$$

Since angles ADB and CAB are obtuse, we must have $b < 2k < 2b$. Since $2x/3 + 1/3x < 1$ for $\frac{1}{2} < x < 1$, equation (1) determines a unique acute angle whenever $1/2 < k/b < 1$. Otherwise there is no equilibrium position except with B on the wall.

Remark. If we calculate the height of the center of gravity of the rod, keeping A directly below C on the wall, we find that the equilibrium position away from the wall is always unstable (i.e., a slight disturbance will tend to lower the center of gravity) while the vertical equilibrium is stable if and only if $b < 2k$.

---

5. If $f(x)$ is a real-valued function defined for $0 < x < 1$, then the formula $f(x) = o(x)$ is an abbreviation for the statement that

$$\frac{f(x)}{x} \to 0 \qquad \text{as } x \to 0.$$

Keeping this in mind, prove the following: if

$$\lim_{x \to 0} f(x) = 0 \quad \text{and} \quad f(x) - f\left(\frac{x}{2}\right) = o(x),$$

then $f(x) = o(x)$.

Solution. Let $\epsilon > 0$ be given. Choose δ so that for all x satisfying $0 < x < \delta$

$$\left| \frac{1}{x}[f(x) - f(x/2)] \right| < \frac{1}{2}\epsilon. \tag{1}$$

Now fix y, $0 < y < \delta$. Then

$$f(y) = \left[f(y) - f\left(\frac{y}{2}\right)\right] + \left[f\left(\frac{y}{2}\right) - f\left(\frac{y}{4}\right)\right] + \cdots + \left[f\left(\frac{y}{2^{n-1}}\right) - f\left(\frac{y}{2^n}\right)\right] + f\left(\frac{y}{2^n}\right).$$

So

$$|f(y)| \le \sum_{i=1}^{n} \left| f\left(\frac{y}{2^{i-1}}\right) - f\left(\frac{y}{2^i}\right)\right| + \left| f\left(\frac{y}{2^n}\right)\right|$$

$$\le \sum_{i=1}^{n} \frac{y}{2^i}\epsilon + \left| f\left(\frac{y}{2^n}\right)\right| = y\epsilon\left(1 - \frac{1}{2^n}\right) + \left| f\left(\frac{y}{2^n}\right)\right|$$

using (1).

Letting $n \to \infty$ we have

$$|f(y)| \le \epsilon y$$

since $f(x) \to 0$ as $x \to 0$.

Thus we have proved: For all $\epsilon > 0$, there is a $\delta > 0$ such that $|f(y)/y| \le \epsilon$ provided $0 < y < \delta$. But, by definition, this is

$$f(x) = o(x).$$

---

6. Suppose that $u_0, u_1, u_2, \ldots$ is a sequence of real numbers such that

(1) $$u_n = \sum_{k=1}^{\infty} u^2_{n+k} \qquad \text{for } n = 0, 1, 2, \ldots.$$

Prove that if Σu_n converges then $u_k = 0$ for all k.

Solution. It is obvious from (1) that the terms u_n are non-increasing and non-negative. Thus, if any one term is zero, so are all its successors, and by induction using (1) so are all its predecessors.

Suppose Σu_n converges. Let p be chosen so that $\Sigma_{n>p}\, u_n < 1$. Then, from (1),

$$u_p = \sum_{n>p} u_n^2 \le \sum_{n>p} u_p u_n = u_p \sum_{n>p} u_n \le u_p$$

with equality at the last step only if $u_p = 0$. But, looking at the first member and the last member, we see that equality must hold throughout. Thus $u_p = 0$ and hence all the u's are zero, as we have shown above.

---

7. Prove that there are no integers x and y for which

$$x^2 + 3xy - 2y^2 = 122. \tag{1}$$

Solution. Multiply (1) by 4 and complete the square to obtain

$$(2x + 3y)^2 - 17y^2 = 488. \tag{2}$$

If (1) has a solution in integers, then so does (2) and hence so does the congruence

$$u^2 \equiv 488 \pmod{17}. \tag{3}$$

But (3) has no solutions because 488 is not a quadratic residue of 17. The most direct way to see this is to check the possibilities $u = 0, 1, \ldots, 8$. (We need not worry about $u = 9, \ldots, 16$ because $u^2 \equiv (17 - u)^2$.) Alternatively we can calculate using the quadratic reciprocity theorem and the Legendre symbol. We have

$$(488/17) = (12/17) = (2/17)^2(3/17) = (3/17) = (17/3) = (2/3) = -1;$$

hence 488 is not a quadratic residue of 17.

Remark. We could equally well argue that a solution of (1) and hence of (2) in integers would imply that 17 is a quadratic residue of 61, since 61 divides 488. But $(17/61) = (61/17) = (10/17) = (2/17)(5/17) = 1 \cdot (17/5) = (2/5) = -1$.

For the quadratic reciprocity theorem and the Legendre symbol see, for example, W. J. LeVeque, *Topics in Number Theory*, vol. 1, Addison-Wesley, Reading, Mass., 1956, page 66 ff.

Afternoon Session

1. Show that the equation $x^2 - y^2 = a^3$ has always integral solutions for x and y whenever a is a positive integer.

Solution. Let $x + y = a^2$, and $x - y = a$. Then $x^2 - y^2 = a^3$, and

$x = \frac{1}{2}(a^2 + a)$ and $y = \frac{1}{2}(a^2 - a)$. Since a^2 and a are both even or both odd, x and y are both integers and a solution exists for every integer a.

REMARK. There are other solutions, for example,

$$x = \frac{a^3 + 1}{2}, \qquad y = \frac{a^3 - 1}{2}, \qquad \text{for } a \text{ odd}$$

and

$$x = \frac{a^3 + 4}{4}, \qquad y = \frac{a^3 - 4}{4}, \qquad \text{for } a \text{ even.}$$

---

2. Assume as known the (true) fact that the alternating harmonic series

$$(1) \qquad 1 - 1/2 + 1/3 - 1/4 + 1/5 - 1/6 + 1/7 - 1/8 + \cdots$$

is convergent, and denote its sum by s. Rearrange the series (1) as follows:

(2)

$$1 + 1/3 - 1/2 + 1/5 + 1/7 - 1/4 + 1/9 + 1/11 - 1/6 + \cdots.$$

Assume as known the (true) fact that the series (2) is also convergent, and denote its sum by S. Denote by s_k, S_k the kth partial sum of the series (1) and (2) respectively. Prove the following statements.

(i) $\quad S_{3n} = s_{4n} + \frac{1}{2} s_{2n},$

(ii) $\quad S \neq s.$

Solution. (i) We have

$$s_{4n} = 1 - \frac{1}{2} + \frac{1}{3} - \frac{1}{4} + \frac{1}{5} - \frac{1}{6} + \cdots + \frac{1}{4n - 1} - \frac{1}{4n}$$

$$\frac{1}{2} s_{2n} = \frac{1}{2} - \frac{1}{4} + \frac{1}{6} + \cdots - \frac{1}{4n}.$$

Adding, we obtain

$$s_{4n} + \frac{1}{2}s_{2n} = 1 + \frac{1}{3} - \frac{1}{2} + \frac{1}{5} + \frac{1}{7} - \frac{1}{4} + \cdots + \frac{1}{4n-3} + \frac{1}{4n-1} - \frac{1}{2n}$$
$$= S_{3n}.$$

(ii) Since (1) is an alternating series with terms that decrease to zero, $\lim_{n\to\infty} s_n = s$ exists. Moreover, $s > 1 - \frac{1}{2} = \frac{1}{2}$, so $s \neq 0$. Therefore

$$S = \lim_{n\to\infty} S_{3n} = \frac{3}{2}s \neq s.$$

REMARK. It is well known that $s = \log_e 2$.

---

3. Let a and b denote real numbers such that $a < b$. The symbol (a, b) will denote the closed interval with the end points a, b. Let there be given a collection of closed intervals $(a_1, b_1), \ldots, (a_n, b_n)$ such that any two of these closed intervals have at least one point in common. Prove that there exists then a point which is contained in every one of these intervals.

Solution. Two closed intervals (c, d) and (e, f) overlap if and only if $c \leq f$ and $e \leq d$.

Since any two of the given intervals overlap we have

$$a_n \leq b_p \tag{1}$$

for all n and all p. Hence $\{a_n\}$ is bounded above (and non-empty) so we can let ξ be the least upper bound of $\{a_n\}$. Since (1) shows that each b_p is an upper bound for $\{a_n\}$, we have $\xi \leq b_p$ for all p, while $a_p \leq \xi$ because ξ is an upper bound for $\{a_n\}$. Thus $\xi \in (a_p, b_p)$ for each p.

REMARKS. We have worded the proof so that it remains valid for an infinite collection of closed intervals that intersect in pairs. This is Helly's theorem in dimension one. See L. Danzer, B. Grunbaum, V. Klee, "Helly's Theorem and Its Relatives," *Proc. Symposium in Pure Mathematics, 7: Convexity,* American Mathematical Society, Providence, R.I., 1963. Also Yaglom and Boltyanskii, *Convex Figures,* Holt, Rinehart and Winston, New York, 1961.

---

4. Given the focus f and the directrix D of a parabola P and a line L, describe (with proof) a Euclidean (i.e., ruler and compass) construction of the point or points of intersection of L and P. Be sure to identify the case for which there are no points of intersection.

Solution. Recall that P is the locus of points equidistant from f and D.

Assume to begin with that L is neither parallel nor perpendicular to D but meets D in a single point o. For convenience of description choose coordinates so that o is the origin, D is the x-axis, f is in the upper half-plane and one ray of L is in the first quadrant with direction angle α, $0 < \alpha < \pi/2$. Let β be the direction angle of $\overrightarrow{of}$; then $0 < \beta < \pi$.

Suppose L meets P at a point q. The circle C with center q and passing through f is then tangent to D at some point of the positive ray. Since $\overrightarrow{of}$ meets C, $\beta \leq 2\alpha$. (See Figure 1.) Thus there will be no point of intersection if $\beta > 2\alpha$.

Fig. 1.

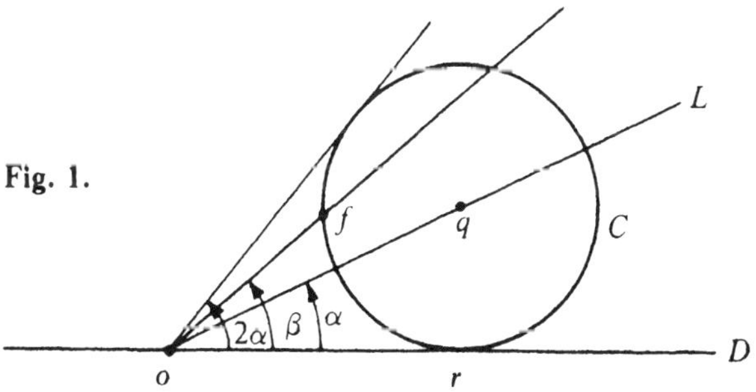

Now suppose $\beta \leq 2\alpha$. We shall construct a point of $P \cap L$ in this case and, in fact, two points if $\beta < 2\alpha$. Choose any point q' on L in the first quadrant and draw the circle C' with center q' and tangent to D. Since $\beta \leq 2\alpha$ the ray $\overrightarrow{of}$ cuts C' at a point f' (and indeed at two points, either of which can be chosen as f', if $\beta < 2\alpha$). If the plane be dilated (or contracted) from o in the ratio $|of|:|of'|$, then f' will be mapped to f, q' will be mapped to q on L, and C' will be mapped on a circle C with center q passing through f and tangent to D. But this implies that q is equidistant from f and D, so $q \in L \cap P$. The Euclidean construction of q is immediate.

Fig. 2.

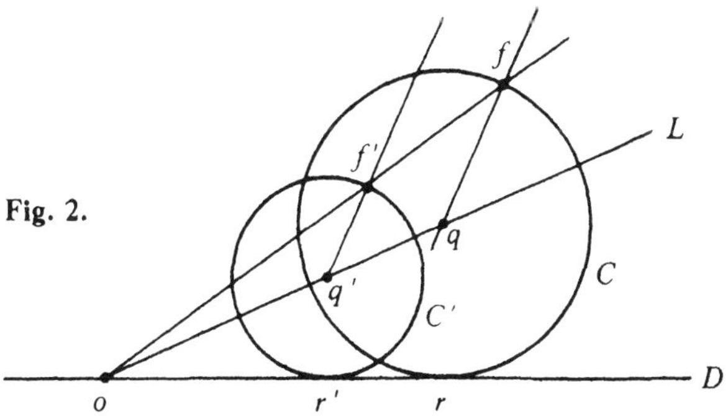

If $f \notin L$, then q is the intersection of L and a line parallel to $\overleftrightarrow{q'f'}$ passing through f. (See Figure 2.) If $f \in L$, a less direct construction is required. For example, let r' be the point at which C' is tangent to D, then draw $\overleftrightarrow{fr}$ parallel to $\overleftrightarrow{f'r'}$ meeting D at r; then q is the intersection of L and a line perpendicular to D at r.

Fig. 3.

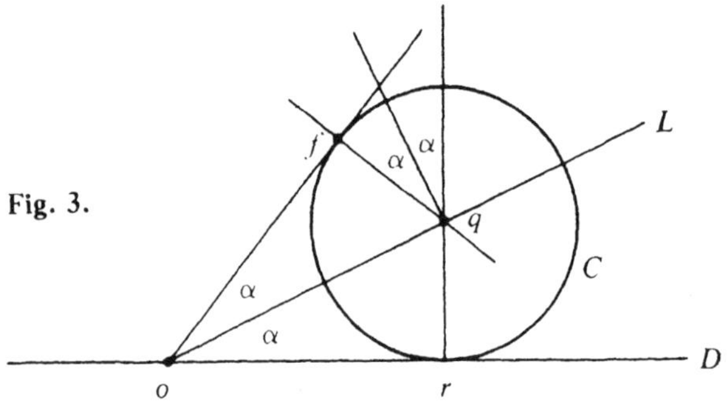

In case $\beta < 2\alpha$, there are two choices for f' and they will lead to two different points of $L \cap P$. If $\beta = 2\alpha$, only one point of $L \cap P$ is found. The analysis of the problem given in the third paragraph shows that every point of $L \cap P$ must be obtained by our construction, so $L \cap P$ consists of only one point and L is tangent to P in this case. (Moreover, Figure 3 shows that the normal to L at q bisects the angle between $\overrightarrow{qf}$ and the vertical ray at q. This is the well-known reflection property of the parabola.)

Now for the omitted cases. If L is perpendicular to D at o, then L must meet P in just one point. [This is a limiting case of the preceding with $\alpha = \pi/2$, so certainly $\beta < 2\alpha$.] The construction is the same as before, but although $\overrightarrow{of}$ meets C' twice, one of the crossings is at o and does not lead to a second point of $L \cap P$. Also the construction given above will fail in this case if $f \in L$. In that case the required point of intersection is the midpoint of of.

Finally suppose L is parallel to (or coincident with) D. Say its equation is $y = a$, and let b be the y-coordinate of f. [We are still assuming D is the x-axis and $b > 0$.] An analogue of the previous construction [i.e., draw any circle with center on L and tangent to D, then translate it along L, if possible, so as to pass through f; the center of the new circle is in $L \cap P$] would serve, but it is easier to draw a circle E with center f and radius equal to the distance from L to D. Then $E \cap L = L \cap P$, obviously. Evidently $L \cap P$ will contain 0, 1, or 2 points according as $b > 2a$, $b = 2a$, or $b < 2a$. Note that these are the limiting forms of the conditions $\beta > 2\alpha$, $\beta = 2\alpha$, $\beta < 2\alpha$ if L is the line $\overleftrightarrow{os}$ where s is kept fixed in the upper half-plane and $o \to \infty$ along the left ray of D.

5. Let $f(x)$ be a real-valued function, defined for $-1 < x < 1$, such that $f'(0)$ exists. Let a_n, b_n be two sequences such that

$$-1 < a_n < 0 < b_n < 1, \quad \lim_{n\to\infty} a_n = 0, \quad \lim_{n\to\infty} b_n = 0.$$

Prove that

$$\lim_{n\to\infty} \frac{f(b_n) - f(a_n)}{b_n - a_n} = f'(0).$$

Solution. Let $\epsilon > 0$ be given. Then we can choose δ, $0 < \delta < 1$, so that for all x with $0 < |x| < \delta$,

$$\left| \frac{f(x) - f(0)}{x} - f'(0) \right| < \epsilon.$$

Let k be chosen so that, for all $n \geq k$,

$$|a_n| < \delta, \quad \text{and} \quad |b_n| < \delta.$$

Then for $m \geq k$ we have

$$(1) \qquad \left| \frac{f(a_m) - f(0)}{a_m} - f'(0) \right| < \epsilon, \qquad \left| \frac{f(b_m) - f(0)}{b_m} - f'(0) \right| < \epsilon.$$

But

$$\begin{aligned} |f(b_m) - f(a_m) - (b_m - a_m)f'(0)| & \\ &= |(f(b_m) - f(0) - b_m f'(0)) - (f(a_m) - f(0) - a_m f'(0))| \\ &\leq |f(b_m) - f(0) - b_m f'(0)| + |f(a_m) - f(0) - a_m f'(0)| \\ &\leq \epsilon|b_m| + \epsilon|a_m| = \epsilon(b_m - a_m). \end{aligned}$$

The penultimate step follows from (1) and the last from $a_m < 0 < b_m$.

Therefore

$$\left| \frac{f(b_m) - f(a_m)}{b_m - a_m} - f'(0) \right| < \epsilon$$

for all $m > k$. Since this is true for any $\epsilon > 0$, we have proved that

$$\lim_{n\to\infty} \frac{f(b_n) - f(a_n)}{b_n - a_n} = f'(0).$$

6. Prove that every positive rational number is the sum of a finite number of distinct terms of the series

$$1 + \frac{1}{2} + \frac{1}{3} + \cdots + \frac{1}{n} + \cdots.$$

Solution. Consider first rational numbers r, $0 \leq r < 1$. We shall prove that such a rational number r can be represented as the sum of a finite number (zero if $r = 0$) of distinct terms of the harmonic series.

The result is obviously true for $r = 0$. If $r = p/q$, where p and q are positive integers, then the result is also true for those cases where $p = 1$. We make the inductive hypothesis that the desired representation is possible for all rational numbers p/q (<1) for which $p < P$. Now consider a rational number $r = P/q < 1$. Let m be the least positive integer for which $1/m \leq P/q$. Then since $P/q < 1$, $m \geq 2$, and we have

$$\frac{1}{m} \leq \frac{P}{q} < \frac{1}{m-1}.$$

Therefore $mP - P < q \leq mP$, so $0 \leq mP - q < P$. Let $R = (P/q) - (1/m)$. Then $R = (mP - q)/qm$, and R is by our inductive hypothesis representable by a finite sum of distinct terms from the harmonic series. Since

$$R < \frac{1}{m-1} - \frac{1}{m} = \frac{1}{m(m-1)} \leq \frac{1}{m}$$

we see that none of the terms used in the expansion of R could be $1/m$. Hence $r = P/q = R + 1/m$ can be expressed as a finite number of distinct terms of the harmonic series. Thus rational numbers less than one with numerator P have the desired representation also. It follows by induction that *all* rational numbers less than one have the desired representation. Now let r be a rational, $r \geq 1$. Let S_n denote the nth partial sum of the harmonic series. Evidently S_n is rational. Since $S_n \to \infty$ as $n \to \infty$, there is an integer $n \geq 1$ such that $S_n \leq r < S_{n+1}$. Then $r - S_n$ is a rational number, and

$$(1) \qquad r - S_n < S_{n+1} - S_n = \frac{1}{n+1} < 1.$$

So $r' = r - S_n$ can be expressed as a finite number of distinct terms of the harmonic series. In view of (1), none of these terms can be in the set $\{1, \frac{1}{2}, \frac{1}{3}, \ldots, \frac{1}{n}\}$. Hence r is the sum of the first n terms of the harmonic series and those additional terms needed to express $r' = r - S_n$ in the desired form, and all these are different.

REMARKS. The proof shows that given r, if we always pick the largest term that will not make the sum too large, we will eventually make the sum exactly r. With a very slight modification the proof shows that we can forbid the use of any fixed finite number of terms. Hence it follows that every positive rational has infinitely many representations of the desired type.

J. C. Owings, *American Mathematical Monthly*, vol. 75 (1968), pages 777-778, gives a different proof of the original problem.

The ancient Egyptians seemed to prefer to represent a rational fraction as a sum of fractions with unit numerators, for example $\frac{4}{7} = \frac{1}{2} + \frac{1}{14}$, $\frac{4}{5} = \frac{1}{2} + \frac{1}{4} + \frac{1}{20}$, etc. For that reason the terms in the harmonic series are frequently called Egyptian fractions. They have been studied by professional and amateur mathematicians alike. One recent problem [Walter Penney, *Journal of Recreational Mathematics*, vol. 3 (July 1970), page 170] asks (a) for a representation using the least number of distinct fractions, and (b) for a representation using the smallest maximum denominator.

EXAMPLE: The minimum number of terms needed for 67/120 is three:

$$\frac{67}{120} = \frac{1}{2} + \frac{1}{18} + \frac{1}{360}$$

is one such representation, and the smallest maximum denominator is ten,

$$\frac{67}{120} = \frac{1}{3} + \frac{1}{8} + \frac{1}{10}.$$

[See Bernhardt Wohlgemuth, "Egyptian Fractions," *Journal of Recreational Mathematics*, vol. 5 (1972) pages 55-58].

---

7. Show that

$$\lim_{n\to\infty} \sum_{s=1}^{n} \left(\frac{a+s}{n}\right)^n \qquad (a > 0)$$

lies between e^a and e^{a+1}.

Solution. We shall evaluate the limit exactly using the facts:

$$(1) \qquad \left(1 + \frac{x}{n}\right)^n \leq e^x$$

if $n > 0$ and $1 + x/n > 0$, and

$$(2) \qquad \lim_{n\to\infty}\left(1+\frac{x}{n}\right)^n = e^x$$

for all real x.

Let

$$S_n = \sum_{s=1}^{n}\left(\frac{a+s}{n}\right)^n = \sum_{r=0}^{n-1}\left(1+\frac{a-r}{n}\right)^n.$$

Recalling that $a > 0$ and using (1), we have

$$S_n \le \sum_{r=0}^{n-1} \exp(a-r) < \sum_{r=0}^{\infty} \exp(a-r) = \frac{e^{a+1}}{e-1}.$$

Therefore

$$(3) \qquad \limsup_{n\to\infty} S_n \le \frac{e^{a+1}}{e-1}.$$

Also, for fixed k and $n > k$

$$S_n \ge \sum_{r=0}^{k}\left(1+\frac{a-r}{n}\right)^n.$$

The limit on the right exists as $n \to \infty$ by (2), so

$$\liminf_{n\to\infty} S_n \ge \sum_{r=0}^{k} \exp(a-r).$$

Since k is arbitrary, we get

$$(4) \qquad \liminf_{n\to\infty} S_n \ge \sum_{r=0}^{\infty} \exp(a-r) = \frac{e^{a+1}}{e-1}.$$

Comparing (3) and (4), we obtain

$$\lim S_n = \frac{e^{a+1}}{e-1}.$$

Evidently

$$1 < \frac{e}{e-1} < e, \quad \text{so} \quad e^a < \lim S_n < e^{a+1},$$

as required.

REMARKS. The inequality (1) follows immediately from the well-known inequality $\log(1 + y) \leq y$, while (2), even for complex values of x, follows from the fact that $\log(1 + y)$ has a power series representation convergent for $|y| < 1$.

THE FIFTEENTH WILLIAM LOWELL PUTNAM MATHEMATICAL COMPETITION

March 5, 1955

Morning Session

1. Prove that there is no set of integers m, n, p except 0, 0, 0 for which $m + n\sqrt{2} + p\sqrt{3} = 0$.

Solution. We recall that (1) if a is a positive integer and $\sqrt{a}$ is not an integer, then $\sqrt{a}$ is irrational.

Suppose that m, n, p are integers such that

$$m + n\sqrt{2} + p\sqrt{3} = 0. \tag{2}$$

If both n and p are zero, so is m. If just one of n and p is zero, we have either $\sqrt{2} = -m/n$ or $\sqrt{3} = -m/p$, both contrary to (1). If neither n nor p is zero, then

$$m^2 = (n\sqrt{2} + p\sqrt{3})^2 = 2n^2 + 3p^2 + 2np\sqrt{6}$$

and

$$\sqrt{6} = (m^2 - 2n^2 - 3p^2)/2np$$

again contrary to (1). So the only integer triplet for which (2) is true is $m = 0, n = 0, p = 0$.

For completeness we include a proof of (1). Suppose a is a positive integer and $\sqrt{a} = b/c$ where b and c are positive integers. We may assume b and c are relatively prime. Then $ac^2 = b^2$. Consider a prime q that divides c. Then q divides b^2 and hence b; so if there is such a prime q, then b and c have a common factor, which is a contradiction. Hence c has no prime factors, so $c = 1$, and $\sqrt{a} = b$, an integer.

---

2. $A_1A_2 \ldots A_n$ is a regular polygon inscribed in a circle of radius r and center O. P is a point on line OA_1 extended beyond A_1. Show that

$$\prod_{i=1}^{n} \overline{PA_i} = \overline{OP}^n - r^n.$$

Solution. We may assume the polygon is in the complex plane with its center at the origin and A_1 on the positive real axis. Then the other vertices are

$$r\omega, r\omega^2, \ldots, r\omega^{n-1}$$

where ω is a primitive nth root of unity.

If P is at the point x, then $\overline{PA_i} = |x - r\omega^{i-1}|$. So

$$\prod_{i=1}^{n} \overline{PA_i} = \left|\prod_{i=1}^{n} (x - r\omega^{i-1})\right| = r^n \left|\prod_{i=1}^{n} \left(\frac{x}{r} - \omega^{i-1}\right)\right| = r^n \left|\left(\frac{x}{r}\right)^n - 1\right|$$

$$= |x^n - r^n| = x^n - r^n = \overline{OP}^n - r^n.$$

At the third step we used the factorization

$$X^n - 1 = \prod_{i=1}^{n} (X - \omega^{i-1}),$$

which is valid because $1, \omega, \omega^2, \ldots, \omega^{n-1}$ are the zeros of $X^n - 1$.

---

3. Suppose that $\sum_{i=1}^{\infty} x_i$ is a convergent series of positive terms which monotonically decrease (that is, $x_1 \geq x_2 \geq x_3 \geq \cdots$). Let P denote the set of all numbers which are sums of some (finite or infinite) subseries of $\sum_{i=1}^{\infty} x_i$. Show that P is an interval if and only if

$$(1) \qquad x_n \leq \sum_{i=n+1}^{\infty} x_i \qquad \text{for every integer } n.$$

Solution. Let $\mathbf{N}$ be the set of positive integers, and let J be a subset of $\mathbf{N}$. We write $S(J)$ for $\sum_{i \in J} x_i$. The problem requires us to show that the range of S is an interval if and only if (1) holds.

Suppose (1) fails for a given sequence. Let p be an index such that

$$x_p > \sum_{i>p} x_i.$$

Choose α so that $\sum_{i>p} x_i < \alpha < x_p$. Then there is no J for which $S(J) = \alpha$; for if $J \cap \{1, 2, \ldots, p\} \neq \emptyset$, then $S(J) \geq x_p$ by the monotonicity of the

x's, while if $J \cap \{1, 2, \ldots, p\} = \emptyset$, then $S(J) \leq \sum_{i>p} x_i$. Since $\sum_{i>p} x_i$ and x_p are both in range (S), we see that range (S) is not an interval. Thus (1) is necessary in order that range (S) be an interval.

Now suppose (1) holds and $0 < y < S(\mathbf{N})$. We shall construct a set L such that $S(L) = y$. We define a sequence $n_1, n_2, n_3, \ldots$ by induction as follows. Let n_1 be the least index for which

$$x_{n_1} < y.$$

(Such an index exists because $x_k \to 0$.) Assuming that $n_1, n_2, \ldots, n_k$ have been chosen so that

$$x_{n_1} + x_{n_2} + \cdots + x_{n_k} < y,$$

let n_{k+1} be the least index exceeding n_k such that

$$x_{n_1} + x_{n_2} + \cdots + x_{n_k} + x_{n_{k+1}} < y.$$

(Again, such an index exists.)

Let $L = \{n_1, n_2, \ldots\}$. Clearly

$$S(L) \leq y. \tag{2}$$

If $p \in N - L$, there is a least index k such that $n_k > p$. In choosing n_k we rejected p, hence

$$x_{n_1} + x_{n_2} + \cdots + x_{n_{k-1}} + x_p \geq y \tag{3}$$

and therefore

$$S(L) + x_p \geq y. \tag{4}$$

We split the remainder of the proof into two cases.

Case 1. The set $N - L$ is finite. Note that $N - L \neq \emptyset$, since in that case $S(L) = S(N) > y$, contradicting (2). Hence $N - L$ has a largest element which we can take to be p in (4). Then

$$L = \{n_1, n_2, \ldots, n_{k-1}\} \cup \{p + 1, p + 2, \ldots\}.$$

Then combining (1) and (3) we see that

$$S(L) = x_{n_1} + x_{n_2} + \cdots + x_{n_{k-1}} + \sum_{i>p} x_i \geq y$$

which together with (2) shows that $S(L) = y$.

Case 2. The set $N - L$ is infinite. Then for any $\epsilon > 0$, we can choose $p \in N - L$ so that $x_p < \epsilon$. Then (4) yields

$$y \leq S(L) + \epsilon.$$

Since ϵ is arbitrary, we obtain $y \leq S(L)$. So again we must have $S(L) = y$.

Since $S(\emptyset) = 0$, it follows that range $(S) = [0, S(\mathbf{N})]$. Thus (1) is both necessary and sufficient in order that range (S) be an interval.

---

4. On a circle, n points are selected and the chords joining them in pairs are drawn. Assuming that no three of these chords are concurrent (except at the endpoints), how many points of intersection are there?

Solution. Any four points on a circle determine just one pair of chords that intersect at an interior point. Since the hypothesis implies that all such points of intersection are distinct, there are $\binom{n}{4}$ points of intersection in the interior of the circle.

---

5. If a parabola is given in the plane, find a geometric construction (ruler and compass) for the focus.

Solution. Analysis: Consider the chords of the parabola determined by a family $\mathfrak{F}$ of parallel lines. It is well known that the midpoints of these chords fall on a line parallel to the axis of the parabola. Such a line is called a diameter of the parabola. A diameter meets the parabola just once. If the diameter determined by $\mathfrak{F}$ meets the parabola at P, then the member of $\mathfrak{F}$ through P is tangent to the parabola. By the focussing property of the parabola, if the diameter through P is reflected in the tangent at P, the resulting line passes through the focus. Hence we are led to the following construction.

Construction: Draw any chord of the parabola and construct another parallel to it. Find the midpoints A and B of these chords and let $\overleftrightarrow{AB}$ meet the parabola at P. Draw $\overleftrightarrow{PQ}$ parallel to the original chord and find C so that $\angle APQ = \angle QPC$, with C on the side of $\overleftrightarrow{PQ}$ opposite to A. The focus is on $\overleftrightarrow{PC}$.

Repeat the construction starting from a different chord to obtain another line through the focus. Assuming the two lines thus constructed are different their intersection is the focus.

To avoid the awkward possibility that the two lines through the focus coincide, be sure that the two starting chords are neither parallel nor perpendicular. To see that this suffices, note that if the starting chords for the two constructions make an angle α, the constructed lines make an angle 2α; hence, if $\alpha \neq 0, \pi/2$, the constructed lines will have different directions.

REMARK. For the actual construction a number of shortcuts can be found. For example, once the first diameter has been found, the axis can quickly be located as the perpendicular bisector of any chord perpendicular to the diameter.

---

6. Find a necessary and sufficient condition on the positive integer n that the equation

$$x^n + (2 + x)^n + (2 - x)^n = 0 \tag{1}$$

have a rational root.

Solution. There can be no real root if n is even, since for real x each term is non-negative and they cannot vanish simultaneously.

If $n = 1$, there is obviously a unique root $x = -4$.

Suppose n is odd and at least 3. When the terms of the equation are expanded and collected, the result is monic with all coefficients non-negative integers and constant term 2^{n+1}. The only possible roots therefore are of the form -2^t. For $x = -1$, all three terms of the given expression are odd, so -1 is not a root. Putting $x = -2$, we find $(-2)^n + 0 + 4^n \neq 0$, so -2 is not a root. If we put $x = -2^{p+1}$ where $p \geq 1$, the left member of (1) becomes

$$2^n[-2^{pn} + (1 - 2^p)^n + (1 + 2^p)^n]$$
$$= 2^n\left[-2^{pn} + 2\left\{1 + \binom{n}{2} 2^{2p} + \binom{n}{4} 2^{4p} + \cdots\right\}\right].$$

The expression in the brackets is $\equiv 2$ [mod 4] (recall we are assuming $n \geq 3$), so -2^{p+1} is not a root for $p \geq 1$. So there are no roots if $n > 1$.

Summarizing, relation (1) has a rational root if and only if $n = 1$.

REMARK. A root with $n > 2$ would give a counterexample to the famous Fermat Conjecture.

---

7. Consider the function f defined by the differential equation

$$f''(x) = (x^3 + ax)f(x)$$

and the initial conditions $f(0) = 1, f'(0) = 0$. Prove that the roots of f are bounded above but unbounded below.

First Solution. We show first that the roots of f are bounded above. Let α be a number such that $x^3 + ax > 0$ for $x \geq \alpha$. We shall prove that no solution of the differential equation

$$y'' - (x^3 + ax)y = 0, \tag{1}$$

except the identically zero solution, has more than one root in $[\alpha, \infty)$. Suppose g is a non-zero solution of (1) and $g(x_1) = g(x_2) = 0$ with $\alpha \leq x_1 < x_2$. Then g is not zero throughout $[x_1, x_2]$, so, changing the sign of g if necessary, we assume that g is somewhere positive on $[x_1, x_2]$. Let g achieve its maximum value on $[x_1, x_2]$ at x_3. Then $g(x_3) > 0$, $g'(x_3) = 0$, and $g''(x_3) \leq 0$ by a standard criterion for a maximum. But

$$g''(x_3) = (x_3{}^3 + ax_3)g(x_3) > 0$$

because $x_3 \geq \alpha$. This contradiction proves that a non-zero solution of (1), in particular f, has at most one root in $[\alpha, \infty)$. Hence the roots of f are bounded above.

We now show that the roots of f are unbounded below. Let β be a number such that $x^3 + ax < -1$ for $x \leq \beta$. We shall prove that every solution of (1) has a root on $(-\infty, x_0]$ for any choice of $x_0 \leq \beta$. Suppose this is false. Then there is a solution h that has constant sign on $(-\infty, x_0]$. Changing the sign of h if necessary, we can assume that h is positive on this interval. For any choice of $x_1 \leq x_0$ we have by the extended mean value theorem

$$h(x) = h(x_1) + (x - x_1)h'(x_1) + \frac{1}{2}(x - x_1)^2 h''(\xi), \tag{2}$$

where ξ is between x and x_1. Then if $x \leq x_0$, we have $\xi < x_0$ and $h''(\xi) = (\xi^3 + a\xi)h(\xi) < 0$ and therefore

$$h(x) < h(x_1) + (x - x_1)h'(x_1).$$

If we could choose x_1 so that $h'(x_1) > 0$, this would show that $h(x) < 0$ for large negative x, contrary to our assumption. So $h'(x) \leq 0$ for all

$x \leq x_0$. But this implies $h(x) \geq h(x_0)$ for $x \leq x_0$. Then (2) yields

$$h(x) < h(x_1) + (x - x_1)h'(x_1) - \frac{1}{2}(x - x_1)^2 h(x_0)$$

since $h''(\xi) = (\xi^3 + a\xi)h(\xi) < -h(x_0)$. But this shows that $h(x)$ is negative for sufficiently large negative values of x. This contradiction proves that every solution of (1), in particular f, has arbitrarily large negative roots.

Second Solution. Again we choose α and β so that $x^3 + ax > 0$ for $x \geq \alpha$, and $x^3 + ax < -1$ for $x \leq \beta$. We shall apply the Sturm comparison theorem.

On the interval $[\alpha, \infty)$ we compare the differential equations

$$u'' + 0 \cdot u = 0 \tag{3}$$

and

$$y'' + (-x^3 - ax)y = 0. \tag{4}$$

Since $0 > -x^3 - ax$ on this interval, we conclude that between any two roots of a non-zero solution of (4) appears a root of any solution of (3). But (3) has a constant solution with no roots at all, so we see that no non-zero solution of (4) has two roots in $[\alpha, \infty)$. Hence f has at most one root in $[\alpha, \infty)$, so its roots are bounded above.

On the interval $(-\infty, \beta]$ we compare (4) with

$$v'' + 1 \cdot v = 0. \tag{5}$$

Since $-x^3 - ax > 1$ on this interval, we conclude that between any two roots of any non-zero solution of (5) there is a root of any solution of (4). For any choice of γ, $\sin(x - \gamma)$ is a non-zero solution of (5) with roots at $\gamma - \pi$ and γ. Hence f has a root in the interval $(\gamma - \pi, \gamma)$ for any $\gamma \leq \beta$. Thus the roots of f are unbounded below.

A proof of the Sturm comparison theorem is given on page 451.

Afternoon Session

1. A sphere rolls along two intersecting straight lines. Find the locus of its center.

First Solution. Let l_1 and l_2 be the given intersecting lines and let π be their plane. Let lines m_1 and m_2 be the bisectors of the angles formed by l_1 and l_2 and let σ_1 and σ_2 be the planes perpendicular to π containing m_1 and m_2, respectively. Then $\sigma_1 \cup \sigma_2$ is the locus of all points equidistant from l_1 and l_2.

Let $\mathcal{C}$ be the right circular cylinder of radius r and axis l_1. Then $\mathcal{C}$ is the locus of all points having distance r from l_1.

A sphere of radius r is tangent to l_1 and l_2 if and only if its center is at distance r from both l_1 and l_2. Hence the desired locus is $\mathcal{C} \cap (\sigma_1 \cup \sigma_2) = (\mathcal{C} \cap \sigma_1) \cup (\mathcal{C} \cap \sigma_2)$. Now the intersection of a right circular cylinder with a plane neither parallel nor perpendicular to its axis is an ellipse, so the locus is the union of two ellipses. These two ellipses have a common minor axis of length $2r$ lying on the line $\sigma_1 \cap \sigma_2$ and major axes lying on the lines m_1 and m_2.

Second Solution. We choose axes so that the given lines lie in the x,y-plane and the x-axis bisects the angle between them. Then the normal forms of the equations of the lines are

$$x \sin\theta - y\cos\theta = 0$$

and

$$x \sin\theta + y\cos\theta = 0,$$

where $0 < \theta < \pi/2$.

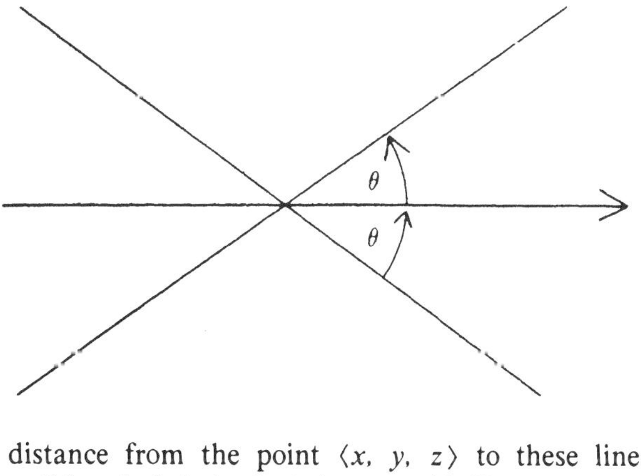

The squared distance from the point $\langle x, y, z\rangle$ to these lines is $z^2 + (x\sin\theta - y\cos\theta)^2$ and $z^2 + (x\sin\theta + y\cos\theta)^2$. The center of the sphere must be at distance r from both lines; hence the desired locus is given by the two equations

$$z^2 + (x\sin\theta - y\cos\theta)^2 = r^2$$

$$z^2 + (x\sin\theta + y\cos\theta)^2 = r^2.$$

Subtracting these equations, we obtain $4xy \sin\theta \cos\theta = 0$. Therefore the locus lies on the union of the planes $x = 0$ and $y = 0$, i.e., the y, z-plane and the x, z-plane. The part on the y, z-plane has the equation $z^2 + y^2 \cos^2 \theta$ r^2, so it is an ellipse having minor axis on the z-axis of length $2r$ a having major axis on the y-axis of length $2r \sec\theta$. The part on the x, z-plane has the equation $z^2 + x^2 \sin^2\theta = r^2$. It is therefore an ellipse havi minor axis of length $2r$ lying on the z-axis and major axis of length $2r \operatorname{cosec}\theta$ lying on the x-axis. The locus is the union of these two ellipses.

---

2. Suppose that f is a function with two continuous derivatives and $f(0) = 0$. Prove that the function g, defined by $g(0) = f'(0)$, $g(x) = f(x)/x$ for $x \neq 0$, has a continuous derivative.

First Solution. It is clear that g has a continuous derivative (even two continuous derivatives) on $\mathbf{R} - \{0\}$ with

$$g'(x) = \frac{xf'(x) - f(x)}{x^2}.$$

Now f is differentiable at 0, so

$$\lim_{x \to 0} \frac{f(x) - f(0)}{x} = \lim_{x \to 0} g(x) = f'(0).$$

Thus g is continuous at 0.

To find $g'(0)$, we must consider

$$\lim_{x \to 0} \frac{g(x) - g(0)}{x} = \lim_{x \to 0} \frac{f(x) - xf'(0)}{x^2}.$$

By the extended mean value theorem, for each x there is a $\theta \in (0, 1)$ such that

$$f(x) = f(0) + xf'(0) + \frac{1}{2} x^2 f''(\theta x),$$

so

$$\lim_{x \to 0} \frac{f(x) - xf'(0)}{x^2} = \lim_{x \to 0} \frac{1}{2} f''(\theta x) = \frac{1}{2} f''(0)$$

since f'' is continuous. Thus g is differentiable at 0 and $g'(0) = f''(0)/2$.

By the ordinary mean value theorem, for each x there is an $\eta \in (0, 1)$ such that

$$f'(x) - f'(0) = xf''(\eta x).$$

Hence we have

$$\lim_{x \to 0} g'(x) = \lim_{x \to 0} \frac{xf'(x) - f(x)}{x^2}$$

$$= \lim_{x \to 0} \left(f''(\eta x) - \frac{1}{2} f''(\theta x) \right) = \frac{1}{2} f''(0),$$

and so g' is continuous at 0.

The explicit references to the mean value theorem can be avoided by using L'Hospital's rule.

Second Solution. Note that

$$g(x) = \int_0^1 f'(xt)\, dt.$$

Since f' has a continuous derivative, we can differentiate under the integral sign and conclude that g' exists and is given by

$$g'(x) = \int_0^1 tf''(xt)\, dt.$$

Then, since f'' is continuous, g' is continuous.

---

3. Prove that there exists no distance-preserving map of a spherical cap into the plane. (Distances on the sphere are to be measured along great circles on the surface.)

Solution. We shall denote the spherical distance between two points A and B by $\rho(A, B)$ and the ordinary Euclidean distance by $|AB|$.

Suppose $X \mapsto X'$ is a distance-preserving map of a spherical cap into the plane. Let A, B, C, D be four points of the cap such that $ABCD$ is a Euclidean square in three-space. Note that a spherical cap must contain such a set. Then $\rho(A, B) = \rho(B, C) = \rho(C, D) = \rho(D, A)$ and $\rho(A, C) = \rho(B, D)$. These relations imply that

$$|A'B'| = |B'C'| = |C'D'| = |D'A'| \quad \text{and} \quad |A'C'| = |B'D'|.$$

Hence $A'B'C'D'$ is a square, and $|A'C'| = \sqrt{2}|A'B'|$. Since distances are preserved by the mapping, $\rho(A, C) = \sqrt{2}\rho(A, B)$. Now $|AC| = \sqrt{2}|AB|$ because $ABCD$ is a square, so

$$\frac{\rho(A, C)}{|AC|} = \frac{\rho(A, B)}{|AB|}. \tag{1}$$

However, $\rho(X, Y)/|XY| = \theta/\sin\theta$, where θ is half the central angle of the arc XY. Now $\theta/\sin\theta$ is a strictly increasing function of θ and hence of $|XY|$ and $|AC| > |AB|$; therefore, (1) is impossible. This contradiction proves that no distance-preserving map of a spherical cap into a plane exists.

---

4. Do there exist 1,000,000 consecutive integers each of which contains a repeated prime factor?

First Solution. We shall prove that there are sequences of consecutive integers of arbitrary length each of which has a repeated prime factor. The proof is by induction on the length. Obviously there are such sequences of length 1.

Suppose $a_1, a_2, \ldots, a_k$ are k consecutive integers (in order), each of which has a repeated prime factor. Let $b = a_1a_2 \cdots a_k$. Then for any integer n

$$nb + a_1, \quad nb + a_2, \quad \ldots, \quad nb + a_k$$

are k consecutive integers, each of which has a repeated prime factor since a_i divides $nb + a_i$. Let p be a prime not dividing b. Then we can choose n so that $nb + a_k + 1$ is divisible by p^2, since this amounts to solving the congruence $bx + a_k + 1 \equiv 0 \pmod{p^2}$ and b is relatively prime to p^2. Then each of the $k + 1$ consecutive integers

$$nb + a_1, \quad nb + a_2, \quad \ldots, \quad nb + a_k, \quad nb + a_k + 1$$

has a repeated prime factor.

It follows that there are sequences of consecutive integers of arbitrary length, and in particular sequences of length 1,000,000, each of which has a repeated prime factor.

Second Solution. Let $p_1, p_2, \ldots, p_s$ be s distinct primes. According to the Chinese Remainder Theorem the simultaneous congruences

$$x \equiv -1 \pmod{p_1^2}$$
$$x \equiv -2 \pmod{p_2^2}$$
$$\cdots\cdots\cdots\cdots\cdots$$
$$x \equiv -s \pmod{p_s^2}$$

have a solution, say n. Then the s consecutive integers

$$n+1, \quad n+2, \quad \ldots, \quad n+s$$

each have a repeated prime factor, for p_i^2 divides $n + i$. Since we may take $s = 1{,}000{,}000$, there do exist sequences of 1,000,000 consecutive integers, each of which contains a repeated prime factor.

For the Chinese Remainder Theorem see, for example, Niven and Zuckerman, *An Introduction to the Theory of Numbers*, 2nd ed., Wiley, New York, 1966, page 33.

---

5. Given an infinite sequence of 0's and 1's and a fixed integer k, suppose that there are no more than k distinct blocks of k consecutive terms. Show that the sequence is eventually periodic. (For example, the sequence 11011010101 followed by alternating 0's and 1's indefinitely, which is periodic beginning with the fifth term.)

Solution. We shall refer to a block of n consecutive terms as an n-block.

The proof is by induction on k. Clearly, if there is at most one 1-block, the sequence is constant, i.e., periodic with period 1.

Now suppose that every infinite sequence with at most k distinct k-blocks is eventually periodic. Then, given an infinite sequence with at most $k + 1$ distinct $(k + 1)$-blocks, consider what happens when each of these blocks is shortened by one on the right. If we obtain in this way fewer than $k + 1$ distinct k-blocks, then the given sequence has at most k distinct k-blocks and is therefore eventually periodic by the inductive hypothesis.

On the other hand, suppose we obtain $k + 1$ distinct k-blocks. Then each k-block has a unique extension to a $(k + 1)$-block. Then in the given sequence, if the ith and the $(i + p)$th k-blocks are the same for some i (where $p > 0$), it follows that the $(i + k)$th and $(i + p + k)$th terms are the same. Hence the $(i + 1)$st and $(i + p + 1)$st k-blocks are the same, and it follows by induction that the sequence is periodic with period p starting with the ith term. Since there are at most $k + 1$ distinct k-blocks in the sequence, a repeated block must appear within the first $k + 2$ blocks, so periodicity must begin not later than the $(k + 1)$st term.

Thus we have shown that every infinite sequence in which there are at most $k + 1$ distinct $(k + 1)$-blocks is eventually periodic. This completes the induction.

REMARKS. It may be shown that the number of terms before the periodic part of the sequence is at most $k - p$. The following sequence shows that this statement is best possible.

$$\overbrace{0,0,\ldots,0}^{k-p},\overbrace{0,0,\ldots,0,1}^{p},\overbrace{0,0,\ldots,0,1}^{p},\overbrace{0,0,\ldots,0,1}^{p},\ \ldots .$$

It does not matter how many distinct characters are permitted in the original sequence.

A related problem appears as Problem E 2307 in the *American Mathematical Monthly,* vol. 79 (1972) page 773.

---

6. Prove: If $f(x) > 0$ for all x and $f(x) \to 0$ as $x \to \infty$, then there exists at most a finite number of solutions of

$$f(m) + f(n) + f(p) = 1 \tag{1}$$

in positive integers m, n, and p.

Solution. We consider first solutions of (1) that satisfy also

$$f(m) \geq f(n) \geq f(p). \tag{2}$$

Since f takes only positive values, for any such solution we have

$$\frac{1}{3} \leq f(m) < 1, \tag{3}$$

$$\frac{1}{2}(1 - f(m)) \leq f(n) < 1 - f(m), \tag{4}$$

and

$$f(p) = 1 - f(m) - f(n). \tag{5}$$

Since $f(x) \to 0$ as $x \to \infty$, there are only finitely many integers m that satisfy (3); for each such m there are only finitely many integers n that satisfy (4); and for each such pair m, n there are only finitely many integers p that satisfy (5).

Thus there are only finitely many integral solutions of (1) that satisfy (2). But all other integer solutions of (1) are obtained from these by permuting m, n, and p. Hence (1) has only finitely many solutions in positive integers.

REMARKS. The proof obviously generalizes to show that for any integer k and number α, the equation

$$f(m_1) + f(m_2) + \cdots + f(m_k) = \alpha$$

has only a finite number of integral solutions.

Problem A.M. 6 of the Twelfth Competition involves a special case of this result. There it was necessary to show that the equation

$$mnp = 2(m-2)(n-2)(p-2)$$

has only a finite number of solutions in positive integers. Since this equation can be rewritten as

$$\log \frac{m}{m-2} + \log \frac{n}{n-2} + \log \frac{p}{p-2} = \log 2,$$

it is an instance of the above.

---

7. Four forces acting on a body are in equilibrium. Prove that, if their lines of action are mutually skew, they are rulings of a hyperboloid.

Solution. Let the forces be F_1, F_2, F_3, F_4 acting along the lines l_1, l_2, l_3, l_4, respectively. Of course we assume that $F_i \neq 0$.

For equilibrium the total moment of the forces about any line m must be zero. Consider a line m that is coplanar with each of the lines l_1, l_2, and l_3. The forces F_1, F_2, and F_3 each have zero moment about m, so F_4 must have zero moment about m as well. This implies that m and l_4 are coplanar.

Thus the mutually skew lines l_1, l_2, l_3, l_4 have the property that any line that is coplanar with each of the first three is also coplanar with the fourth. This implies that they are rulings of a non-degenerate quadric surface. We sketch a proof of this below.

Non-degenerate ruled quadric surfaces always meet some planes in hyperbolas, and consequently they are sometimes generically referred to as hyperboloids. (See James and James, *Mathematical Dictionary*, Van Nostrand, New York, 1949.) More commonly, however, they are divided into two classes, hyperboloids and hyperbolic paraboloids. The former are central quadrics and with suitable choice of coordinates have an equation of the form

$$\frac{x^2}{a^2} + \frac{y^2}{b^2} - \frac{z^2}{c^2} = 1.$$

The latter are non-central and in appropriate coordinates have an equation of the form

$$z = \frac{x^2}{a^2} - \frac{y^2}{b^2}.$$

The distinction can easily be made projectively. Hyperbolic paraboloids are those ruled non-degenerate quadrics that are tangent to the plane at infinity. In affine terms we can say that the hyperbolic paraboloids are those ruled non-degenerate quadrics for which the members of each family of rulings are all parallel to a fixed plane.

It is possible that four forces in equilibrium should act along the rulings of a hyperbolic paraboloid. For example, let $F_1 = \mathbf{j}$ acting at $\langle 0, 0, 0\rangle$, $F_2 = -3\mathbf{j} - 3\mathbf{k}$ acting at $\langle 1, 0, 0\rangle$, $F_3 = 3\mathbf{j} + 6\mathbf{k}$ acting at $\langle 2, 0, 0\rangle$, and $F_4 = -\mathbf{j} - 3\mathbf{k}$ acting at $\langle 3, 0, 0\rangle$, where $\mathbf{i}, \mathbf{j}, \mathbf{k}$ are unit vectors in the directions of the x, y, and z axes. These forces are in equilibrium and their lines of action are rulings of the hyperbolic paraboloid $z = xy$.

We sketch the proof of the result mentioned above concerning the rulings of a quadric surface. For simplicity we treat the problem projectively. In this context two coplanar lines always intersect. If l is a line and p is a point not on l, then $p \vee l$ is the plane containing p and l.

Let l_1, l_2, and l_3 be three mutually skew lines. Let $\mathfrak{M}$ be the set of lines that meet each of l_1, l_2, and l_3. Suppose p is a point of l_1. Then $p \vee l_2$ and $p \vee l_3$ are distinct planes (since l_2 and l_3 are skew), and any line that passes through p and meets both l_2 and l_3 lies in both of them. Hence $(p \vee l_2) \cap (p \vee l_3)$ is the unique line through p that meets both l_2 and l_3. Thus we see that through each point of l_1 passes a unique member of $\mathfrak{M}$. The union of the lines in $\mathfrak{M}$ is a non-degenerate quadric surface $\mathfrak{Q}$. and we see that l_1, l_2, and l_3 lie wholly on $\mathfrak{Q}$.

Now let l_4 be any line distinct from l_1, l_2, and l_3 that meets every member of $\mathfrak{M}$. It is easy to see that l_1, l_2, l_3, and l_4 are mutually skew. We shall show that l_4 lies entirely on $\mathfrak{Q}$. If $q \in l_4$, then $n = (q \vee l_1) \cap (q \vee l_2)$ is a line through q meeting l_1 and l_2. Say it meets l_1 at r. There is a line $m \in \mathfrak{M}$ through r and it meets l_4 (by our assumption on l_4), say at s. Now if $q \neq s$, then there would be two lines m and n through r meeting both l_2 and l_4; but there is only one, namely, $(r \vee l_2) \cap (r \vee l_4)$. So $q = s \in m \subseteq \mathcal{Q}$. Thus l_4 lies on $\mathcal{Q}$.

We have shown that if four mutually skew lines in projective three-space have the property that any line meeting the first three meets also the fourth, then these lines are rulings of a quadric surface.

For the theory of quadric surfaces, see A. Seidenberg, *Lectures in Projective Geometry*, Van Nostrand, Princeton, N.J., 1962, pages 208 ff.

THE SIXTEENTH WILLIAM LOWELL PUTNAM MATHEMATICAL COMPETITION

March 3, 1956

Morning Session

1. Evaluate

$$\lim_{x\to\infty}\left[\frac{1}{x}\,\frac{a^x-1}{a-1}\right]^{1/x}$$

where $a > 0$, $a \neq 1$.

Solution. Let

$$f(x)=\left[\frac{1}{x}\,\frac{a^x-1}{a-1}\right]^{1/x}.$$

Then for $x > 0$ and $a > 1$, we have

$$\log f(x) = -\frac{\log x}{x} - \frac{\log(a-1)}{x} + \frac{\log(a^x-1)}{x}.$$

As $x \to +\infty$,

$$\frac{\log x}{x} \to 0,\ \frac{\log(a-1)}{x} \to 0$$

and

$$\frac{\log(a^x-1)}{x} = \frac{\log(1-a^{-x})}{x} + \log a \to \log a.$$

Hence $\log f(x) \to \log a$.

On the other hand, if $0 < a < 1$ and $x > 0$, then

$$\log f(x) = -\frac{\log x}{x} - \frac{\log(1-a)}{x} + \frac{\log(1-a^x)}{x},$$

and it is clear that all three terms approach zero as $x \to +\infty$.

Since exp is a continuous function, we have

$$\lim_{x\to+\infty} f(x) = \exp \lim_{x\to+\infty} \log f(x) = \begin{cases} \exp \log a = a & \text{if } a > 1 \\ \exp 0 = 1 & \text{if } 0 < a < 1. \end{cases}$$

More concisely, $\lim_{x\to+\infty} f(x) = \max(a, 1)$.

REMARK. It is of some interest to see what happens if $\lim_{x\to\infty}$ is replaced by $\lim_{x\to-\infty}$. To do this, note that for $x < 0$ and $a > 1$, we have

$$\log f(x) = -\frac{\log|x|}{x} - \frac{\log(a-1)}{x} + \frac{\log(1-a^x)}{x},$$

and all three terms have limit zero as $x \to -\infty$; since in this case $1 - a^x \to 1$.

For $x < 0$ and $0 < a < 1$, on the other hand,

$$\log f(x) = -\frac{\log|x|}{x} - \frac{\log(1-a)}{x} + \frac{\log(1-a^{-x})}{x} + \log a,$$

so $\lim_{x\to-\infty} f(x) = \log a$.

Hence we have

$$\lim_{x\to-\infty} f(x) = \begin{cases} 1 & \text{if } a > 1, \\ a & \text{if } 0 < a < 1. \end{cases}$$

In other words, $\lim_{x\to-\infty} f(x) = \min(a, 1)$. In particular $\lim_{x\to\infty} f(x) \neq \lim_{x\to-\infty} f(x)$, so that $\lim_{|x|\to\infty} f(x)$ does not exist.

---

2. Prove that every positive integer has a multiple whose decimal representation involves all ten digits.

Solution. If n is a positive integer and p is any other positive integer, then one of the integers

$$p + 1, p + 2, \ldots, p + n$$

is a multiple of n. Given n, choose $p = 1{,}234{,}567{,}890 \times 10^k$, where k is so large that $10^k > n$. Then all of the integers $p + 1, p + 2, \ldots, p + n$ have decimal representations beginning with 1234567890 ..., and one of these is a multiple of n.

---

3. A particle falls in a vertical plane from rest under the influence of gravity and a force perpendicular to and proportional to its velocity. Obtain the equations of the trajectory and identify the curve.

Solution. Suppose the position of the particle at time t is $\langle x(t), y(t)\rangle$. Then the velocity vector is given by $\langle x'(t), y'(t)\rangle$ and the acceleration vector by $\langle x''(t), y''(t)\rangle$. We choose the coordinate system so that the particle starts at the origin with x running horizontally and y vertically as usual.

Since the vector $\langle y'(t), -x'(t)\rangle$ is perpendicular to and of the same length as the velocity vector the differential equation governing the motion is

$$\langle x''(t), y''(t)\rangle = c\langle y'(t), -x'(t)\rangle + \langle 0, -g\rangle \tag{1}$$

where c is a constant incorporating the mass of the given particle. The initial conditions are $x(0) = y(0) = x'(0) = y'(0) = 0$. Separating (1) into two equations we have

$$x''(t) = cy'(t)$$

$$y''(t) = -cx'(t) - g.$$

Differentiating the first of these equations, we get

$$x'''(t) = cy''(t) = -c^2x'(t) - cg,$$

which in standard form is

$$x'''(t) + c^2x'(t) = -cg. \tag{2}$$

The corresponding homogeneous differential equation, $x'''(t) + c^2x'(t) = 0$, has the three linearly independent solutions 1, $\sin ct$, $\cos ct$. Since the right member of (2) is a solution of the homogeneous differential equation, we look for a particular solution of the form kt, and find that $-gt/c$ is such a solution. Hence the general solution of (2) is

$$x(t) = -gt/c + \alpha + \beta \cos ct + \gamma \sin ct.$$

The initial conditions are $x(0) = x'(0) = x''(0) = 0$ (the latter from $y'(0) = 0$). Hence

$$x(t) = -\frac{gt}{c} + \frac{g}{c^2}\sin ct. \tag{3}$$

Therefore $y' = x''/c = -g/c \sin ct$ and

$$y(t) = \frac{g}{c^2}(-1 + \cos ct), \tag{4}$$

using the initial condition $y(0) = 0$.

Thus equations (3) and (4) describe the motion. They are the parametric equations of a cycloid traced by a point on the rim of a circle of radius g/c^2 rolling along the underside of the x axis with velocity $-g/c$.

REMARK. We have assumed throughout, of course, that $c \neq 0$. If $c = 0$, the motion is just free fall.

---

4. Suppose the n times differentiable real function $f(x)$ has at least $n + 1$ distinct zeros in the closed interval $[a, b]$ and that the polynomial $P(z) \equiv z^n + C_{n-1}z^{n-1} + \cdots + C_0$ has only real zeros. Show that $(D^n + C_{n-1}D^{n-1} + \cdots + C_0)f(x)$ has at least one zero in the interval $[a, b]$ where D^n denotes, as usual, d^n/dx^n.

Solution. We first prove a lemma.

LEMMA. *Suppose f is differentiable on $[a, b]$ and has $m + 1$ distinct zeros there. Then for any real number λ, $(D - \lambda)f$ has at least m distinct zeros on $[a, b]$.*

Proof. Consider the identity

$$(D - \lambda)f(x) = e^{\lambda x}D(e^{-\lambda x}f(x)).$$

Applying Rolle's theorem to the right member of this identity we see that there is a zero of $(D - \lambda)f$ between any two consecutive zeros of f. Therefore there are at least m distinct zeros on $[a, b]$. ■

We can now prove the result stated in the problem by induction on n. If $n = 1$, this is just the lemma with $m = 1$.

Assume the result is true for $n = k$. Suppose P has degree $k + 1$ and that f is $k + 1$ times differentiable and has $k + 2$ distinct zeros on $[a, b]$. Since P has all real roots, we can write $P(z) = Q(z)(z - \lambda)$, where λ is real and Q is a polynomial of degree k with all real roots. Then $g = (D - \lambda)f$ is k times differentiable on $[a, b]$ and has at least $k + 1$ distinct zeros on $[a, b]$, by the lemma with $m = k + 1$. Hence by the inductive hypothesis $Q(D)g$ has at least one zero on $[a, b]$. But

$$Q(D)g = Q(D)(D - \lambda)f = P(D)f.$$

Thus the result is true for $n = k + 1$. This completes the induction.

---

5. Given n objects arranged in a row. A subset of these objects is called unfriendly if no two of its elements are consecutive. Show that the number of unfriendly subsets each having k elements is

$$\binom{n - k + 1}{k}.$$

Solution. For each subset S of an ordered set of n objects we form a linear arrangement of A's and B's by writing an A in positions corresponding to members of S and B's elsewhere. We describe such an arrangement as unfriendly if no two A's are consecutive. Then an unfriendly subset corresponds to an unfriendly arrangement, and we must show that the number of unfriendly arrangements of k A's and $n - k$ B's is $\binom{n-k+1}{k}$.

We shall establish a bijective correspondence between all unfriendly arrangements of k A's and $n - k$ B's and all arrangements of k A's and $n - 2k + 1$ B's.

Given an unfriendly arrangement of k A's and $n - k$ B's, remove the B standing immediately to the right of each of the first $k - 1$ A's (i.e., all but the last A). We obtain in this way an arrangement of k A's and $n - 2k + 1$ B's.

Conversely, given an arrangement of k A's and $n - 2k + 1$ B's, insert a B after each of the A's but the last. We obtain in this way an unfriendly arrangement of k A's and $n - k$ B's.

Obviously these two transformations are inverses of one another. Therefore the number of unfriendly arrangements of k A's and $n - k$ B's is the same as the number of arrangements of k A's and $n - 2k + 1$ B's, namely

$$\binom{n - k + 1}{k}.$$

This problem was treated by Irving Kaplansky, "Solution of the problème des ménages, "*Bulletin of the American Mathematical Society,* vol. 49 (1943), pages 784-785. The problem has been generalized by H. D. Abramson, "On Selecting Separated Objects from a Row," *American*

Mathematical Monthly, vol. 76 (1969), pages 1130–1131. The generalization requires greater separation between the elements in the unfriendly subset.

---

6. (i) A transformation of the plane into itself preserves all rational distances. Prove that it preserves all distances.

(ii) Show that the corresponding theorem for the line is false.

Solution. (i) Suppose T is a transformation of the plane into itself that preserves all rational distances. Denote the distance between any two points P, Q by $d(P, Q)$.

Let A and B be any two distinct points in the plane. Given any positive number $\epsilon < d(A, B)$, choose a rational number r such that

$$d(A, B) - \epsilon < r < d(A, B),$$

and choose a second rational number s such that $s < \epsilon$ and $r + s > d(A, B)$. Then the circle of radius r about A and the circle of radius s about B intersect; let C be one of the intersection points. Then $d(A, C) = r$, and $d(B, C) = s$. Hence $d(TA, TC) = r$ and $d(TB, TC) = s$, so we have

$$\begin{aligned} r - s &= d(TA, TC) - d(TB, TC) \\ &\leq d(TA, TB) \leq d(TA, TC) + d(TC, TB) = r + s. \end{aligned}$$

Hence

$$d(A, B) - 2\epsilon < d(TA, TB) < d(A, B) + \epsilon.$$

Since ϵ can be taken arbitrarily small, this shows that

$$d(TA, TB) = d(A, B).$$

Since A and B were chosen arbitrarily, this proves that T is distance preserving.

(ii) Identify the line with $\mathbf{R}$, as usual. Consider the transformation $S(x) = x$ if x is rational, and $S(x) = x + 1$ if x is irrational. Then S preserves all rational distances, but not all distances. For example $d(0, \sqrt{2}) = \sqrt{2}$, but $d(S(0), S(\sqrt{2})) = d(0, \sqrt{2} + 1) = \sqrt{2} + 1$.

Remarks. The result and the argument given above remain valid for Euclidean n-dimensional space E^n for any $n \geq 2$. A stronger result is true:

If $n \geq 2$ and T maps E^n into itself so that, for *one* positive number α, $d(TA, TB) = \alpha$ whenever $d(A, B) = \alpha$, then T is an isometry. (See F. S. Beckman and D. A. Quarles, Jr., "On Isometries of Euclidean Spaces," *Proc. Amer. Math. Soc.*, vol. 4 (1953), pp. 810–815.)

Every isometric map of E^n into itself can be represented as the product of not more than $n + 1$ reflections. Hence, when expressed in Cartesian coordinates, it is given by a (possibly) inhomogeneous linear transformation where the coefficient matrix is orthogonal. (See P. B. Yale, *Geometry and Symmetry*, Holden-Day, San Francisco, 1968, p. 60. The proof there generalizes immediately to E^n.)

---

7. Prove that the number of odd binomial coefficients in any finite binomial expansion is a power of 2.

Solution. First we note that $(1 + x)^2 \equiv 1 + x^2 \pmod 2$ and hence, by induction $(1 + x)^\beta \equiv 1 + x^\beta \pmod 2$ if β is any power of 2.

Let the exponent of the binomial be n, a positive integer, and represent n in the form

$$n = 2^{\alpha_1} + 2^{\alpha_2} + \cdots + 2^{\alpha_s}$$

where the α's are integers and $0 \leq \alpha_1 < \alpha_2 < \cdots < \alpha_s$. [Note that in effect this is the dyadic expansion of n.] For convenience let $\beta_i = 2^{\alpha_i}$. Then we write the finite binomial expansion in the form

$$\begin{aligned}(1 + x)^n &= (1 + x)^{\beta_1}(1 + x)^{\beta_2} \cdots (1 + x)^{\beta_s} \\ &\equiv (1 + x^{\beta_1})(1 + x^{\beta_2}) \cdots (1 + x^{\beta_s}) \pmod 2.\end{aligned}$$

When the latter expression is multiplied out, there are clearly 2^s terms, each involving a different power of x because each non-negative integer has exactly one representation as a sum (possibly empty) of distinct powers of 2. Hence exactly 2^s terms of $(1 + x)^n$ have odd integers as coefficients, that is, exactly 2^s of the binomial coefficients $\binom{n}{i}$, $i = 0, 1, \ldots, n$, are odd. The proof shows that s is the number of unit digits which appear in the dyadic expansion of n. The result is also valid for $n = 0$.

REMARKS. This problem appeared as Problem E 1288, *American Mathematical Monthly*, vol. 65 (1958), pages 368–369. A generalization appeared as Problem 4723, *American Mathematical Monthly*, vol. 65 (1958), page 48: "Given a non-negative integer n and a prime p, obtain an expression for the number of binomial coefficients $\binom{n}{r}$ which are not divisible by p." If n is written in the base p, let $\{n_i\}$ be the "digits" which appear. Then the required number of binomial coefficients not divisible by p is given by $\Pi\,(n_i + 1)$.

The same process used to establish the original result can be used to prove the following: Suppose p is a prime and n and k are integers, $0 \le k \le n$. Let

$$n = n_0 + n_1 p + n_2 p^2 + \cdots + n_t p^t$$
$$k = k_0 + k_1 p + k_2 p^2 + \cdots + k_t p^t,$$

be the p-adic expansions of n and k. Then

$$\binom{n}{k} \equiv \binom{n_0}{k_0}\binom{n_1}{k_1}\binom{n_2}{k_2}\cdots\binom{n_t}{k_t} \pmod{p}.$$

Afternoon Session

1. Show that if the differential equation

$$M(x, y)\,dx + N(x, y)\,dy = 0$$

is both homogeneous and exact then the solution $y = f(x)$ satisfies $xM + yN = C$ (constant).

Solution. If M and N are homogeneous of degree k, then by Euler's theorem

$$x\frac{\partial M}{\partial x} + y\frac{\partial M}{\partial y} = kM$$

and

$$x\frac{\partial N}{\partial x} + y\frac{\partial N}{\partial y} = kN.$$

If the given differential equation is exact, then

$$\frac{\partial M}{\partial y} = \frac{\partial N}{\partial x}.$$

Hence, if both conditions are satisfied,

$$d(xM + yN) = Mdx + Ndy + x\left(\frac{\partial M}{\partial x}\,dx + \frac{\partial M}{\partial y}\,dy\right)$$

$$+\, y\left(\frac{\partial N}{\partial x}\,dx + \frac{\partial N}{\partial y}\,dy\right)$$

$$= Mdx + Ndy + \left(x\,\frac{\partial M}{\partial x} + y\,\frac{\partial M}{\partial y}\right)dx$$

$$+ \left(x\,\frac{\partial N}{\partial x} + y\,\frac{\partial N}{\partial y}\right)dy$$

$$= (k + 1)(Mdx + Ndy),$$

a relation valid on the entire domain of M and N.

Now if $y = f(x)$, where f is defined on an interval, is a solution of the given differential equation (that is, the substitution of $f(x)$ for y makes $Mdx + Ndy$ zero) then this substitution in $xM + yN$ gives a function g defined on an interval with $dg = 0$. Hence g is a constant, say $g(x) = C$. Then $y = f(x)$ satisfies $xM + yN = C$.

---

2. Suppose that each set X of points in the plane has an associated set $\overline{X}$ of points called its cover. Suppose further that

(1) $\overline{X \cup Y} \supset \overline{\overline{X}} \cup \overline{Y} \cup Y$, where $\cup$ designates point set sum (or union) and $\supset$ denotes set inclusion.

Prove: (i) $\overline{X} \supset X$, (ii) $\overline{\overline{X}} = \overline{X}$, (iii) $X \supset Y$ implies $\overline{X} \supset \overline{Y}$. Prove conversely that (i), (ii) and (iii) imply (1).

Solution. In (1) let $Y = X$: then

$$\overline{X} \supset \overline{\overline{X}} \cup \overline{X} \cup X$$

from which it is clear that $\overline{X} \supset X$, which is (i).

We note from the above that $\overline{X} \supset \overline{\overline{X}}$. In (i) we replace X by $\overline{X}$ to get $\overline{\overline{X}} \supset \overline{X}$. These two relations imply that $\overline{\overline{X}} = \overline{X}$, which is (ii).

Suppose $Y \subset X$. Then $X \cup Y = X$, and (1) reduces to

$$\overline{X} \supset \overline{\overline{X}} \cup \overline{Y} \cup Y$$

whence $\overline{X} \supset \overline{Y}$, which is (iii).

Now suppose that (i), (ii), and (iii) hold. For any sets X and Y we have

$$X \subset X \cup Y.$$

By (iii),

$$\overline{X} \subset \overline{X \cup Y}.$$

Hence by (ii)

$$\overline{\overline{X}} \subset \overline{X \cup Y}. \tag{2}$$

Also

$$Y \subset X \cup Y$$

and by (iii)

$$\overline{Y} \subset \overline{X \cup Y}. \tag{3}$$

From (i)

$$X \cup Y \subset \overline{X \cup Y}$$

so

$$Y \subset \overline{X \cup Y}. \tag{4}$$

Now (2), (3) and (4) together yield

$$\overline{X \cup Y} \supset \overline{\overline{X}} \cup \overline{Y} \cup Y,$$

which is the required condition (1).

REMARK. See Garrett Birkhoff, *Lattice Theory, Amer. Math. Soc. Colloquium Pub.*, vol. 25, Providence, R.I., 1967, page 113.

---

3. A sphere is inscribed in a tetrahedron and each point of contact of the sphere with the four faces is joined to the vertices of the face containing the point. Show that the four sets of three angles so formed are identical.

Solution. Let the vertices of the tetrahedron be P_i, $i = 1, 2, 3, 4$, and let Q_i be the point of contact of the sphere with the face opposite P_i. We adopt the convention that when i, j, k, l appear in a statement, they represent distinct elements of $\{1, 2, 3, 4\}$.

Now P_iQ_j and P_iQ_k are tangents to the sphere from the same external point; hence $|P_iQ_j| = |P_iQ_k|$. Similarly, $|P_lQ_j| = |P_lQ_k|$. Hence, $\triangle P_iQ_jP_l \cong \triangle P_iQ_kP_l$ by s.s.s. Therefore, $\angle P_iQ_jP_l = \angle P_iQ_kP_l$. We denote this angle by $|il|$. Clearly we have

$$(1) \qquad |il| = |li|.$$

Since the angles at Q_i add to 2π, we have

$$|23| + |34| + |42| = 2\pi,$$
$$|34| + |41| + |13| = 2\pi,$$
$$|41| + |12| + |24| = 2\pi,$$
$$|12| + |23| + |31| = 2\pi.$$

If we add the first two of these equations, subtract the third and fourth, and use (1), we obtain $2 \cdot |34| - 2 \cdot |12| = 0$. Hence, $|12| = |34|$, and by symmetry

$$(2) \qquad |ij| = |kl|.$$

The angles at Q_1 are $|23|$, $|34|$, and $|42|$, and these are respectively equal to the angles at Q_2, namely, $|41|$, $|34|$, and $|13|$, by (2). By symmetry, the central angles are the same in all four faces.

Remark. The angles appear with the same orientation in each face. Thus, when viewed from outside the tetrahedron, the angles $|23|$, $|34|$, and $|42|$ at Q_1 appear in the same order (clockwise or counterclockwise) as the corresponding angles $|41|$, $|34|$, and $|13|$ at Q_2. This is hard to visualize, but if it were not so, the faces of the tetrahedron would be asymmetrically partitioned into two classes, which is clearly impossible in a symmetrical situation.

---

4. Prove that if A, B, and C are angles of a triangle measured in radians then $A \cos B + \sin A \cos C > 0$.

Solution. We distinguish two cases.

Case 1. $0 < B < \pi/2$. Then $\cos B > 0$. Since $A > 0$, $A > \sin A$;

hence $A \cos B > \sin A \cos B$. Also $C = \pi - A - B < \pi - B$, and $0 < C < \pi$, so $\cos C > \cos(\pi - B) = -\cos B$. Therefore $\cos B + \cos C > 0$. Hence

$$A \cos B + \sin A \cos C > \sin A (\cos B + \cos C) > 0;$$

and the desired inequality is established in Case 1.

Case 2. $\frac{\pi}{2} \leq B < \pi$. Then $\cos B \leq 0$, $0 < A < \frac{\pi}{2}$, so $A < \tan A$, and $\tan A > 0$. Hence

$$A \cos B \geq \tan A \cos B.$$

Also $B = \pi - A - C$. So

$$\cos B + \cos A \cos C = -\cos(A + C) + \cos A \cos C = \sin A \sin B > 0.$$

Therefore

$$\begin{aligned} A \cos B + \sin A \cos C &\geq \tan A \cos B + \sin A \cos C \\ &= \tan A (\cos B + \cos A \cos C) > 0. \end{aligned}$$

Thus the required inequality is proved in Case 2 also.

---

5. Consider a set of $2n$ points in space, $n > 1$. Suppose they are joined by at least $n^2 + 1$ segments. Show that at least one triangle is formed. Show that for each n it is possible to have $2n$ points joined by n^2 segments without any triangles being formed.

Solution. Let T_n be the statement: If $2n$ points are joined by at least $n^2 + 1$ segments, then a triangle is formed. We shall prove T_n by induction.

Suppose A, B, C, D are four points joined by at least five segments. Then at most one segment is missing, say AB, and BCD form a triangle. Thus T_2 is true.

Now assume T_k is true. Let $2(k + 1)$ points be given connected by at least $(k + 1)^2 + 1$ segments. Let A and B be two points that are joined and let $\mathcal{S}$ be the set of $2k$ other points. Suppose p points of $\mathcal{S}$ are joined to A and q points of $\mathcal{S}$ are joined to B. If $p + q > 2k$, then some point C of $\mathcal{S}$ is joined to both A and B, and ABC is a triangle. If $p + q \leq 2k$, then at most $2k + 1$ segments have A or B as endpoints and at least

$$(k + 1)^2 + 1 - (2k + 1) = k^2 + 1$$

connect points of $\mathcal{S}$. By T_k some three points of $\mathcal{S}$ form a triangle. Thus T_{k+1} is true.

Therefore, T_n is true for all $n > 1$ (and for $n = 1$ in a vacuous way).

To show that $2n$ points can be joined by n^2 segments without any triangles being formed, divide the points into two sets S, T of n elements each. If every point in S is joined to every point in T, then n^2 segments are used and no triangle is formed.

REMARK. If $2n$ points are joined by $n^2 + 1$ segments, then at least n triangles are formed. For a discussion of this and related problems, see P. Turan, "On the Theory of Graphs," *Colloquium Mathematica,* vol. 3 (1954, pages 19–30.

---

6. Given $T_1 = 2$, $T_{n+1} = T_n^2 - T_n + 1$, $n > 0$, Prove:

(i) If $m \neq n$, T_m and T_n have no common factor greater than 1.

(ii) $$\sum_{i=1}^{\infty} \frac{1}{T_i} = 1.$$

Solution. The first few members of the sequences are $T_1 = 2$, $T_2 = 3$, $T_3 = 7$. We shall prove by induction that

(1) $$T_{n+1} = 1 + \prod_{i=1}^{n} T_i \quad \text{for } n \geq 1.$$

This is true for $n = 1$. Suppose it is true for $n = k$. Then

$$\begin{aligned} T_{k+2} &= 1 + T_{k+1}(T_{k+1} - 1) \\ &= 1 + T_{k+1}\left[\prod_{i=1}^{k} T_i\right] \\ &= 1 + \prod_{i=1}^{k+1} T_i. \end{aligned}$$

(The first step by the given recursion, the second by the inductive hypothesis.) This completes the inductive proof of (1).

Now suppose $m \neq n$, say $m < n$. Then (1) shows that T_m divides $T_n - 1$, so T_m and T_n are relatively prime. This is (i).

Next we prove by induction that

(2) $$\sum_{i=1}^{n} \frac{1}{T_i} = 1 - \frac{1}{T_{n+1} - 1}$$

for all n. This is true for $n = 1$ and, if it is true for $n = k$, we have

$$\sum_{i=1}^{k+1} \frac{1}{T_i} = 1 - \frac{1}{T_{k+1} - 1} + \frac{1}{T_{k+1}}$$

$$= 1 - \frac{1}{T_{k+1}(T_{k+1} - 1)}$$

$$= 1 - \frac{1}{T_{k+2} - 1}.$$

Thus (2) is established.

Since $T_n \to \infty$ as $n \to \infty$, it follows from (2) that

$$\sum_{i=1}^{\infty} \frac{1}{T_i} = 1,$$

as required.

REMARK. It has been proved by P. Erdös and E. G. Straus (*Indian Journal of Mathematics*, vol. 27 (1963), pp. 129–133) that if $a_1, a_2, \ldots$ are positive integers such that

$$a_{n+1} \geq \prod_{k=1}^{n} a_k \quad \text{and} \quad \sum_{n=1}^{\infty} \frac{1}{a_n}$$

is rational, then $a_{n+1} = a_n^2 - a_n + 1$ for all sufficiently large n.

---

7. The polynomials $P(z)$ and $Q(z)$ with complex coefficients have the same set of numbers for their zeros but possibly different multiplicities. The same is true of the polynomials

$$P(z) + 1 \quad \text{and} \quad Q(z) + 1.$$

Prove that $P(z) \equiv Q(z)$.

Solution. The desired conclusion is false for polynomials of degree zero, so we shall assume that at least one of the polynomials is not constant. Suppose P has degree m and Q has degree n. We may assume by symmetry that $m \geq n$. Let the distinct zeros of P be $\{\lambda_1, \lambda_2, \ldots, \lambda_r\}$, and let the distinct zeros of $P + 1$ be $\{\mu_1, \mu_2, \dots, \mu_s\}$. These sets are clearly disjoint. Counting multiplicities, P', the derivative of both P and $P + 1$, must have at least $(m - r)$ zeros in $\{\lambda_1, \lambda_2, \dots, \lambda_r\}$ and $(m - s)$ zeros in $\{\mu_1, \mu_2, \dots, \mu_s\}$. Therefore $(m - r) + (m - s) \leq m - 1$, the degree of P'. (It is here that the assumption $m > 0$ enters.) Hence $r + s > m$. But each of the $r + s$ numbers $\lambda_1, \lambda_2, \dots, \lambda_r, \mu_1, \mu_2, \dots, \mu_s$ is a zero of $P - Q$, a polynomial of degree at most m. It follows that $P(z) \equiv Q(z)$ as required.

REMARK. This problem is discussed in a more general setting by W. W. Adams and E. G. Straus, "Non-Archimedean Analytic Functions Taking the Same Value at the Same Points," *Illinois Journal of Mathematics*, vol. 15 (1971), pages 418–424.

THE SEVENTEENTH WILLIAM LOWELL PUTNAM MATHEMATICAL COMPETITION

March 2, 1957

Morning Session

1. The normals to a surface all intersect a fixed straight line. Show that the surface is a portion of a surface of revolution.

Solution. The problem is not properly stated. A glance at the figure depicting the union of a cutaway right circular cylinder and a portion of a spherical cap shows that it is possible to have a connected C^1-surface which satisfies the hypothesis of the problem but which requires a very liberal interpretation of the term *surface* to accommodate the conclusion. The example can be made much more complicated. In fact, there is a C^ω-surface S satisfying the hypothesis such that the smallest rotationally invariant set containing S includes the union of any prescribed countable collection of C^∞-surfaces of revolution.

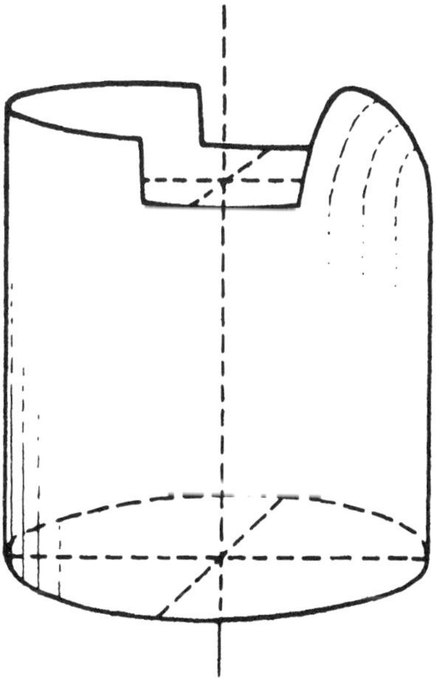

By changing the hypothesis and the conclusion we shall obtain two valid variants of the problem. Since the subject is necessarily rather technical, as the above example suggests, we begin with the definition of a surface. Let E be Euclidean three-space.

DEFINITION. A subset S of E is a C^1-surface if and only if, for each point $p \in S$, there is an open neighborhood N of p in E and a C^1-function $f: N \to \mathbf{R}$ with non-vanishing gradient such that $S \cap N$ is the zero set of f.

We shall need the following facts. If S is a C^1-surface and $p \in S$, then there is a unique plane tangent to S at p, and if π is any other plane through p, there is a neighborhood U of p in E such that $\pi \cap S \cap U$ is a curve with a non-singular C^1-parametrization that passes through p (i.e., p is not an endpoint.) This follows from the implicit function theorem. (See, for example, R. C. Buck, *Advanced Calculus*, New York, McGraw-Hill, 1956, Chap. 5.)

We consider throughout a C^1-surface S in E with the property that all of its normals intersect a fixed line l. If $p \in E - l$, then $C(p)$ is the circle through p in a plane perpendicular to l with its center on l. We note that a surface of revolution with axis l is a surface which, with any point p not on l, contains the entire circle $C(p)$.

LEMMA 1. *Suppose a connected C^1-curve A lies in a plane π and all the normals to A in π pass through a fixed point $b \notin A$. Then A lies on a circle with center b.*

Proof. Choose rectangular coordinates in π with origin at b. At any point p of A the line pb is perpendicular to the tangent to A. This means that the tangents to A all lie in the direction field of the differential equation

$$x\,dx + y\,dy = 0; \tag{1}$$

that is, A is an integral curve of (1). Since (1) is non-singular on $\pi - \{b\}$, any connected integral curve of (1) lying in $\pi - \{b\}$ lies on a unique maximal integral curve. The maximal integral curves of (1) are evidently circles with center b, so the lemma follows. ■

LEMMA 2. *Let π be a plane perpendicular to l. Then any connected C^1-curve in $\pi \cap S - l$ lies on a circle $C(p)$ for some p.*

Proof. Let A be a connected C^1-curve in $\pi \cap S - l$. We shall prove that all of the normals to A pass through $\pi \cap l$.

Let q be any point of A and let τ be the plane tangent to S at q. Then $\tau \neq \pi$ since the normal to π at q does not meet l. The line tangent to A at q lies in both π and τ; hence it is $\pi \cap \tau$. Because π and τ are perpendicular respectively to two intersecting lines in the plane σ of q and l (namely, l and the normal to S at q), $\pi \cap \tau$ is perpendicular to σ. Therefore $\pi \cap \sigma$ is the normal to A at q in π. It passes through $\pi \cap l$, and thus we have shown that all the normals to A pass through the point $\pi \cap l$. Then it follows from Lemma 1 that A lies on a circle $C(p)$. ∎

LEMMA 3. *Let $p \in S - l$. Then $C(p) \cap S$ is open relative to $C(p)$.*

Proof. Let π be the plane of $C(p)$ and let q be any point of $C(p) \cap S$. Now π is not tangent to S at q since the normal to π at q does not meet l; hence $\pi \cap S$ contains a connected C^1-curve A that passes through q. By Lemma 2, A lies on $C(q) = C(p)$. Hence A is an arc of $C(p)$ that passes through q and lies in $C(p) \cap S$. Since q was chosen arbitrarily in $C(p) \cap S$, it follows that $C(p) \cap S$ is open relative to $C(p)$. ∎

THEOREM A. *S is locally a surface of revolution; that is, for every point p of S there is a neighborhood N of p in E and a surface of revolution S^* such that $S \cap N = S^* \cap N$.*

Proof. We introduce cylindrical coordinates r, θ, z with axis l.

Suppose $p \in S - l$. We take an open neighborhood N of p as in the definition of a surface. Cutting N down if necessary, we may assume that

$$N = \{\langle r, \theta, z\rangle : |r - r(p)| < \epsilon, |\theta - \theta(p)| < \epsilon, |z - z(p)| < \epsilon\}$$

for some positive ϵ, which we take small enough to ensure that $N \cap l = \emptyset$.

Suppose $q \in S \cap N$. Since $S \cap N$ is closed relative to N, $C(q) \cap S \cap N$ is closed relative to $C(q) \cap N$. Since $C(q) \cap S$ is open relative to $C(q)$, $C(q) \cap S \cap N$ is open relative to $C(q) \cap N$. Thus, $C(q) \cap S \cap N$ is not empty and both open and closed relative to $C(q) \cap N$. The latter being connected, $C(q) \cap S \cap N = C(q) \cap N$. It follows that

$$S \cap N = \cup \{C(q) \cap N : q \in S \cap N\}.$$

Hence $S \cap N = S^* \cap N$, where

$$S^* = \cup \{C(q) : q \in S \cap N\},$$

a surface of revolution. (It is a C^1-surface because any point of S^* has a neighborhood N_1 obtained by rotating N about l, and $S^* \cap N_1$ is obtained by rotating $S \cap N$ about l.)

Now suppose $p \in S \cap l$. We take a neighborhood N of p as in the definition of a surface. We may assume that

$$N = \{\langle r, \theta, z\rangle : r < \epsilon, |z - z(p)| < \epsilon\}$$

for some positive ϵ.

Let $q \in S \cap N - l$. As before, $C(q) \cap S \cap N$ is open and closed relative to $C(q) \cap N$, but in this case $C(q) \cap N = C(q)$, so $S \cap N \supseteq C(q)$. Hence $S \cap N$ is a surface of revolution. ∎

THEOREM B. *If S is closed, then S is a surface of revolution.*

Proof. Under this hypothesis $C(q) \cap S$ is both open and closed relative to $C(q)$ for any $q \notin l$. Since $C(q)$ is connected, $C(q) \cap S$ is either empty or $C(q)$. Hence S is a surface of revolution. ∎

---

2. A uniform wire is bent into a form coinciding with the portion of the curve $y = e^x$, $0 \leq x \leq a$, $a > 1$, and the line segment $a - 1 \leq x \leq a$, $y = e^a$. The wire is then suspended from the point $(a - 1, e^a)$ and a horizontal force F is applied at the point $(0, 1)$ to hold the wire in coincidence with the curve and segment. Assuming the x axis is horizontal, show that the force F is directed to the right.

Solution. The following three statements are evidently equivalent.

The force F is directed to the right.

The moment of the force of gravity on the wire is clockwise about the point of support.

The centroid of the wire falls to the right of the point of support.

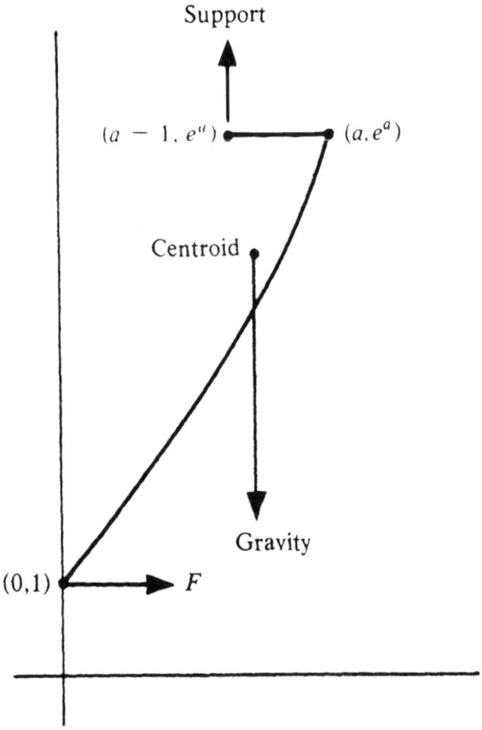

Hence if $\bar{x}$ is the x-coordinate of the centroid of the wire, we must prove that

$$(1) \qquad \bar{x} > a - 1.$$

Without loss of generality, we may take the constant linear density of the wire to be 1. Then the mass of the curved portion of the wire is

$$\int_0^a \sqrt{1 + e^{2x}}\, dx,$$

and its x-moment is

$$\int_0^a x\sqrt{1 + e^{2x}}\, dx.$$

The straight segment of the wire has mass 1 and x-moment $a - \frac{1}{2}$. Hence (1) is equivalent to

$$a - \frac{1}{2} + \int_0^a x\sqrt{1 + e^{2x}}\, dx > (a - 1)\left(1 + \int_0^a \sqrt{1 + e^{2x}}\, dx\right)$$

which is in turn equivalent to

$$(2) \qquad \int_0^a x\sqrt{1 + e^{2x}}\, dx - (a - 1)\int_0^a \sqrt{1 + e^{2x}}\, dx > -\frac{1}{2}.$$

So we must prove (2) for all $a > 1$.

Let $f(a)$ be given by the left member of (2). Then

$$f'(a) = \sqrt{1 + e^{2a}} - \int_0^a \sqrt{1 + e^{2x}}\, dx.$$

Since $1 + e^{2x} < (\frac{1}{2}e^{-x} + e^x)^2$,

$$\int_0^a \sqrt{1 + e^{2x}}\, dx < \int_0^a \left(\frac{1}{2}e^{-x} + e^x\right) dx$$

$$= e^a - \frac{1}{2} - \frac{1}{2}e^{-a} < e^a < \sqrt{1 + e^{2a}}$$

for $a > 0$. Hence $f'(a) > 0$ for $a > 0$. This implies $f(a) > f(0) = 0$ for $a > 0$. Then (2) follows immediately and we are done.

3. A and B are real numbers and k a positive integer. Show that

$$\left|\frac{\cos kB \cos A - \cos kA \cos B}{\cos B - \cos A}\right| < k^2 - 1$$

whenever the left side is defined.

Solution. Let $x = (A - B)/2$, $y = (A + B)/2$. The numerator of the expression on the left is

$$\frac{1}{2}[\cos(kB + A) + \cos(kB - A) - \cos(kA + B) - \cos(kA - B)]$$

$$= \frac{1}{2}[\cos(kB + A) - \cos(kA + B)] + \frac{1}{2}[\cos(kB - A) - \cos(kA - B)]$$

$$= \sin(k - 1)x \sin(k + 1)y + \sin(k + 1)x \sin(k - 1)y.$$

The denominator is $2 \sin x \sin y$, and we may assume this is not zero. Hence we have

$$\left|\frac{\cos kB \cos A - \cos kA \cos B}{\cos B - \cos A}\right| \le \frac{1}{2}\left|\frac{\sin(k - 1)x}{\sin x}\right| \cdot \left|\frac{\sin(k + 1)y}{\sin y}\right| + \frac{1}{2}\left|\frac{\sin(k + 1)x}{\sin x}\right| \cdot \left|\frac{\sin(k - 1)y}{\sin y}\right|.$$

Now since $|\sin nz| \le n|\sin z|$ for all z and all positive integers n with strict inequality unless $n = 1$ or $\sin z = 0$ (this is proved below), both terms are less than $(k^2 - 1)/2$ provided $k > 1$. Therefore

$$\left|\frac{\cos kB \cos A - \cos kA \cos B}{\cos B - \cos A}\right| < k^2 - 1$$

provided $k > 1$ and $\cos B \ne \cos A$. Obviously, we have equality if $k = 1$, so the problem is not accurately phrased.

We now prove that $|\sin nz| \le n|\sin z|$ for all positive integers n. If $\sin z = 0$, equality holds for every integer n, so we assume from now on that $\sin z \ne 0$; then $|\cos z| < 1$. We continue by induction on n. Clearly there is equality if $n = 1$. For $n = 2$, we have $|\sin 2z| = 2|\cos z| \cdot |\sin z| < 2|\sin z|$. Suppose we have strict inequality for $n = k$. Then

$$\begin{aligned} |\sin(k + 1)z| &= |\sin kz \cos z + \cos kz \sin z| \\ &\le |\sin kz| + |\sin z| < (k + 1)|\sin z|. \end{aligned}$$

Thus we have strict inequality for $n = k + 1$. Hence we have proved

$$|\sin nz| \leq |\sin z|$$

for all real z and all positive integers n with strict inequality unless $n = 1$ or $\sin z = 0$.

REMARK. In the report on the Competition (*American Mathematical Monthly*, vol. 64 (1957), pages 649–654) the error in the statement of the problem was explained as follows. "The questions committee originally had a '$\leq$' in this question, but, in some manner, the '$=$' was lost before the question reached the director. The omission was discovered in reading the proofs, but the chairman of the committee decided that it was just as well to give the contestants the opportunity to discover the error and correct it."

---

4. $P(z)$ is a complex polynomial whose roots (as points in the Argand plane) can be covered by a closed circular disc of radius R. Show that the roots of $nP(z) - kP'(z)$ can be covered by a closed circular disc of radius $R + |k|$, where n is the degree of $P(z)$, k is any complex number, and $P'(z)$ is the derivative of $P(z)$.

Solution. Suppose that the roots of $P(z)$ all lie in the closed disc, D_1, $|z - c| \leq R$. We shall prove that the roots of $nP(z) - kP'(z)$ lie in the closed disc, D_2, $|z - c| \leq R + |k|$.

If the roots of $P(z)$ are $\lambda_1, \lambda_2, \ldots, \lambda_n$, then $P(z) = A(z - \lambda_1)(z - \lambda_2) \cdots (z - \lambda_n)$ where A is a constant. Taking the logarithmic derivative, we find

$$\frac{P'(z)}{P(z)} = \sum_{i=1}^{n} \frac{1}{z - \lambda_i}.$$

Suppose $u \notin D_2$. Then

$$|u - \lambda_i| \geq |u - c| - |\lambda_i - c| > R + |k| - R = |k|$$

since $\lambda_i \in D_1$. Therefore, for $i = 1, 2, \ldots, n$, $\dfrac{|k|}{|u - \lambda_i|} < 1$, and

$$|nP(u) - kP'(u)| = |P(u)| \cdot \left| n - k \sum_{i=1}^{n} \frac{1}{u - \lambda_i} \right|$$

$$\geq |P(u)| \cdot \left(n - \sum_{i=1}^{n} \frac{|k|}{|u - \lambda_i|} \right) > 0.$$

Thus u is not a root of $nP(z) - kP'(z)$.

This proves that all the roots of $nP(z) - kP'(z)$ lie in D_2 as claimed.

---

5. Given n points in the plane, show that the largest distance determined by these points cannot occur more than n times.

Solution. Suppose $\mathcal{S}$ is a set in the plane. By a *diameter* of $\mathcal{S}$ we mean a segment connecting two points of $\mathcal{S}$ that are as far apart as any two points of $\mathcal{S}$. We are asked to prove (1) a set of n points in a plane can have at most n diameters. We shall use induction on n. Clearly, (1) is true for $n = 2$ or 3.

Assume that (1) is true for $n = k$. Let $\mathcal{E}$ be a set in a plane with $k + 1$ points.

Suppose some point of $\mathcal{E}$, say X, is the endpoint of at most one diameter of $\mathcal{E}$. Then $\mathcal{E} - \{X\}$ is a set of k points in the plane and has at most k diameters by the inductive hypothesis. Then $\mathcal{E}$ has at most $k + 1$ diameters.

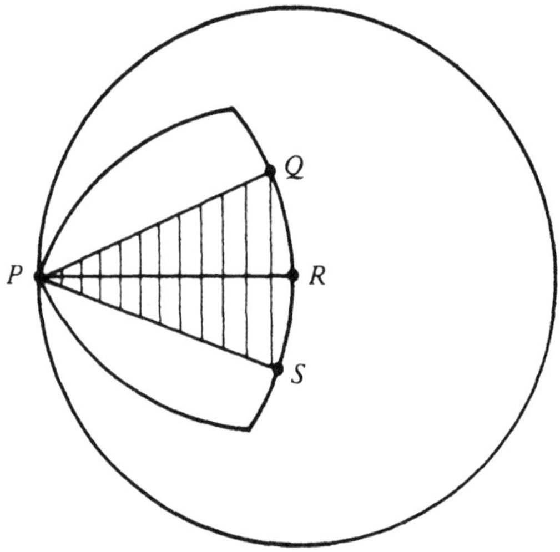

Suppose some point of $\mathcal{E}$, say P, is the endpoint of at least three diameters, PQ, PR, and PS. Let $r = |PQ|$. Then Q, R, and S lie on a circle of radius r about P and, in fact, on a minor arc of that circle. We choose the notation so that R is between Q and S on that arc. Since every two points of $\mathcal{E}$ are at most r apart, $\mathcal{E}$ lies in the intersection $\mathcal{I}$ of three closed circular disks of radius r and centers P, Q, and S. Now, except for the point P, $\mathcal{I}$ is inside the circle of radius r about R, and therefore R is the end point of just one diameter of $\mathcal{E}$, namely, RP. Therefore, the previous paragraph applies, and $\mathcal{E}$ has at most $k + 1$ diameters.

Finally, suppose each point of $\mathcal{E}$ is the endpoint of exactly two diameters. Then the diameters have altogether $2(k + 1)$ endpoints, so there are exactly $k + 1$ of them.

Thus, in any case, $\mathcal{E}$ has at most $k + 1$ diameters. Hence (1) is true for $n = k + 1$. It follows by induction that (1) is true for all n.

REMARK. The geometric part of the argument can be completely formalized as follows:

An endpoint of a diameter of $\mathcal{E}$ is an extreme point of the convex hull of $\mathcal{E}$. Therefore, P, Q, R, S are vertices of a convex quadrilateral, and we can choose notation so that $\overleftrightarrow{PR}$ separates Q and S. Suppose RT is a diameter of $\mathcal{E}$ with $T \neq P$. Then T does not lie on $\overleftrightarrow{PR}$, so we may assume T and S are on opposite sides of $\overleftrightarrow{PR}$. Then P, R, S, T are vertices of a convex quadrilateral and PR must be a diagonal. Let PR and ST intersect at U. Then

$$|PR| + |ST| = |PU| + |US| + |RU| + |UT| > |PS| + |RT|.$$

Since $|PR| = |PS| = |RT| = r$, this shows that $|ST| > r$, which is impossible. Hence there is no such diameter RT, and R is the endpoint of just one diameter RP, as claimed.

For other results of a similar nature see P. Erdos, "On Sets of Distances of n Points," *American Mathematical Monthly*, vol. 53 (1956), pages 248–250.

---

6. $S_1 = \ln a$ and $S_n = \sum_{i=1}^{n-1} \ln (a - S_i)$, $n > 1$.

Show that

$$\lim_{n \to \infty} S_n = a - 1.$$

Solution. The given recursion can be written

$$S_{n+1} = S_n + \ln (a - S_n).$$

The polygonal representation of this recursion is shown in the figure [see p. 223]. It is clear that, with any choice of $S_1 < a$, we have

$$S_2 \leq S_3 \leq S_4 \leq \cdots \leq a - 1, \tag{1}$$

and that the sequence converges to $a - 1$.

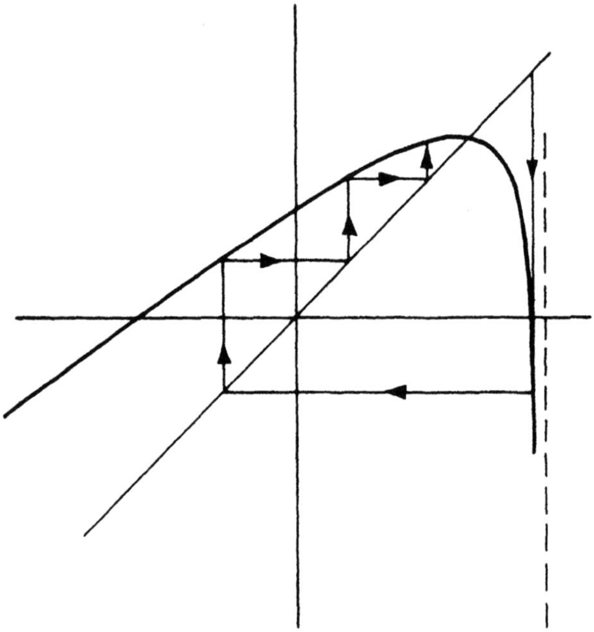

$x = a$

To prove this analytically, we let

$$f(x) = x + \ln(a - x)$$

for $x < a$. Then $f'(x) = 1 - 1/(a - x)$, which is positive for $x < a - 1$ and negative for $x > a - 1$. Since $f(a - 1) = a - 1$, it follows that $f(x) \leq a - 1$ for all $x < a$. Also, if $x \leq a - 1$, then $\ln(a - x) \geq 0$, so $f(x) \geq x$. Then (1) follows immediately, so the sequence $\{S_n\}$ has a limit, say T. Clearly, $T \leq a - 1$, so T is a point of continuity for f. Hence

$$f(T) = f(\lim S_n) = \lim f(S_n) = \lim S_{n+1} = T.$$

This gives $\ln(a - T) = 0$, and therefore $T = a - 1$.

We have proved $\lim S_n = a - 1$, as required.

---

7. Each member of a set of circles in the xy plane is tangent to the x axis and no two of the circles intersect. Show that:

(i) the points of tangency can include all the rational points on the axis, but

(ii) the points of tangency cannot include all the irrational points.

First Solution. (i) Suppose p, q, r, and s are integers such that $p/q \neq r/s$. Choose $k > 2$. Consider the two circles with centers at $\langle p/q, 1/kq^2 \rangle$ and $\langle r/s, 1/ks^2 \rangle$ and tangent to the x-axis. We claim these circles do not intersect. If they did, the distance between their centers would not exceed the sum of their radii; squaring, we would have

$$\left(\frac{p}{q} - \frac{r}{s}\right)^2 + \left(\frac{1}{kq^2} - \frac{1}{ks^2}\right)^2 \leq \left(\frac{1}{kq^2} + \frac{1}{ks^2}\right)^2$$

which implies

$$(ps - rq)^2 \leq 4/k^2.$$

But this is impossible since $k > 2$ and $(ps - rq)^2$ is a positive integer. This establishes our claim.

Now, given a rational number r, write it in its lowest terms as a quotient of two integers, $r = p/q$, and let C_r be the circle with center $\langle p/q, 1/kq^2 \rangle$ and radius $1/kq^2$. Then C_r is tangent to the x-axis at $\langle r, 0 \rangle$, and, as we showed above, C_r and C_s do not intersect if $r \neq s$. Hence $\{C_r\colon r \text{ rational}\}$ is a set of circles each of which is tangent to the x-axis, for which the points of tangency include all rational points on the x-axis, and no two of which intersect.

(ii) Since any circular region contains a point both of whose coordinates are rational, and since the number of such points is countable, it is impossible to have an uncountable family of circles in the plane whose interiors are disjoint. Any two disjoint circles having a common tangent have also disjoint interiors, so it is impossible that uncountably many disjoint circles should all be tangent to the x-axis. Since the set of irrational points on the x-axis is uncountable, (ii) follows.

Second Solution. (i) We can construct a set of circles as required in (i) by induction. Let the rational numbers be enumerated $r_1, r_2, \ldots$. Let C_1 be a circle tangent to the x-axis at the point $\langle r_1, 0 \rangle$ with radius 1. Suppose disjoint circles $C_1, C_2, \ldots, C_n$ have been constructed so that C_i is tangent to the x-axis at $\langle r_i, 0 \rangle$ for $i = 1, 2, \ldots, n$. Since $\langle r_{n+1}, 0 \rangle$ is outside all of these circles, it is at positive distance, say δ, from their union. Let C_{n+1} be a circle of radius $\delta/3$ tangent to the x-axis at $\langle r_{n+1}, 0 \rangle$. Then C_{n+1} is disjoint from $C_1, C_2, \ldots, C_n$. We obtain in this way a sequence of circles that includes one tangent to the x-axis at each rational point.

(ii) Let $\mathcal{C}$ be a set of disjoint circles in the plane all tangent to the x-axis. For each positive number α let $\mathcal{C}_\alpha$ be the set of circles in $\mathcal{C}$ having radius $\geq \alpha$. For a fixed α, two members of $\mathcal{C}_\alpha$ that fall on the same side of the x-axis have points of tangency separated by at least 2α. Hence there are at most countably many such circles. Thus, $\mathcal{C}_\alpha$ is the union of two countable sets, so it is countable. Now the union of countably many countable sets is countable, and $\mathcal{C} = \bigcup_{n=1}^{\infty} \mathcal{C}_{1,n}$, so $\mathcal{C}$ is countable. The set of points at which members of $\mathcal{C}$ are tangent to the x-axis is therefore countable, so it cannot include all irrational points.

REMARK. See L. R. Ford, "Fractions," *American Mathematical Monthly*, vol. 45 (1938), pages 586–601, for more information concerning the family of circles constructed in our first solution.

Afternoon Session

1. Consider the determinant $|a_{ij}|$ of order 100 with $a_{ij} = i \times j$. Prove that if the absolute value of each of the 100! terms in the expansion of this determinant is divided by 101 then the remainder in each case is 1.

Solution. Each term in the expansion of the given determinant is, except for sign, the product of all possible row indices and all possible column indices, that is, $(100!)^2$, and this is the absolute value of every term.

Now 101 is a prime, so by Wilson's theorem $100! \equiv -1 \pmod{101}$. Hence $(100!)^2 \equiv (-1)^2 \equiv 1 \pmod{101}$, as required.

For Wilson's theorem see A. H. Beiler, *Recreations in the Theory of Numbers*. Dover, New York, 1964, or any text on number theory.

---

2. If facilities for division are not available, it is sometimes convenient in determining the decimal expansion of $1/A$, $A > 0$ to use the iteration $X_{k+1} = X_k(2 - AX_k)$, $k = 0, 1, 2, \ldots$, where X_0 is a selected "starting" value. Find the limitations, if any, on the starting value X_0 in order that the above iteration converges to the desired value $1/A$.

First Solution. The polygonal representation of this recursion is shown in the figure. (See p. 223 for an explanation.) It is clear that if X_0 lies in $(0, 2/A)$ then

$$0 < X_1 \le X_2 \le X_3 \le \cdots \le \frac{1}{A}, \tag{1}$$

and the sequence converges to $1/A$. If $X_0 = 0$ or $2/A$, then obviously

$$X_1 = X_2 = X_3 = \cdots = 0, \tag{2}$$

and if X_0 lies outside $0, 2/A$, then

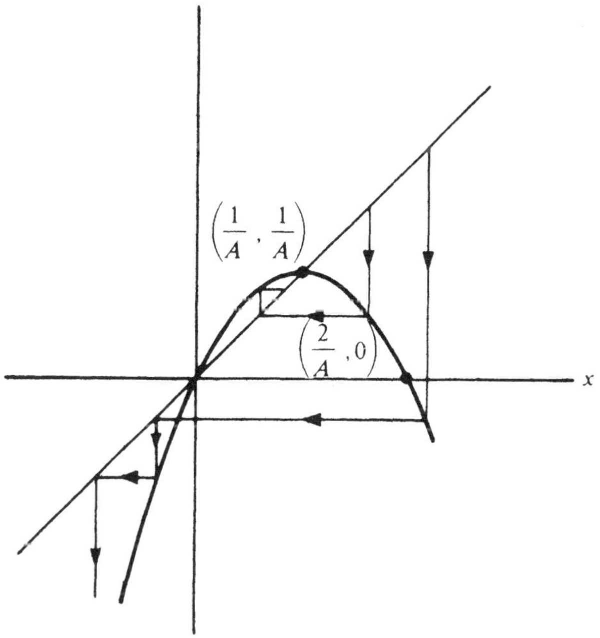

(3) $$0 > X_1 > X_2 > X_3 > \cdots$$

and the sequence diverges to $-\infty$.

To make this rigorous, we define $f(x) = x(2 - Ax)$. We find that f achieves its maximum value $1/A$ for $x = 1/A$. Moreover, $f(x) > x$ for $0 < x < 1/A$ and $f(x) < x$ for $x < 0$. Then (1), (2), and (3) follow immediately. In case (1) the sequence must be convergent and the limit must be a positive root of $f(x) = x$, and there is only one, namely, $x = 1/A$. In case (2) the sequence obviously converges to 0. In case (3) it either diverges or converges to a negative root of $f(x) = x$, but there is no such root.

Therefore, the sequence converges to $1/A$ if and only if $0 < X_0 < 2/A$.

Second Solution. We can find an explicit formula for X_n from which the limiting behavior is obvious. We have

$$1 - AX_{n+1} = (1 - AX_n)^2;$$

therefore

$$1 - AX_n = (1 - AX_0)^{2^n}.$$

It is now obvious that $1 - AX_n \to 0$, that is $X_n \to 1/A$, if and only if $|1 - AX_0| < 1$, that is $0 < X_0 < 2/A$.

3. For $f(x)$ a positive, monotone decreasing function defined in $0 \le x \le 1$ prove that

$$\frac{\int_0^1 xf^2(x)\,dx}{\int_0^1 xf(x)\,dx} \le \frac{\int_0^1 f^2(x)\,dx}{\int_0^1 f(x)\,dx}.$$

Solution. The desired inequality is equivalent to

$$\int_0^1 f^2(x)dx \int_0^1 yf(y)dy - \int_0^1 yf^2(y)dy \int_0^1 f(x)dx \ge 0$$

which can be rewritten

$$\int_0^1 \int_0^1 f(x)f(y)y[f(x) - f(y)]\,dx\,dy \ge 0. \tag{1}$$

Denote the left member of (1) by I. Then

$$I = \int_0^1 \int_0^1 f(x)f(y)x[f(y) - f(x)]\,dx\,dy.$$

(We have interchanged the variables of integration and then the order of integration.) Hence

$$2I = \int_0^1 \int_0^1 f(x)f(y)(y - x)[f(x) - f(y)]\,dx\,dy.$$

Because f is decreasing, $(y - x)[f(x) - f(y)] \ge 0$ for all x and y. Then since f is everywhere positive, it is clear that $2I \ge 0$. This proves (1).

REMARK. The argument given generalizes to prove

$$\int_0^1 f(x)g(x)w(x)dx \int_0^1 w(x)dx \le \int_0^1 f(x)w(x)dx \int_0^1 g(x)w(x)dx$$

whenever f is decreasing, g is increasing, and w is a non-negative weight function (assuming, of course, that the integrals exist). Moreover, the inequality is strict unless either f or g is constant over the support of w (i.e., the closure of the set $\{x\colon w(x) > 0\}$). The present problem is the special case $g(x) = x$, $w(x) = f(x)$.

4. Let $a(n)$ be the number of representations of the positive integer n as the sums of 1's and 2's taking order into account. For example, since

$$4 = 1 + 1 + 2 = 1 + 2 + 1 = 2 + 1 + 1$$
$$= 2 + 2 = 1 + 1 + 1 + 1,$$

then $a(4) = 5$. Let $b(n)$ be the number of representations of n as the sum of integers greater than 1, again taking order into account and counting the summand n. For example, since $6 = 4 + 2 = 2 + 4 = 3 + 3 = 2 + 2 + 2$, we have $b(6) = 5$. Show that for each n, $a(n) = b(n + 2)$.

First Solution. By direct counting we can find the first few values of the two sequences:

n	1	2	3	4	5	6
$a(n)$	1	2	3	5	8	13
$b(n)$	0	1	1	2	3	5

Thus it appears that the sequences are the well-known Fibonacci sequence satisfying the recursion

$$a(n + 1) = a(n) + a(n - 1), \qquad n = 2, 3, \ldots. \tag{1}$$

To prove that this is indeed the case, note that any representation of $n + 1$ as an ordered sum of 1's and 2's becomes, on removal of the last summand, either a representation of n (if the removed term is a 1) or a representation of $n - 1$ (if the removed term is a 2). Conversely, any representation of either n or $n - 1$ becomes a representation of $n + 1$ on adjoining either a 1 or 2, as appropriate. It follows that the a-sequence does indeed satisfy the recursion (1).

Next we show that the b's satisfy the same recursion. Delete the last term from a representation of $n + 1$ as a sum of integers greater than 1. We obtain then either a representation of $n - 1$, $n - 2$, ..., 2, or a vacuous sum. Conversely, any such representation extends to a representation of $n + 1$. Hence for $n \geq 1$

$$b(n + 1) = b(n - 1) + b(n - 2) + \cdots + b(2) + 1. \tag{2}$$

If n is at least 2, we have also

$$b(n) = b(n - 2) + \cdots + b(2) + 1. \tag{3}$$

Comparing (2) and (3), we find

$$b(n+1) = b(n-1) + b(n), \qquad n \geq 2.$$

So the b-sequence satisfies the same recursion as the a-sequence.

Then since $b(3) = a(1)$, $b(4) = a(2)$, it follows by induction that $b(n+2) = a(n)$ for all $n \geq 1$.

Second Solution. We can find the power-series generating function for $a(n)$. The number of ways that n can be represented as an ordered sum of k 1's and 2's is clearly the coefficient of x^n in $(x + x^2)^k$. Hence

$$1 + \sum_{n=1}^{\infty} a(n)x^n = \sum_{k=0}^{\infty} (x + x^2)^k = \frac{1}{1 - x - x^2}.$$

On the other hand, the number of ways that n can be represented as an ordered sum of k integers greater than 1 is the coefficient of x^n in

$$(x^2 + x^3 + \cdots)^k = \left(\frac{x^2}{1-x}\right)^k.$$

So

$$1 + \sum_{n=2}^{\infty} b(n)x^n = \sum_{k=0}^{\infty} \left(\frac{x^2}{1-x}\right)^k = \left(1 - \frac{x^2}{1-x}\right)^{-1}$$
$$= \frac{1-x}{1-x-x^2}$$
$$= 1 + \frac{x^2}{1-x-x^2}.$$

Thus

$$\sum_{n=2}^{\infty} b(n)x^n = x^2 + x^2 \sum_{n=1}^{\infty} a(n)x^n$$

and the result follows immediately.

[The formal manipulation of the series can be justified either by considering only small values of x, for example $|x| < \frac{1}{10}$, which makes each series involved absolutely convergent, or by considering the ring of formal power series in which convergence means only the ultimate stabilization of terms of each degree.]

Third Solution. A one-to-one correspondence between representations of n as a sum of 1's and 2's and representations of $n + 2$ as a sum of integers greater than 1 can be established as follows. Representations of n as an ordered sum of 1's and 2's are obviously in one-to-one correspondence with representations of $n + 2$ as a sum of 1's and 2's ending with a 2 (and having one more summand). If in such a representation written in linear order we bracket each 2 with the longest string of 1's immediately preceding it, we obtain a representation of $n + 2$ as an ordered sum of integers greater than 1. Conversely, in any representation of $n + 2$ as such an ordered sum, we can replace each summand by a string of 1's followed by a single 2.

Example. The following representation of 9 as an ordered sum of 1's and 2's,

$$9 = 1 + 1 + 2 + 2 + 2 + 1$$

corresponds to the representation of $11(=9 + 2)$

$$11 = 1 + 1 + 2 + 2 + 2 + 1 + 2,$$

which corresponds to

$$\begin{aligned} 11 &= (1 + 1 + 2) + (2) + (2) + (1 + 2) \\ &= 4 + 2 + 2 + 3, \end{aligned}$$

and also conversely.

It follows that $b(n + 2) = a(n)$.

---

5. With each subset X of a set is associated a second subset $f(X)$. The association is such that whenever X contains Y then $f(X)$ contains $f(Y)$. Show that for some set A, $f(A) = A$.

Solution. Let the given set be S. Define

$$\mathcal{C} = \{X \subseteq S: X \subseteq f(X)\},$$

and let A be the union of all members of $\mathcal{C}$. We shall prove that $A = f(A)$.

Suppose $X \in \mathcal{C}$; then $X \subseteq f(X)$ and $X \subseteq A$. Therefore, $f(X) \subseteq f(A)$, by hypothesis, so $X \subseteq f(A)$. By the definition of union

$$A \subseteq f(A). \tag{1}$$

By the hypothesis, $f(A) \subseteq f(f(A))$, so $f(A) \in \mathcal{C}$. Again by definition of union,

$$f(A) \subseteq A. \tag{2}$$

Comparing (1) and (2), we see that $A = f(A)$, as claimed.

REMARKS. By essentially the same argument one can prove the Knaster-Tarski fixed point theorem—namely: Every order-preserving mapping of a complete lattice into itself has a fixed element. See G. Szasz, *Introduction to Lattice Theory,* Academic Press, New York, 1963.

Fraenkel (in *Abstract Set Theory,* North Holland Publishing Co., 1953) ascribes the result to Dedekind, whose work however was not published until 1932, and independently to Peano and Zermelo.

---

6. The curve $y = f(x)$ passes through the origin with a slope of 1. It satisfies the differential equation $(x^2 + 9)y'' + (x^2 + 4)y = 0$. Show that it crosses the x axis between

$$x = \frac{3}{2}\pi \quad \text{and} \quad x = \sqrt{\frac{63}{53}}\,\pi.$$

Solution. We shall use the Sturm comparison theorem. We state the theorem now and give the proof later.

STURM COMPARISON THEOREM. *Let I be an interval in $\mathbf{R}$ and suppose the functions u and v satisfy*

$$u''(x) + A(x)u(x) = 0$$
$$v''(x) + B(x)v(x) = 0$$

for $x \in I$, where A and B are continuous functions such that $A(x) \geq B(x)$ for $x \in I$. Assume v is not identically zero on I and let α and β be zeros of v with $\alpha < \beta$. Then there is a zero of u in the open interval (α, β) unless $A(x) = B(x)$ for $\alpha \leq x \leq \beta$ and u and v are proportional in this interval.

For the stated problem we first compare the differential equations

$$y'' + \frac{x^2 + 4}{x^2 + 9}\,y = 0 \tag{1}$$

and

$$v'' + \frac{4}{9} v = 0. \tag{2}$$

As a solution of (2) we take $v(x) = \sin \frac{2}{3}x$, which has zeros at 0 and $\frac{3}{2}\pi$. Since

$$\frac{x^2 + 4}{x^2 + 9} \geq \frac{4}{9}$$

for all x with strict inequality for $x \neq 0$, any solution of (1) must have a zero in $(0, \frac{3}{2}\pi)$. Hence there exists a $\xi \in (0, \frac{3}{2}\pi)$ such that $f(\xi) = 0$. Moreover, the graph of f must cross the x-axis at ξ because otherwise $f'(\xi) = 0$, and then the uniqueness theorem for solutions of (1) would imply that $f(x) = 0$ for all x.

Now for $0 \leq x \leq \frac{3}{2}\pi$, we have

$$\frac{x^2 + 4}{x^2 + 9} < \frac{53}{63}. \tag{3}$$

To see this, note that (3) is equivalent to $10x^2 < 225$. Since $\pi^2 < 10$, we see that $10(\frac{3\pi}{2})^2 < 225$, so (3) follows.

If we set

$$u(x) = \sin \sqrt{\frac{53}{63}}\, x,$$

then

$$u''(x) + \frac{53}{63} u(x) = 0.$$

Applying the Sturm comparison theorem again (with $I = [0, \frac{3}{2}\pi]$), we conclude that u has a zero on $(0, \xi)$. But the first positive zero of u is at $\sqrt{63/53}\pi$, so $\sqrt{63/53}\pi < \xi$. Thus we have

$$\sqrt{\frac{63}{53}}\pi < \xi < \frac{3}{2}\pi,$$

as required.

Proof of the Sturm Comparison Theorem. The zeros of v are isolated, hence we may assume that β is the next zero after α; i.e., that v has a fixed sign on (α, β). Suppose u has no zero in (α, β). Then u also has a fixed sign on (α, β). Changing the signs of u and/or v if necessary (which does not affect the location of the zeros), we may assume that both u and v are positive on (α, β). It is then clear that $v'(\alpha) \geq 0$ and $v'(\beta) \leq 0$. A non-zero solution of a non-singular second-order linear differential equation and its derivative cannot both vanish at the same point, so we must have

$$v'(\alpha) > 0, \qquad v'(\beta) < 0. \tag{4}$$

Consider $w(x) = u(x)v'(x) - u'(x)v(x)$. We have

$$\begin{aligned} w'(x) &= u(x)v''(x) - u''(x)v(x) \\ &= (A(x) - B(x))u(x)v(x) \geq 0 \end{aligned} \tag{5}$$

for all $x \in [\alpha, \beta]$. Therefore, w is non-decreasing on $[\alpha, \beta]$, and in particular $w(\alpha) \leq w(\beta)$. Then from the definition of w we obtain

$$u(\alpha)v'(\alpha) \leq u(\beta)v'(\beta).$$

Comparing this with (4) and remembering that $u(\alpha) \geq 0$, $u(\beta) \geq 0$, we see that $u(\alpha) = u(\beta) = 0$ and $w(\alpha) = w(\beta) = 0$. But then, since w is monotone on $[\alpha, \beta]$, both w and its derivative must vanish throughout this interval. Thus,

$$v(x)u'(x) - u(x)v'(x) = 0 \tag{6}$$

while (5) gives $A(x) = B(x)$ for $\alpha \leq x \leq \beta$.

For $\alpha < x < \beta$, we may divide (6) by $v(x)^2$ and conclude $(u/v)' = 0$, whence u/v is constant on (α, β). Thus u and v are proportional on (α, β). By continuity they are also proportional on $[\alpha, \beta]$. ∎

REMARK. Using essentially the same argument, we reach the same conclusion if we drop the assumption that $v(\alpha) = 0$ and instead assume

$$u(\alpha) = v(\alpha), \qquad u'(\alpha) = v'(\alpha).$$

This variation of the theorem is used in Problem P.M. 6 of the Twenty-second Competition.

For a more general treatment of the Sturm comparison theorem, see Coddington and Levinson, *Theory of Ordinary Differential Equations*, McGraw-Hill, New York, 1955, page 208 ff.

---

7. Let C be a closed convex planar disc bounded by a regular polygon. Show that for each positive integer n there exists a set of points $S(n)$ in the

plane such that each n points of $S(n)$ can be covered by C, but $S(n)$ itself cannot be covered by C.

Solution. Suppose C is a regular polygon of k-sides and r is the radius of the inscribed circle. For a given positive integer n, let $S = S(n)$ be a circle of radius $r \sec(\pi/2kn)$. We must show that

(i) C cannot be placed so as to cover S, and

(ii) if $P_1, P_2, \ldots, P_n$ are any n points of S, then C can be placed so as to cover $\{P_1, \ldots, P_n\}$.

Keeping C fixed, we prove that no circular disk of radius exceeding r can be placed inside of C. From this (i) follows immediately. For a fixed radius r_1, the set E of points that are centers of disks of radius r_1 lying inside C is convex. Moreover, E is invariant under rotations about the center O of C of angle $2\pi/k$. Hence, E is either void or contains O. If $r_1 > r$, the circle of radius r_1 about O does not lie inside C, so E is void in this case; that is, no circular disk of radius r_1 can be placed inside C.

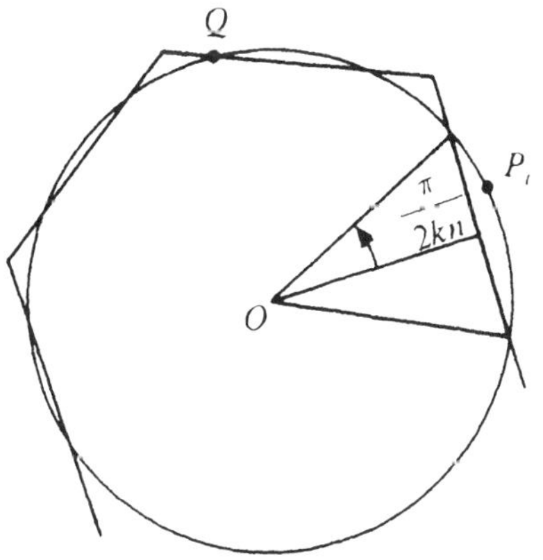

Now we attack (ii). Let $P_1, P_2, \ldots, P_n$ be points of S and fix a reference point Q in C at distance $r \sec(\pi/2kn)$ from O; place C so that O coincides with the center of S, and rotate C about O so that Q describes the circle S. Consider one of the P's, say P_i. As Q describes S, P_i will be on or outside of C when Q is in a set A_i that is the union of k closed arcs each of length π/kn radians (precisely when $2\pi m/k \leq \angle P_iOQ \leq 2\pi m/k + \pi/kn$ for some integer m, if Q is chosen as shown). The length of A_i is π/n and the total length of $\bigcup_{i=1}^{n} A_i$ is at most π. Therefore $S - \bigcup_{i=1}^{n} A_i$ has length at least π and so is not void. If C is rotated so that $Q \in S - \bigcup A_i$, then each of the points $P_1, \ldots, P_n$ is inside of C. Thus (ii) is proved.

THE EIGHTEENTH WILLIAM LOWELL PUTNAM MATHEMATICAL COMPETITION

February 8, 1958

Morning Session

1. If $a_0, a_1, \ldots, a_n$ are real numbers satisfying

$$\frac{a_0}{1} + \frac{a_1}{2} + \cdots + \frac{a_n}{n+1} = 0,$$

show that the equation $a_0 + a_1 x + a_2 x^2 + \cdots + a_n x^n = 0$ has at least one real root.

Solution. If $f(x) = a_0 + a_1 x + \cdots + a_n x^n$, then

$$\int_0^1 f(x)dx = \frac{a_0}{1} + \frac{a_1}{2} + \cdots + \frac{a_n}{n+1} = 0.$$

Hence, by the mean value theorem for integrals, there exists a number ξ between 0 and 1 such that

$$f(\xi) = \int_0^1 f(x)dx = 0.$$

REMARK. This problem appears in G. H. Hardy, *A Course in Pure Mathematics*, 7th ed., Cambridge University Press, 1938, page 243. It is stated there that the problem appeared in the Cambridge Mathematical Tripos for 1929.

---

2. Two uniform solid spheres of equal radii are so placed that one is directly above the other. The bottom sphere is fixed, and the top sphere, initially at rest, rolls off. At what point will contact between the two spheres be "lost"? Assume the coefficient of friction is such that no slipping occurs.

Solution. Let S be the rolling sphere; let a be its radius and M its mass. Recall that the moment of inertia of a uniform solid sphere is $I = (2/5)Ma^2$.

We assume that the motion is essentially two-dimensional. Then the state of the system is defined by the angle θ between the line of centers and the vertical as shown in the figure. We denote the time derivative of θ by $\dot{\theta}$. Let v be the translational speed of the center of S and let ω be the speed of rotation of S.

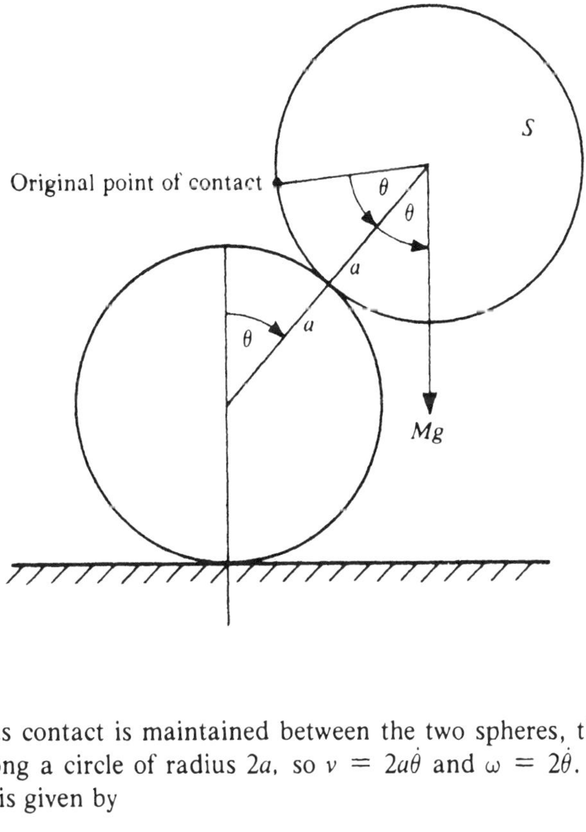

As long as contact is maintained between the two spheres, the center of S moves along a circle of radius $2a$, so $v = 2a\dot{\theta}$ and $\omega = 2\dot{\theta}$. The kinetic energy of S is given by

$$\frac{1}{2} Mv^2 + \frac{1}{2} I\omega^2 = \frac{14}{5} Ma^2\dot{\theta}^2$$

and the potential energy by

$$2Mag \cos \theta$$

relative to the level of the center of the lower sphere. The total energy is constant so

$$2Mag \cos \theta + \frac{14}{5} Ma^2\dot{\theta}^2 = 2Mag.$$

(The right member is the left evaluated for $\theta = 0$.) Therefore we have

$$\frac{7}{5} a\dot{\theta}^2 = g(1 - \cos\theta). \tag{1}$$

To keep S in a circular orbit of radius $2a$, a force toward the center of magnitude $2aM\dot{\theta}^2$ is required. This force is supplied by the component of the gravitational force along the line of centers, which is $Mg\cos\theta$. As long as this component exceeds the required centripetal force, the excess is opposed by the force of contact between the two spheres. When the component of gravity becomes inadequate to supply the necessary force, contact is lost and S goes into a parabolic path while rotating at constant speed. The loss of contact occurs when

$$2aM\dot{\theta}^2 = Mg\cos\theta.$$

Combining this with (1), we obtain

$$\frac{7}{10}\cos\theta = 1 - \cos\theta;$$

Hence contact is lost when $\cos\theta = 10/17$, i.e., when $\theta = \arccos(10/17)$.

REMARKS. It is interesting to verify that contact is lost at the same point even if the two spheres are of different sizes.

The motion described is not realizable, for equation (1) shows that the time required for the upper sphere to roll to angular position α would be

$$\int_0^\alpha \frac{1}{\sqrt{1 - \cos\theta}}\, d\theta$$

and this improper integral diverges since

$$\frac{1}{\sqrt{1 - \cos\theta}} \sim \frac{\sqrt{2}}{\theta} \qquad \text{as } \theta \to 0.$$

Hence it would be more accurate to phrase the question as follows: The upper sphere is displaced slightly and allowed to roll off. Find the limit of the position at which contact is lost as the initial displacement approaches zero.

For a more general treatment of this problem, see A. S. Ramsey, *Dynamics*, Cambridge University Press, 1929, page 210.

3. Real numbers are chosen at random from the interval $(0 \le x \le 1)$. If after choosing the nth number the sum of the numbers so chosen first exceeds 1, show that the expected or average value for n is e.

First Solution. We assume that the phrase "real numbers are chosen at random" means that the x's are independent and each has the uniform distribution on $[0, 1]$. Then the probability that $(x_1, x_2, \ldots, x_n)$ falls in a region S of the cube $[0, 1]^n$ is the n-dimensional content of S.

Let p_n be the probability that $x_1 + x_2 + \cdots + x_n \le 1$. The probability that $x_1 + x_2 + \cdots + x_n > 1$ but $x_1 + x_2 + \cdots + x_{n-1} \le 1$ is then

$$q_n = p_{n-1} - p_n.$$

It is proved below that $p_n = 1/n!$ Hence the expected number of choices required to make the sum exceed one is

$$E = \sum_{n=1}^{\infty} nq_n = \sum_{n=1}^{\infty} n\left(\frac{1}{(n-1)!} - \frac{1}{n!}\right) = \sum_{n=1}^{\infty} \frac{1}{(n-1)!}(n-1)$$

$$= \sum_{n=2}^{\infty} \frac{1}{(n-2)!} = e.$$

LEMMA. *The n-dimensional content $V_n(a)$ of the region in R^n determined by the inequalities $x_i \ge 0$, $i = 1, 2, \ldots, n$ and $x_1 + x_2 + \cdots + x_n \le a$ is $a^n/n!$*

Proof. This is evidently true for $n = 1$. Suppose it is true for $n = k$. Then $V_{k+1}(a)$ can be formed by "slicing" perpendicular to the x_{k+1}-axis. The slice for $x_{k+1} = b$ is the k-dimensional region determined by the inequalities $x_i \ge 0$, $i = 1, 2, \ldots, k$ and $x_1 + x_2 + \cdots + x_k \le a - b$ for $0 \le b \le a$. By the inductive hypothesis its k-dimensional content is $V_k(a - b) = (a - b)^k/k!$. Hence

$$V_{k+1}(a) = \int_0^a V_k(a - x_{k+1})dx_{k+1} = \int_0^a \frac{(a - x_{k+1})^k}{k!} dx_{k+1}$$

$$= \frac{a^{k+1}}{(k+1)!}.$$

Thus the formula $V_n(a) = a^n/n!$ is established for all positive integers n. ■

Evidently $p_n = V_n(1) = 1/n!$.

Second Solution. Let the expected number of trials required to obtain a score of c or more be $E(c)$. Suppose $0 < c \le 1$. If the first draw is in the interval $[x, x + \Delta x]$ where $x < c$, then the expected number of draws will be about $1 + E(c - x)$. Hence for $0 < c \le 1$,

$$E(c) = 1 + \int_0^c E(c - x)\,dx = 1 + \int_0^c E(u)\,du. \tag{1}$$

(The integral exists since E is increasing.) We see from (1) that E is continuous, so by the fundamental theorem of calculus,

$$E'(c) = E(c).$$

Therefore

$$E(c) = \lambda e^c,$$

for some λ. In order to satisfy (1), we must have $\lambda = 1$. Hence $E(1) = e$.

REMARK. D. J. Newman and M. S. Klamkin [*American Mathematical Monthly*, vol. 66 (1959), pages 50–51] have found the expected value of the least n for which $x_1{}^k + \cdots + x_n{}^k > 1$. Klamkin and J. H. van Lint (*Statistica Nederlandica*, vol. 26 (1972), pages 191–196) have extended the result to more general functions than powers.

---

4. If $a_1, a_2, \ldots, a_n$ are complex numbers such that

$$|a_1| = |a_2| = \cdots = |a_n| = r \neq 0,$$

and if ${}_nT_s$ denotes the sum of all products of these n numbers taken s at a time, prove that

$$\left|\frac{{}_nT_s}{{}_nT_{n-s}}\right| = r^{2s-n}$$

whenever the denominator of the left-hand side is different from zero.

Solution. For any non-zero complex number, z, we have $z^{-1} = \bar{z}/|z|^2$ where the bar denotes complex conjugation. If J is a set of s indices selected from $\{1, 2, \ldots, n\}$, then

$$\prod_{i \notin J} a_i = a_1 a_2 \cdots a_n \prod_{i \in J} a_i^{-1} = a_1 a_2 \cdots a_n \prod_{i \in J} \frac{\bar{a}_i}{r^2}$$

$$= \frac{a_1 a_2 \cdots a_n}{r^{2s}} \prod_{i \in J} \bar{a}_i.$$

Therefore

$$_nT_{n-s} = \sum_J \prod_{i \notin J} a_i = \frac{a_1 a_2 \cdots a_n}{r^{2s}} \sum_J \prod_{i \in J} \overline{a_i}$$

$$= \frac{a_1 a_2 \cdots a_n}{r^{2s}} \, {}_n\overline{T_s}.$$

Hence

$$|{}_nT_{n-s}| = \frac{r^n}{r^{2s}} |{}_n\overline{T_s}| = r^{n-2s} |{}_nT_s|,$$

whence the required formula follows at once.

---

5. Show that the integral equation

$$f(x, y) = 1 + \int_0^x \int_0^y f(u, v)\, du\, dv$$

has at most one solution continuous for $0 \leq x \leq 1, 0 \leq y \leq 1$.

Solution. Suppose there are two continuous solutions and let g be their difference. Then g is continuous and

$$g(x, y) = \int_0^x \int_0^y g(u, v)\, du\, dv.$$

Since g is continuous it is bounded on the given square. Let M be a bound. Then

$$|g(x, y)| \leq \int_0^x \int_0^y |g(u, v)|\, du\, dv \leq \int_0^x \int_0^y M\, du\, dv = Mxy$$

for $0 \leq x \leq 1, 0 \leq y \leq 1$.

We now prove that

$$(1) \qquad |g(x, y)| \leq M \frac{x^n}{n!} \frac{y^n}{n!}$$

for any positive integer n. This has been proved for $n = 1$. Assume that it is true for $n = k$; then

$$|g(x, y)| \leq \int_0^x \int_0^y |g(u, v)|\, du\, dv$$

$$\leq \int_0^x \int_0^y M \frac{u^k}{k!} \frac{v^k}{k!}\, du\, dv = M \frac{x^{k+1}}{(k+1)!} \frac{y^{k+1}}{(k+1)!}.$$

Thus (1) is established by mathematical induction. But for any fixed x and y

$$\lim_{n \to \infty} M \frac{x^n}{n!} \frac{y^n}{n!} = 0.$$

Hence $|g(x, y)| \leq 0$, that is $g(x, y) = 0$. Thus there cannot be two different continuous solutions.

REMARK. There is a solution to the given integral equation. It is readily found by the power series method to be given by

$$f(x, y) = 1 + xy + \frac{x^2}{2!} \frac{y^2}{2!} + \frac{x^3}{3!} \frac{y^3}{3!} + \cdots + \frac{x^n}{n!} \frac{y^n}{n!} + \cdots$$

$$= J_0(\sqrt{-4xy}) = J_0(2i\sqrt{xy})$$

where J_0 is the Bessel function of the first kind of order zero. See Whittaker and Watson, *Modern Analysis*, 4th ed., Cambridge University Press, 1940, page 372.

The problem generalizes immediately to n-dimensions. See Problem 4885, *American Mathematical Monthly*, vol. 67 (1960), page 87; Solution, vol. 68 (1961), page 73.

---

6. What is the smallest amount that may be invested at interest rate i, compounded annually, in order that one may withdraw 1 dollar at the end of the first year, 4 dollars at the end of the second year, $\ldots$, n^2 dollars at the end of the nth year, in perpetuity?

Solution. The present value of one dollar to be paid after n years is $(1 + i)^{-n}$ dollars. Hence the value in dollars of the given annuity is

$$\sum_{n=1}^{\infty} n^2(1 + i)^{-n}.$$

Since

$$\frac{1}{1-x} = \sum_{n=0}^{\infty} x^n,$$

we have

$$\frac{x}{(1-x)^2} = x\frac{d}{dx}\left(\frac{1}{1-x}\right) = \sum_{n=1}^{\infty} nx^n$$

and

$$\frac{x+x^2}{(1-x)^3} = x\frac{d}{dx}\left(\frac{x}{(1-x)^2}\right) = \sum_{n=1}^{\infty} n^2x^n$$

for $|x| < 1$. Putting $x = 1/(1+i)$, we obtain

$$\sum_{n=1}^{\infty} n^2(1+i)^{-n} = \frac{(1+i)(2+i)}{i^3}.$$

(At 6 percent interest, the cost of the annuity would be \$10,109.26.)

---

7. Show that ten equal-sized squares cannot be placed on a plane in such a way that no two have an interior point in common and the first touches each of the others.

Solution. Suppose that such a set of unit squares exists. Let S, S_1, S_2 be three of the non-overlapping squares with centers respectively at C, C_1, C_2 and suppose S touches both S_1 and S_2.

Let $\theta = \angle C_1CC_2$, $a = |C_1C|$, $b = |CC_2|$, $c = |C_2C_1|$. Then $1 \le a \le \sqrt{2}$, $1 \le b \le \sqrt{2}$, and $c \ge 1$. By the law of cosines

$$\cos\theta = \frac{a^2+b^2-c^2}{2ab} \le \frac{a^2+b^2-1}{2ab} = f(a,b).$$

$$\frac{\partial f}{\partial a} = \frac{a^2-b^2+1}{2a^2b} \quad\text{and}\quad \frac{\partial f}{\partial b} = \frac{b^2-a^2+1}{2ab^2}.$$

Both of these derivatives are non-negative throughout the allowable (a, b)

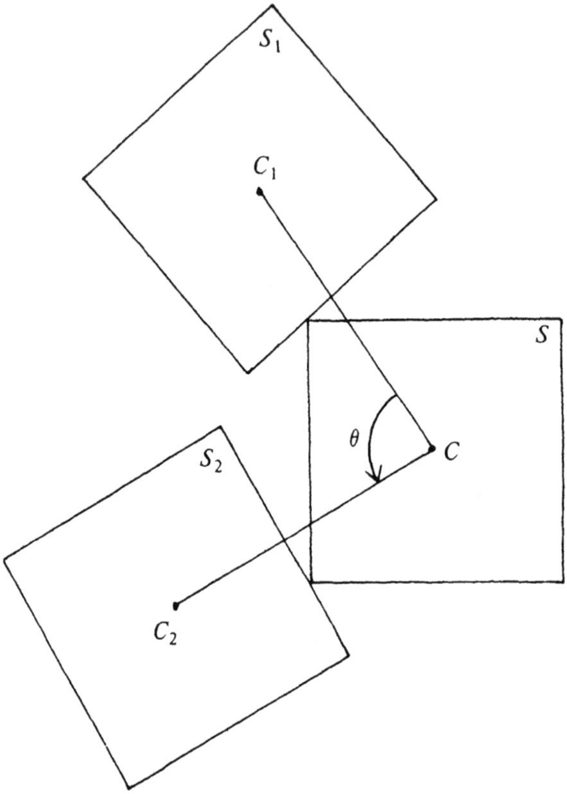

domain, so it follows that the maximum value of f on that domain is $f(\sqrt{2}, \sqrt{2}) = 3/4$. Thus $\cos\theta \leq 3/4$ and $\theta \geq \arccos 3/4$. Let $\alpha = \arccos 3/4$. Then

$$\cos 3\alpha = 4\cos^3\alpha - 3\cos\alpha = -\frac{9}{16} < -\frac{1}{2} = \cos\frac{2\pi}{3}.$$

Hence $3\alpha > 2\pi/3$, and so $\angle C_1CC_2 = \theta \geq \alpha > 2\pi/9$.

Now suppose that $S, S_1, S_2, S_3, \ldots, S_9$ are the non-overlapping squares, with S touching each of the others. Let $C, C_1, C_2, C_3, \ldots, C_9$ be the respective centers. Choose polar coordinates with center at C, and number the squares so that the angular coordinates of $C_1, C_2, \ldots, C_9$ increase monotonically between 0 and 2π. Then

$$\angle C_1CC_2 + \angle C_2CC_3 + \cdots + \angle C_9CC_1 = 2\pi,$$

so at least one of these nine angles is no greater than $2\pi/9$, contrary to what we have proved above. Hence the configuration is impossible.

REMARK. Suppose S^* is a square of side 2 concentric with S as shown.

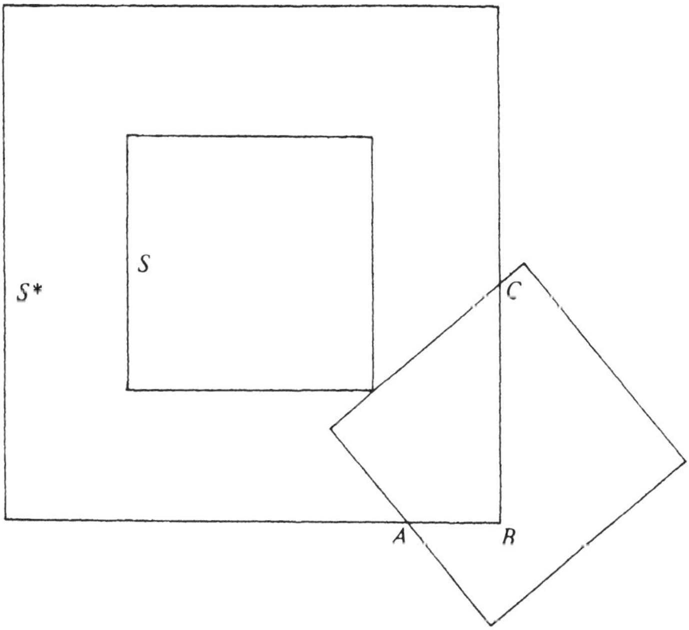

By a rather long analysis of cases, it can be shown that any unit square which touches S but contains no interior point of S must cut off from S^* an arc (for example ABC in the figure) of length at least 1. Since the perimeter of S^* is 8, at most 8 squares with disjoint interiors can touch S. This proof has the advantage that it extends easily to prove that the only configuration of eight unit squares touching S is the familiar checkerboard arrangement.

Afternoon Session

1. (i) Given line segments A, B, C, D, with A the longest, construct a quadrilateral with these sides and with A and B parallel, when possible.

Solution. Construct a triangle PQR with $|PQ| = A - B$, $|PR| = C$, and $|QR| = D$. This is possible if and only if these lengths satisfy the strict triangle inequality, i.e., the sum of the two shorter exceeds the longest. Extend PQ to S so that $|QS| = B$, and construct a parallelogram $SQRT$. Then $PSTR$ is a quadrilateral satisfying the conditions of the problem.

A second solution can be obtained by extending PQ in the opposite direction, but it is congruent to the first.

If $A = B$ (it is not clear if this is supposed to be ruled out by the hypothesis "A the longest"), the triangle constructed above will be degenerate

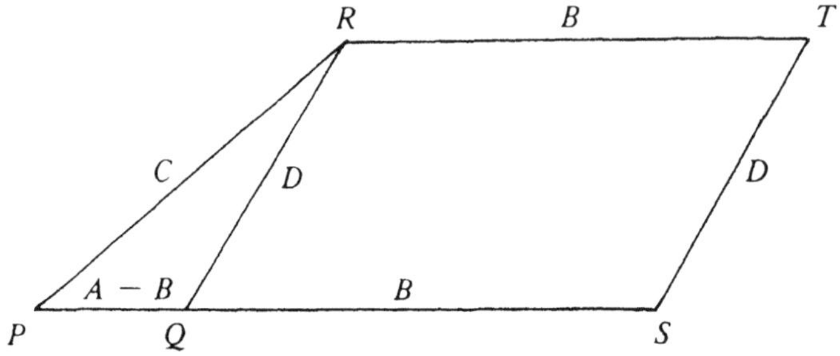

and the triangle inequality becomes $C = D$. In this case there are infinitely many non-congruent solutions: any parallelogram with adjacent sides of lengths A and C will do.

---

1. (ii) Given any acute-angled triangle ABC and one altitude AH, select any point D on AH, then draw BD and extend until it intersects AC in E, and draw CD and extend until it intersects AB in F. Prove angle AHE = angle AHF.

First Solution. Draw l through A parallel to BC. Since angles ABC and ACB are acute, the foot H of the altitude falls between B and C. Assuming $D \neq A, H$, which are trivial cases, complete the diagram as shown.

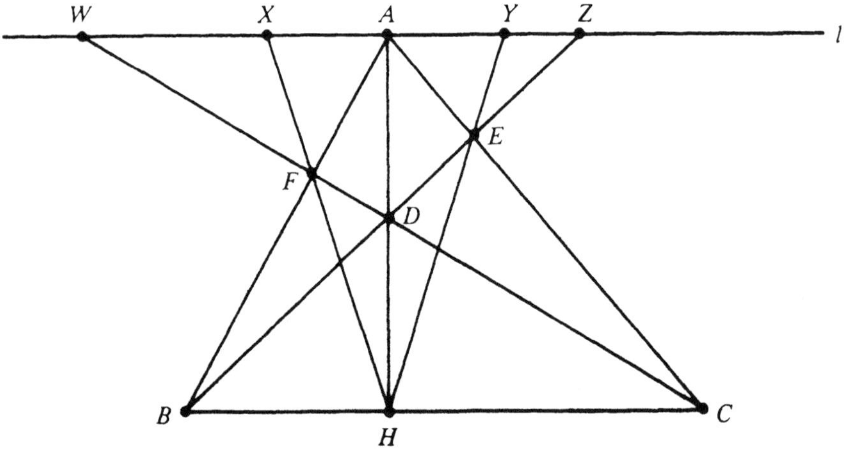

Considering several pairs of similar triangles, we see that

$$\frac{|AX|}{|BH|} = \frac{|AF|}{|BF|} = \frac{|AW|}{|BC|}, \qquad \frac{|AY|}{|CH|} = \frac{|AE|}{|CE|} = \frac{|AZ|}{|CB|},$$

$$\frac{|AW|}{|HC|} = \frac{|AD|}{|HD|} = \frac{|AZ|}{|HB|}.$$

Therefore, $|AX| \cdot |BC| = |AW| \cdot |BH| = |AZ| \cdot |HC| = |AY| \cdot |BC|$, whence $|AX| = |AY|$. So right triangles AHX and AHY are congruent and $\angle AHX = \angle AHY$, as required.

Second Solution. Choose BC and AH as cartesian axes. Let $A = (0, a)$, $B = (b, 0)$, $C = (c, 0)$, $D = (0, d)$. Then the equation of BD is $dx + by = bd$, and the equation of AC is $ax + cy = ac$. These lines meet at

$$E = \left(\frac{bc(a-d)}{ab-cd}, \frac{ad(b-c)}{ab-cd}\right).$$

Hence the slope of HE is

$$\frac{ad(b-c)}{bc(a-d)}.$$

Interchanging b and c, we find

$$\text{slope } HF = \frac{ad(c-b)}{bc(a-d)} = -\text{slope } HE.$$

Therefore, $\angle AHE = \angle AHF$, as required.

The acute angle hypothesis shows that b and c have opposite signs, while a and d have the same sign (or $d = 0$). Hence $ab - cd \neq 0$, so E exists. Also $ac - bd \neq 0$, so F exists. If $a = d$, the lines HE and HF both coincide with AH, so $\angle AHE = \angle AHF$ anyway. The proof shows that we can take D anywhere on the line AH as long as neither $ab - cd$ nor $ac - bd$ is zero.

Third Solution. The diagonal BD of the complete quadrilateral BF, FD, DH, HB is divided harmonically by the other diagonals HF and AC at G and E. Therefore, HF, HE; HB, HD is a harmonic pencil. But HB and HD are perpendicular and hence they bisect the angles formed by HF and HE. See N, A. Court, *College Geometry*, 2nd ed., Barnes and Noble, New York, 1952.

REMARK. It is interesting to compare the special cases that arise in the different proofs. The case $D = H$ is exceptional in the synthetic and projective arguments, but not in the analytic argument, while the case $BD \| AC$ is exceptional for the synthetic and analytic arguments, but not the pro-

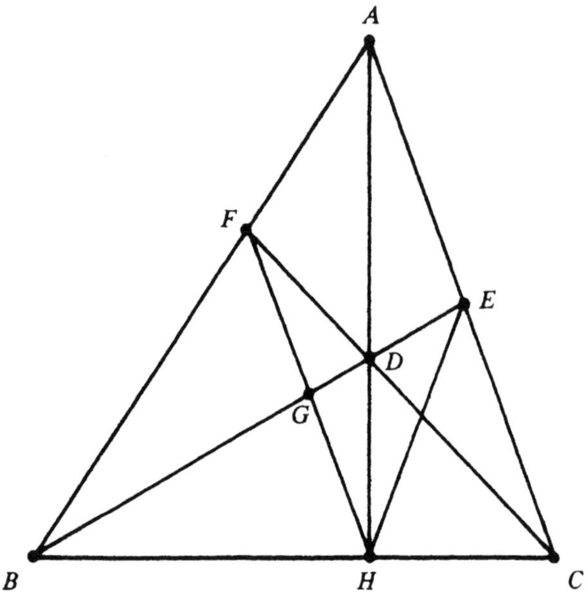

jective one. In the synthetic proof, the figure changes its appearance markedly if D is outside AH.

---

2. Prove that the product of four consecutive positive integers cannot be a perfect square or cube.

Solution. Let the four consecutive integers be $x - 1, x, x + 1, x + 2$. Their product is

$$P = (x - 1)(x + 2)x(x + 1) = (x^2 + x - 2)(x^2 + x) = (x^2 + x - 1)^2 - 1.$$

If P were a perfect square, then P and $(x^2 + x - 1)^2$ would be consecutive positive integers, both perfect squares, which is impossible. Hence P is not a perfect square.

Now suppose P is a perfect cube. Then $x > 2$, since $x - 1$ is positive and $x = 2$ gives $P = 24$. One of the integers x and $x + 1$ must be odd. If x is odd, it is relatively prime to $x - 1$, $x + 1$, and $x + 2$, so $(x - 1)(x + 1)(x + 2) = x^3 + 2x^2 - x - 2$ is also a perfect cube. But

$$x^3 < x^3 + 2x^2 - x - 2 < (x + 1)^3$$

for $x > 2$, so we would have a perfect cube between the cubes of two consecutive integers, which is impossible. If $x + 1$ is odd, then it is relatively prime to $(x - 1)x(x + 2) = x^3 + x^2 - 2x$ which again must be a perfect cube. But

$$x^3 < x^3 + x^2 - 2x < (x + 1)^3$$

for $x > 2$, and again we have a contradiction. So P cannot be a perfect cube.

REMARKS. The second part is the Diophantine equation

$$y^3 = z^2 - 1, \tag{1}$$

where $z = x^2 + x - 1$. It has been shown by Mordell that the only solution of (1) in positive integers is $y = 2$, $z = 3$. Since 8 is not the product of four consecutive positive integers, P cannot be a perfect cube. See L. J. Mordell, "The Diophantine Equation $y^2 = ax^3 + bx^2 + cx + d$, or Fifty Years After," *Proceedings of the London Mathematical Society*, Series 3, vol. 38 (1963), pages 454–458.

Erdös and Selfridge (*Illinois J. Math.* vol. 19 (1975), pp. 292-301) have proved that the product of two or more consecutive positive integers is never an nth power ($n > 1$).

---

3. In a round-robin tournament with n players (each pair of players plays one game) in which there are no draws, the numbers of wins scored by the players are $s_1, s_2, \ldots, s_n$. Prove that a necessary and sufficient condition for the existence of 3 players, A, B, C, such that A beat B, B beat C, and C beat A is

$$s_1^2 + s_2^2 + \cdots + s_n^2 < (n - 1)(n)(2n - 1)/6.$$

Solution. For any tournament outcome T, let

$$U(T) = s_1^2 + s_2^2 + \cdots + s_n^2.$$

The outcome of a round-robin tournament will be called *transitive* if and only if there are no examples of three players A, B, and C such that A beat B, B beat C, and C beat A. In this case, "beat" is a transitive linear-order relation on the set of players, so we can number the players $P_1, P_2, \ldots, P_n$ so that P_i beat P_j if and only if $i > j$. Then the final scores of the players are, respectively,

$$0, 1, \ldots, n - 1.$$

Hence in the transitive case we have

$$U(T) = 0^2 + 1^2 + \cdots + (n-1)^2 = (n-1)n(2n-1)/6.$$

This proves that the given condition is sufficient.

Now consider a non-transitive tournament outcome, that is, suppose there are three players, A, B, and C, such that A beat B, B beat C, and C beat A. We may assume that s_A is the least of the numbers s_A, s_B, and s_C; then $s_A \le s_B$. Now reverse the outcome of the match between A and B. We get a new tournament outcome in which A's score is $s_A - 1$, B's score is $s_B + 1$, and every other player's score remains the same. Then U is increased by

$$(s_A - 1)^2 - s_A{}^2 + (s_B + 1)^2 - s_B{}^2 = 2(s_B - s_A) + 2 > 0.$$

Thus any non-transitive tournament outcome can be changed so as to increase U. But it is impossible to do this indefinitely because the number of possible outcomes is finite. So starting with a non-transitive outcome, after a finite number of such changes the outcome becomes transitive and U increases to $(n-1)n(2n-1)/6$. Hence for a non-transitive outcome

$$U(T) < (n-1)n(2n-1)/6.$$

Thus the given condition is necessary.

REMARK. Tournaments have been extensively analyzed. See J. W. Moon, *Topics on Tournaments*, Holt, Rinehart and Winston, New York, 1969.

---

4. What is the average straight line distance between two points on a sphere of radius 1?

Solution. We take the radius of the sphere to be a to help keep our dimensions in order. By symmetry we need only compute the average length of all chords emanating from some fixed point which we take to be the north pole N of a spherical coordinate system. Slice the sphere into zones by parallels of co-latitude. The zone between co-latitude θ and co-latitude $\theta + \Delta\theta$ has area approximately

$$(2\pi a \sin\theta)(a\,\Delta\theta)$$

and all the chords from N to points in this zone have length approximately

$$2a \sin(\theta/2).$$

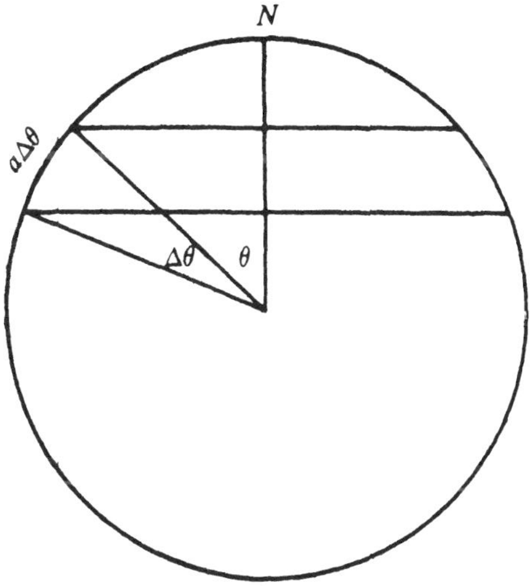

Therefore, the average length is

$$L = \frac{4\pi a^3}{S} \int_0^\pi \sin\frac{\theta}{2}\sin\theta\, d\theta,$$

where $S = 4\pi a^2$ is the surface area of the sphere. Since $a = 1$, we have

$$L = \int_0^\pi \sin\frac{\theta}{2}\left(2\sin\frac{\theta}{2}\cos\frac{\theta}{2}\right) d\theta = \frac{4}{3}\sin^3\frac{\theta}{2}\Big|_0^\pi = \frac{4}{3}.$$

---

5. Given an infinite number of points in a plane, prove that if all the distances determined between them are integers then the points are all in a straight line.

Solution. Suppose the given set contains three non-collinear points, say A, B, and C, such that $|AB| = r$ and $|AC| = s$ where r and s are integers. If P is any point at integral distance from both A and B, then by the triangle inequality, $|PA - PB|$ is one of the integers $0, 1, 2, \ldots, r$. Hence P must fall on one of the hyperbolas

$$H_i = \{X: |XA - XB| = i\}, \qquad i = 1, 2, \ldots, r-1$$

or on the union H_0 of $\overleftrightarrow{AB}$ and the perpendicular bisector of AB.

Similarly, a point P at integral distance from both A and C must be on one of the hyperbolas

$$K_i = \{X: |XA - XC| = i\}, \qquad i = 1, 2, \ldots, s-1$$

or on the union K_0 of $\overleftrightarrow{AC}$ and the perpendicular bisector of AC. Any point of our given set must be in one of the sets $H_i \cap K_j$. Because $\overleftrightarrow{AB} \neq \overleftrightarrow{AC}$, none of the sets H_i is the same as any of the sets K_j. Unless both i and j are 0, $H_i \cap K_j$ is the intersection of two second-degree curves not both degenerate; hence $H_i \cap K_j$ contains at most four points. But $H_0 \cap K_0$ also contains at most four points since H_0 and K_0 do not share a common line. Therefore the given set contains at most $4(r + 1)(s + 1)$ points, contrary to the hypothesis that it is infinite. This contradiction shows that the given points are collinear.

REMARK. It is known that for any positive integer n, there exist n non-collinear points in the plane such that each set of two are at integral distances. In fact, we can construct such a set lying on a circle. If θ is an angle such that both $\sin\theta$ and $\cos\theta$ are rational, the points $e^{2mi\theta}$, m an integer, are all at rational distances from one another. By a change of scale we can make any finite number of these distances integral.

References. N. H. Anning and P. Erdős, "Integral Distances," *Bulletin of the American Mathematical Society*, vol. 51 (1945), pages 598–600; A. S. Besicovitch, "Rational Polygons," *Mathematika*, vol. 6 (1959), page 98; D. E. Daykin, "Rational Polygons," *Mathematika*, vol. 10 (1963), pages 125–131; and H. Hadwiger, H. Debrunner, and V. Klee, *Combinatorial Geometry in the Plane*, Holt, Rinehart and Winston, New York, 1964, pages 4–6.

---

6. A projectile moves in a resisting medium. The resisting force is a function of the velocity and is directed along the velocity vector. The equation $x = f(t)$ gives the horizontal distance in terms of the time t. Show that the vertical distance y is given by

$$y = -gf(t)\int \frac{dt}{f'(t)} + g\int \frac{f(t)}{f'(t)}\,dt + Af(t) + B$$

where A and B are constants and g is the acceleration due to gravity.

Solution. In vector form the differential equation of the motion is

$$m\langle x'', y''\rangle = -R\langle x', y'\rangle - m\langle 0, g\rangle, \tag{1}$$

where primes denote differentiation with respect to time and R is the magnitude of the resistance.

Since the x-component of the motion is given by $x = f(t)$, it follows from (1) that

$$R = -m \frac{f''(t)}{f'(t)}$$

assuming $f(t)$ does not vanish. Then y satisfies the differential equation

$$y'' = +\frac{f''(t)}{f'(t)} y' - g.$$

Dividing by $f'(t)$ we get the exact differential form

$$\frac{y''}{f'(t)} - \frac{f''(t)}{[f'(t)]^2} y' = \left(\frac{y'}{f'(t)}\right)' = -\frac{g}{f'(t)}.$$

Thus

$$\frac{y'}{f'(t)} = \frac{y'(0)}{f'(0)} - g \int_0^t \frac{dr}{f'(r)},$$

and so

$$y(t) = y(0) + \frac{y'(0)}{f'(0)} [f(t) - f(0)] - g \int_0^t f'(s) \int_0^s \frac{dr}{f'(r)} ds.$$

The last integral can be integrated by parts,

$$u = \int_0^s \frac{dr}{f'(r)}, \qquad dv = f'(s)ds$$

to give

$$y(t) = \frac{y'(0)}{f'(0)} f(t) + y(0) - \frac{y'(0)f(0)}{f'(0)}$$

$$- g \cdot f(t) \int_0^t \frac{dr}{f'(r)} + g \int_0^t \frac{f(s)}{f'(s)} ds,$$

and this is in the form required.

The preceding work depends on the assumption that $f'(t) \neq 0$ for any t. Let us consider this assumption.

Suppose that the dependence of the resistance on the velocity is given by

$$R = \phi(x', y')$$

where $\phi : \mathbf{R}^2 \to \mathbf{R}$ is a function satisfying the Lipschitz condition

$$|\phi(u_1, v_1) - \phi(u_2, v_2)| \le K\{|u_1 - u_2| + |v_1 - v_2|\}$$

for some constant K. The original differential equation is then

$$(2) \qquad \begin{aligned} x'' &= -\frac{1}{m}\phi(x', y')x' \\ y'' &= -\frac{1}{m}\phi(x', y')y' - g \end{aligned}$$

and this system has a unique solution for any initial conditions (even if they are set at a time other than 0). It is clear that, if $x'(t_0) = 0$, then a solution of (2) can be found by solving

$$y'' = -\frac{1}{m}\phi(0, y')y' - g$$

and letting x be constant. Hence from the uniqueness it follows that, if $x'(t_0) = 0$ for any t_0, then x is constant. In terms of the given function f, this means that f' does not vanish anywhere unless f is constant. Thus our previous work is justified except in the trivial case of constant f. Obviously, we get no information about the nature of R in this case, so there is no way to compute y from f.

---

7. Prove that if $f(x)$ is continuous for $a \le x \le b$ and $\int_a^b x^n f(x)\,dx = 0$ for $n = 0, 1, 2, \ldots$ then $f(x)$ is identically zero on $a \le x \le b$.

First Solution. The hypothesis evidently implies that

$$(1) \qquad \int_a^b f(x)p(x)\,dx = 0$$

for any polynomial p. The Weierstrass approximation theorem [see R. C. Buck, *Advanced Calculus*, McGraw-Hill, New York, 1956, p. 39] guarantees that, given $\epsilon > 0$, there exists a polynomial p such that

$$|f(x) - p(x)| < \epsilon \quad \text{for all } x \text{ in } [a, b].$$

Let M be a bound for $|f(x)|$ on $[a, b]$. Then using (1) we find

$$\left| \int_a^b f(x)^2\, dx \right| = \left| \int_a^b f(x)\{ f(x) - p(x)\}\, dx \right|$$

$$\leq \int_a^b |f(x)| \cdot |f(x) - p(x)|\, dx \leq \int_a^b M\epsilon\, dx = M\epsilon(b - a).$$

Since ϵ can be chosen arbitrarily small, we have

$$\int_a^b f(x)^2\, dx = 0.$$

Since f is continuous we conclude that $f(x) = 0$ for all x in $[a, b]$.

Second Solution. Since the hypothesis is invariant under translations and changes of scale, we may assume that $a = 0$, $b = 2\pi$.

According to Parseval's theorem a continuous function is everywhere zero if all of its Fourier coefficients are zero. [See Georgi P. Tolstov, *Fourier Series*, Russian tr., Richard A. Silverman, Prentice-Hall, Englewood Cliffs, N.J., 1962, pp. 119-122. This reference also supplies a proof of the Weierstrass theorem (used in the first solution) via Fourier series]. Hence it suffices to prove

$$\int_0^{2\pi} f(x) \sin nx\, dx = 0 \tag{2}$$

and

$$\int_0^{2\pi} f(x) \cos nx\, dx = 0 \tag{3}$$

for $n = 0, 1, 2, \ldots$.

Now

$$\sin nx = nx - \frac{n^3}{3!} x^3 + \frac{n^5}{5!} x^5 - \cdots,$$

where the series converges uniformly on any finite interval. Multiplying by $f(x)$ and integrating termwise, we get (2). The proof of (3) is similar.

Third Solution. As we remarked above, it makes no difference what interval we consider; for this proof it is convenient to assume that $a = 0$, $b = 1$.

LEMMA. *Suppose g is a non-negative continuous function on $[0, 1]$ which attains its maximum value at a unique point x_0. Suppose f is any real continuous function on $[0, 1]$ such that $f(x_0) > 0$. Then for all sufficiently large integers n*

$$\int_0^1 f(x) g(x)^n \, dx > 0.$$

Proof. We assume $0 < x_0 < 1$. (If $x_0 = 0$ or 1, the proof requires some trivial modifications.)

By continuity there is an open interval (c, d) containing x_0 and a positive number α such that

$$f(x) > \alpha \qquad \text{for } c < x < d.$$

Let $\lambda = \sup\{g(x): x \notin (c, d)\}$. Since this maximum is attained at some point $x_1 \neq x_0$, we have $\lambda < g(x_0)$. Choose μ so that $\lambda < \mu < g(x_0)$ and let (h, k) be an interval such that

$$g(x) > \mu \qquad \text{for } h < x < k.$$

Finally let M be a bound for $|f(x)|$ on $[0, 1]$. Then

$$\int_0^1 f(x) g(x)^n \, dx = \left(\int_0^c + \int_c^h + \int_h^k + \int_k^d + \int_d^1\right) f(x) g(x)^n \, dx.$$

Now

$$\left|\int_0^c + \int_d^1\right| \leq M\lambda^n,$$

$$\int_c^h + \int_k^d \geq 0,$$

and

$$\int_h^k \geq \alpha\mu^n (k - h).$$

Hence

$$\int_0^1 f(x) g(x)^n \, dx \geq \alpha\mu^n(k - h) - M\lambda^n$$

$$= \left[\alpha(k - h) - M\left(\frac{\lambda}{\mu}\right)^n\right]\mu^n.$$

Since $\lambda/\mu < 1$, this is positive for all sufficiently large n, and the lemma is proved. ■

We now attack the original question. If f is positive at any point, it is (by continuity) positive at some rational point, say p/q where p and q are integers, $0 \leq p \leq q \neq 0$. Take $g(x) = x^p(1 - x)^{q-p}$ in the lemma. Evidently g is non-negative on $[0, 1]$ and it is easily checked that it attains its maximum at the unique point p/q. Hence the lemma asserts

$$\int_0^1 f(x)\, g(x)^n\, dx > 0$$

for sufficiently large n. But $\{g(x)\}^n$ is a polynomial for any choice of n, so the hypothesis on f implies that the integral is zero. This contradiction shows that f is nowhere positive.

The same argument shows that $-f$ is nowhere positive. Therefore f is everywhere zero.

REMARK. It was reported by the paper grader that no contestant made any significant progress with this problem except by using either the Weierstrass theorem or the theory of Fourier series.

THE NINETEENTH WILLIAM LOWELL PUTNAM MATHEMATICAL COMPETITION

November 22, 1958

Morning Session

1. Let $f(m, 1) = f(1, n) = 1$ for $m \geq 1$, $n \geq 1$, and let $f(m, n) = f(m-1, n) + f(m, n-1) + f(m-1, n-1)$ for $m > 1$ and $n > 1$. Also let

$$S(n) = \sum_{a+b=n} f(a, b), \qquad a \geq 1 \text{ and } b \geq 1.$$

Prove that

$$S(n+2) = S(n) + 2S(n+1) \qquad \text{for } n \geq 2.$$

Solution. If we write the value of $f(m, n)$ at the point $\langle m, n\rangle$ in the plane and border the resulting array with zeros as in the diagram,

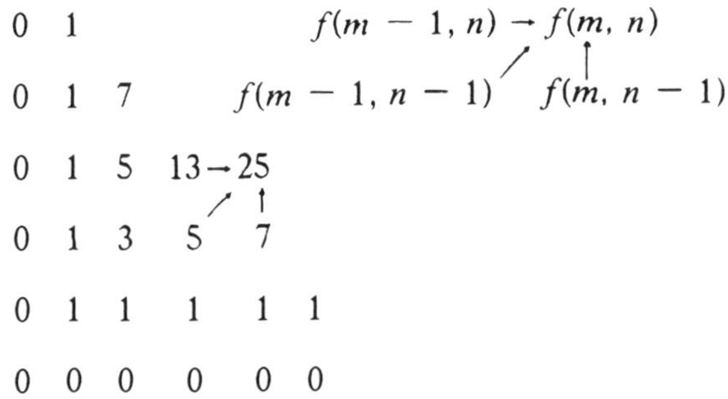

we see that the recursion relation together with the given values for $f(1, n)$ and $f(m, 1)$ amount to the assertion that every non-zero entry in this array (except $f(1, 1)$) is the sum of the entry immediately to its left, the entry just below it, and the entry diagonally below it to the left.

Now $S(n+2)$ is the sum of the terms on the $(n+2)$nd diagonal, $x + y = n + 2$, and it is clear from the diagram that each non-zero term on the $(n+1)$st diagonal enters this sum twice while each term on the nth diagonal enters once; hence, $S(n+2) = 2S(n+1) + S(n)$.

This argument can be carried out formally as follows:

$$S(n+2) = \sum_{j=1}^{n+1} f(n+2-j, j)$$

$$= f(n+1, 1) + \sum_{j=2}^{n} \{f(n+1-j, j) + f(n+2-j, j-1)$$

$$+ f(n+1-j, j-1)\} + f(1, n+1)$$

$$= \left\{ f(n, 1) + \sum_{j=2}^{n} f(n+1-j, j) \right\}$$

$$+ \left\{ \sum_{k=1}^{n-1} f(n+1-k, k) + f(1, n) \right\} + \sum_{k=1}^{n-1} f(n-k, k)$$

$$= S(n+1) + S(n+1) + S(n).$$

In the third step we set $j = k + 1$ in two of the sums and used the facts that $f(n+1, 1) = f(n, 1)$ and $f(1, n+1) = f(1, n)$.

REMARK. This recursion is studied in greater detail by R. G. Stanton and D. D. Cowan, "Note on a 'Square' Functional Equation," *SIAM Review*, vol. 12, no. 2 (April 1970), pages 277-279. This problem occurs as Lemma 4 in the given paper.

---

2. Let

$$R_1 = 1, \qquad R_{n+1} = 1 + n/R_n, \qquad n \geq 1.$$

Show that for $n \geq 1$,

$$\sqrt{n} \leq R_n \leq \sqrt{n} + 1.$$

Solution. We shall use induction on n. Evidently $\sqrt{1} = 1 = R_1 \leq \sqrt{1} + 1$, so the given formula is true for $n = 1$. Suppose we know that it is true for $n = k$, i.e., for k a positive integer

$$\sqrt{k} \leq R_k \leq \sqrt{k} + 1.$$

Then

$$\sqrt{k+1}-1=\frac{k}{\sqrt{k+1}+1}<\frac{k}{\sqrt{k}+1}\le\frac{k}{R_k}\le\frac{k}{\sqrt{k}}=\sqrt{k}<\sqrt{k+1}.$$

Hence

$$\sqrt{k+1}<1+\frac{k}{R_k}=R_{k+1}<\sqrt{k+1}+1,$$

and the induction is complete.

Note that the proof shows that the inequalities are strict for all integers, $n>1$.

REMARK. This problem was considered by Leo Moser and Max Wyman, "On Solutions of $x^d = 1$ in Symmetric Groups," *Canadian Journal of Mathematics*, vol. 7 (1955), pages 159–168.

They obtained the continued fraction expansion

$$R_{n+1}=1+\frac{n}{1+}\,\frac{n-1}{1+}\,\frac{n-2}{1+}\cdots\frac{1}{1},$$

which follows directly from the recurrence, $R_{n+1}=1+n/R_n$ and proved that

$$\lim_{n\to\infty}(R_n-\sqrt{n})=\tfrac{1}{2}.$$

---

3. Under the assumption that the following set of relations has a unique solution for $u(t)$, determine it.

$$\frac{du(t)}{dt}=u(t)+\int_0^1 u(s)\,ds,$$

$$u(0)=1.$$

Solution. Put

$$b=\int_0^1 u(s)ds.$$

Then b is a constant and the given equation becomes the familiar linear differential equation

$$u'(t) = u(t) + b$$

with the general solution

$$u(t) = -b + ce^t.$$

There remains the problem of determining b and c. We have

$$b = \int_0^1 (-b + ce^s)ds = -b + c(e - 1)$$

and

$$u(0) = -b + c = 1.$$

These two equations imply that

$$b = (e - 1)/(3 - e) \quad \text{and} \quad c = 2/(3 - e)$$

so

$$u(t) = \frac{1}{3 - e}(2e^t - e + 1).$$

It is readily checked that this function is indeed a solution.

The above derivation proves the uniqueness of the solution function u, so this assumption in the statement of the problem is unnecessary.

---

4. In assigning dormitory rooms, a college gives preference to pairs of students in this order:

$$AA,\ AB,\ AC,\ BB,\ BC,\ AD,\ CC,\ BD,\ CD,\ DD,$$

in which AA means two seniors, AB means a senior and a junior, etc. Determine numerical values to assign to A, B, C, D so that the set of numbers $A + A$, $A + B$, $A + C$, $B + B$, etc., corresponding to the order above will be in descending magnitude. Find the general solution and the solution in least positive integers.

Solution. The inequalities to be solved are

$$(1)\qquad 2A > A + B > A + C > 2B > B + C > A + D > 2C > B + D > C + D > 2D.$$

Evidently we must have $A > B > C > D$, so put

$$A = \alpha + B,\ B = \beta + C,\ C = \gamma + D,$$

where α, β and γ are positive. The first, second, fourth, eighth, and ninth inequalities in (1) are now satisfied, while the third, fifth, sixth, and seventh inequalities become, respectively

$$\alpha > \beta, \qquad \gamma > \alpha, \qquad \alpha + \beta > \gamma, \qquad \text{and} \quad \gamma > \beta.$$

The inequality $\gamma > \beta$ is a consequence of $\gamma > \alpha$ and $\alpha > \beta$, so it can be dropped from this system. Now put $\alpha = \beta + \delta$, $\gamma = \alpha + \epsilon$ where δ and ϵ are positive. Then $\alpha + \beta > \gamma$ becomes $\beta > \epsilon$, so we can put $\beta = \epsilon + \zeta$ with ζ positive. Then

$$\alpha = \delta + \epsilon + \zeta$$
$$\beta = \epsilon + \zeta$$
$$\gamma = \delta + 2\epsilon + \zeta,$$

and finally,

$$(2)\qquad \begin{aligned} A &= 2\delta + 4\epsilon + 3\zeta + D \\ B &= \delta + 3\epsilon + 2\zeta + D \\ C &= \delta + 2\epsilon + \zeta + D. \end{aligned}$$

Here D can be chosen arbitrarily, while δ, ϵ, and ζ must be positive. It follows either from the derivation or from a direct check that if A, B, C, D satisfy (2), where δ, ϵ and ζ are positive, then they satisfy the inequalities (1).

To find integral solutions, we note that if we start with integral A, B, C, D satisfying (1), then all the numbers α, β, γ, δ, ϵ, ζ will be positive integers. Hence the least solution in positive integers is obtained by choosing $\delta = \epsilon = \zeta = D = 1$. Hence $A = 10$, $B = 7$, $C = 5$, $D = 1$ is the least solution in positive integers.

5. Show that the number of non-zero terms in the expansion of the nth order determinant having zeros in the main diagonal and ones elsewhere is

$$n!\left|1-\frac{1}{1!}+\frac{1}{2!}-\frac{1}{3!}+\cdots+\frac{(-1)^n}{n!}\right|.$$

Solution. Recall that the determinant of the $n \times n$ matrix $M = (m_{ij})$ is

$$\sum_{\pi} \epsilon(\pi)\, m_{1j_1} m_{2j_2} \cdots m_{nj_n}$$

where

$$\pi = \begin{pmatrix} 1 & 2 & \dots & n \\ j_1 & j_2 & \dots & j_n \end{pmatrix}$$

runs through all the permutations of the set $\{1, 2, \dots, n\}$ and $\epsilon(\pi)$ is $+1$ or -1 according as π is even or odd. For the matrix in question with zeros on the main diagonal the term corresponding to the permutation π will be zero if and only if π has a fixed point (i.e., there is an index i such that $j_i = i$). Hence the problem calls for finding how many permutations of the set $\{1, 2, \dots, n\}$ have no fixed point. Let this number be A_n. We shall give three ways to evaluate A_n.

First method. We shall derive the relation

$$(1) \qquad A_n = (n-1)(A_{n-1} + A_{n-2}),$$

and from this deduce the required formula.

Let B_n be the number of permutations π of $\{1, 2, \dots, n\}$ with no fixed point, and such that $\pi(1) = 2$. It is clear that $A_n = (n-1)B_n$ (since 2 could be replaced by any element of $\{2, \dots, n\}$). To enumerate B_n we consider separately the cases

$$\text{(i)} \qquad \pi(2) = 1.$$

and

$$\text{(ii)} \qquad \pi(2) > 2.$$

In case (i), π corresponds to a unique fixed point free permutation of $\{3, \dots, n\}$; hence, the number of such permutations is A_{n-2}.

In case (ii), π has the form

$$\pi = \begin{pmatrix} 1 & 2 & \dots & k & \dots \\ 2 & j & \dots & 1 & \dots \end{pmatrix} \qquad \text{where } j \neq 1.$$

To this π we associate the permutation

$$\pi^* = \begin{pmatrix} 2 & \dots & k & \dots \\ j & \dots & 2 & \dots \end{pmatrix},$$

obtained by deleting the first column of π, and replacing the entry 1 in the kth column by the symbol 2. Then π^* is a fixed point free permutation on $\{2, \dots, n\}$ and the correspondence between π and π^* is one to one. Hence, the number of permutations in case (ii) is A_{n-1}.

Combining the two cases we get

$$A_n = (n - 1)B_n = (n - 1)(A_{n-1} + A_{n-2}),$$

as asserted.

To complete the derivation of the required formula for A_n we set $A_n = n!\, C_n$ in (1) to get

$$n!\, C_n = (n - 1)(n - 1)!\, C_{n-1} + (n - 1)!\, C_{n-2}.$$

Dividing by $(n - 1)!$ we obtain

$$nC_n = (n - 1)\, C_{n-1} + C_{n-2}$$

or, equivalently,

$$C_n - C_{n-1} = -\frac{1}{n}(C_{n-1} - C_{n-2}).$$

By iteration this yields

$$\text{(2)} \qquad C_n - C_{n-1} = \left(-\frac{1}{n}\right)\left(-\frac{1}{n-1}\right) \cdots \left(-\frac{1}{3}\right)(C_2 - C_1) = \frac{(-1)^n}{n!}.$$

Since obviously $A_1 = 0$, $A_2 = 1$ and therefore $C_1 = 0$, $C_2 = \frac{1}{2}$. Now sum equation (2) to get

$$C_n = \sum_{m=0}^{n} \frac{(-1)^m}{m!}$$

from which we get

$$A_n = n!\left(1 - \frac{1}{1!} + \frac{1}{2!} \cdots + \frac{(-1)^n}{n!}\right).$$

Second method. Note that A_k is the number of fixed-point free permutations of any set having k elements.

Let us classify the permutations of $\{1, 2, \ldots, n\}$ according to how many points they leave fixed. A_n permutations have no fixed points. For any particular point a in $\{1, 2, \ldots, n\}$ there are A_{n-1} permutations that fix a but move all other points; hence there are altogether nA_{n-1} that fix exactly one point. For any two distinct points a and b there are A_{n-1} permutations that fix both a and b but move all other points, and there are $\binom{n}{2} A_{n-2}$ permutations that have exactly two fixed points. Continuing this reasoning, we see that there are $\binom{n}{r}A_{n-r}$ permutations with exactly r fixed points. We make this formula valid for $r = n$ by defining $A_0 = 1$. Since each permutation has some number of fixed points, we have

$$n! = A_n + \binom{n}{1} A_{n-1} + \binom{n}{2} A_{n-2} + \cdots + \binom{n}{r} A_{n-r} + \cdots + A_0.$$

Now the system of equations

$$(3) \qquad B_n = \sum_{i=0}^{n} \binom{n}{i} A_{n-i}, \qquad n = 0, 1, \ldots, k$$

can be solved for A_n, giving

$$(4) \qquad A_n = \sum_{j=0}^{n} (-1)^j \binom{n}{j} B_{n-j}.$$

To see this we note that if (3) holds, then the right member of (4) is

$$\sum_{j=0}^{n} (-1)^j \binom{n}{j} \sum_{i=0}^{n-j} \binom{n-j}{i} A_{n-j-i}.$$

The coefficient of A_{n-k} in this double sum is

$$\sum_{j=0}^{k} (-1)^j \binom{n}{j}\binom{n-j}{k-j} = \binom{n}{k} \sum_{j=0}^{k} (-1)^j \binom{k}{j}$$

$$= 0 \quad \text{if } k > 0$$

$$= 1 \quad \text{if } k = 0,$$

since the last sum is the binomial expansion of $(1 - 1)^k$. Thus the right member reduces to A_n as claimed in (4). Conversely, one can prove that (4) implies (3).

In the present case, $B_n = n!$, so

$$A_n = \sum_{j=0}^{n} (-1)^j \binom{n}{j} (n-j)! = n! \sum_{j=0}^{n} (-1)^j \frac{1}{j!},$$

as required.

REMARK. The explicit inversion of equation (3) to the form (4) is often important in combinatorial analysis.

Third method. We can deduce the formula for A_n quite directly from what is known as the principle of inclusion and exclusion. This says, that if X is any finite set and $Y_1, Y_2, \ldots, Y_n$ are subsets of X, then

$$|X - (Y_1 \cup Y_2 \cup \cdots \cup Y_n)| = |X| - \Sigma |Y_i| + \Sigma |Y_i \cap Y_j|$$
$$- \Sigma |Y_i \cap Y_j \cap Y_k| + \cdots + (-1)^r \Sigma |Y_{i_1} \cap Y_{i_2} \cap \cdots \cap Y_{i_r}| + \cdots;$$

the sums being over all distinct sets of r indices from $\{1, 2, \ldots, n\}$. (Here $|A|$ denotes the number of members of A.) In the present instance we take X to be the set of permutations of $\{1, 2, \ldots, n\}$ and Y_i to be the subset of those that fix i (and possibly other points). Then

$$|Y_{i_1} \cap Y_{i_2} \cap \cdots \cap Y_{i_r}| = (n-r)!$$

for each of the r-element subsets $\{i_1, i_2, \ldots, i_r\}$ of $\{1, 2, \ldots, n\}$, so

$$\begin{aligned} A_n &= |X - (Y_1 \cup Y_2 \cup \cdots Y_n)| \\ &= n! - \binom{n}{1}(n-1)! + \binom{n}{2}(n-2)! - \cdots \\ &\quad + (-1)^r \binom{n}{r}(n-r)! + \cdots \\ &= n! \sum_{r=0}^{n} (-1)^r \frac{1}{r!} \end{aligned}$$

as before.

REMARKS. The problem is a famous one, known as the "problème des rencontres." It is often posed in the form: What is the probability that, if n letters are placed at random in their envelopes, no letter is put in the correct envelope? The answer is, of course,

$$A_n/n! = \sum_{r=0}^{n} (-1)^r \frac{1}{r!}.$$

This is very near to e^{-1} if n is at all large.

The "problème des rencontres" was formulated by P. R. Montmort about 1708. See the bibliography provided with the solution to Problem 4146, *American Mathematical Monthly*, vol. 53 (1946), pages 107-110, for additional historical and other references.

The name "sub-factorial" is sometimes given to the number A_n, and although not standardized, the notations $n!!$ and $\lfloor\!\underline{\;n}$ have been used for n sub-factorial. See *Mathematical Gazette*, vol. 34 (1950), pages 302-303.

---

6. Let $a(x)$ and $b(x)$ be continuous functions on $0 \le x \le 1$ and let $0 \le a(x) \le a < 1$ on that range. Under what other conditions (if any) is the solution of the equation for u,

$$u = \underset{0 \le x \le 1}{\text{maximum}} \; [b(x) + a(x) \cdot u],$$

given by

$$u = \underset{0 \le x \le 1}{\text{maximum}} \left| \frac{b(x)}{1 - a(x)} \right| ?$$

Solution. Since the functions involved are continuous, the maximum values must be attained, and since we are given that $a(x)$ is bounded and less than unity, we see that the following statements are all equivalent:

(1) $u = \max\limits_{0 \le x \le 1} [b(x) + a(x)u]$

(2) $(\forall x) \quad u \ge b(x) + a(x) \cdot u$ with equality for at least one value of x.

(3) $(\forall x) \quad (1 - a(x))\, u \ge b(x)$ with equality for at least one value of x.

(4) $(\forall x) \quad u \ge \dfrac{b(x)}{1 - a(x)}$ with equality for at least one value of x.

(5) $u = \max\limits_{0 \le x \le 1} \dfrac{b(x)}{1 - a(x)}$.

Thus no additional conditions are required. Equation (1) always has a unique solution for u, given by (5).

---

7. Let a and b be relatively prime positive integers, b even. For each positive integer q let $p = p(q)$ be chosen so that

$$\left| \frac{p}{q} - \frac{a}{b} \right|$$

is a minimum. Prove that

$$\lim_{n \to \infty} \sum_{q=1}^{n} \frac{q \left| \frac{p}{q} - \frac{a}{b} \right|}{n} = \frac{1}{4}.$$

Solution. Rewrite $q\,|p/q - a/b|$ in the form

$$\frac{1}{b}\,|pb - qa|.$$

For each q we are to choose p to minimize this; then $pb - qa$ will be the absolutely least residue of qa modulo b. Since a is relatively prime to b, as q varies through a complete set of residues modulo b, so will qa, and therefore $pb - qa$ will take the values

$$-C + 1, -C + 2, \ldots -1, 0, 1, \ldots, C - 1, C$$

where $b = 2C$ (recall that b is even) and the contribution to the sum will be

$$\frac{1}{b}(0 + 1 + 2 + \cdots + C - 1 + C + C - 1 + \cdots + 1)$$

$$= \frac{C^2}{2C} = b/4.$$

Thus if $n = b \cdot r + s$ where $0 \le s < b$, we have

$$\sum_{q=1}^{n} q \left| \frac{p}{q} - \frac{a}{b} \right| = r\frac{b}{4} + \sum_{q=br+1}^{br+s} q \left| \frac{p}{q} - \frac{a}{b} \right|$$

since q runs through r complete residue systems (mod b) and then the integers $br + 1, br + 2, \ldots, br + s$. Since $p(q)$ is the integer nearest to qa/b,

$$q \left| \frac{p}{q} - \frac{a}{b} \right| = \left| p - \frac{qa}{b} \right| \le \frac{1}{2}.$$

We have therefore

$$\left|\frac{1}{4}n - \sum_{q=1}^{n} q\left|\frac{p}{q} - \frac{a}{b}\right|\right| = \left|\frac{s}{4} - \sum_{q=br+1}^{br+s} q\left|\frac{p}{q} - \frac{a}{b}\right|\right|$$

$$\leq \frac{s}{4} + s \sup_{q} q\left|\frac{p}{q} - \frac{a}{b}\right|$$

$$\leq \frac{3}{4}s < \frac{3}{4}b.$$

Dividing by n, we get

$$\left|\frac{1}{4} - \frac{1}{n}\sum_{q=1}^{n} q\left|\frac{p}{q} - \frac{a}{b}\right|\right| < \frac{3b}{4n}.$$

Letting $n \to \infty$, we see that

$$\lim_{n\to\infty} \frac{1}{n}\sum_{q=1}^{n} q\left|\frac{p}{q} - \frac{a}{b}\right| = \frac{1}{4}.$$

Afternoon Session

1. Given

$$b_n = \sum_{k=0}^{n} \binom{n}{k}^{-1}, \qquad n \geq 1,$$

prove that

$$b_n = \frac{n+1}{2n} b_{n-1} + 1, \qquad n \geq 2,$$

and hence, as a corollary,

$$\lim_{n\to\infty} b_n = 2.$$

Solution. We have

$$n!\, b_n = \sum_{k=0}^{n} k!(n-k)! = \sum_{k=1}^{n+1} (k-1)!(n-k+1)!$$

so

$$\begin{aligned}2n!b_n &= 0!n! + \sum_{k=1}^{n} [k!(n-k)! + (k-1)!(n-k+1)!] + n!0! \\ &= 2n! + (n+1)\sum_{k=1}^{n} (k-1)!(n-k)! \\ &= 2n! + (n+1)[(n-1)!b_{n-1}].\end{aligned}$$

Dividing by $2n!$ we get

$$b_n = 1 + \frac{n+1}{2n} b_{n-1}$$

as required.

Let $b_n = 2 + C_n$. Then

$$nC_n = 1 + \frac{n+1}{2(n-1)}(n-1)C_{n-1}.$$

We shall prove by induction that $nC_n \le 6$ for all n. This is clearly true for $n = 1, 2, 3$ ($C_1 = 0$, $C_2 = \frac{1}{2}$, $C_3 = \frac{2}{3}$). Assume it is true for $n = k - 1$ where $k \ge 4$. Then

$$kC_k \le 1 + \frac{k+1}{2(k-1)} 6 \le 1 + \frac{5}{2 \cdot 3} 6 = 6,$$

where we used the fact that $(x+1)/(x-1) = 1 + 2/(x-1)$ increases as x decreases for $x > 1$. This completes the induction.

From the inequality $C_n \le 6/n$ and the obvious fact that $C_n \ge 0$, it follows that $C_n \to 0$, and therefore $b_n \to 2$.

REMARK. One can establish $\lim b_n = 2$ directly since

$$\begin{aligned}b_n &= 2 + \frac{2}{n} + \sum_{k=2}^{n-2} \frac{1}{\binom{n}{k}} \\ &\le 2 + \frac{2}{n} + \frac{n-3}{\binom{n}{2}} = 2 + \frac{4}{n} \cdot \frac{n-2}{n-1}\end{aligned}$$

for $n \ge 3$.

2. Given a set of $n + 1$ positive integers, none of which exceeds $2n$, show that at least one member of the set must divide another member of the set.

Solution. Every positive integer can be written uniquely in the form $2^p q$ where p is a non-negative integer and q is a positive odd integer, the odd part of n. The odd part of an integer in the set $S = \{1, 2, 3, \ldots, 2n\}$ must be one of the n integers $1, 3, 5, \ldots, (2n - 1)$. Given $(n + 1)$ integers in S, at least two must have the same odd part, that is they must be of the form

$$2^{p_1}q \quad \text{and} \quad 2^{p_2}q,$$

where $p_1 \neq p_2$. We can choose the notation so that $p_1 < p_2$; then $2^{p_1}q$ divides $2^{p_2}q$.

---

3. If a square of unit side be partitioned into two sets, then the diameter (least upper bound of the distances between pairs of points) of one of the sets is not less than $\sqrt{5}/2$. Show also that no larger number will do.

Solution. Suppose the square is $ABCD$ (with unit side) and the mid-points of the sides AB and BC are E and F, respectively. Then $|AF| = |DF| = |DE| = |CE| = \sqrt{5}/2$.

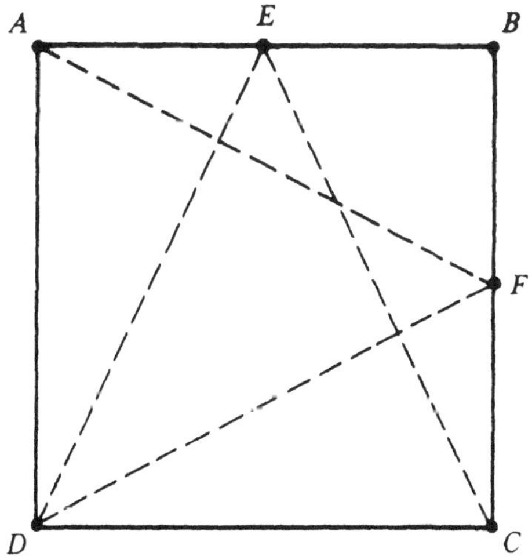

Suppose that the square is partitioned into two sets S and T of diameter less than $\sqrt{5}/2$, and choose the notation so that $A \in S$. Then $F \in T$, $D \in S$, $E \in T$, $C \in S$, since A and F, for example, are too far apart to be both members of S. Thus A and C are in same subset S, but $|AC| = \sqrt{2} > \sqrt{5}/2$, contradicting the fact that the diameter of S is $<\sqrt{5}/2$.

On the other hand, one can clearly partition the square into two rectangular sets of diameter $\sqrt{5}/2$, as indicated.

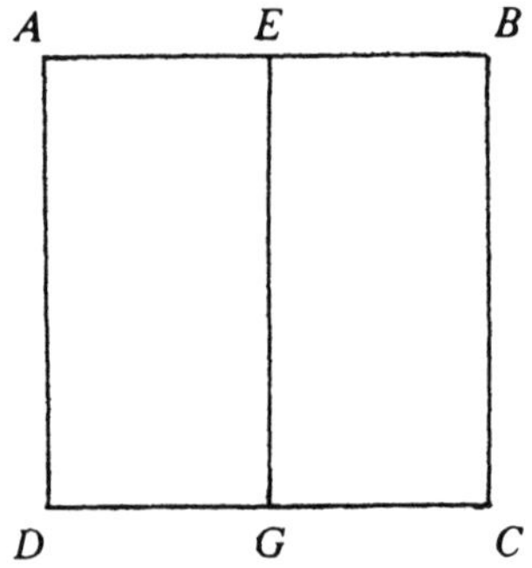

The points of the dividing segment EG can be apportioned to either subset. (It is obvious that the diameter of a rectangle is the length of its diagonal, but an analytic proof can easily be given.)

---

4. Let C be a real number, and let f be a function such that

$$\lim_{x \to \infty} f(x) = C, \qquad \lim_{x \to \infty} f'''(x) = 0.$$

Prove that

$$\lim_{x \to \infty} f'(x) = 0 \quad \text{and} \quad \lim_{x \to \infty} f''(x) = 0,$$

where superscripts denote derivatives.

Solution. By Taylor's theorem (finite form, or extended law of the mean), for any x there exist numbers $\xi(x)$ and $\eta(x)$ between 0 and 1 such that

$$f(x+1) = f(x) + f'(x) + \frac{1}{2}f''(x) + \frac{1}{6}f'''(x + \xi(x)) \tag{1}$$

$$f(x-1) = f(x) - f'(x) + \frac{1}{2}f''(x) - \frac{1}{6}f'''(x - \eta(x)). \tag{2}$$

Adding these relations and transposing terms, we get

(3)

$$f''(x) = f(x+1) - 2f(x) + f(x-1) - \frac{1}{6}f'''(x+\xi(x)) + \frac{1}{6}f'''(x-\eta(x)).$$

Subtracting (2) from (1) and transposing terms, we get

(4)

$$2f'(x) = f(x+1) - f(x-1) - \frac{1}{6}f'''(x+\xi(x)) - \frac{1}{6}f'''(x-\eta(x)).$$

As $x \to \infty$, so do $x - \eta(x)$ and $x + \xi(x)$, so all terms on the right of (3) and (4) have limits, and

$$\lim_{x\to\infty} f''(x) = C - 2C + C - \frac{1}{6}\cdot 0 + \frac{1}{6}\cdot 0 = 0$$

and

$$\lim_{x\to\infty} f'(x) = \frac{1}{2}\left[C - C - \frac{1}{6}\cdot 0 - \frac{1}{6}\cdot 0\right] = 0.$$

REMARKS. The hypothesis can be weakened; it is sufficient to assume that f''' is bounded. Moreover, if we assume $\lim_{x\to\infty} f(x)$ exists and $f^{(n)}$ is bounded, we can prove that $\lim_{x\to\infty} f^{(k)}(x) = 0$ for $0 < k < n$. This was published by Littlewood in "The Converse of Abel's Theorem," *Proceedings of the London Mathematical Society* (2), vol. 9 (1910–11), pages 434–448. For results concerning precise bounds for $f^{(k)}(x)$ given bounds for $f(x)$ and $f^{(n)}(x)$ where $0 < k < n$, see I. J. Schoenberg, "The Elementary Cases of Landau's Problem of Inequalities Between Derivatives," *American Mathematical Monthly*, vol. 80 (1973), pages 121–158.

---

5. The lengths of successive segments of a broken line are represented by the successive terms of the harmonic progression $1, 1/2, 1/3, \ldots, 1/n, \ldots$. Each segment makes with the preceding segment a given angle θ. What is the distance and what is the direction of the limiting point (if there is one) from the initial point of the first segment?

First Solution. We may identify the plane with the complex number plane in such a way that the first segment extends from 0 to 1. Then the next segment extends from 1 to $1 + \frac{1}{2} e^{i\theta}$, since this segment represents the complex number $\frac{1}{2} e^{i\theta}$.

The nth segment of the path represents the complex number $(1/n) e^{i(n-1)\theta}$, and when added to the previous $(n - 1)$ segments, the sum ends at the point $\Sigma_{p=1}^{n} (1/p)\, e^{i(p-1)\theta}$. Thus the question really concerns the convergence and evaluation of

$$\sum_{p=1}^{\infty} \frac{1}{p} e^{i(p-1)\theta}. \tag{1}$$

This becomes the harmonic series if $\theta = 0$ (or a multiple of 2π) and does not converge in this case. We shall show that the series (1) converges in all other cases.

A theorem of Abel asserts: If $b_1, b_2, b_3, \ldots$ is a real sequence that decreases monotonically to zero, and if the partial sums of the series $\Sigma_{p=1}^{\infty} a_p$ are bounded, then $\Sigma_{p=1}^{\infty} a_p b_p$ converges. A proof of this theorem was given in the discussion of Problem A.M. 7 of the Eleventh Competition (p. 325). It is also proved in a number of advanced calculus texts, for example, Louis Brand, *Advanced Calculus: An Introduction to Classical Analysis*, Wiley, New York, 1955, pages 418–420; or T. M. Apostol, *Mathematical Analysis*, Addison-Wesley, Reading, Mass., 1957, page 365.

For the present case, take $b_p = 1/p$ and $a_p = e^{i(p-1)\theta}$. Since the a's form a geometric progression, and since we are assuming $e^{i\theta} \neq 1$, we have

$$\left| \sum_{p=1}^{n} a_p \right| = \left| \frac{1 - e^{in\theta}}{1 - e^{i\theta}} \right| \leq \frac{2}{|1 - e^{i\theta}|}.$$

Thus the partial sums are bounded, Abel's theorem applies, and the series (1) converges whenever $e^{i\theta} \neq 1$.

To evaluate (1) we can use another theorem of Abel. If $\Sigma_{p=1}^{\infty} c_p$ converges, then

$$\lim_{r \to 1-} \sum_{p=1}^{\infty} c_p r^p = \sum_{p=1}^{\infty} c_p$$

where the limit is taken for r increasing to 1 through real values. See Brand, pages 423–424, or Apostol, page 421.

In the present case,

$$\Sigma \frac{1}{p} e^{i(p-1)\theta} = e^{-i\theta} \lim_{r \to 1-} \Sigma \frac{1}{p} (re^{i\theta})^p.$$

Putting $z = re^{i\theta}$, we recognize the series on the right as the Taylor's series expansion for the principal value of $-\log(1 - z)$, valid for $|z| < 1$.

Hence

$$\Sigma \frac{1}{p} e^{i(p-1)\theta} = -e^{i\theta} \lim_{r \to 1} \log (1 - re^{i\theta}) = -e^{i\theta} \log (1 - e^{i\theta}).$$

Now

$$1 - e^{i\theta} = 2 \sin \frac{1}{2} \theta \, e^{(1/2)i(\theta - \pi)} \quad \text{for } 0 < \theta < 2\pi$$

and here $2 \sin \frac{1}{2}\theta > 0$ and $\frac{1}{2}(\theta - \pi)$ is the principal value of the argument (since $-\pi < \frac{1}{2}(\theta - \pi) < \pi$). Therefore

$$\log (1 - e^{i\theta}) = \log \left(2 \sin \frac{1}{2}\theta \right) + \frac{1}{2} i \, (\theta - \pi)$$

and

$$\Sigma \frac{1}{p} e^{i(p-1)\theta} = -e^{-i\theta} \left\{ \log \left(2 \sin \frac{1}{2}\theta \right) + \frac{1}{2} i(\theta - \pi) \right\}.$$

Hence the limit point is at a distance

$$(2) \qquad \left| \Sigma \frac{1}{p} e^{i(p-1)\theta} \right| = \sqrt{\left(\log \left(2 \sin \frac{\theta}{2} \right) \right)^2 + \frac{1}{4}(\theta - \pi)^2}$$

from the origin; its argument is

$$(3) \qquad \pi - \theta + \arg \left[\log \left(2 \sin \frac{\theta}{2} \right) + \frac{i}{2}(\theta - \pi) \right].$$

REMARK. If $\theta = \pi$, the series to be summed is the well-known alternating series

$$\sum_{n=1}^{\infty} \frac{(-1)^{n-1}}{n} = \log 2.$$

In this case, (2) above reduces to log 2 and (3) becomes 0.

Second Solution. One can prove the relation

$$(4) \qquad \sum_{p=1}^{\infty} \frac{1}{p} e^{ip\theta} = -\log (1 - e^{i\theta}) = -\log \left(2 \sin \frac{1}{2}\theta \right) + \frac{i}{2}(\pi - \theta)$$

using the theory of Fourier series. The Fourier series of $\frac{1}{2}(\pi - \theta)$ on the interval $[0, 2\pi]$ is readily found to be

$$\sum_{p=1}^{\infty} \frac{1}{p} \sin p\theta.$$

With slightly more difficulty the Fourier series of $\log(2 \sin \frac{1}{2}\theta)$ is found to be

$$-\sum_{p=1}^{\infty} \frac{1}{p} \cos p\theta.$$

Since both functions on the right side of (4) have continuous derivatives on the open interval $(0, 2\pi)$, they are represented by their Fourier series on this interval, i.e., the series converge and

$$\sum \frac{1}{p} \cos p\theta = -\log\left(2 \sin \frac{1}{2}\theta\right),$$

$$\sum \frac{1}{p} \sin p\theta = \frac{1}{2}(\pi - \theta)$$

for $0 < \theta < 2\pi$.

Combining these relations and using $e^{ip\theta} = \cos p\theta + i \sin p\theta$ we obtain (4).

---

6. Let a complete oriented graph on n points be given, i.e., a set of n points $1, 2, 3, \ldots, n$, and between any two points i and j a direction, $i \to j$. Show that there exists a permutation of the points, $[a_1, a_2, a_3, \ldots, a_n]$, such that $a_1 \to a_2 \to a_3 \to \cdots \to a_n$.

First Solution. We shall prove this by induction on n. It is obviously true for $n = 2$ (and vacuously true for $n = 1$). We assume the result for $n = 1, 2, \ldots, k$ and consider a complete oriented graph on $k + 1$ points. Pick any one of these points, say b, and consider two subsets of the remaining k points

$$A = \{x: x \to b\},\ C = \{x: b \to x\}.$$

Then A and C, with the given directions for their pairs, are complete oriented graphs having, say, p and q points, respectively; $p + q = k$. By the induction hypothesis A can be enumerated so that $a_1 \to a_2 \to \cdots \to a_p$, and C can be enumerated so that $c_1 \to c_2 \to \cdots \to c_q$. Then the required chain is given by

$$a_1 \to a_2 \to \cdots \to a_p \to b \to c_1 \to c_2 \to \cdots c_q.$$

[Note that either A or C might be empty, but this creates no difficulty.]

Second Solution. Again assume that the result is true for $n = k$ points, and let a complete oriented graph G on $k + 1$ points be given. Pick any point b of G. By the inductive hypothesis, the remaining k points can be labeled $a_1, a_2, \ldots, a_k$ so that

$$a_1 \to a_2 \to \cdots \to a_k.$$

Then b can be fitted into this sequence either just before a_i, where i is the least index with $b \to a_i$, or at the end of the sequence if no such index exists.

---

7. Let $a_1, a_2, \ldots, a_n$ be a permutation of the integers $1, 2, \ldots, n$. Call a_i a "big" integer if $a_i > a_j$ for all $j > i$. Find the mean number of "big" integers over all permutations on the first n positive integers.

First Solution. If σ is a permutation, let $N_i(\sigma)$ be the number of "big" integers occurring at position i. Then $N_i(\sigma) = 0$ or 1. The average value of $N_i(\sigma)$ over all the permutations is $1/(n - i + 1)$ because after $a_1, a_2, \ldots, a_{i-1}$ have been selected, the question of whether or not a_i will be a big integer is whether or not a_i is the greatest among the $(n - i + 1)$ integers left.

Let the number of big integers in σ be $N(\sigma)$. Then $N(\sigma) = N_1(\sigma) + N_2(\sigma) + \cdots + N_n(\sigma)$, and the average value of $N(\sigma)$ over all the $n!$ permutations will be the sum of the average values of the separate terms, $N_i(\sigma)$, for $i = 1, 2, \ldots, n$. Hence this average is

$$1/n + 1/(n-1) + 1/(n-2) + \cdots + 1.$$

Second Solution. The average number of big integers in a permutation of n distinct integers is the same no matter what these integers are. Call this number A_n.

Given a permutation of $\{1, 2, \ldots n\}$, remove the element 1 and close up to obtain a permutation of $\{2, 3, \ldots, n\}$. This defines an n-to-1 mapping ξ of permutations of $\{1, 2, \ldots, n\}$ into permutations of $\{2, 3, \ldots, n\}$. If the element 1 appeared at the end of the original permutation σ, then $\xi(\sigma)$

has one big integer fewer than σ. In all other cases $\xi(\sigma)$ has the same number of big integers as σ. If $N(\sigma)$ denotes the number of big integers in σ, this means that

$$N(\sigma) = N(\xi(\sigma)) + 1 \quad \text{for } (n-1)! \text{ permutations } \sigma$$

and

$$N(\sigma) = N(\xi(\sigma)) \quad \text{for the remaining permutations.}$$

Hence

$$\sum_{\sigma} N(\sigma) = \sum_{\sigma} N(\xi(\sigma)) + (n-1)!.$$

Dividing through by $n!$ we see that

$$A_n = \frac{1}{n!} \sum_{\sigma} N(\xi(\sigma)) + 1/n.$$

Since the mapping ξ is always exactly n-to-1,

$$\frac{1}{n!} \sum_{\sigma} N(\xi(\sigma)) = A_{n-1}.$$

This gives the recurrence

$$A_n = A_{n-1} + 1/n.$$

Since $A_1 = 1$, it follows that

$$A_n = 1 + 1/2 + 1/3 + \cdots + 1/n.$$

THE TWENTIETH WILLIAM LOWELL PUTNAM MATHEMATICAL COMPETITION

November 21, 1959

Morning Session

1. Let n be a positive integer. Prove that $x^n - (1/x^n)$ is expressible as a polynomial in $x - (1/x)$ with real coefficients if and only if n is odd.

Solution. Let $z = x - 1/x$. If the desired representation

$$x^n - \left(\frac{1}{x}\right)^n = P_n(z)$$

exists, the coefficient of z^n in P_n must be one. Then equating the terms in $1/x^n$, we see that $-1/x^n = (-1/x)^n$, which implies that n is odd.

To show conversely that the representations exist for all odd n, we use induction. Clearly they exist for $n = 1, 3$ with $P_1(z) = z$, $P_3(z) = z^3 + 3z$. Suppose representations exist for $n = 1, 3, \ldots, 2k - 1$, where $k \geq 2$. Then

$$\left(x^2 + \frac{1}{x^2}\right)\left(x^{2k-1} - \frac{1}{x^{2k-1}}\right) = x^{2k+1} - \frac{1}{x^{2k+1}} + x^{2k-3} - \frac{1}{x^{2k-3}};$$

that is,

$$(z^2 + 2)P_{2k-1}(z) = x^{2k+1} - 1/x^{2k+1} + P_{2k-3}(z).$$

So we define

$$P_{2k+1}(y) = (y^2 + 2)P_{2k-1}(y) - P_{2k-3}(y)$$

and we have

$$P_{2k+1}(z) = x^{2k+1} - 1/x^{2k+1}$$

and there is such a polynomial for $n = 2k + 1$. Therefore $x^n - 1/x^n$ can be written as a polynomial in $x - 1/x$ with integer coefficients for all odd positive integers n.

REMARKS. For n a positive even integer, say $n = 2k$, there exists no function Q, polynomial or otherwise, such that

$$x^n - \frac{1}{x^n} = Q\left(x - \frac{1}{x}\right).$$

If such a function did exist, by putting $x = \frac{1}{2}$ and $x = -2$ successively we would have

$$\frac{1}{2^{2k}} - 2^{2k} = Q\left(-\frac{3}{2}\right) = 2^{2k} - \frac{1}{2^{2k}},$$

a contradiction.

It is easy to see that $x^n + 1/x^n$ can be represented as a polynomial in $x + 1/x$ for all positive integers n. Putting $x = e^{i\theta}$, this shows that $\cos n\theta$ is a polynomial in $\cos\theta$ for all n. The same substitution in the result of the problem shows that $\sin n\theta$ is a polynomial in $\sin\theta$ for odd n; but, of course, it is not for even n.

---

2. Prove that if the points in the complex plane corresponding to two distinct complex numbers z_1 and z_2 are two vertices of an equilateral triangle, then the third vertex corresponds to $-\omega z_1 - \omega^2 z_2$, where ω is an imaginary cube root of unity.

Solution. If the complex numbers z_1, z_2, z_3 are the vertices of an equilateral triangle, then

$$\frac{z_3 - z_1}{z_2 - z_1} = \alpha = e^{\pm \pi i/3}.$$

Hence $z_3 = (1 - \alpha)z_1 + \alpha z_2 = -\alpha^2 z_1 - \alpha^4 z_2$, because $\alpha^2 - \alpha + 1 = 0$ and $\alpha^3 = -1$ for either determination of α. Thus

$$z_3 = -\omega z_1 - \omega^2 z_2$$

where $\omega = \alpha^2 = e^{+2\pi i/3}$ is one of the two complex cube roots of unity.

Note that given z_1 and z_2 there are two ways to complete an equilateral triangle, corresponding to the two choices for ω.

---

3. Find all complex-valued functions f of a complex variable such that $f(z) + zf(1 - z) = 1 + z$ for all z.

Solution. We have

$$f(z) + zf(1 - z) = 1 + z \tag{1}$$

for all z. Replace z by $1 - z$ in this equation to get

$$f(1 - z) + (1 - z)f(z) = 2 - z. \tag{2}$$

Eliminating $f(1 - z)$ from (1) and (2), we get

$$(1 - z + z^2)f(z) = 1 - z + z^2.$$

Hence $f(z) = 1$ for all z except possibly the two values $e^{\pm \pi i/3}$ for which $1 - z + z^2 = 0$.

Let $\alpha = e^{i\pi/3}$, $\bar{\alpha} = e^{-i\pi/3}$. Then $\alpha + \bar{\alpha} = 1$, $\alpha\bar{\alpha} = 1$. Let $f(\alpha) = 1 + \beta$, $f(\bar{\alpha}) = 1 + \gamma$. Then equation (1) with $z = \alpha$ becomes

$$\beta + \alpha\gamma = 0,$$

so we choose β arbitrarily and set $\gamma = -\bar{\alpha}\beta$. With these values (1) is readily checked for $z = \alpha, \bar{\alpha}$. Hence a function f satisfies the given equation if and only if it has the form

$$f(\alpha) = 1 + \beta$$

$$f(\bar{\alpha}) = 1 - \bar{\alpha}\beta$$

$$f(z) = 1 \quad \text{for all other values of } z,$$

where β is any complex number.

---

4. If f and g are real-valued functions of one real variable, show that there exist numbers x and y such that $0 \le x \le 1$, $0 \le y \le 1$, and $|xy - f(x) - g(y)| \ge 1/4$.

Solution. Since

$$1 = (1 - f(1) - g(1)) + (f(1) + g(0)) + (f(0) + g(1)) - (f(0) + g(0))$$

one of the numbers

$$|1 - f(1) - g(1)|, \; |f(1) + g(0)|, \; |f(0) + g(1)|, \; |f(0) + g(0)|$$

is at least $\frac{1}{4}$. Thus relation (1) holds for at least one of the points (1, 1), (1, 0), (0, 1), or (0, 0).

---

5. A sparrow, flying horizontally in a straight line, is 50 feet directly below an eagle and 100 feet directly above a hawk. Both hawk and eagle fly directly toward the sparrow, reaching it simultaneously. The hawk flies twice as fast as the sparrow. How far does each bird fly? At what rate does the eagle fly?

Solution. Although not specifically stated, it is clearly intended that each of the three birds flies at a uniform speed. So we consider the general problem of a predator flying with speed v in pursuit of a prey flying along a straight path with speed rv, $r < 1$.

Let the prey start at (0, 0) and move along the x-axis so that his position at time t is $(rvt, 0)$. Let the predator start at $(0, h)$, $h > 0$. If at time t the predator is at (x, y), then the condition that he always flies directly toward his prey is that

(1) $$\frac{dx/dt}{dy/dt} = \frac{rvt - x}{-y}.$$

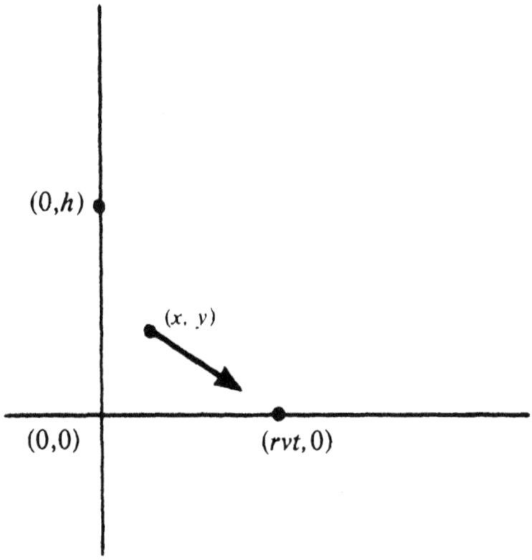

As long as y remains positive, $dy/dt < 0$, so we may choose y as the independent variable. Then (1) becomes

(2) $$y\frac{dx}{dy} = x - rvt.$$

The condition that the predator has constant speed v is

$$\sqrt{\left(\frac{dx}{dt}\right)^2 + \left(\frac{dy}{dt}\right)^2} = v,$$

which we can write as

$$-\sqrt{\left(\frac{dx}{dy}\right)^2 + 1} = v\,\frac{dt}{dy}, \tag{3}$$

where we have introduced a minus sign because dt/dy is negative.

We differentiate (2) to get

$$y\,\frac{d^2x}{dy^2} + \frac{dx}{dy} = \frac{dx}{dy} - rv\,\frac{dt}{dy},$$

and using (3) we have

$$y\,\frac{d^2x}{dy^2} = r\sqrt{\left(\frac{dx}{dy}\right)^2 + 1}.$$

We can write this as the first-order differential equation

$$y\,\frac{dz}{dy} = r\sqrt{1 + z^2},$$

where $z = dx/dy$. Separating the variables we have

$$r\,\frac{dy}{y} = \frac{dz}{\sqrt{1 + z^2}},$$

whence

$$r \log y = \log\left(z + \sqrt{1 + z^2}\right) + C. \tag{4}$$

Since $z = dx/dy = 0$ when $y = h$, we find $C = r \log h$. Then (4) can be rewritten

$$\left(\frac{y}{h}\right)^r = z + \sqrt{1 + z^2}$$

and solved for $z(= dx/dy)$ to get

$$(5) \qquad 2\frac{dx}{dy} = \left(\frac{y}{h}\right)^r - \left(\frac{y}{h}\right)^{-r}.$$

This can be integrated again using the initial condition $x = 0$ when $y = h$ to give

$$(6) \qquad 2x = \frac{h}{1+r}\left(\frac{y}{h}\right)^{1+r} - \frac{h}{1-r}\left(\frac{y}{h}\right)^{1-r} + \frac{2rh}{1-r^2}.$$

This is valid only for $y > 0$. Recalling that $0 < r < 1$, we see from (6) that

$$x \to \frac{rh}{1-r^2}$$

as $y \to 0$. We see from (5) that $y(dx/dy) \to 0$ and then from (2) that

$$x - rvt \to 0$$

as $y \to 0$. It follows that the predator catches his prey at the point $(rh/(1-r^2), 0)$ when $t = h/(1-r^2)v$. At this time the prey has flown $rh/(1-r^2)$ and the predator, $h/(1-r^2)$.

Now for the hawk and the sparrow, $r = \frac{1}{2}$, $h = 100$ ft., so the sparrow flies 200/3 ft., and the hawk, 400/3 ft. [Note that since the sparrow is above the hawk we would have to choose a coordinate system with y positive in the downward direction.]

In the case of the eagle, $h = 50$ ft. and r is unknown, but

$$\frac{r50}{1-r^2} = \frac{200}{3};$$

hence $4r^2 + 3r - 4 = 0$ and

$$r = \frac{-3+\sqrt{73}}{8}$$

since r must be positive.

The speed of the eagle is $1/r$ times the speed of the sparrow, or

$$\frac{v(3+\sqrt{73})}{16} \approx .721\, v$$

where v is the speed of the hawk.

Since the time of capture is $400/3v$, the eagle flies a distance of

$$\frac{400}{3}\left(\frac{3+\sqrt{73}}{16}\right) \approx 96.2 \text{ ft.}$$

REMARK. The problem appears as Advanced Problem 3573 in the *American Mathematical Monthly*, vol. 39 (1932), page 454.

An entertaining paper by Arthur Bernhart, "Curves of Pursuit," *Scripta Mathematica*, 20, nos. 3-4 (1954), outlines some of the history of the pursuit problem.

---

6. Let m and n be integers greater than 1, and $a_1, \ldots, a_{m+1}$ real numbers. Prove that there exist real n by n matrices $A_1, \ldots, A_m$ such that (i) Det $(A_j) = a_j$ for $j = 1, \ldots, m$, and (ii) Det $(A_1 + \cdots + A_m) = a_{m+1}$.

Solution. Let

$$A_1 = \begin{pmatrix} a_1 & 0 & & \\ 1 & 1 & & 0 \\ & & 1 & \\ & 0 & & \ddots \\ & & & & 1 \end{pmatrix}, \qquad A_2 = \begin{pmatrix} a_2 & b & & \\ 0 & 1 & & 0 \\ & & 1 & \\ & 0 & & \ddots \\ & & & & 1 \end{pmatrix},$$

$$A_i = \begin{pmatrix} a_i & & & \\ & 1 & & 0 \\ & & 1 & \\ & 0 & & \ddots \\ & & & & 1 \end{pmatrix}, \qquad i = 3, 4, \ldots, m,$$

where b is to be determined. Evidently, det $(A_j) = a_j$ for $j = 1, \ldots, m$. Then

$$A_1 + A_2 + \cdots + A_m = \begin{pmatrix} s & b & & \\ 1 & m & & 0 \\ & & m & \\ & 0 & & \ddots \\ & & & & m \end{pmatrix},$$

where $s = a_1 + a_2 + \cdots + a_m$.

We have

$$\det(A_1 + A_2 + \cdots + A_m) = (sm - b)m^{n-2}.$$

Hence we set $b = sm - a_{m+1}m^{-n+2}$ to get the desired value a_{m+1}.

REMARK. There are many other possibilities. The difficulty of the problem is due largely to its highly underdetermined character.

---

7. If f is a real-valued function of one real variable which has a continuous derivative on the closed interval $[a, b]$ and for which there is no $x \in [a, b]$ such that $f(x) = f'(x) = 0$, then prove that there is a function g with continuous first derivative on $[a, b]$ such that $fg' - f'g$ is positive on $[a, b]$.

Solution. Let f: $[a, b] \to \mathbf{R}$ be a function with a continuous derivative such that f and f' do not vanish at the same point. Let $S = \{x : f(x) = 0\}$. Then S is a finite set. For if not, S would have an accumulation point $x_0 \in [a, b]$, and there would exist a sequence $\{x_n\}$ in S such that $x_n \to x_0$ and $x_n \neq x_0$. Then $f(x_0) = 0$ by continuity and

$$f'(x_0) = \lim_{n \to \infty} \frac{f(x_n) - f(x_0)}{x_n - x_0} = 0$$

contrary to our hypothesis that f and f' do not vanish at the same point.

Since f' does not vanish on S, there is a polynomial h such that $f'(x)h(x) = -1$ for all $x \in S$. For each positive number c define

$$g_c(x) = xf(x) + ch(x)$$

and

$$\begin{aligned} w_c(x) &= f(x)g_c'(x) - f'(x)g_c(x) \\ &= f(x)^2 + c(f(x)h'(x) - f'(x)h(x)). \end{aligned}$$

We shall prove that, for some sufficiently small c, w_c is positive throughout $[a, b]$. Then g_c is the required function.

By construction $fh' - f'h$ is positive (indeed equal to 1) for the points of S. By continuity $fh' - f'h$ remains positive on a set $T \supseteq S$, where T is open relative to $[a, b]$. Moreover, $|fh' - f'h|$ is bounded on the interval $[a, b]$; say by M. Since f^2 is continuous and positive on the compact set

$[a, b] - T$, it is bounded away from zero on this set; say by ϵ. Then, if $0 < c < \epsilon/M$,

$$w_c(x) \geq f(x)^2 - c\,|f(x)h'(x) - f'(x)h(x)| > \epsilon - cM > 0$$

for $x \in [a, b] - T$, while

$$w_c(x) \geq c(f(x)h'(x) - f'(x)h(x)) > 0,$$

if $x \in T$. Thus for $0 < c < \epsilon/M$, w_c is positive on all of $[a, b]$ and the corresponding function g_c has the desired property.

REMARK. If we are willing to assume that f is of class C^2, (i.e., f'' exists and is continuous) then the problem is much easier. There are then continuous functions p and q such that f is a solution of

$$y'' + p(x)y' + q(x)y = 0.$$

For example, $p = \dfrac{-f''f'}{(f')^2 + (f)^2}$, $q = \dfrac{-f''f}{(f')^2 + (f)^2}$.

This second-order differential equation has a second solution linearly independent of f; call it g. Then the Wronskian

$$fg' - f'g$$

cannot vanish on $[a, b]$. Changing the sign of g if necessary, we can make $fg' - f'g > 0$ on $[a, b]$.

Afternoon Session

1. Let each of m distinct points on the positive part of the X-axis be joined to n distinct points on the positive part of the Y-axis. Obtain a formula for the number of intersection points of these segments (exclusive of endpoints), assuming that no three of the segments are concurrent.

Solution. Each pair of points on the X-axis together with each pair of points on the Y-axis determine a convex quadrilateral whose diagonals meet somewhere in the first quadrant. Conversely, each intersection point arises in this way. Since no three segments are concurrent, except at the endpoints, each intersection point arises uniquely. There are, therefore,

$$\binom{m}{2}\binom{n}{2} = mn(m-1)(n-1)/4$$

points of intersection. (Compare Problem A.M. 4 of the Fifteenth Competition.)

---

2. Let c be a positive real number. Prove that c can be expressed in infinitely many ways as a sum of infinitely many distinct terms selected from the sequence

$$1/10,\ 1/20,\ \ldots,\ 1/10n,\ \ldots.$$

Solution. We shall prove a more general result: Suppose c is a positive number and $a(1), a(2), \ldots$ is any sequence of positive numbers such that $a(n) \to 0$ as $n \to 0$ and

$$\sum_{n=1}^{\infty} a(n) \tag{1}$$

diverges. Then there exist infinitely many strictly increasing sequences of positive integers, $n_1, n_2, \ldots$ such that

$$\sum_{i=1}^{\infty} a(n_i) = c.$$

Let k be any integer such that $a(k) < c$. Define the sequence $n_1, n_2, \ldots$ recursively by $n_1 = k$, n_2 is the least integer exceeding n_1 such that $a(n_1) + a(n_2) < c, \ldots$, in general n_t is the least integer exceeding n_{t-1} such that

$$a(n_1) + a(n_2) + \cdots + a(n_{t-1}) + a(n_t) < c.$$

Since $a(n) \to 0$ as $n \to \infty$ there will always be such an integer. Now

$$\sum_{i=1}^{\infty} a(n_i) \tag{2}$$

converges to a number not greater than c because the terms are positive and c is an upper bound for the partial sums. We shall prove, in fact, that (2) does converge to c.

Let $\epsilon > 0$ be given. Choose $p > k$ so that $a(n) < \epsilon$ for all $n > p$. Since (1) diverges, there must be infinitely many terms of (1) which do not appear in (2). Suppose that q is an omitted index exceeding p; i.e., $n_i \neq q$ for any i and $q > p$. Since $n_1 < q$ and $n_i \to \infty$, we can choose r so that $n_{r-1} < q < n_r$. In choosing n_r we rejected the choice of q, hence

$$a(n_1) + a(n_2) + \cdots + a(n_{r-1}) + a(q) \geq c.$$

Therefore,

$$c - \sum_{i=1}^{r-1} a(n_i) \leq a(q) < \epsilon.$$

It follows that (2) converges to c.

Since we have constructed a sequence starting with k for any choice of k such that $a(k) < c$, and since there are infinitely many such choices, it follows that there are infinitely many sequences of the required type.

In the given problem $a(n) = 1/10n$, and the above result is applicable since $\Sigma\ 1/10n$ diverges.

REMARK. We can always represent c in infinitely many ways so that the different representations use disjoint subsets of the original terms. For suppose we have found m representations using all different terms. Removing all terms already used from the a sequence leaves a new sequence tending to zero with divergent sum, and from this sequence we can construct an $(m + 1)$st representation.

---

3. Give an example of a continuous real-valued function f from $[0, 1]$ to $[0, 1]$ which takes on every value in $[0, 1]$ an infinite number of times.

First Solution. It is well known that there exists a continuous surjective map, g: $[0, 1] \to [0, 1] \times [0, 1]$; for example the Peano space filling curve. If π denotes the projection of the unit square on its first coordinate, we can take $f = \pi \circ g$ to get a continuous function from $[0, 1]$ to $[0, 1]$ that takes each value uncountably often.

To obtain a more explicit function with this property, we may proceed as follows: Let X be the space of sequences of 0's and 1's with the product topology, $\{0, 1\}$ being taken as a discrete space. There is a continuous surjective map Φ: $X \to X$ such that the inverse image of every point is uncountable; for example, let $\Phi(x_1, x_2, x_3, x_4, x_5, \ldots) = x_1, x_3, x_5, \ldots$. There is an injective continuous map, $\eta: X \to [0, 1]$; for example, let

$$\eta(x_1, x_2, x_3, \ldots) = 2 \sum_n \frac{1}{3^n} x_n.$$

There is a surjective continuous map $\theta: X \to [0, 1]$; for example, let $\theta(x_1, x_2, x_3, \ldots) = \Sigma_n\ 1/2^n x_n$.

Now consider the diagram

$$\begin{array}{ccc} X & \xrightarrow{\Phi} & X \\ \eta\downarrow & & \downarrow\theta \\ [0,1] & \xrightarrow{f} & [0,1]. \end{array}$$

Since X is a compact set and since η is injective, $\eta(X)$ is a compact, and therefore closed, subset of $[0, 1]$ and η^{-1} is a continuous map from $\eta(X)$ to X. We note that $\eta(X)$ is the well-known middle third set of Cantor. By the Tietze extension theorem, there is a continuous extension f of $\theta\Phi\eta^{-1}$ over $[0, 1]$. This map, f, makes the above diagram commutative. If $\alpha \in [0, 1]$, then $\theta^{-1}(\alpha)$ is non-empty, so $\Phi^{-1}\theta^{-1}(\alpha)$ is uncountable. Since η is injective, $\eta\Phi^{-1}\theta^{-1}(\alpha)$ is uncountable. Finally

$$f^{-1}(\alpha) \supseteq \eta\Phi^{-1}\theta^{-1}(\alpha),$$

so $f^{-1}(\alpha)$ is uncountable.

To make the function f completely explicit, we may define f on the complement of $\eta(X)$ by making it linear on each maximal interval in $[0, 1] - \eta(X)$.

The Tietze Extension Theorem may be stated as follows: *Let A be a closed set in $\mathbf{R}^n$ and let f be a continuous, bounded, real-valued function defined on A. Then there exists a continuous, real-valued g defined on $\mathbf{R}^n$ which is an extension of f and is such that*

$$\text{lub}\,\{g(x) \mid x \in \mathbf{R}^n\} = \text{lub}\,\{f(x) \mid x \in A\}$$

and

$$\text{glb}\,\{g(x) \mid x \in \mathbf{R}^n\} = \text{glb}\,\{f(x) \mid x \in A\}.$$

See Haaser, Lasalle and Sullivan, *Mathematical Analysis*, vol. 2, Blaisdell, Waltham, Mass., 1964, pages 354–356.

The theorem remains valid for any closed set A in a normal (i.e., T_1 and T_4) topological space.

Second Solution. A function that almost satisfies the conditions can be defined by an infinite trigonometric series and then modified slightly to meet all the requirements.

Choose p so that $\frac{1}{9} < p < 1$ and put

$$h(x) = (1 - p) \sum_{k=1}^{\infty} p^{k-1} \cos(3^{k^2} \pi x) \quad \text{for } 0 \leq x \leq 1.$$

Since the terms of this series are uniformly dominated by those of the geometric series $(1 - p)\,\Sigma p^{k-1}$, h is a continuous function and $|h(x)| \leq 1$ for all x. Moreover, h achieves this bound only for $x = 0, \frac{1}{3}, \frac{2}{3}, 1$, where it takes the values $+1$ and -1 alternately. We shall prove that h takes every other value in the interval $[-1, +1]$ infinitely often.

Let $\alpha \in (-1, +1)$. Put

$$h(x) = h_n(x) + R_n(x)$$

where

$$h_n(x) = (1 - p) \sum_{k=1}^{n-1} p^{k-1} \cos(3^{k^2} \pi x)$$

and

$$R_n(x) = (1 - p) \sum_{k=n}^{\infty} p^{k-1} \cos(3^{k^2} \pi x).$$

There is an integer q such that, for all $n \geq q$,

$$h_n(0) > \alpha > h_n(1).$$

Fix an integer $n \geq q$. Then h_n takes the value α somewhere, say at t_n.

Since

$$|h_n'(x)| \leq (1 - p) \sum_{k=1}^{n-1} p^{k-1} 3^{k^2} \pi \leq 3^{(n-1)^2} \pi$$

for any x, we know that

$$|h_n(x) - \alpha| < p^n(1 - 2p)$$

for

$$|x - t_n| < \frac{1}{\pi} p^n(1 - 2p) 3^{-(n-1)^2}. \tag{1}$$

Let I_n be that part of this interval that lies in $[0, 1]$. The length of I_n is at least the right member of (1).

The term $(1 - p)p^{n-1} \cos(3^{n^2}\pi x)$ oscillates from $-(1 - p)p^{n-1}$ to $+(1 - p)p^{n-1}$ on every interval of length $2 \cdot 3^{-n^2}$, while $|R_{n+1}(x)|$ is uniformly bounded by p^n. Hence $R_n(x)$ oscillates at least between

$-(1 - 2p)p^{n-1}$ and $(1 - 2p)p^{n-1}$ on every interval of length $2\cdot 3^{-n^2}$. The interval I_n contains A_n non-overlapping intervals of this length, where

$$A_n = \left[\frac{\text{length } I_n}{2\cdot 3^{-n^2}}\right] \geq \left[\left(\frac{1-2p}{6\pi}\right)(9p)^n\right].$$

([ ] denotes the greatest integer function.) On each of these little intervals, $h = h_n + R_n$ takes the value α at least once. Thus h takes the value α at least A_n times. Since we can take n as large as we please and $(9p)^n \to \infty$, h takes the value α infinitely often, as claimed.

There are many ways to convert h to a function with range $[0, 1]$ taking every value infinitely often. For example,

$$f(x) = \sin^2(2h(x))$$

defines such a function.

The construction of h exemplifies a standard way to define a pathological function—a rapidly convergent sum of even more rapid oscillating terms. It is not hard to prove, for example, that h is nowhere differentiable.

---

4. Given the following matrix of 25 elements

$$\begin{pmatrix} 11 & 17 & 25 & 19 & 16 \\ 24 & 10 & 13 & 15 & 3 \\ 12 & 5 & 14 & 2 & 18 \\ 23 & 4 & 1 & 8 & 22 \\ 6 & 20 & 7 & 21 & 9 \end{pmatrix},$$

choose five of these elements, no two coming from the same row or column, in such a way that the minimum of these five elements is as large as possible. Prove that your answer is correct.

Solution. Since the set of border elements of the matrix is the union of two rows and two columns, we may choose at most four elements from the border and must choose at least one element from the central 3×3 submatrix. Since the largest element in this central submatrix is 15, there is no admissible choice for which the minimum exceeds 15. But the choice 25, 15, 18, 23, 20 satisfies the conditions and has minimum 15.

REMARK. It is easy to check that there is just one choice that realizes the minimum value of 15.

The problem is a variation on what is called the assignment problem: Given an $n \times n$ matrix, choose n elements, one from each row and one from each column, so that the sum of these elements is as large as possible. The name derives from the following example. Suppose a company has n employees and n jobs. If a_{ij} is the utility of assigning the ith employee to the jth job, then the most desirable assignment is found by solving the assignment problem for the matrix (a_{ij}).

For the given matrix, it happens that the same choice is also the unique solution to the assignment problem.

For more information on the assignment problem, see G. Dantzig, *Linear Programming and Extensions*, Princeton University Press, 1963, page 316 ff., or Reinfeld and Vogel, *Mathematical Programming*, Prentice-Hall, Englewood Cliffs, N.J., 1958, page 238 ff.

---

5. Find the equation of the smallest sphere which is tangent to both of the lines: (i) $x = t + 1$, $y = 2t + 4$, $z = -3t + 5$, and (ii) $x = 4t - 12$, $y = -t + 8$, $z = t + 17$.

First Solution. Let the given lines be l and m. Then there is a unique segment PQ perpendicular to both lines with P on l and Q on m. The required sphere has PQ as its diameter.

Suppose l and m are given in terms of a parameter t by $\mathbf{a} + t\mathbf{v}$ and $\mathbf{b} + t\mathbf{w}$, respectively, where $\mathbf{a}$, $\mathbf{b}$, $\mathbf{v}$, and $\mathbf{w}$ are vectors. If the lines are not parallel, $\mathbf{v}$ and $\mathbf{w}$ are linearly independent. Since PQ is perpendicular to both lines, it has the direction of $\mathbf{v} \times \mathbf{w}$, say $\overrightarrow{PQ} = \rho\mathbf{v} \times \mathbf{w}$, where ρ is a scalar. Let P and Q be the points $\mathbf{a} + \sigma\mathbf{v}$ and $\mathbf{b} + \tau\mathbf{w}$, respectively. Then $\overrightarrow{PQ} = \mathbf{b} - \mathbf{a} - \sigma\mathbf{v} + \tau\mathbf{w}$ and

$$\mathbf{a} - \mathbf{b} = -\rho(\mathbf{v} \times \mathbf{w}) - \sigma\mathbf{v} + \tau\mathbf{w}.$$

Hence we can calculate ρ, σ, and τ by expressing $\mathbf{a} - \mathbf{b}$ in terms of the independent vectors $\mathbf{v} \times \mathbf{w}$, $\mathbf{v}$ and $\mathbf{w}$. Then the center of the required sphere is at

$$\mathbf{a} + \sigma\mathbf{v} + \frac{1}{2}\rho(\mathbf{v} \times \mathbf{w})$$

and its radius is

$$\frac{1}{2}|\rho|\,\|\mathbf{v} \times \mathbf{w}\|.$$

For the example in question, $\mathbf{a} = \langle 1, 4, 5\rangle$ $\mathbf{b} = \langle -12, 8, 17\rangle$, $\mathbf{v} = \langle 1, 2, -3\rangle$, $\mathbf{w} = \langle 4, -1, 1\rangle$ and $\mathbf{v} \times \mathbf{w} = \langle -1, -13, -9\rangle$. Then ρ, σ and τ are found from the equations

$$13 = \rho - \sigma + 4\tau$$

$$-4 = 13\rho - 2\sigma - \tau$$

$$-12 = 9\rho + 3\sigma + \tau$$

which give

$$\rho = \frac{-147}{251}, \qquad \sigma = \frac{-782}{251}, \qquad \tau = \frac{657}{251}.$$

The center of the sphere is therefore at

$$\langle 1, 4, 5\rangle - \frac{782}{251}\langle 1, 2, -3\rangle - \frac{147}{502}\langle -1, -13, -9\rangle$$

$$= \frac{1}{502}\langle -915, 791, 8525\rangle.$$

The square of the radius is

$$\frac{1}{4}\rho^2 \|\vec{\mathbf{v}} \times \vec{\mathbf{w}}\|^2 = \left(\frac{147}{502}\right)^2 (251) = \frac{147^2}{1004}.$$

The equation of the sphere is

$$(502x + 915)^2 + (502y - 791)^2 + (502z - 8525)^2 = 251(147)^2.$$

Second Solution. Let P and Q be chosen on l and m, respectively, so that PQ is as short as possible. Then PQ is perpendicular to each of the lines l and m. The sphere with PQ as diameter is the required sphere, for it is tangent to l and m and no smaller sphere intersects both l and m.

If P is the point $\langle s + 1, 2s + 4, -3s + 5\rangle$ and Q is the point $\langle 4t - 12, -t + 8, t + 17\rangle$, then

$$|PQ|^2 = (s - 4t + 13)^2 + (2s + t - 4)^2 + (-3s - t - 12)^2$$

$$= 14s^2 + 2st + 18t^2 + 82s - 88t + 329.$$

The quadratic terms of this function are positive definite, so the minimum is achieved at the unique point at which both partial derivatives vanish.

Hence the equations

$$28s + 2t + 82 = 0$$

$$2s + 36t - 88 = 0$$

determine the desired values of s and t, which turn out to be $-782/251$ and $657/251$, respectively.

We can now find P and Q and continue as in the first solution.

REMARK. The complicated arithmetic was not intended by the examination committee. The second equation of (i) was intended to be $y = 2t - 4$. Then in the first solution, we find $\rho = -1$, $\sigma = -2$, $\tau = +3$. In the second solution the parameters turn out to be $s = -2$ and $t = 3$.

---

6. Prove that, if x and y are positive irrationals such that $1/x + 1/y = 1$, then the sequences $[x], [2x], \ldots, [nx], \ldots$ and $[y], [2y], \ldots, [ny], \ldots$ together include every positive integer exactly once. (The notation $[x]$ means the largest integer not exceeding x.)

Solution. Evidently $x > 1$, and therefore the numbers $0, x, 2x, 3x, \ldots$ all differ by more than one. Hence the integers

$$[x], [2x], [3x], \ldots \tag{1}$$

are all positive and different. Similarly the integers

$$[y], [2y], [3y], \ldots \tag{2}$$

are all positive and different.

Suppose some integer p appears in both (1) and (2). Say $p = [ax] = [by]$ where a and b are positive integers. Then $p < ax < p + 1$ and $p < by < p + 1$. (The possibilities $p = ax$ and $p = by$ are excluded here, and in similar places elsewhere, because x and y are irrational). Then

$$p = \frac{p}{x} + \frac{p}{y} < a + b < \frac{p + 1}{x} + \frac{p + 1}{y} = p + 1$$

and we have found an integer, $a + b$, between the integers p and $p + 1$. This is impossible. Hence no integers appear in both (1) and (2).

Suppose some positive integer p is not included in either (1) or (2). Then we can find positive integers a and b such that

$$ax < p < p + 1 < (a + 1)x,$$

and

$$by < p < p + 1 < (b + 1)y.$$

Then

$$\frac{a}{p} + \frac{b}{p} < \frac{1}{x} + \frac{1}{y} = 1,$$

so $a + b < p$. Also

$$\frac{a+1}{p+1} + \frac{b+1}{p+1} > \frac{1}{x} + \frac{1}{y} = 1,$$

so $a + b + 2 > p + 1$. Thus we have found two integers, p and $p + 1$, between the integers $a + b$ and $a + b + 2$, which is impossible. We conclude that every integer appears either in (1) or in (2).

REMARK. This is sometimes called Beatty's problem, after Samuel Beatty (1881-1970). In a slightly different form it appeared as Problem 3117, *American Mathematical Monthly*, vol. 34 (1927), pages 158-159. Howard Grossman, "A Set Containing All Integers," *American Mathematical Monthly*, vol. 69 (1962), pages 532-533, gives a proof by analyzing lattice points. A. S. Fraenkel, "The Bracket Function and Complementary Sets of Integers," *Canadian Journal of Mathematics*, vol. 21 (Jan. 1969), pages 6-27, gives a history, a bibliography, and a generalization of the problem.

---

7. For each positive integer n, let f_n be a real-valued symmetric function of n real variables. Suppose that for all n and for all real numbers $x_1, \ldots, x_{n+1}, y$, it is true that

(1) $f_n(x_1 + y, \ldots, x_n + y) = f_n(x_1, \ldots, x_n) + y$,

(2) $f_n(-x_1, \ldots, -x_n) = -f_n(x_1, \ldots x_n)$,

(3) $f_{n+1}(f_n(x_1, \ldots, x_n), \ldots, f_n(x_1, \ldots, x_n), x_{n+1}) = f_{n+1}(x_1, \ldots, x_{n+1})$.

Prove that

(4) $f_n(x_1, \ldots, x_n) = (x_1 + \cdots + x_n)/n$.

Solution. Since $f_1(0) = -f_1(0)$ by (2), we see

that (5) $$f_1(0) = 0.$$

By (1) and (5) we have

$$f_1(x) = f_1(0) + x = x,$$

and this relation is (4) for the case $n = 1$.

For any integer $n > 1$ and any number c

$$\begin{aligned} f_n(c, 0, \ldots, 0, -c) &= f_n(-c, 0, \ldots, 0, c) \\ &= -f_n(c, 0, \ldots, 0, -c) \end{aligned}$$

by symmetry and by (2); hence

(6) $$f_n(c, 0, \ldots, 0, -c) = 0.$$

We now make the inductive hypothesis that (4) is valid for $n = k$. Then equation (3) says that $f_{k+1}(x_1, x_2, \ldots, x_{k+1})$ depends only on x_{k+1} and the sum of $x_1, \ldots, x_k$. Hence, if $a_1 + a_2 + \cdots + a_{k+1} = 0$, then

(7) $$f_{k+1}(a_1, a_2, \ldots, a_{k+1}) = f_{k+1}(-a_{k+1}, 0, \ldots, 0, a_{k+1}) = 0,$$

using (6).

Now let $x_1, x_2, \ldots, x_{k+1}$ be any numbers and let u be their average. Set $a_i = x_i - u$. Then $a_1 + a_2 + \cdots + a_{k+1} = 0$, and

$$f_{k+1}(x_1, x_2, \ldots, x_{k+1}) = f_{k+1}(a_1, a_2, \ldots, a_{k+1}) + u = u$$

by (1) and (7). This is (4) for $n = k + 1$. Therefore (4) is true for all integers by induction.

THE TWENTY-FIRST WILLIAM LOWELL PUTNAM MATHEMATICAL COMPETITION

December 3, 1960

Morning Session

1. Let n be a given positive integer. How many solutions are there in ordered positive integer pairs (x, y) to the equation

$$\frac{xy}{x+y} = n?$$

Solution. The given equation is equivalent to

$$(x-n)(y-n) = n^2. \tag{1}$$

Evidently, either both $x > n$ and $y > n$, or $x < n$ and $y < n$. There are no solutions satisfying both $0 < x < n$, $0 < y < n$ because then $|x - n||y - n| < n \cdot n = n^2$. Hence we need only look for integral solutions of (1) with $x - n > 0$, $y - n > 0$. It is clear that there are as many solutions as there are ordered factorizations of n^2 into two factors.

If the prime factorization of n is $p_1^{\alpha_1} p_2^{\alpha_2} \cdots p_k^{\alpha_k}$, then

$$n^2 = p_1^{2\alpha_1} p_2^{2\alpha_2} \cdots p_k^{2\alpha_k}$$

and the number of ordered factorizations of n^2 is

$$(2\alpha_1 + 1)(2\alpha_2 + 1) \cdots (2\alpha_k + 1).$$

---

2. Show that if three points are inside a closed square of unit side, then some pair of them are within $\sqrt{6} - \sqrt{2}$ units apart.

Solution. If three points are in a closed square region, it is possible to translate them together so that one becomes a corner while the others remain inside the given square. (There is a smallest rectangle R (possibly degenerate) that contains the given points and has sides parallel to those of the square. Each side of R contains at least one of the given points, so one of the given points must fall on two sides of R, i.e., it is a vertex of R. Then all of R can be translated inside the square so that this vertex becomes a corner.) The translation does not affect the mutual distances between the three points, so we may assume that one of the three given points is O in the square $OABC$. We choose axes as shown.

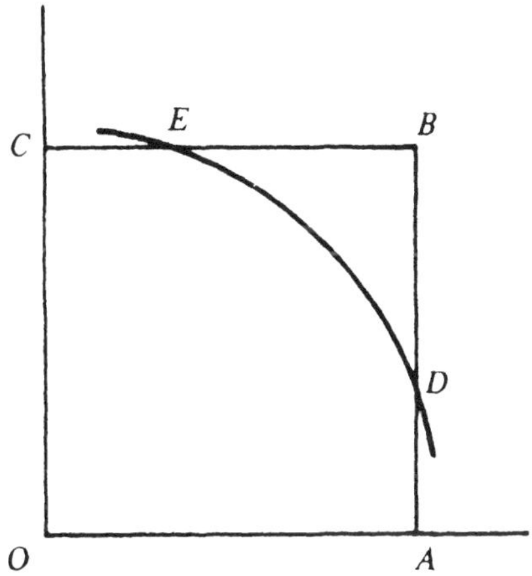

Assume that the three points O, P, Q are all separated by more than $\alpha = \sqrt{6} - \sqrt{2}$. Then P and Q lie outside the circle $x^2 + y^2 = \alpha^2$. This circle crosses AB and BC at $D = \langle 1, \sqrt{\alpha^2 - 1}\rangle$ and $E = \langle \sqrt{\alpha^2 - 1}, 1\rangle$, respectively. Now P and Q, being in the right triangle DBE, are no farther apart than D and E. But

$$\begin{aligned} |DE|^2 &= 2(1 - \sqrt{\alpha^2 - 1})^2 \\ &= 2(1 - \sqrt{7 - 4\sqrt{3}})^2 = 2(1 - (2 - \sqrt{3}))^2 \\ &= 2(\sqrt{3} - 1)^2 = (\sqrt{6} - \sqrt{2})^2 = \alpha^2. \end{aligned}$$

So $|PQ| \leq \alpha$, a contradiction.

This proves that, of any three points in a closed unit square, some two are no farther than α apart. If the three points are strictly inside the square, they are also inside a square of side $s < 1$, and hence some two of them are at distance at most $\alpha s < \alpha$; that is, some two are less than α apart.

REMARKS. An interesting discussion of this and related problems, together with some history and bibliography, appears in an article by Benjamin L. Schwartz, "Separating Points in a Rectangle," *Mathematics Magazine*, vol. 46 (1973), pages 62–70.

3. Show that if t_1, t_2, t_3, t_4, t_5 are real numbers, then

$$\sum_{j=1}^{5} (1 - t_j)\exp\left(\sum_{k=1}^{j} t_k\right) \leq e^{e^{e^e}}.$$

Solution. We first show that

$$(1) \qquad (1 - s + a)e^s \leq e^a$$

for any choice of s. The derivative of the function on the left with respect to s is

$$(a - s)e^s$$

which vanishes only for $s = a$, and this critical point is easily seen to be a maximum point (since $\lim_{s\to-\infty} (1 - s + a)e^s = 0$ and

$$\lim_{s\to+\infty} (1 - s + a)e^s = -\infty).$$

The claimed inequality (1) follows.

Then, taking $a = 0$, $s = t_5$ in (1), we have

$$(1 - t_5)e^{t_5} \leq e^0 = 1.$$

Multiplying by the positive number e^{t_4} and adding $(1 - t_4)e^{t_4}$, we find

$$(1 - t_4)e^{t_4} + (1 - t_5)e^{t_4+t_5} \leq (1 - t_4)e^{t_4} + e^{t_4} \leq e,$$

the last step by (1) with $a = 1$.

We continue this process.

$$(1 - t_3)e^{t_3} + (1 - t_4)e^{t_3+t_4} + (1 - t_5)e^{t_3+t_4+t_5} \leq (1 - t_3)e^{t_3} + e \cdot e^{t_3} \leq e^e.$$

Similarly

$$\begin{aligned}(1 - t_2)e^{t_2} + (1 - t_3)e^{t_2+t_3} + (1 - t_4)e^{t_2+t_3+t_4} + (1 - t_5)e^{t_2+t_3+t_4+t_5} \\ \leq (1 - t_2)e^{t_2} + e^e e^{t_2} \leq e^{e^e}\end{aligned}$$

and finally

$$\sum_{j=1}^{5} (1 - t_j) \exp\left(\sum_{k=1}^{j} t_k\right) \leq e^{e^{e^e}}.$$

4. Given two points in the plane, P and Q, at fixed distances from a line L, and on the same side of the line, as indicated, the problem is to find a third point R so that $PR + RQ + RS$ is a minimum, where RS is perpendicular to L. Consider all cases.

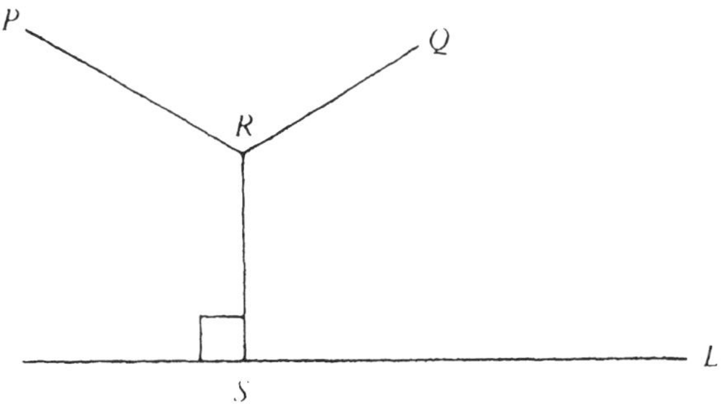

Solution. We take axes in the plane so that L is the x-axis, $P = (0, a)$, and $Q = (c, b)$. Without loss of generality we may assume $c \geq 0$, $0 < b \leq a$. Let $Q^* = (c, -b)$ be the reflection of Q in L, and let K be the intersection of line PQ^* with L. For any point X on L, we have

$$PX + QX = PX + XQ^* \geq PQ^* = PK + KQ^* = PK + QK$$

with strict inequality unless $X = K$.

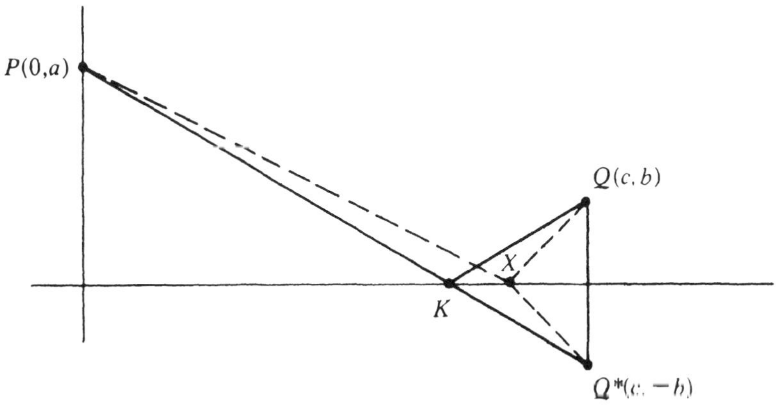

Consider the infinite strip in the plane given by $x \geq 0$, $0 < y \leq a$, and divide it into three parts as shown.

We will consider three cases of the problem.

Case 1. $c \leq \sqrt{3}(a - b)$; i.e., Q falls in the closed triangle POA.

Case 2. $\sqrt{3}(a - b) < c < \sqrt{3}(a + b)$; i.e., Q falls in the open triangle PAB or on the open segment PB.

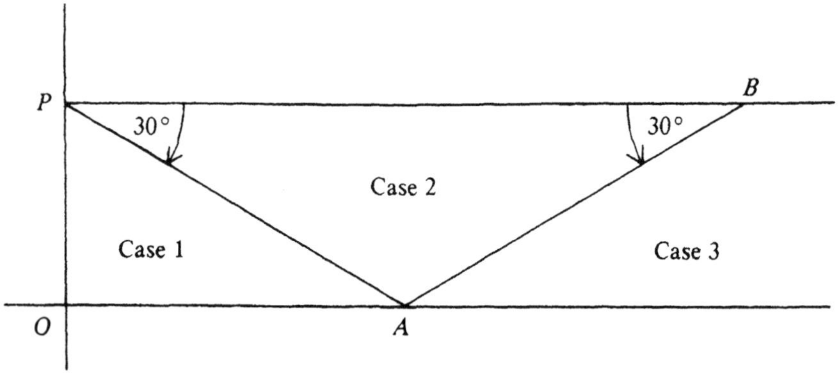

Case 3. $c \geq \sqrt{3}\,(a+b)$; i.e., Q falls on or to the right of AB.

Let $\sigma(X) = PX + QX + \tau(X)$, where $\tau(X)$ is the distance from X to the line L. We seek the point at which σ is minimum. We have seen that the minimum of σ along L is uniquely achieved at K.

As the sum of convex functions, σ is convex. (A function is said to be *convex* if and only if the chords of its graph lie on or above the graph; analytically, $\sigma(\lambda X + (1-\lambda)Y) \leq \lambda\sigma(X) + (1-\lambda)\sigma(Y)$ for any two points X and Y and any number λ with $0 < \lambda < 1$, where we treat the points X and Y as vectors to form $\lambda X + (1-\lambda)Y$. See, for example, M. R. Hestenes, *Calculus of Variations and Optimal Control Theory*, Wiley, New York, 1966, page 45 ff.)

It is clear that $\sigma(X) \to \infty$ as $X \to \infty$ (in any direction). Since σ is continuous, it achieves its minimum value somewhere. A minimum point must either be a critical point for σ (i.e., a point at which grad σ vanishes) or a point of non-differentiability. Moreover, it follows from convexity that if σ has a unique critical point, it is the unique minimum point.

Now σ is differentiable except at P, Q, and along L, and at any other point X,

$$(\text{grad } \sigma)(X) = u + v + w$$

where u, v, and w are unit vectors in the direction from P to X, from Q to X, and perpendicular to L in the direction away from L, respectively. The sum of three unit vectors is zero if and only if they are at angles of 120° to one another. So $(\text{grad } \sigma)(X) = 0$ if and only if X is a point in the upper half-plane (other than P or Q) such that $\overrightarrow{QX}$ and $\overrightarrow{PX}$ have directions 210° and 330°, respectively (since w has angle 90°). It is clear that such a point exists if and only if we are in Case 2, and then R, the unique minimum point, is found as the intersection of PA with the ray from Q having angle 210°.

In Cases 1 and 3 σ does not have a critical point, so the minimum must occur at P, Q, or some point of L. Since the minimum of σ for points on L has been shown to be K, the minimum occurs at K, P, or Q.

In Case 1, we claim $R = Q$. Since $\sigma(Q) < \sigma(P)$ in this case, except for the degenerate case $P = Q$, it will be sufficient to prove $\sigma(Q) < \sigma(K)$ to establish our claim.

In Case 3, we claim $R = K$. Since $\sigma(Q) \leq \sigma(P)$, it will be sufficient to prove $\sigma(Q) > \sigma(K)$ in this case.

Now $\sigma(Q) = PQ + \tau(Q) = \sqrt{(a - b)^2 + c^2} + b$ and $\sigma(K) = PK + QK = PQ^* = \sqrt{(a + b)^2 + c^2}$. Therefore,

$$\begin{aligned}\sigma(Q)^2 - \sigma(K)^2 &= (a - b)^2 + c^2 + b^2 + 2b\sqrt{(a - b)^2 + c^2} \\ &\qquad - [(a + b)^2 + c^2] \\ &= b[b - 4a + 2\sqrt{(a - b)^2 + c^2}\,].\end{aligned}$$

In Case 1 we have $c \leq \sqrt{3}\,(a - b)$ and therefore

$$\sigma(Q)^2 - \sigma(K)^2 \leq b[b - 4a + 2\sqrt{4(a - b)^2}\,] = -3b^2 < 0,$$

and in Case 3, $c \geq \sqrt{3}\,(a + b)$, so

$$\begin{aligned}\sigma(Q)^2 - \sigma(K)^2 &\geq b[b - 4a + 2\sqrt{(a - b)^2 + 3(a + b)^2}\,] \\ &= b[b - 4a + 2\sqrt{4(a^2 + ab + b^2)}\,] > b^2 > 0.\end{aligned}$$

This completes the proof.

To summarize:

In Case 1, $R = Q$.

In Case 2, R is the point at which a ray from Q at angle 210° meets PA.

In Case 3, R is the point at which L meets the line joining P to Q^*, the reflection of Q in L.

---

5. Consider a polynomial $f(x)$ with real coefficients having the property $f(g(x)) = g(f(x))$ for every polynomial $g(x)$ with real coefficients. Determine and prove the nature of $f(x)$.

Solution. Consider a constant function g, say $g(x) = a$. Then $f(g(x)) = g(f(x))$ becomes $f(a) = a$. Since this is true for all real a, f is the identity function, i.e., $f(x) = x$.

REMARK. The proof doesn't require the hypothesis that f be a polynomial function.

---

6. A player throwing a die scores as many points as on the top face of the die and is to play until his score reaches or passes a total n. Denote by $p(n)$ the probability of making exactly the total n, and find the value of $\lim_{n\to\infty} p(n)$.

REMARK. Since the score of a dice player will increase by an average of $3\frac{1}{2}$ per throw, it seems clear that his score will ultimately assume about $1/3\frac{1}{2}$ or $2/7$ of the possible values. Hence the probability that he achieves some large score n precisely must be very nearly $2/7$. On the basis of this heuristic argument we expect that

$$\lim_{n\to\infty} p(n) = 2/7.$$

We shall show that this is indeed true.

First Solution. If by definition $p(0) = 1$ and $p(n) = 0$, $n < 0$ then the following relations are easy to verify.

$$\begin{aligned} p(0) &= 1 \\ p(1) &= \frac{1}{6}\, p(0) \\ p(2) &= \frac{1}{6}\, [p(1) + p(0)] \\ p(3) &= \frac{1}{6}\, [p(2) + p(1) + p(0)] \\ p(n) &= \frac{1}{6}\, [p(n-1) + p(n-2) + \cdots + p(n-6)], \quad n > 0. \end{aligned} \tag{1}$$

Adding these equations and then canceling, we obtain

$$p(n) + \frac{5}{6}\, p(n-1) + \frac{4}{6}\, p(n-2) + \cdots + \frac{1}{6} p(n-5) = 1.$$

Now if it is assumed that $p(n) \to p$ as $n \to \infty$, then it follows that $(21/6)p = 1$ and hence that $p = 2/7$.

We shall now show that, regardless of the initial conditions, a sequence $\{p(n)\}$ that satisfies the recursion (1) is convergent.

Let

$$\underline{p}(n) = \min \{p(n - i) : i = 1, 2, \ldots, 6\}$$
$$\bar{p}(n) = \max \{p(n - i) : i = 1, 2, \ldots, 6\}$$

From (1) we obtain

$$\frac{1}{6}\,[5\,\underline{p}(n) + \bar{p}(n)] \le p(n) \le \frac{1}{6}\,[5\,\bar{p}(n) + \underline{p}(n)].$$

Hence $\underline{p}(n) \le \underline{p}(n + 1) \le \bar{p}(n + 1) \le \bar{p}(n)$, so the sequence $\{\underline{p}(n)\}$ is non-decreasing and $\{\bar{p}(n)\}$ is non-increasing. Let

$$\bar{q} = \lim_{n\to\infty} \bar{p}(n)$$

$$\underline{q} = \lim_{n\to\infty} \underline{p}(n).$$

Then $\underline{q} \le \bar{q}$.

Now suppose $\bar{q} > \underline{q}$ and set $\epsilon = \frac{1}{5}(\bar{q} - \underline{q})$. Since $\epsilon > 0$, for some N and all $k \ge N$

$$\underline{q} - \epsilon < \underline{p}(k) \le \underline{q}$$

so

$$\frac{1}{6}\,[5(\underline{q} - \epsilon) + \bar{q}] < \frac{1}{6}\,[5\,\underline{p}(k) + \bar{p}(k)] \le p(k),$$

i.e., $\underline{q} < p(k)$. From this follows $\underline{q} < \underline{p}(k)$ for $k > N + 5$, a contradiction since $\underline{p}(k)$ increases to $\underline{q}$.

We conclude $\bar{q} = \underline{q}$. Then it follows immediately that $\lim_{n\to\infty} p(n) = \bar{q} = \underline{q}$.

Second Solution. Define $p(0) = 1$, $p(n) = 0$ for $n < 0$, and for each $n = 0, 1, 2, \ldots$ let

$$\alpha_n = (p(n - 5), p(n - 4), p(n - 3), p(n - 2), p(n - 1), p(n))^T.$$

Then for all n we have

$$\alpha_{n+1} = A\alpha_n$$

where

$$A = \begin{bmatrix} 0 & 1 & 0 & 0 & 0 & 0 \\ 0 & 0 & 1 & 0 & 0 & 0 \\ 0 & 0 & 0 & 1 & 0 & 0 \\ 0 & 0 & 0 & 0 & 1 & 0 \\ 0 & 0 & 0 & 0 & 0 & 1 \\ \frac{1}{6} & \frac{1}{6} & \frac{1}{6} & \frac{1}{6} & \frac{1}{6} & \frac{1}{6} \end{bmatrix}.$$

Hence $\alpha_n = A^n \alpha_0$.

Now A is row-stochastic (i.e., it has non-negative entries and each row sums to one), and it is easy to see that A^6 has all positive entries. It is shown in the theory of Markov processes that, if some power of a row-stochastic matrix has all positive entries, its powers converge to a row-stochastic matrix having all rows the same. Therefore $\lim A^n = B$ where B has all rows the same, and

$$\lim \alpha_n = (\lim A^n)\alpha_0 = B\alpha_0,$$

a vector with all components the same, say p. We have

$$\lim_n p(n) = p.$$

If $\beta = (1, 2, 3, 4, 5, 6)$, then $\beta A = \beta$. Therefore $\beta A^n = \beta$, and taking limits $\beta B = \beta$. Then

$$21p = \beta(B\alpha_0) = \beta\alpha_0 = 6,$$

so $p = 2/7$.

For the theorem on row-stochastic matrices see, for example, P. A. P. Moran, *Introduction to Probability Theory*, Clarendon Press, Oxford, 1968, page 112.

Third Solution. We now roll up the really heavy artillery and argue as follows.

The probability of obtaining a total of n in k throws (exactly) is the coefficient of x^n in

$$\left[\frac{1}{6}(x + x^2 + \cdots + x^6)\right]^k.$$

Since obtaining n in k throws and obtaining n in l throws for $k \neq l$ are mutually exclusive events, the probability of ever obtaining a total of n is the coefficient of x^n in

$$\sum_{k=0}^{\infty} \left[\frac{1}{6}(x + x^2 + \cdots + x^6) \right]^k .$$

Hence

$$\begin{aligned} \sum_{n=0}^{\infty} p(n)x^n &= \frac{6}{6 - (x + x^2 + \cdots + x^6)} \\ &= \frac{6}{(1 - x)(6 + 5x + 4x^2 + 3x^3 + 2x^4 + x^5)} \\ &= \frac{2}{7(1 - x)} + \frac{2}{7}\,\frac{15 + 10x + 6x^2 + 3x^3 + x^4}{6 + 5x + 4x^2 + 3x^3 + 2x^4 + x^5}. \end{aligned}$$

Thus

$$\sum_{n=0}^{\infty} \left(p(n) - \frac{2}{7} \right) x^n = \frac{2}{7}\,\frac{15 + 10x + 6x^2 + 3x^3 + x^4}{6 + 5x + 4x^2 + 3x^3 + 2x^4 + x^5}.$$

Thus far the argument has been formally combinatoric in character. If we now regard x as a complex variable and show that the denominator of the last fraction does not vanish inside or on the unit circle in the complex plane, then it follows that for some $x > 1$, $\sum_{n=0}^{\infty} (p(n) - (2/7))x^n$ converges and hence $\lim_{n \to \infty} p(n) = 2/7$.

Let

$$D = 6 + 5x + 4x^2 + 3x^3 + 2x^4 + x^5.$$

If $x = 1$, $D \neq 0$. Assume now that $x \neq 1$ but $|x| \leq 1$. Then

$$D(1 - x) = 6 - x - x^2 - x^3 - x^4 - x^5 - x^6,$$

so

$$|D||1 - x| > 6 - 6|x| \geq 0.$$

Thus the zeros of D are outside the unit circle.

Remark. This is one of the problems criticized by L. J. Mordell in his article "The Putnam Competition," *American Mathematical Monthly*, vol. 70 (1963), pages 481-490. The published solution in the *Monthly* is a condensed version of our third solution, and Mordell felt it was too sophisticated for an undergraduate competition. The examiners no doubt envisioned something like our first solution. See page 623 in Appendix.

---

7. Let $N(n)$ denote the smallest positive integer N such that $x^N = 1$ for every permutation x on n symbols, where 1 denotes the identity permutation. Prove that if $n > 1$,

$$\frac{N(n)}{N(n-1)} = 1 \text{ if } n \text{ is divisible by 2 distinct primes,}$$

$$= p \text{ if } n \text{ is a power of a prime } p.$$

Solution. Let $L(n)$ be the least common multiple of the integers $1, 2, \ldots, n$. Then $L(n-1) \mid L(n)$ for $n = 2, 3, \ldots$. We shall prove that $N(n) = L(n)$ for all n.

In any group an element of order i satisfies $x^N = 1$ if and only if $i \mid N$. Since the permutation group Σ on n symbols contains elements of each of the orders $1, 2, \ldots, n$, we have $i \mid N(n)$ for $i = 1, 2, \ldots, n$ and therefore $L(n) \mid N(n)$. Conversely, every permutation in Σ is the product of commuting cycles, each of which is of order at most n, so $x^{L(n)} = 1$ for all $x \in \Sigma$. Hence $L(n) \geq N(n)$, and therefore $L(n) = N(n)$.

Suppose n is an integer greater than one but not a prime power. Then n can be written as the product of two smaller integers that are relatively prime, say $n = ab$. Then a and b both divide $L(n-1)$. Since they are relatively prime, $ab = n$ also divides $L(n-1)$. Therefore, $L(n) \mid L(n-1)$ and hence $L(n) = L(n-1)$ in this case. On the other hand, suppose n is a power of a prime p, say $n = p^a$. Then $n/p \mid L(n-1)$, so $n \mid pL(n-1)$, and therefore $L(n) \mid pL(n-1)$. Now any integer less than n is divisible by a power of p no greater than p^{a-1}, so $p^a \nmid L(n-1)$ and hence $L(n) \neq L(n-1)$. Thus $L(n)/L(n-1)$ divides p, but $L(n)/L(n-1) \neq 1$. Since p is a prime, we have $L(n)/L(n-1) = p$ in this case.

Since $L(n) = N(n)$, we have proved

$$\frac{N(n)}{N(n-1)} = 1 \quad \text{if } n \text{ is divisible by two distinct primes}$$

$$= p \quad \text{if } n \text{ is a power of a prime } p.$$

Afternoon Session

1. Find all solutions of $n^m = m^n$ in integers n and m ($n \neq m$). Prove that you have obtained all of them.

Solution. First we consider positive integer solutions. Then the given equation is equivalent to

$$\frac{1}{m}\log m = \frac{1}{n}\log n. \tag{1}$$

The function $(\log x)/x$ is strictly increasing for $0 < x \leq e$ and strictly decreasing for $e \leq x$, as we see by considering its derivative $(1 - \log x)/x^2$. Hence a solution of (1) with $m < n$ must have $m < e$, so $m = 1$ or 2. If $m = 1$, there are no values of $n > 1$ which satisfy (1). If $m = 2$, then $n = 4$ is the unique solution of (1) with $n > 2$. Thus $m = 2$, $n = 4$ is the only solution of the given equation in positive integers with $m < n$. Clearly, there is just one other solution in positive integers with $m \neq n$, namely, $m = 4$, $n = 2$.

In the original equation neither m nor n can be zero, since $0^n = n^0$ has no non-zero solution n. If m is negative, say $m = -k$, and n is positive, the requirement becomes $(-k)^n = 1/n^k$. But this has no positive integral solution n, since the left member would be an integer but the right would not, unless $n = 1$, in which case the left member is negative.

If both m and n are negative, say $m = -k$, $n = -l$, then $(-k)^{-l} = (-l)^{-k}$ giving

$$(-1)^l = (-1)^k \quad \text{and} \quad k^l = l^k$$

with k and l unequal positive integers. As we have seen, this implies $k = 2$, $l = 4$, or vice versa. Both of these solutions satisfy the sign condition.

Thus there are four solutions to the original equation in unequal integers:

$$(m, n) = (2, 4), (4, 2), (-2, -4), (-4, -2).$$

The problem is related to Problem A.M. 1 of the Twelfth Competition. It has attracted the attention of mathematicians at least since the eighteenth century. In Solomon Hurwitz, "On the Rational Solutions of $m^n = n^m$ with $m \neq n$," *American Mathematical Monthly*, vol. 74 (1967), pages 298–300, it is shown that all rational solutions of the equation with $|m| > |n|$ are given by

$$m = \left(1 + \frac{1}{s}\right)^{s+1} \qquad n = \left(1 + \frac{1}{s}\right)^{s}$$

where s is a positive integer, and

$$m = -\left(1 + \frac{1}{s}\right)^{s+1} \qquad n = -\left(1 + \frac{1}{s}\right)^{s}$$

where s is a negative odd integer.

---

2. Evaluate the double series

$$\sum_{j=0}^{\infty} \sum_{k=0}^{\infty} 2^{-3k-j-(k+j)^2}.$$

Solution. If we write out a few terms of the double sequence of terms, we see that each of the terms 2^{-2m} occurs exactly once. Since all the terms

$j \backslash k$	0	1	2	3
0	2^0	2^{-4}	2^{-10}	2^{-18}
1	2^{-2}	2^{-8}	2^{-16}	
2	2^{-6}	2^{-14}		
3	2^{-12}			

are positive, the double series may be rearranged to make the simple series

$$\sum_{m=0}^{\infty} 2^{-2m}$$

which sums to 4/3. The sum of the original double series is therefore 4/3. To formalize this argument one must prove that

$$(j, k) \mapsto (k + j)^2 + 3k + j$$

is a bijection from the set of ordered pairs of non-negative integers to the set of non-negative even integers. This is straightforward but long.

A direct analytic argument can be given as follows: Sum the double series along the diagonals defined by $j + k = n$, and then sum on n.

We find

$$\sum_{j=0}^{\infty}\sum_{k=0}^{\infty} 2^{-3k-j-(j+k)^2} = \sum_{n=0}^{\infty}\sum_{k=0}^{n} 2^{-2k-n-n^2}$$

$$= \sum_{n=0}^{\infty}\left[\frac{4}{3}(1 - 2^{-2n-2})2^{-n-n^2}\right]$$

$$= \frac{4}{3}\left[\sum_{n=0}^{\infty} 2^{-n-n^2} - \sum_{n=0}^{\infty} 2^{-(n+1)(n+2)}\right]$$

$$= \frac{4}{3}\left[\sum_{n=0}^{\infty} 2^{-n(n+1)} - \sum_{m=1}^{\infty} 2^{-m(m+1)}\right]$$

$$= 4/3.$$

At the third step, the separation of the sum into two sums is permissible because the new sums are convergent.

---

3. The motion of the particles of a fluid in the plane is specified by the following components of velocity

$$\frac{dx}{dt} = y + 2x(1 - x^2 - y^2),$$

(1)

$$\frac{dy}{dt} = -x.$$

Sketch the shape of the trajectories near the origin. Discuss what happens to an individual particle as $t \to +\infty$, and justify your conclusion.

Solution. This is an autonomous differential system (i.e., the variable t does not appear explicitly), so the trajectories form a smooth covering of the plane by curves, except at the points where both dx/dt and dy/dt vanish. The only such point is the origin.

To find the structure of the trajectories near the origin, we drop the terms of degree greater than one and solve instead the system

$$\frac{dx}{dt} = 2x + y \qquad \frac{dy}{dt} = -x. \tag{2}$$

If we set $v = x + y$, we find $dv/dt = v$, whence $v = ae^t$, and we obtain

$$x = (at + b)e^t \qquad y = (-at - b + a)e^t \tag{3}$$

as solutions of (2), where a and b are arbitrary constants.

[The system (2) can also be solved by differentiating the first equations and eliminating dy/dt with the aid of the second. Or we could immediately write the solution in the vector form

$$\begin{pmatrix} x \\ y \end{pmatrix} = e^{At} \begin{pmatrix} c_1 \\ c_2 \end{pmatrix},$$

where A is the coefficient matrix $\begin{pmatrix} 2 & 1 \\ -1 & 0 \end{pmatrix}$ and c_1, c_2 are arbitrary constants. The evaluation of e^{At} involves a change of basis to get A in canonical form and that is what we did above. The variable $v = x + y$ is chosen because $(1, 1)$ is the row eigenvector for A. (It happens that A has only one one-dimensional eigen-subspace.)]

To find the nature of the curves (3), consider instead

$$x_1 = at + b \qquad y_1 = -at - b + a. \tag{4}$$

Since $x_1 + y_1$ is constant, these equations represent uniform motion along straight lines as shown. All points are stationary along the line $x_1 + y_1 = 0$, and other points move in the directions indicated.

Restoring the factor e^t causes the trajectories to "pull-in" rapidly toward the origin as $t \to -\infty$. All are tangent to the line $x + y = 0$ at the origin as $t \to -\infty$, as shown. This configuration is called a *node*. The trajectories of the original system (1) have a similar appearance in a small neighborhood of the origin. (See, for example, F. G. Tricomi, *Differential Equations*, Hefner, New York, 1961.)

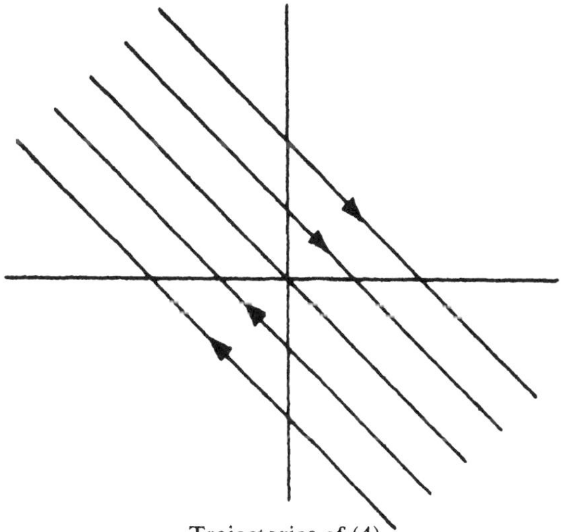

Trajectories of (4).

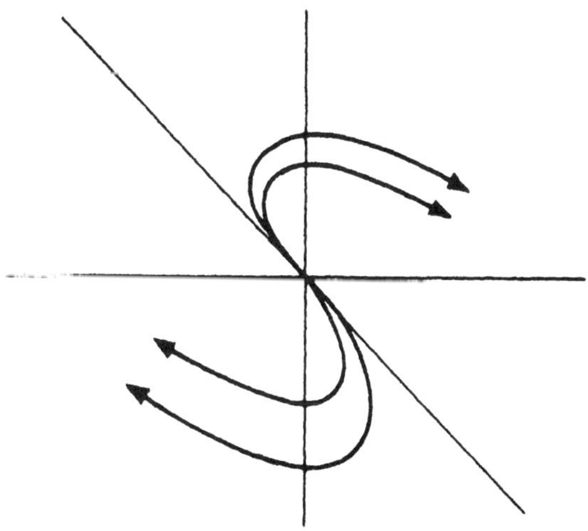

Trajectories of (3).

Returning to the original system (1), we see that

$$x = \cos t, \qquad y = -\sin t$$

a solution. This is the unit circle described uniformly clockwise. We shall how that all other trajectories (save the degenerate $x = y = 0$) are asymptotic to the unit circle as $t \to \infty$. Curves starting inside the unit circle spiral out towards it, while curves starting outside the unit circle spiral in towards it.

To prove this we switch to polar coordinates. We have

$$r\frac{dr}{dt} = x\frac{dx}{dt} + y\frac{dy}{dt} = 2x^2(1 - r^2). \tag{5}$$

For a trajectory inside the unit circle (5) shows that r is non-decreasing. Since $x = 0$ is not a trajectory (we exclude the degenerate trajectory), every interval of a trajectory contains points at which $dr/dt \neq 0$, so r is strictly increasing along any trajectory inside the unit circle. Moreover, r is not bounded above by a number $\rho < 1$ because then the circle of radius ρ would be a trajectory, which is not the case as (5) shows. Along trajectories outside the unit circle r is strictly decreasing and there is no limit circle of radius greater than one. To verify the spiraling we consider

$$\frac{d\theta}{dt} = \frac{1}{r^2}\left(x\frac{dy}{dt} - y\frac{dx}{dt}\right) = -1 + (r^2 - 1)\sin 2\theta. \tag{6}$$

This shows that $d\theta/dt < -1/2$ in the annulus $3/4 < r < 6/5$, so the trajectories wind around the origin in the clockwise direction in this region.

The trajectories are therefore as pictured on page 533. The dotted curve has the equation $y + 2x(1 - x^2 - y^2) = 0$; it is the locus of points where trajectories have vertical tangents. Horizontal tangents occur along the y-axis.

---

4. Consider the arithmetic progression $a, a + d, a + 2d, \ldots$, where a and d are positive integers. For any positive integer k, prove that the progression has either no exact kth powers or infinitely many.

Solution. Suppose the given arithmetic progression contains n^k (i.e., it contains a perfect kth power). Then the equation

$$(n + d)^k = n^k + d\left[\binom{k}{1}n^{k-1} + \binom{k}{2}n^{k-2}d + \cdots + d^{k-1}\right]$$

shows that $(n + d)^k$ is in the progression. By induction, $(n + 2d)^k$, $(n + 3d)^k, \ldots$ are also in the progression. Thus, if the progression contains one perfect kth power, it contains infinitely many.

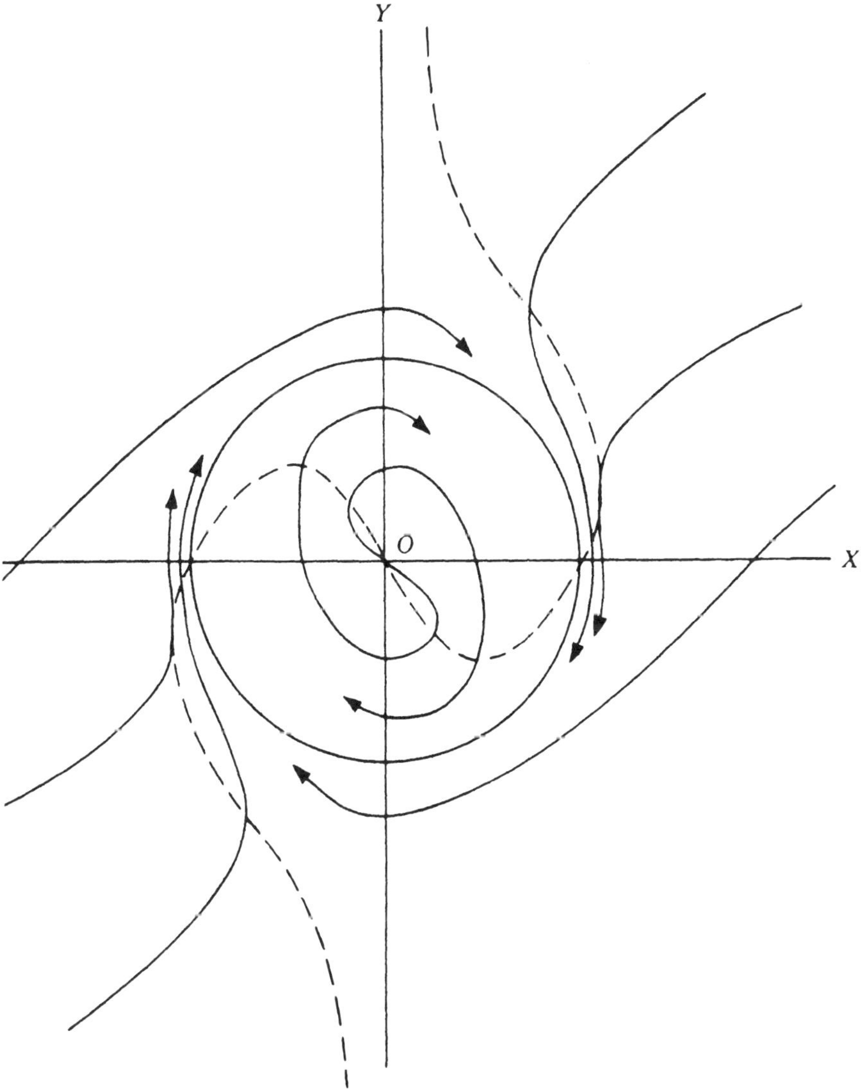

5. Define a sequence as follows:

$$a_0 = 0$$

$$a_1 = 1 + \sin(-1)$$

$$\vdots$$

$$a_n = 1 + \sin(a_{n-1} - 1)$$

$$\vdots$$

Evaluate

$$\lim_{n\to\infty} \frac{1}{n} \sum_{k=1}^{n} a_k .$$

Solution. Let $b_n = a_n - 1$. Then $b_0 = -1$ and

(1) $$b_n = \sin b_{n-1}, \qquad n = 1, 2, 3, \ldots .$$

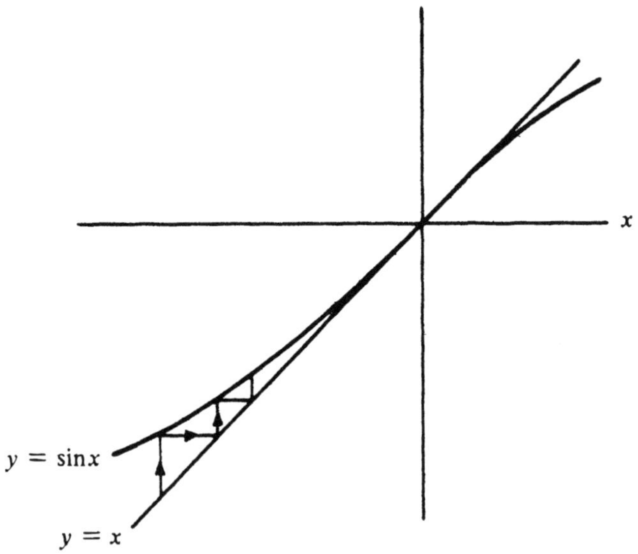

The polygonal representation of this recursion (see page 223) suggests that b_n increases to 0 as $n \to \infty$. Analytically, we note that for $-\pi < x < 0$, $x < \sin x < 0$. So from $-1 \le b_{n-1} < 0$ there follows $b_{n-1} < b_n < 0$. Hence

$$-1 = b_0 < b_1 < b_2 < \cdots < 0.$$

Therefore $\lim b_n = c$ exists. Passing to the limit in (1) we obtain $\sin c = c$, so $c = 0$.

We next prove that

(2) $$\lim_{n\to\infty} \frac{1}{n} \sum_{k=1}^{n} b_k = 0.$$

Let $\epsilon > 0$ be given and choose p so that $b_k > -\epsilon$ for $k > p$. Choose m

so that $m\epsilon > p$ (and $m > p$). Then if $n > m$, we have

$$\sum_{k=1}^{n}(-b_k) = \sum_{k=1}^{p}(-b_k) + \sum_{k=p+1}^{n}(-b_k) < p + n\epsilon < 2n\epsilon,$$

and hence

$$0 > \frac{1}{n}\sum_{k=1}^{n} b_k > -2\epsilon.$$

Since ϵ was arbitrary, (2) follows.

Finally, we have

$$\lim_{n\to\infty}\frac{1}{n}\sum_{k=1}^{n} a_k = \lim_{n\to\infty}\left(1 + \frac{1}{n}\sum_{k=1}^{n} b_k\right) = 1.$$

---

6. Any positive integer may be written in the form $n = 2^k(2l + 1)$. Let $a_n = e^{-k}$ and $b_n = a_1 a_2 a_3 \cdots a_n$. Prove that Σb_n converges.

Solution. It is clear that $a_n = e^0 = 1$ if n is odd and $a_n \le e^{-1}$ if n is even. Therefore

$$b_{2k} = a_1 a_2 \cdots a_{2k} \le e^{-k},$$

and

$$b_{2k+1} \le e^{-k}.$$

Therefore,

$$b_1 + b_2 + \cdots b_{2k} < b_1 + b_2 + \cdots + b_{2k+1}$$

$$\le 1 + 2e^{-1} + 2e^{-2} + \cdots + 2e^{-2k} < 1 + \frac{2e^{-1}}{1 - e^{-1}}.$$

Thus the partial sums of $\Sigma\, b_n$ are bounded. Since the series has positive terms, it converges.

---

7. Let $g(t)$ and $h(t)$ be real, continuous functions for $t \geq 0$. Show that any function $v(t)$ satisfying the differential inequality

$$\frac{dv}{dt} + g(t)v \geq h(t), \qquad v(0) = c,$$

satisfies the further inequality $v(t) \geq u(t)$ where

$$\frac{du}{dt} + g(t)u = h(t), \qquad u(0) = c.$$

From this, conclude that for sufficiently small $t > 0$, the solution of

$$\frac{dv}{dt} + g(t)v = v^2, \qquad v(0) = c_1$$

may be written

$$v = \max_w \left[c_1 e^{-\int_0^t [g(s) - 2w(s)]ds} - \int_0^t e^{-\int_0^t [g(s_1) - 2w(s_1)]ds_1} w^2(s)\, ds \right]$$

where the maximization is over all continuous functions $w(t)$ defined over some t-interval $[0, t_0]$.

Solution. Let

$$G(t) = \int_0^t g(s)ds.$$

Then $G' = g$. Put $x = ue^G$, $y = ve^G$. Then

$$\frac{dx}{dt} = \left(\frac{du}{dt} + gu\right)e^G = he^G$$

$$\frac{dy}{dt} = \left(\frac{dv}{dt} + gv\right)e^G \geq he^G,$$

since $e^G > 0$ for all t. Thus $(d/dt)(y - x) \geq 0$. Now $x(0) = y(0) = c$, and we conclude $x(t) \leq y(t)$ and therefore $u(t) \leq v(t)$ for all $t \geq 0$.

For the second part of the problem, we suppose that v satisfies

$$\frac{dv}{dt} + gv = v^2 \text{ for } t \in [0, t_0], \quad v(0) = c_1,$$

and we let w be any continuous function on $[0, t_0]$. Then

$$(v - w)^2 \geq 0 \quad \text{for all } t,$$

$$\frac{dv}{dt} + (g - 2w)v = v^2 - 2wv \geq -w^2.$$

If $W(t) = \int_0^t w(s)ds$ and $y = ve^{G-2W}$ then

$$\frac{dy}{dt} = \left(\frac{dv}{dt} + (g - 2w)v\right)e^{G-2W} \geq -w^2e^{G-2W}$$

$$y(t) \geq c_1 - \int_0^t w^2(s) \exp[G(s) - 2W(s)]ds$$

$$v(t) \geq c_1 \exp[-G(t) + 2W(t)] - \int_0^t \exp[G(s) - G(t) - 2(W(s) - W(t))]w^2(s)ds.$$

So

$$v(t) \geq \max_w \left\{c_1 \exp[-G(t) + 2W(t)] - \int_0^t \exp[G(s) - G(t) - 2(W(s) - W(t))]w^2(s)ds\right\}.$$

Now if we take $w = v$, then all of the preceding inequalities become equalities so we conclude

$$v(t) = \max_w \left\{c_1 \exp[-G(t) + 2W(t)] - \int_0^t \exp[G(s) - G(t) - 2(W(s) - W(t))]w^2(s)ds\right\},$$

which is equivalent to the form given.

The solution of the equation

$$\frac{dv}{dt} + gv = v^2,$$

subject to the initial condition $v(0) = c_1$, may be unbounded on a finite interval, say $v \to \infty$ as $t \uparrow t_1$. Then the preceding analysis is valid only for $t < t_1$. If we pick any t_0 with $0 < t_0 < t_1$, then it is valid for $[0, t_0]$.

THE TWENTY-SECOND WILLIAM LOWELL PUTNAM MATHEMATICAL COMPETITION

December 2, 1961

Morning Session

1. The graph of the equation $x^y = y^x$ in the first quadrant (i.e., the region where $x > 0$ and $y > 0$) consists of a straight line and a curve. Find the coordinates of the intersection point of the line and the curve.

Solution. In the first quadrant the given equation is equivalent to

$$\frac{1}{y}\log y = \frac{1}{x}\log x. \tag{1}$$

Consider the function given by

$$f(t) = \frac{1}{t}\log t \quad \text{for } t > 0.$$

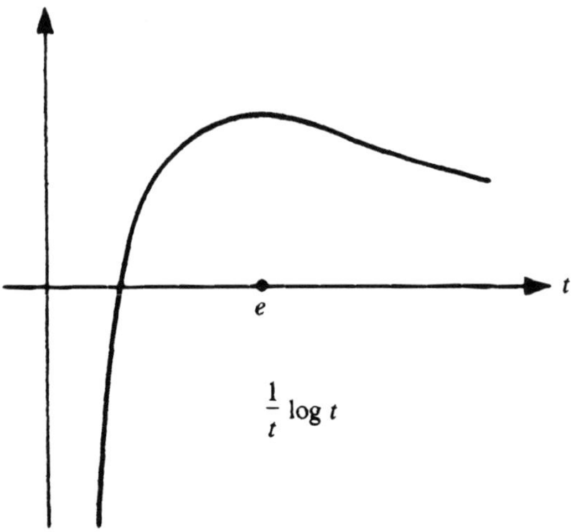

Since $f'(t) = (1 - \log t)/t^2$, it is clear that f is strictly increasing for $t \le e$, is strictly decreasing for $t \ge e$, and achieves its maximum value e^{-1} for t

$= e$. Moreover, $f(t) \to -\infty$ as $t \to 0$, and $f(t) \to 0$ as $t \to \infty$. It follows that for $\alpha \in (0, e^{-1})$ the equation $f(t) = \alpha$ has two solutions, one in $(1, e)$, the other in (e, ∞). For α near 0, the lower solution is just above 1 and the upper solution is large. As α increases to e^{-1}, the lower solution increases to e and the upper solution decreases to e. Therefore, the locus (1) consists of the line $y = x$ and a curve M lying in the quadrant $x > 1$, $y > 1$ and asymptotic to the line $x = 1$ and $y = 1$, as shown. M is evidently symmetric in the line $y = x$ and crosses that line at $\langle e, e \rangle$.

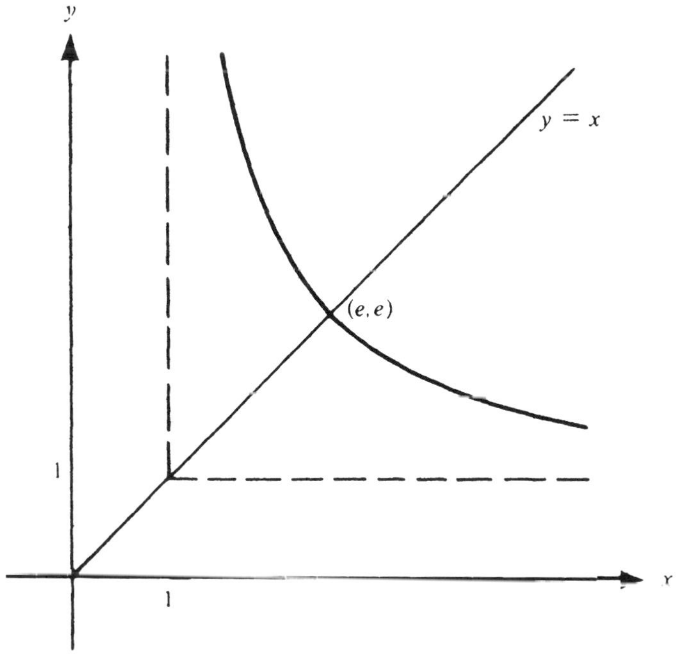

We can establish the smoothness of the curve M analytically. If F is a smooth function (say C^∞) defined on an open set in $\mathbf{R}^2$ then the level sets (contour lines) of F are smooth curves except possibly at critical points of F (i.e., points where both partial derivatives of F vanish). At a critical point the structure of the level sets is the same as that of the level sets of the second Taylor polynomial of F, provided the second degree terms are a non-degenerate quadratic form.

Let

$$F = \frac{1}{y}\log y - \frac{1}{x}\log x \text{ for } x > 0,\ y > 0.$$

Then

$$F_1' = \frac{-1}{x^2}(1 - \log x), \qquad F_2' = \frac{1}{y^2}(1 - \log y),$$

and the only critical point of F is at $\langle e, e\rangle$. At this point the second Taylor polynomial of F is

$$\frac{1}{2e^3}((x - e)^2 - (y - e)^2).$$

This is a non-degenerate quadratic form that vanishes along the lines of slopes $+1$ and -1 through the point $\langle e, e\rangle$. It follows that the level sets of F are everywhere smooth curves except the one through $\langle e, e\rangle$ which is locally the union of two smooth curves having slopes $+1$ and -1 at that point. Since this level set is the required locus, this proves that the curve M described above is smooth.

REMARKS. The problem was discussed at length by E. J. Moulton, in "The Real Function Defined by $x^y = y^x$, *American Mathematical Monthly*, vol. 23 (1916), pages 233–237. R. Robinson Rowe (*Journal of Recreational Mathematics*, vol. 3 (1970), pages 176–178) calls the curve M the "mutuabola."

For a discussion of level sets, see R. C. Buck, *Advanced Calculus*, McGraw-Hill, New York, 1965, page 349 ff.

---

2. For a real-valued function $f(x, y)$ of the two positive real variables x and y, define $f(x, y)$ to be *linearly bounded* if and only if there exists a positive number K such that $|f(x, y)| < (x + y)K$ for all positive x and y. Find necessary and sufficient conditions on the real numbers α and β such that $x^\alpha y^\beta$ is linearly bounded.

Solution. Suppose $x^\alpha y^\beta < (x + y)K$ for all positive x, y. Setting $x = y = t$, we find $t^{\alpha+\beta-1} < 2K$ for all positive t. It follows that $\alpha + \beta = 1$. Now set $x = s$, $y = 1 - s$. Then $s^\alpha(1 - s)^\beta < K$ for $0 < s < 1$. Letting $s \to 0$, we see that $\alpha \geq 0$, letting $s \to 1$, we find $\beta \geq 0$. Thus, in order that $x^\alpha y^\beta$ be linearly bounded it is necessary that $\alpha \geq 0$, $\beta \geq 0$, and $\alpha + \beta = 1$.

Conversely, suppose $\alpha \geq 0$, $\beta \geq 0$, $\alpha + \beta = 1$. Then for $0 < t < 1$, it is obvious that

$$t^\alpha(1 - t)^\beta < 1^\alpha \cdot 1^\beta = 1.$$

Hence

$$x^\alpha y^\beta = \left(\frac{x}{x+y}\right)^\alpha \left(1 - \frac{x}{x+y}\right)^\beta (x+y) < x+y$$

for any positive x and y, so $x^\alpha y^\beta$ is linearly bounded with $K = 1$. Thus the conditions stated are both necessary and sufficient.

---

3. Evaluate

$$\lim_{n\to\infty} \sum_{j=1}^{n^2} \frac{n}{n^2 + j^2}.$$

Solution. We write the sum in the form

$$S_n = \frac{1}{n} \sum_{j=1}^{n^2} \frac{1}{1 + \left(\frac{j}{n}\right)^2}.$$

Since

$$\int_{j/n}^{(j+1)/n} \frac{dx}{1+x^2} < \frac{1}{n} \frac{1}{1 + \left(\frac{j}{n}\right)^2} < \int_{(j-1)/n}^{j/n} \frac{dx}{1+x^2},$$

we get

(1) $$\int_{1/n}^{(n^2+1)/n} \frac{dx}{1+x^2} < S_n < \int_0^n \frac{dx}{1+x^2}.$$

Now

$$\int_0^n \frac{dx}{1+x^2} = \text{arc tan } n, \qquad \lim_{n\to\infty} \text{arc tan } n = \frac{\pi}{2}$$

$$\int_{1/n}^{n+(1/n)} \frac{dx}{1+x^2} = \text{arc tan}\left(n + \frac{1}{n}\right) - \text{arc tan}\left(\frac{1}{n}\right),$$

and

$$\lim_{n\to\infty}\left(\arctan\left(n+\frac{1}{n}\right)-\arctan\left(\frac{1}{n}\right)\right)=\frac{\pi}{2}.$$

Thus both the left and right members of (1) have the limit $\pi/2$, so

$$\lim_{n\to\infty} S_n = \frac{\pi}{2}.$$

---

4. Define a function f over the domain of positive integers as follows: $f(1) = 1$, and for $n > 1, f(n) = (-1)^k$ where k is the total number of prime factors of n. For example $f(9) = (-1)^2, f(20) = (-1)^3$. Define $F(n)$ as $\sum f(d)$ where the sum ranges over all positive integer divisors of n. Prove that for every positive integer n, $F(n) = 0$ or $F(n) = 1$. For which integers n is $F(n) = 1$?

Solution. Suppose m and n are relatively prime positive integers. Then every divisor of mn is uniquely a product d_1d_2 where $d_1 | m$, $d_2 | n$, and conversely. Also $f(d_1d_2) = f(d_1)f(d_2)$.

$$F(mn) = \sum_{d|mn} f(d) = \sum_{d_1|m}\sum_{d_2|n} f(d_1)f(d_2)$$

$$= F(m)F(n).$$

Thus F is a *multiplicative* numerical function and it suffices to evaluate F on prime powers. Evidently for p a prime

$$F(p^\alpha) = f(1) + f(p) + f(p^2) + \cdots + f(p^\alpha)$$

$$= 1 - 1 + 1 - \cdots + (-1)^\alpha$$

$$= \begin{cases} 1 & \text{if } \alpha \text{ is even,} \\ 0 & \text{if } \alpha \text{ is odd.} \end{cases}$$

Let n be any positive integer and suppose

$$n = p_1^{\alpha_1}p_2^{\alpha_2}\cdots p_k^{\alpha_k}$$

is its canonical factorization into primes. Then $F(n) = F(p_1^{\alpha_1})F(p_2^{\alpha_2})\cdots F(p_k^{\alpha_k})$ and we see that $F(n) = 0$ if some prime appears with odd exponent in the prime factorization of n, and $F(n) = 1$ if all primes appear with even exponents. In other words,

$$F(n) = 1$$

if n is a perfect square, and

$$F(n) = 0$$

if n is not a perfect square.

---

5. Let Ω be a set of n points, where $n > 2$. Let Σ be a nonempty subcollection of the 2^n subsets of Ω that is closed with respect to unions, intersections, and complements (that is, if A and B are members of Σ, then so are $A \cup B$, $A \cap B$, $\Omega - A$ and $\Omega - B$, where $\Omega - B$ denotes all points in Ω but not in B). If k is the number of members of Σ, what are the possible values of k? Give a proof.

First Solution. Since Σ is not empty, say it contains A. Then Σ contains also $\Omega - A$ and $A \cap (\Omega - A) = \emptyset$. Hence also $\Omega = \Omega - \emptyset \in \Sigma$.

Among the non-empty members of Σ certain are *minimal* in the sense that they do not contain any other member of Σ except $\emptyset$. Any non-empty member A of Σ contains a minimal element, for example, a set B of least cardinal satisfying $B \subseteq A$, $B \in \Sigma$, $B \neq \emptyset$.

Let $B_1, B_2, \ldots, B_p$ be a proper enumeration of all the minimal elements of Σ. These elements are mutually disjoint. Suppose $i \neq j$, then $B_i \cap B_j \in \Sigma$ and $B_i \cap B_j \subseteq B_i$. Hence by the minimality of B_i either $B_i \cap B_j = \emptyset$ or $B_i \cap B_j = B_i$. The latter is impossible since it implies $B_j \supseteq B_i$, hence $B_i = B_j$ or $B_i = \emptyset$, both contradictions.

Now if $\Omega - (B_1 \cup B_2 \cup \cdots \cup B_p) \neq \emptyset$, then it would contain some minimal element of Σ, say B_i, and we would have

$$B_i = B_i \cap [\Omega - (B_1 \cup B_2 \cup \cdots \cup B_p)] = \emptyset$$

a contradiction. So $B_1 \cup B_2 \cup \cdots \cup B_p = \Omega$.

Suppose $C \in \Sigma$. Then

$$C = C \cap \Omega = C \cap (B_1 \cup \cdots \cup B_p)$$
$$= (C \cap B_1) \cup (C \cap B_2) \cup \cdots \cup (C \cap B_p).$$

Now each of the sets $C \cap B_i$ is either $\emptyset$ or B_i (by the minimality of B_i). Thus we have shown: Every element of Σ is the union of some subcollection of the sets $\{B_i\}$. Conversely every such union is a member of Σ. Moreover, since the sets $\{B_i\}$ are non-empty and mutually disjoint, distinct subcollections of $\{B_i\}$ have distinct unions. There are therefore exactly 2^p elements of Σ, one for each subset of $\{B_1, B_2, \ldots, B_p\}$. Thus $k = 2^p$.

Every power of 2 from 2 to 2^n is possible.

If p is an integer $1 \le p \le n$, choose any partition of Ω into p disjoint subsets $B_1, \ldots, B_p$. Then the 2^p unions of the various subsets of $\{B_1, \ldots, B_p\}$ form a collection Σ with the required properties containing 2^p members.

Second Solution. For the subsets of any set Ω define the operation $\oplus$ (called the *symmetric difference*) by

$$A \oplus B = (A \cup B) - (A \cap B)$$

$$= (A \cup B) \cap ((\Omega - A) \cup (\Omega - B)).$$

With this operation the set $P(\Omega)$ of all subsets of Ω becomes a group in which $\emptyset$ is the identity and every element is its own inverse. The second way of writing the definition of $\oplus$ shows that the collection Σ of the problem is closed with respect to $\oplus$. Since every element is its own inverse, Σ is a subgroup of $P(\Omega)$. Hence Σ is a finite group with every element of order 2, so it is a 2-group and has size 2^t for some t. Thus $k = 2^t$ for some t, and here $t \ge 1$ since $\emptyset, \Omega \in \Sigma$ and $t \le n$ since $|P(\Omega)| = 2^n$.

We cannot show by purely group theoretic methods that every power of two within this range can occur, because a subgroup of $P(\Omega)$ need not be closed with respect to unions and intersections, even those subgroups that contain Ω. It is, of course, true that, if $\{B_1, B_2, \ldots, B_p\}$ is a partition of Ω into p disjoint sets, then the subgroup of $P(\Omega)$ generated by the B's is closed with respect to unions, intersections, and complements and has size 2^p, but this requires calculations similar to those in the first solution.

REMARK. A collection of sets closed with respect to unions, intersections, and complements is called a *Boolean algebra* of sets.

---

6. If $J_2 = \{0, 1\}$ is the field of integers modulo 2, and if $J_2[x]$ is the integral domain of polynomials in one indeterminate with coefficients in J_2, prove that $p(x) = 1 + x + x^2 + \cdots + x^n$ is reducible (factorable) in case $n + 1$ is composite. Is the converse true? That is, if $n + 1$ is prime, is $p(x)$ irreducible?

Solution. If $n + 1 = r \cdot s$, $r > 1$, $s > 1$ is a factorization of $n + 1$, then

$$1 + x + \cdots + x^n = (1 + x + \cdots + x^{r-1})(1 + x^r + x^{2r} + \cdots + x^{(s-1)r})$$

is valid over the integers and, a fortiori, when the coefficients are taken modulo 2.

The converse is not true, since

$$(1 + x + x^3)(1 + x^2 + x^3) = 1 + x + x^2 + 3x^3 + x^4 + x^5 + x^6$$
$$\equiv 1 + x + x^2 + x^3 + x^4 + x^5 + x^6 \pmod 2).$$

and $6 + 1$ is a prime.

---

7. Let S be a nonempty closed set in the Euclidean plane for which there is a closed disk D (a circle together with its interior) containing S such that D is a subset of *every* closed disk that contains S. Prove that every point inside D is the midpoint of a segment joining two points of S.

Solution. We claim that S contains the entire circumference of D. If not, there is a point P on the circumference of D not in S, and since S is closed there is disk E of radius $\epsilon > 0$ about P containing no point of S. Then there is a disk F that contains all of $D - E$ and hence all of S, but not P. But this contradicts the fact that any disk containing S contains all of D. [We give an analytic proof of the existence of F below.]

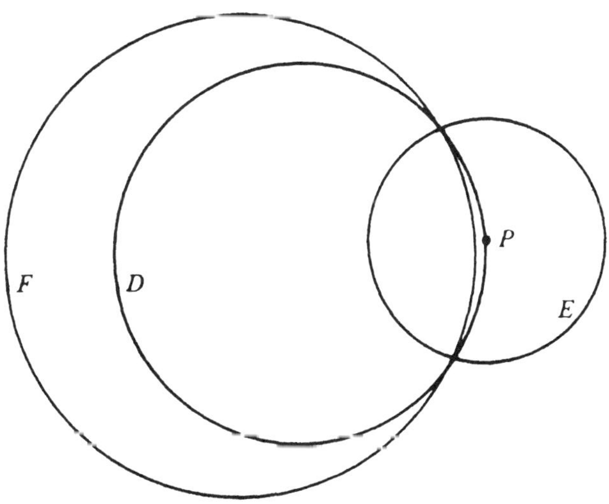

Obviously, every point inside D is the midpoint of a chord of the circumference of D, and hence the midpoint of a segment joining two points of S.

Analytic proof of the existence of F: We may suppose that D is bounded by the circle $x^2 + y^2 - a^2 = 0$ and that P$\langle a, 0\rangle$ is not in S. Then E is bounded by $(x - a)^2 + y^2 - \epsilon^2 = 0$. Let F be the disk bounded by the circle

$$\phi(x, y) = (x^2 + y^2 - a^2) - \frac{1}{2}((x - a)^2 + y^2 - \epsilon^2) = 0.$$

Then $F = \{\langle x, y\rangle : \phi(x, y) \le 0\}$. Since $(x^2 + y^2 - a^2) \le 0$ for $\langle x, y\rangle \in D$ and $(x - a)^2 + y^2 - \epsilon^2 > 0$ for $\langle x, y\rangle \notin E$, we have $\phi(x, y) < 0$ for $\langle x, y\rangle \in D - E$, so $D - E \subseteq F$. But $\phi(a, 0) = \frac{1}{2}\epsilon^2$, so $P \notin F$.

Afternoon Session

1. Let $\alpha_1, \alpha_2, \alpha_3, \ldots$ be a sequence of positive real numbers; define s_n as $(\alpha_1 + \alpha_2 + \cdots + \alpha_n)/n$ and r_n as $(\alpha_1^{-1} + \alpha_2^{-1} + \cdots + \alpha_n^{-1})/n$. Given that $\lim s_n$ and $\lim r_n$ exist as $n \to \infty$, prove that the product of these limits is not less than 1.

Solution. It is clearly sufficient to prove that $r_n s_n \ge 1$ for all n. Let $\beta_i = \alpha_i^{1/2}$ and $\gamma_i = \alpha_i^{-1/2}$. Then by the Cauchy-Schwarz inequality

$$\begin{aligned} n^2 = \left(\sum_{i=1}^{n} \beta_i\gamma_i\right)^2 &\le \left(\sum_{i=1}^{n} \beta_i^2\right)\left(\sum_{i=1}^{n} \gamma_i^2\right) \\ &= \left(\sum_{i=1}^{n} \alpha_i\right)\left(\sum_{i=1}^{n} \alpha_i^{-1}\right) \\ &= (ns_n)(nr_n) \end{aligned}$$

and it follows that

$$r_n s_n \ge 1.$$

---

2. Let α and β be given positive real numbers, with $\alpha < \beta$. If two points are selected at random from a straight line segment of length β, what is the probability that the distance between them is at least α?

Solution. We interpret "at random" to mean that the pair of points x, y is chosen so that the probability that $\langle x, y\rangle$ falls in any region in the square $[0, \beta] \times [0, \beta]$ is proportional to the area of that region. Then the favorable region is evidently the union of the two triangular regions shown and the probability of a favorable outcome is

$$\frac{(\beta - \alpha)^2}{\beta^2} = \left(1 - \frac{\alpha}{\beta}\right)^2.$$

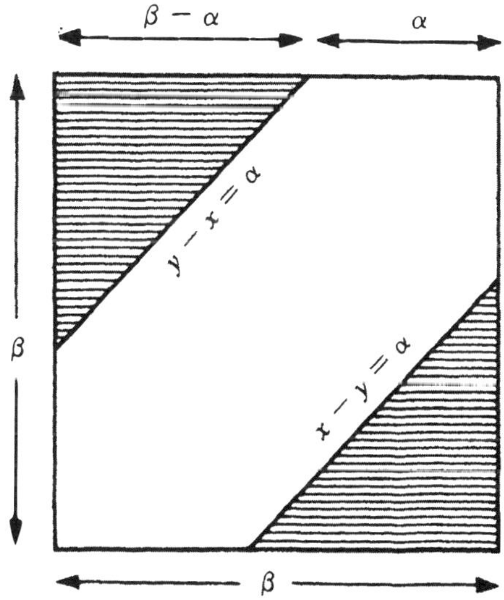

---

3. Consider four points in a plane, no three of which are collinear, and such that the circle through three of them does not pass through the fourth. Prove that one of the four points can be selected having the property that it lies inside the circle determined by the other three.

First Solution. We give first a heuristic argument. Start with a large circle surrounding all four points and gradually shrink it, keeping the points on or inside it until it can no longer be reduced without allowing one of the points to escape to the exterior. Then either three of the given points lie on the circumference and the fourth inside, in which case we are through, or two of the given points are opposite on the circle and two are interior points. In this case, consider the one-parameter family of circles that contain the first two points. As we run through this family we come to a first one that contains another of the given points, and that one will have the fourth point in its interior.

This argument can be formalized as follows: First, the smallest closed disk containing the given points exists by the limit argument given on page 281. Then the appropriate member of the one-parameter family of circles can be found as in our second solution, below.

Second Solution. We can number the points so that P_3 and P_4 are on the same side of the line $\overleftrightarrow{P_1P_2}$. (This can be accomplished by taking P_1P_2 to be an edge of the convex hull of P_1, P_2, P_3, P_4.) We choose coordinate axes so that P_1, P_2 are on the y-axis and P_3 and P_4 have positive x-coordinates.

Let

$$\phi(x, y) = x^2 + y^2 + \alpha x + \beta y + \gamma = 0$$

be the equation of the circle through P_1, P_2, and P_3. Since the given points are not concyclic, $\phi(P_4) \neq 0$. If $\phi(P_4) < 0$, then P_4 is interior to this circle and we have the desired result.

If $\phi(P_4) > 0$, then we set

$$\psi(x, y) = \phi(x, y) - \lambda x$$

and choose λ so that $\psi(P_4) = 0$, i.e., $\lambda = \phi(P_4)/x(P_4)$. Then $\psi(x, y) = 0$ is the equation of the circle through P_1, P_2, and P_4. P_3 is interior to this circle since

$$\psi(P_3) = -\lambda x(P_3) < 0, \quad \text{since } \lambda \text{ and } x(P_3)$$

are each positive.

Note. In this argument we are, in effect, considering the one-parameter family of circles that contain P_1 and P_2.

Third Solution. Let R be a point on (the surface of) a sphere $\mathcal{S}$ in three-space, with R not in the plane of the given points P_1, P_2, P_3, P_4. By stereographic projection from R we transform the given points into four points Q_1, Q_2, Q_3, Q_4 of $\mathcal{S}$. Since no three P's are collinear, no three Q's are coplanar with R. Since the P's are not concyclic, the Q's are not coplanar. Thus Q_1, Q_2, Q_3, Q_4, R are five points on $\mathcal{S}$, no four of which are coplanar.

Suppose each of the four planes determined by three Q's fails to separate R from the fourth Q. Then R would be in the interior of the tetrahedron $Q_1Q_2Q_3Q_4$. But this is impossible since Q_1, Q_2, Q_3, Q_4, and R are cospherical. Hence there is a plane π determined by three Q's, say Q_1, Q_2, Q_3, which separates R from Q_4. Then P_4 is separated from infinity by the stereographic projection on the original plane of $\pi \cap \mathcal{S}$. But the latter is a circle through P_1, P_2, P_3; hence P_4 is inside the circle determined by P_1, P_2, P_3.

4. For a fixed positive integer n let $x_1, x_2, \ldots, x_n$ be real numbers satisfying $0 \le x_k \le 1$ for $k = 1, 2, \ldots, n$. Determine the maximum value, as a function of n, of the sum of the $n(n-1)/2$ terms:

$$\sum_{\substack{i,j=1 \\ i<j}}^{n} |x_i - x_j|.$$

Solution. If we keep all but one of the x's, say $x_2, x_3, \ldots, x_n$, fixed, then the sum in question is a convex function of x_1, since it is a sum of convex functions. Hence the maximum value is achieved either for $x_1 = 0$ or for $x_1 = 1$; conceivably it could be achieved for both and even for all intermediate values.

Since the set of n-tuples being considered is a bounded closed set in $\mathbf{R}_n$, a maximum is achieved at some point. We can then change any one of the x's not already either a 0 or a 1 into either a 0 or a 1 without decreasing the sum. Hence among the points where the maximum is achieved there is at least one for which every x is either a 0 or a 1. Consider this point.

If p of the x's are zero and $(n - p)$ are one, then the sum is clearly

$$p(n-p) = \left(\frac{n}{2}\right)^2 - \left(\frac{n}{2} - p\right)^2$$

and this will be largest when $p = n/2$ if n is even, or $p = (n \pm 1)/2$ if n is odd.

The maximum value of the sum is therefore

$$\frac{1}{4}n^2 \quad \text{if } n \text{ is even,}$$

$$\frac{1}{4}(n^2 - 1) \quad \text{if } n \text{ is odd.}$$

REMARK. If n is even, then the maximum is achieved only if half the x's are at one end and half at the other. But if n is odd, $(n-1)/2$ x's must be at each end, while the last x may be taken anywhere on the interval.

---

5. Let k be a positive integer, and n a positive integer greater than 2. Define

$$f_1(n) = n,\ f_2(n) = n^{f_1(n)},\ \ldots,\ f_{j+1}(n) = n^{f_j(n)},\ \text{etc.}$$

Prove either part of the inequality

$$f_k(n) < n!!! \cdots ! < f_{k+1}(n),$$

where the middle term has k factorial symbols.

Solution. Define $g_1(n) = n!$, $g_{k+1}(n) = [g_k(n)]!$. Then the required inequalities are

$$f_k(n) < g_k(n) < f_{k+1}(n), \qquad n > 2.$$

Proof of the lower inequality. If $t \geq 2n^2$, then

(1) $$t! > (n^2)^{t-n^2} = n^t \cdot n^{t-2n^2} \geq n^t.$$

Also, if $n = 3$ or 4, $k \geq 2$, then

$$g_k(n) \geq g_2(3) = 6! = 720 > 32 \geq 2n^2$$

and if $n \geq 5$, $k \geq 1$,

$$g_k(n) \geq n! \geq n(n-1)(n-2) > 2n^2.$$

Now obviously $f_1(n) < g_1(n)$ for any integer $n \geq 3$ and $f_2(3) < f_2(4) = 256 < 720 = g_2(3) < g_2(4)$. We proceed by induction on n. Assume $f_k(n) < g_k(n)$ where $n \geq 3$, and $k \geq 2$ if $n = 3$ or 4. Then $g_k(n) > 2n^2$, so by (1)

$$f_{k+1}(n) = n^{f_k(n)} < n^{g_k(n)} < [g_k(n)]! = g_{k+1}(n).$$

This completes the inductive proof of the lower inequality. ■

Proof of the upper inequality. We note first that $g_{k+1}(n) = g_k(n!)$ and put $g_0(n) = n$.

We shall prove by induction on k that

(2) $$g_0(n)g_1(n)g_2(n) \cdots g_k(n) < n^{g_0(n)g_1(n)g_2(n) \cdots g_{k-1}(n)}$$

for $k \geq 1$ and $n > 2$.

First, we consider $k = 1$ and $n > 2$. We have

$$n \cdot n! = n \cdot n(n-1) \cdots 3 \cdot 2 < n \cdot n \cdot n \cdots n \cdot n = n^n,$$

i.e.,

$$g_0(n)g_1(n) < n^{g_0(n)}.$$

Now assume that (2) holds for a fixed k and all $n > 2$. Then

$$\begin{aligned} g_0(n)g_1(n)g_2(n)\cdots g_{k+1}(n) &= g_0(n)g_0(n!)g_1(n!)\cdots g_k(n!) \\ &< n(n!)^{g_0(n!)\cdots g_{k-1}(n!)} \\ &\le n^{g_1(n)\cdots g_k(n)}\cdot(n!)^{g_1(n)\cdots g_k(n)} \\ &= (nn!)^{g_1(n)\cdots g_k(n)} \\ &< (n^{g_0(n)})^{g_1(n)\cdots g_k(n)} \\ &= n^{g_0(n)g_1(n)\cdots g_k(n)}. \end{aligned}$$

This completes the proof of (2) by induction on k. ■

Now we shall prove

(3) $$g_0(n)g_1(n)\cdots g_k(n) < f_{k+1}(n)$$

for $k \ge 1$ and $n \ge 2$. (Obviously there is equality for $k = 0$). For $k = 1$, this is $n \cdot n! < n^n$, which we have already established. If we assume (3) for a fixed k and use (2) we obtain

$$g_0(n)g_1(n)\cdots g_{k+1}(n) < n^{g_0(n)g_1(n)\cdots g_k(n)} < n^{f_{k+1}(n)} = f_{k+2}(n),$$

which is (3) for $k + 1$. Hence (3) for all k follows by induction. Obviously (3) implies the required upper inequality

$$g_k(n) < f_{k+1}(n).$$

---

6. Consider the function $y(x)$ satisfying the differential equation $y'' = -(1 + \sqrt{x})y$ with $y(0) = 1$ and $y'(0) = 0$. Prove that $y(x)$ vanishes exactly once on the interval $0 < x < \pi/2$, and find a positive lower bound for the zero.

Solution. We shall apply the Sturm comparison theorem to the three functions determined by the following differential equations with initial conditions:

$$\begin{aligned} u'' + 3u &= 0, & u(0) &= 1, & u'(0) &= 0, \\ y'' + (1+\sqrt{x})y &= 0, & y(0) &= 1, & y'(0) &= 0, \\ v'' + v &= 0, & v(0) &= 1, & v'(0) &= 0. \end{aligned}$$

We see that $u(x) = \cos\sqrt{3}x$ and $v(x) = \cos x$.

For $0 < x < \pi/2$, we have

$$3 > 1 + \sqrt{x} > 1;$$

hence by the Sturm theorem the first zero of u, namely $\pi/2\sqrt{3}$, occurs before the first zero of y, say ξ, and the first zero of y occurs before the first zero of v, namely, $\pi/2$. So we have $\pi/2\sqrt{3} < \xi < \pi/2$.

Suppose y had a second zero, say η, in $[0, \pi/2]$. Then by the Sturm theorem a zero of u would appear in $(\xi, \eta) \subseteq (\pi/2\sqrt{3}, \pi/2)$. But u has no such zero, so y has but one zero in $[0, \pi/2]$.

REMARK. A proof of the Sturm comparison theorem is given on page 451. See the remark on page 452 for the version used in the first part of the proof.

---

7. Given a sequence $\{a_n\}$ of non-negative real numbers such that $a_{n+m} \leq a_n a_m$ for all pairs of positive integers, m and n, prove that the sequence $\{\sqrt[n]{a_n}\}$ has a limit as $n \to \infty$.

Solution. If $a_p = 0$ for some p, then from $a_{p+m} \leq 0 \cdot a_m$ we have $a_{p+1} = 0$, $a_{p+2} = 0, \ldots,$ and indeed $a_n = 0$ for all $n \geq p$. In this event $\lim a_n{}^{1/n} = 0$.

If no $a_p = 0$, then $a_n > 0$ for all n, and we can put

$$b_n = \log a_n.$$

Then the given inequality becomes

$$b_{m+n} \leq b_m + b_n$$

and it follows that

$$b_{km+t} \leq kb_m + b_t$$

for any non-negative integers k, m, t.

We fix the positive integer m, and for any n, let $n = k(n)m + t(n)$ where $0 \leq t(n) < m$. Let $c = \max(b_0, b_1, \ldots, b_{m-1})$

$$\begin{aligned} b_n = b_{k(n)m+t(n)} &\leq k(n)b_m + b_{t(n)} \\ &\leq k(n)b_m + c \end{aligned}$$

and hence

$$\frac{b_n}{n} < \frac{k(n)}{n} b_m + \frac{c}{n}.$$

As $n \to \infty$, $\dfrac{k(n)}{n} \to \dfrac{1}{m}$, hence

$$\text{(1)} \qquad \limsup \frac{1}{n} b_n \le \frac{1}{m} b_m.$$

If α is a number exceeding $\liminf b_n/n$, we can choose m so that $b_m/m < \alpha$. Then (1) shows that

$$\limsup \frac{1}{n} b_n \le \alpha.$$

Hence we conclude

$$\limsup \frac{1}{n} b_n \le \liminf \frac{1}{n} b_n.$$

Thus, $\lim b_n/n$ exists, but possibly improperly. It is less than b_1, for we may take $m = 1$ in (1), but it may be $-\infty$. In any case

$$\lim_{n\to\infty} a_n{}^{1/n} = \lim_{n\to\infty} \exp\left(\frac{1}{n} b_n\right)$$

exists.

THE TWENTY-THIRD WILLIAM LOWELL PUTNAM MATHEMATICAL COMPETITION

December 1, 1962

Morning Session

1. Given five points in a plane, no three of which lie on a straight line, show that some four of these points form the vertices of a convex quadrilateral.

Solution. Consider the convex hull K of the five points. If K is a pentagon or a quadrilateral, the result is obvious. Since the points do not lie on a line, the only remaining possibility is that K is a triangle ABC with vertices among the given points. The other two points D and E are in the interior of ABC and the line DE meets the triangle twice, not at a vertex. By relettering if necessary, we may suppose that DE meets AB and AC but not BC. Then B, C, D, and E are the vertices of a convex quadrilateral.

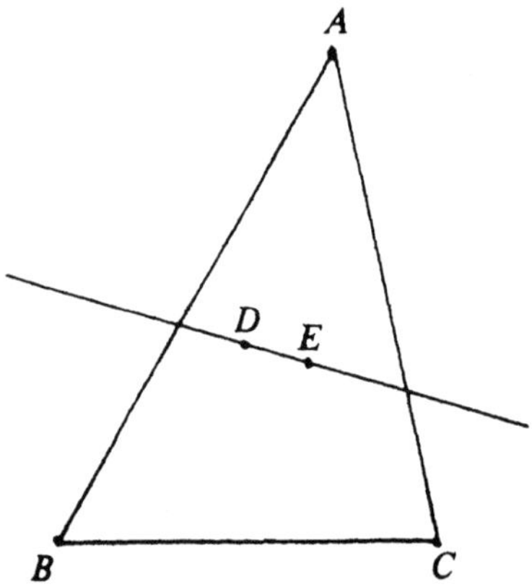

REMARK. This problem and its generalizations to more points were first studied by Esther Klein Szekeres. A proof that every set of nine points with no three collinear contains the vertices of a convex pentagon, and further references, will be found in W. E. Bonnice, "On Convex Planar Polygons Determined by a Finite Planar Set," *American Mathematical Monthly*, vol. 81 (1974), pages 749–752.

---

2. Find every real-valued function f whose domain is an interval I (finite or infinite) having 0 as a left-hand endpoint, such that for every positive member x of I the average of f over the closed interval $[0, x]$ is equal to the geometric mean of the numbers $f(0)$ and $f(x)$.

Solution. Since the geometric mean is defined only for positive numbers, such a function must be everywhere positive; moreover, it must be integrable on $[0, x]$ for every x in its domain.

Put $f(0) = a$ for convenience and let

$$F(x) = \int_0^x f(t)\, dt. \tag{1}$$

Then the condition of the problem is equivalent to

$$af(x) = \left(\frac{1}{x} F(x)\right)^2 \tag{2}$$

for all positive $x \in I$.

Now (1) shows that F is continuous, so for $x > 0$, f is continuous by (2) and F is then differentiable by (1). Then (2) becomes

$$aF'(x) = \frac{1}{x^2} F(x)^2.$$

We can integrate this differential equation by separating the variables, and we find

$$F(x) = \frac{ax}{1 - cx},$$

where c is the constant of integration. Therefore

$$f(x) = F'(x) = \frac{a}{(1 - cx)^2}.$$

for $x > 0$. We note that this formula remains valid for $x = 0$. If $c > 0$, then f is not integrable on $[0, 1/c]$, hence we see that every solution is given by

$$f(x) = \frac{a}{(1 - cx)^2} \begin{cases} \text{for } 0 \le x < \dfrac{1}{c}, & \text{if } c > 0 \\ \text{for } 0 \le x < \infty, & \text{if } c \le 0, \end{cases} \tag{3}$$

where $a > 0$.

It is readily checked that all functions defined by (3) are solutions of the problem.

REMARK. If $a < 0$, then (3) gives solutions of (2) that are not solutions of the problem. If $a = 0$, then (2) has many more solutions, namely, any function that is zero almost everywhere on $[0, \infty)$.

---

3. In a triangle ABC in the Euclidean plane, let A' be a point on the segment from B to C, B' a point on the segment from C to A, and C' a point on the segment from A to B such that

$$\frac{AB'}{B'C} = \frac{BC'}{C'A} = \frac{CA'}{A'B} = k,$$

where k is a positive constant. Let Δ be the triangle formed by parts of the segments obtained by joining A and A', B and B', and C and C'. Prove that the areas of the triangles Δ and ABC are in the ratio.

$$\frac{(k-1)^2}{k^2+k+1}.$$

First Solution. Use barycentric coordinates for the triangle ABC.

$$A = \langle 1, 0, 0\rangle$$

$$B = \langle 0, 1, 0\rangle$$

$$C = \langle 0, 0, 1\rangle$$

$$A' = \frac{1}{k+1}\langle 0, k, 1\rangle$$

$$B' = \frac{1}{k+1}\langle 1, 0, k\rangle$$

$$C' = \frac{1}{k+1}\langle k, 1, 0\rangle.$$

The equations of BB' and CC' are $z = kx$ and $x = ky$, respectively. If their intersection P is given by $\langle \lambda, \mu, \nu\rangle$, then $\lambda = k\mu$, $\nu = k\lambda$, and $\lambda + \mu + \nu = 1$. So

$$P = \frac{1}{k^2+k+1}\langle k, 1, k^2\rangle.$$

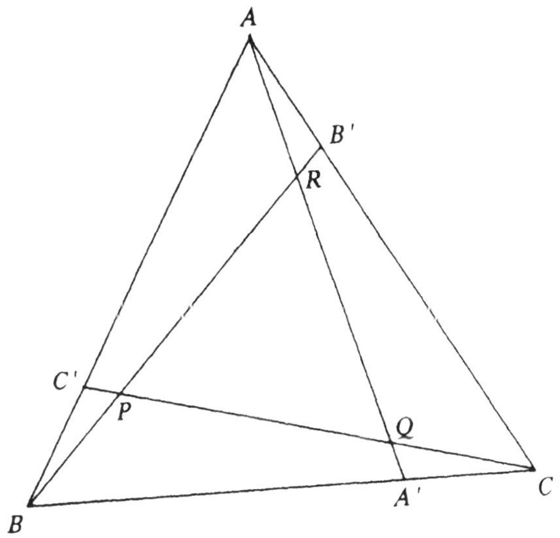

By symmetry

$$Q = \frac{1}{k^2 + k + 1} \langle k^2, k, 1 \rangle$$

$$R = \frac{1}{k^2 + k + 1} \langle 1, k^2, k \rangle.$$

Then

$$\frac{\text{area } PQR}{\text{area } ABC} = \frac{1}{(k^2 + k + 1)^3} \begin{vmatrix} k & 1 & k^2 \\ k^2 & k & 1 \\ 1 & k^2 & k \end{vmatrix}$$

$$= \frac{k^6 - 2k^3 + 1}{(k^2 + k + 1)^3} = \frac{(k^3 - 1)^2}{(k^2 + k + 1)^3} = \frac{(k - 1)^2}{k^2 + k + 1}.$$

Second Solution. A synthetic solution may be of interest.

$$\frac{\Delta\, AQC}{\Delta\, AQC'} = \frac{QC}{QC'} = \frac{\Delta\, BQC}{\Delta\, BQC'}$$

$$\therefore \quad \frac{\Delta\, BQC}{\Delta\, AQC} = \frac{\Delta\, BQC'}{\Delta\, AQC'} = \frac{BC'}{AC'} = k.$$

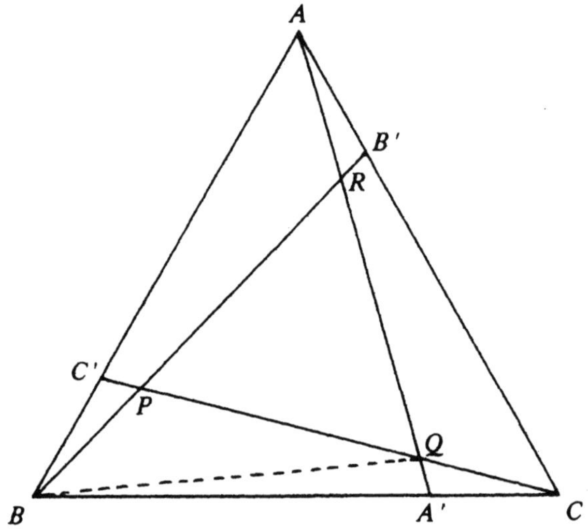

Similarly

$$\frac{\Delta\, AQC}{\Delta\, A'QC} = \frac{AQ}{A'Q} = \frac{\Delta\, BAQ}{\Delta\, BA'Q}$$

$$\therefore \quad \frac{\Delta\, AQB}{\Delta\, AQC} = \frac{\Delta\, A'QB}{\Delta\, A'QC} = \frac{A'B}{A'C} = \frac{1}{k}.$$

Thus

$$\frac{\Delta\, AQB + \Delta\, BQC + \Delta\, AQC}{\Delta\, AQC} = \frac{\Delta\, ABC}{\Delta\, AQC} = 1 + k + \frac{1}{k} = \frac{1 + k + k^2}{k}.$$

By cyclic symmetry we have

$$\frac{\Delta\, AQC}{\Delta\, ABC} = \frac{\Delta\, BRA}{\Delta\, ABC} = \frac{\Delta\, CPB}{\Delta\, ABC} = \frac{k}{1 + k + k^2}.$$

Now $\Delta\, PQR = \Delta\, ABC - \Delta\, AQC - \Delta\, BRA - \Delta\, CPB$, so

$$\frac{\Delta\, PQR}{\Delta\, ABC} = 1 - 3\,\frac{k}{1 + k + k^2} = \frac{(1 - k)^2}{1 + k + k^2}.$$

4. Assume that $|f(x)| \le 1$ and $|f''(x)| \le 1$ for all x on an interval of length at least 2. Show that $|f'(x)| \le 2$ on the interval.

First Solution. We may suppose without loss of generality that the interval in question is $[-1, +1]$. Using Taylor's formula to expand about the point $x \in [-1, 1]$, we find

$$f(1) = f(x) + (1 - x)f'(x) + \frac{1}{2}(1 - x)^2 f''(\xi),$$

$$f(-1) = f(x) + (-1 - x)f'(x) + \frac{1}{2}(-1 - x)^2 f''(\eta),$$

where $\xi \in (x, 1)$ and $\eta \in (-1, x)$. Hence

$$f(1) - f(-1) = 2f'(x) + \frac{1}{2}(1 - x)^2 f''(\xi) - \frac{1}{2}(1 + x)^2 f''(\eta).$$

Using the given bounds for f and f'', we get

$$2|f'(x)| \le |f(1)| + |f(-1)| + \frac{1}{2}(1 - x)^2 |f''(\xi)| + \frac{1}{2}(1 + x)^2 |f''(\eta)|$$

$$\le 2 + \frac{1}{2}(1 - x)^2 + \frac{1}{2}(1 + x)^2 = 3 + x^2 \le 4.$$

Therefore

$$|f'(x)| \le 2.$$

Second Solution. Essentially the same argument, but less elegantly phrased, is the following.

Suppose there is an x in $[-1, 1]$ such that

$$f'(x) = 2 + \epsilon, \qquad \text{where } \epsilon > 0.$$

Since $|f''(t)| \le 1$ for all $t \in [-1, 1]$, we have

$$\begin{aligned} f'(t) &\ge 2 + \epsilon + t - x \quad \text{for } -1 \le t \le x \\ &\ge 2 + \epsilon - t + x \quad \text{for } x \le t \le 1. \end{aligned}$$

Hence

$$f(1) - f(-1) = \int_{-1}^{+1} f'(t)\,dt$$

$$\geq 4 + 2\epsilon - \frac{1}{2}(1 + x)^2 - \frac{1}{2}(1 - x)^2 = 3 + 2\epsilon - x^2$$

$$\geq 2 + 2\epsilon.$$

But $|f(1) - f(-1)| \leq |f(1)| + |f(-1)| \leq 2$. Hence there can be no such x.

Similarly we cannot have $f'(x) < -2$.

REMARKS. The inequality is the best possible, for, if $f(x) = \frac{1}{2}(x + 1)^2 - 1$, the hypothesis is satisfied and $f'(1) = 2$. The first solution shows that $|f'|$ can attain the value 2 only for $x = \pm 1$, and then it follows easily that the function just described and three others obtained from it by reflection are the only extremal functions.

The result was first established by Landau in "Einige Ungleichungen fur zweimal differentzierbaren Funktionen," *Proceedings of the London Mathematical Society* (2), vol. 13 (1914), pages 43-49. For a treatment of several similar inequalities, see I. J. Schoenberg, "The Elementary Cases of Landau's Problem of Inequalities between Derivatives," *American Mathematical Monthly*, vol. 80 (1973), pages 121-158.

---

5. Evaluate in closed form

$$\sum_{k=1}^{n} \binom{n}{k} k^2.$$

Note:

$$\binom{n}{k} = \frac{n(n-1)\cdots(n-k+1)}{1 \cdot 2 \cdots k}.$$

First Solution.

$$\sum_{k=1}^{n} \binom{n}{k} k^2 = \sum_{k=1}^{n} \frac{n!}{(n-k)!k!}[k(k-1) + k]$$

$$= \sum_{k=2}^{n} \frac{n(n-1)(n-2)!}{(n-k)!(k-2)!} + \sum_{k=1}^{n} \frac{n(n-1)!}{(n-k)!(k-1)!}$$

$$= n(n-1)\sum_{j=0}^{n-2}\binom{n-2}{j} + n\sum_{j=0}^{n-1}\binom{n-1}{j}$$

$$= n(n-1)2^{n-2} + n2^{n-1} = n(n+1)2^{n-2}.$$

Second Solution. If we apply the operator $x(d/dx)$ to

$$(1+x)^n = \sum_{k=0}^{n}\binom{n}{k}x^k,$$

we obtain the identity

$$nx(1+x)^{n-1} = \sum_{k=1}^{n}\binom{n}{k}kx^k.$$

Applying it again, we find

$$n(n-1)x^2(1+x)^{n-2} + nx(1+x)^{n-1} = \sum_{k=1}^{n}\binom{n}{k}k^2x^k.$$

Put $x = 1$ and we have

$$\sum_{k=1}^{n}\binom{n}{k}k^2 = n(n-1)2^{n-2} + n2^{n-1} = n(n+1)2^{n-2}.$$

---

6. Let S be a set of rational numbers such that whenever a and b are members of S, so are $a+b$ and ab, and having the property that for every rational number r exactly one of the following three statements is true:

$$r \in S, \quad -r \in S, \quad r = 0.$$

Prove that S is the set of all positive rational numbers.

Solution. If $r \neq 0$, then either $r \in S$ or $-r \in S$. Since $r^2 = (-r)^2$, in either case we have $r^2 \in S$.

In particular, $1 \in S$. Then from the sum property it follows that every positive integer is in S.

If p and q are positive integers, then $1/q^2 \in S$ by our first result, and

$$\frac{p}{q} = pq\left(\frac{1}{q^2}\right) \in S$$

by the product property. Thus every positive rational is in S. Now the hypothesis implies that no negative rational is in S and $0 \notin S$, so S is just the set of positive rational numbers.

REMARKS. The given properties of the set S characterize the set of positive elements in an abstract ordered field. Hence the result can be interpreted as meaning that there is only one way to make the field of rational numbers into an ordered field. See, for example, "Birkhoff and MacLane, "A Survey of Modern Algebra, Macmillan, New York, 1941, page 48.

If the product property is dropped from the hypothesis, then we can prove that S is either the set of positive rationals or the set of negative rationals. For each positive integer q, choose $\epsilon_q = \pm 1$ so that $\epsilon_q/q \in S$. The sum rule shows that $\epsilon_q(p/q) \in S$ for all positive integers p. Then the choice $p = q$ gives $\epsilon_q = \epsilon_1$ and therefore $\epsilon_1 r \in S$ for all positive rationals r.

Afternoon Session

1. Let $x^{(n)} = x(x-1)\cdots(x-n+1)$ for n a positive integer and let $x^{(0)} = 1$. Prove that

$$(x+y)^{(n)} = \sum_{k=0}^{n} \binom{n}{k} x^{(k)} y^{(n-k)}.$$

Note:

$$\binom{n}{k} = \frac{n(n-1)\cdots(n-k+1)}{1\cdot 2\cdots k}.$$

First Solution. We use induction on n. The required equation is clearly valid for $n = 0$. Suppose it is true for $n = p$. Then

$$(x+y)^{(p+1)} = (x+y)^{(p)}(x+y-p)$$

$$= \sum_{k=0}^{p} \binom{p}{k} x^{(k)} y^{(p-k)}(x-k+y-(p-k))$$

$$= \sum_{k=0}^{p} \binom{p}{k} x^{(k+1)} y^{(p-k)} + \sum_{k=0}^{p} \binom{p}{k} x^{(k)} y^{(p-k+1)}$$

$$= \sum_{k=1}^{p+1} \binom{p}{k-1} x^{(k)} y^{(p+1-k)} + \sum_{k=0}^{p} \binom{p}{k} x^{(k)} y^{(p+1-k)}$$

$$= \sum_{k=0}^{p+1} \left[\binom{p}{k-1} + \binom{p}{k} \right] x^{(k)} y^{(p+1-k)}$$

$$= \sum_{k=0}^{p+1} \binom{p+1}{k} x^{(k)} y^{(p+1-k)}.$$

Thus it is also true for $n = p + 1$. This completes the induction.

Second Solution. We have

$$(1 + t)^x = \sum_{k=0}^{\infty} \frac{x^{(k)}}{k!} t^k,$$

$$(1 + t)^y = \sum_{k=0}^{\infty} \frac{y^{(k)}}{k!} t^k$$

for $|t| < 1$ and all x, y. Multiplying these series we obtain

$$(1 + t)^{x+y} = \sum_{n=0}^{\infty} \sum_{k=0}^{n} \frac{x^{(k)} y^{(n-k)}}{k!(n-k)!} t^n,$$

for $|t| < 1$. But we know

$$(1 + t)^{x+y} = \sum_{n=0}^{\infty} \frac{(x+y)^{(n)}}{n!} t^n.$$

Since power series representations are unique, we have

$$\frac{(x+y)^{(n)}}{n!} = \sum_{k=0}^{n} \frac{x^{(k)} y^{(n-k)}}{k!(n-k)!},$$

which is equivalent to the required identity.

---

2. Let R be the set of all real numbers and S the set of all subsets of the positive integers. Construct a function f whose domain is R and whose range is in S, such that $f(a)$ is a proper subset of $f(b)$ whenever $a < b$.

Solution. Let ϕ be an enumeration of the rationals, i.e., a bijective map from **N** to **Q**, where **N** is the set of positive integers and **Q** is the set of rational numbers.

Define a function f as follows

$$f(r) = \{n : \phi(n) < r\}$$

where r is a real number. Then the domain of f is $\mathbf{R}$ and the values of f are subsets of $\mathbf{N}$.

If $a < b$ and $n \in f(a)$, then $\phi(n) < a < b$, so $n \in f(b)$. Thus $f(a) \subseteq f(b)$. Moreover, the inclusion is proper because there is a rational number between a and b and therefore an integer p such that $a < \phi(p) < b$, so $p \in f(b)$ but $p \notin f(a)$.

REMARK. For the proof it is sufficient that $\phi: \mathbf{N} \to \mathbf{R}$ have dense range.

---

3. Let S be a convex region in the Euclidean plane containing the origin. Assume that every ray (that is, half-line) from the origin has at least one point outside S. Prove that S is bounded. (A region in the plane is defined to be convex if and only if the line segment joining every pair of its points lies entirely within the region.)

Solution. The examiners presumably intended *region* to mean *open set*, for the statement is false without some topological hypothesis on S. For example, the strip $0 < x < 1$ in the xy-plane together with the origin forms an unbounded convex set S such that no ray from the origin lies wholly in S.

We shall prove that a ray-bounded convex set S is bounded provided either the origin is an interior point of S or S is closed.

We shall use polar coordinates (ρ, θ). For any θ the set $\{\rho : (\rho, \theta) \notin S\}$ is not empty by hypothesis and bounded below by zero. Let

$$f(\theta) = \inf \{\rho : (\rho, \theta) \notin S\}.$$

The intersection of S with a ray is an interval, so we have

$$(1) \qquad 0 \leq \rho < f(\theta) \Rightarrow (\rho, \theta) \in S$$

and

$$(2) \qquad \rho > f(\theta) \Rightarrow (\rho, \theta) \notin S.$$

Hence, if M is an upper bound for f, then S lies in the closed disk of radius M about O.

Assume O is an interior point of S. Let D be a disk centered at O with

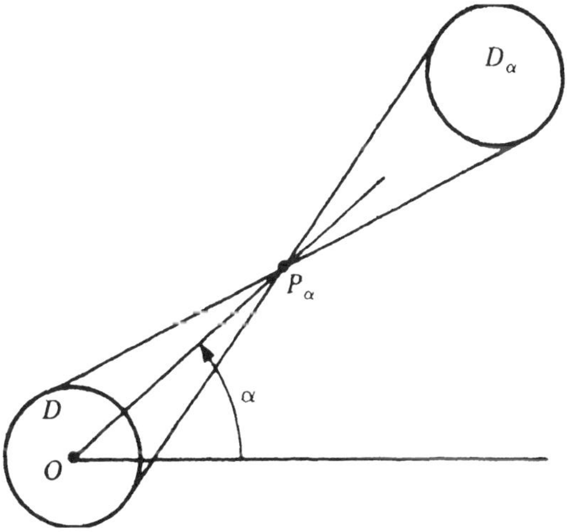

$D \subseteq S$. For any $\alpha \in [0, 2\pi]$, let P_α be the point $(1 + f(\alpha), \alpha)$ and let D_α be the disk obtained by reflecting D through P_α. Suppose $Q \in D_\alpha$, and let Q^* be its reflection through P_α. Then $Q^* \in D \subseteq S$, so if $Q \in S$, then P_α would lie between two points of S, which is impossible. So $D_\alpha \cap S = \emptyset$. Now D_α subtends a positive angle, say 2ϵ, at the origin, and every ray from the origin having direction angle in $I_\alpha = (\alpha - \epsilon, \alpha + \epsilon)$ meets D_α at a distance less than $2(1 + f(\alpha))$. Therefore

$$f(\theta) \leq 2(1 + f(\alpha))$$

for $\theta \in I_\alpha$. By the Heine-Borel theorem, some finite number of these intervals, say $I_{\alpha_1}, I_{\alpha_2}, \ldots, I_{\alpha_n}$, cover $[0, 2\pi]$. Then the largest of the numbers $2(1 + f(\alpha_1))$, $2(1 + f(\alpha_2))$, $\ldots$, $2(1 + f(\alpha_n))$, is an upper bound for f. This proves that S is bounded if O is an interior point.

Now assume that S is closed but unbounded. Then we can choose angles θ_n in $[0, 2\pi]$ so that $f(\theta_n) \to \infty$. By the Bolzano-Weierstrass theorem some subsequence of $\{\theta_n\}$ is convergent, and we may as well assume that $\{\theta_n\}$ itself is convergent. Say $\theta_n \to \beta$. Let $\rho_0 = 1 + f(\beta)$. Then for all large n, $f(\theta_n) > \rho_0$, so by (1), $(\rho_0, \theta_n) \in S$. Now S is closed and $(\rho_0, \theta_n) \to (\rho_0, \beta)$, so $(\rho_0, \beta) \in S$. But we know from (2) that $(\rho_0, \beta) \notin S$. This contradiction proves that S is bounded if it is closed.

---

4. The Euclidean plane is divided into regions by drawing a finite number of circles. Show that it is possible to color each of these regions either red or blue in such a way that no two adjacent regions have the same color. (Two such regions are said to be adjacent if and only if their boundaries have an arc of a circle in common.)

Solution. For each region into which the plane is divided, count the number of circles within which that region lies. If this number is odd color the region red, while if the number is even color it blue.

Then, if two regions share a common boundary arc, one of these regions is interior to one more of the given circles than is the other region. Hence these two regions have different colors.

REMARK. A number of coloring problems (from the simple up to the four-color problem) are discussed in the pamphlet by E. B. Dynkin and V. A. Uspenskii, *Multicolor Problems,* trans. from the Russian, Heath, Boston, 1963.

---

5. Prove that for every integer n greater than 1:

$$\frac{3n+1}{2n+2} < \left(\frac{1}{n}\right)^n + \left(\frac{2}{n}\right)^n + \cdots + \left(\frac{n}{n}\right)^n < 2.$$

Solution. For $n > 1$ and $x > 0$ the function x^n is convex (i.e., its graph is concave up), so its integral is less than the approximation given by the trapezoidal rule. Hence

$$\frac{1}{n+1} = \int_0^1 x^n\,dx < \frac{1}{n}\left[\frac{1}{2}\left(\frac{0}{n}\right)^n + \left(\frac{1}{n}\right)^n + \left(\frac{2}{n}\right)^n + \cdots + \left(\frac{n-1}{n}\right)^n + \frac{1}{2}\left(\frac{n}{n}\right)^n\right].$$

Multiply by n and add $1/2$ to get

$$\frac{3n+1}{2n+2} = \frac{n}{n+1} + \frac{1}{2} < \left[\left(\frac{1}{n}\right)^n + \left(\frac{2}{n}\right)^n + \cdots + \left(\frac{n-1}{n}\right)^n + \left(\frac{n}{n}\right)^n\right].$$

The required upper approximation can be obtained by considering the lower Riemann sum over the same subdivision (and adding 1 to both sides), but the argument below gives a much sharper estimate.

Since $1 - x \le e^{-x}$ for any x, we have

$$\left(1 - \frac{i}{n}\right)^n \le e^{-i}$$

for $0 \leq i \leq n$. Hence

$$\left(\frac{n}{n}\right)^n + \left(\frac{n-1}{n}\right)^n + \cdots + \left(\frac{2}{n}\right)^n + \left(\frac{1}{n}\right)^n \leq 1 + e^{-1} + \cdots + e^{-(n-1)}$$

$$\leq \frac{1}{1-e^{-1}} = \frac{e}{e-1} < 2.$$

Since $\lim_{n\to\infty}(1 - i/n)^n = e^{-i}$, it follows easily that

$$\lim_{n\to\infty}\left[\left(\frac{n}{n}\right)^n + \left(\frac{n-1}{n}\right)^n + \cdots + \left(\frac{2}{n}\right)^n + \left(\frac{1}{n}\right)^n\right] = \frac{e}{e-1}.$$

---

6. Let

$$f(x) = \sum_{k=0}^{n} a_k \sin kx + b_k \cos kx,$$

where a_k and b_k are constants. Show that, if $|f(x)| \leq 1$ for $0 \leq x \leq 2\pi$ and $|f(x_i)| = 1$ for $0 \leq x_1 < x_2 < \cdots < x_{2n} < 2\pi$, then $f(x) = \cos(nx + a)$ for some constant a.

Solution. By writing $\cos kx = \frac{1}{2}(e^{ikx} + e^{-ikx})$, $\sin kx = (1/2i)(e^{ikx} - e^{-ikx})$ and $z = e^{ix}$, any trigonometric polynomial $f(x)$ of degree less than or equal to n (i.e., a function of the given form) can be written in the alternative form

$$f(x) = z^{-n}P(z)$$

where $P(z)$ is an ordinary polynomial (with complex coefficients) of degree less than or equal to $2n$.

Each zero b of the trigonometric polynomial corresponds to a zero e^{ib} of $P(z)$ on the unit circle in the complex plane. Differentiating, and noting that $dz/dx = iz$, we find

$$f'(x) = iz^{-n}(zP'(z) - nP(z)).$$

Hence if b is a double root of f (i.e., $f(b) = f'(b) = 0$), then P has a double zero at e^{ib}. [And similarly for triple zeros, etc., but we shall not need that result.] Hence a non-zero trigonometric polynomial of degree n can have no more than $2n$ zeros in one period counting multiplicities (because, if so, the ordinary polynomial P would have more zeros than its degree). Also if it has $2n$ zeros, then it is determined by them, up to a constant multiple.

Now consider the given trigonometric polynomial f. We must assume that f

has real coefficients since for complex coefficients the result fails; e.g., for $f(x) = \cos x + i \sin x$. It also seems clear that the intent of the problem is that f is not a constant. Note that either $f(x) \equiv 1$ or $f(x) \equiv -1$ would satisfy the condition of the problem but not the conclusion if $n > 0$.

We see that

$$1 - f(x)^2 \quad \text{and} \quad f'(x)^2$$

are both non-negative trigonometric polynomials of degree $2n$ with double zeros at $2n$ distinct points $x_1, x_2, \ldots, x_{2n}$. Since we are assuming that $1 - [f(x)]^2$ is not identically zero, there is a constant $m \geq 0$ such that

$$f'(x)^2 = m^2[1 - f(x)^2]. \tag{1}$$

The possibility $m = 0$ leads to the conclusion that f is a constant, so we assume $m \neq 0$.

This differential equation has solutions of the form

$$f(x) = \cos(mx + a) \tag{2}$$

pieced together with segments of the form $f(x) = \pm 1$. [The differential equation is singular for $f = \pm 1$, and splitting of solutions may indeed occur along these lines; e.g.,

$$\begin{aligned} f(x) &= 1 \quad \text{for } x \leq 0 \\ &= \cos mx \quad \text{for } 0 < x \leq \frac{\pi}{m} \\ &= -1 \quad \text{for } x \geq \frac{\pi}{m} \end{aligned}$$

is a solution.]

We are interested only in solutions that are trigonometric polynomials of degree $\leq n$, and since these are analytic functions of period 2π there can be no piecing and hence they must have the form (2) with m an integer. When f is of the form (2), $|f(x)| = 1$ for exactly $2m$ values of x. Hence $m = n$ and we have established the desired conclusion.

THE TWENTY-FOURTH WILLIAM LOWELL PUTNAM MATHEMATICAL COMPETITION

December 7, 1963

Morning Session

1. (i) Show that a regular hexagon, six squares, and six equilateral triangles can be assembled without overlapping to form a regular dodecagon.

(ii) Let $P_1, P_2, \ldots, P_{12}$ be the successive vertices of a regular dodecagon. Explain how the three diagonals P_1P_9, P_2P_{11}, and P_4P_{12} intersect.

Solution. (i) Let squares be drawn on the outside of each side of a regular hexagon of side s. Then the wedge-shaped indentations have vertex angle $60°$ $(=360° - 90° - 120° - 90°)$, and adjacent sides of length s. Hence they can be filled in with equilateral triangles of side s. The resulting dodecagon has all sides of length s and all angles $150°$ $(=90° + 60°)$; hence it is regular.

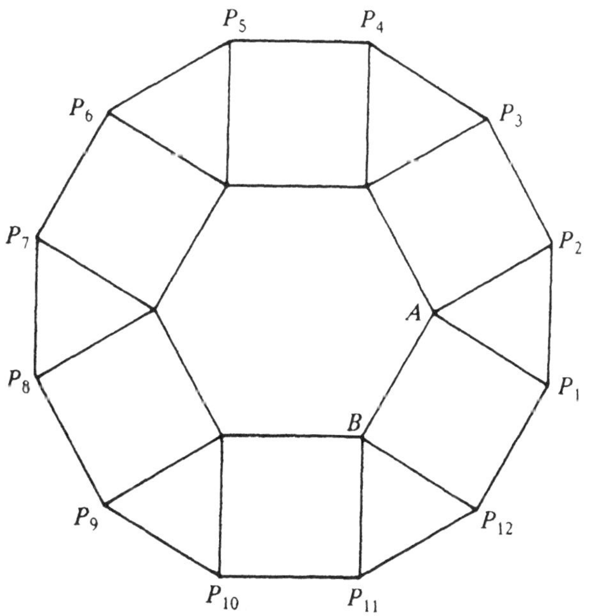

(ii) Letter the vertices as shown in the diagram. Since the regular dodecagon can be inscribed in a circle, $\angle P_1P_{12}P_4$ is half the central angle of arc P_1P_4, which is clearly a quarter circle. Hence $\angle P_1P_{12}P_4 = 45°$, so $P_{12}P_4$ is the diagonal of the square $P_1\ ABP_{12}$ and passes through A. Similarly P_1P_9 passes through B. Hence P_1P_9 and P_4P_{12} both pass through the center of the square P_1ABP_{12}. Now the hexagon $P_1P_2ABP_{11}P_{12}$ is evidently symmetric about the line P_2P_{11}, so P_2P_{11} also passes through the midpoint of the square P_1ABP_{12}. Thus the three diagonals mentioned are concurrent.

---

2. Let $\{f(n)\}$ be a strictly increasing sequence of positive integers such that $f(2) = 2$ and $f(mn) = f(m)f(n)$ for every relatively prime pair of positive integers m and n (the greatest common divisor of m and n is equal to 1). Prove that $f(n) = n$ for every positive integer n.

Solution. Since f is strictly increasing and integer-valued, $f(n + p) \geq f(n) + p$ for any integers n and p. Let $f(3) = a$. Then

$$f(5) \geq a + 2$$

$$f(15) = f(3)f(5) \geq a(a + 2) = a^2 + 2a$$

$$f(18) \geq a^2 + 2a + 3.$$

Also

$$f(6) = f(2)f(3) = 2a$$

$$f(5) \leq 2a - 1$$

$$f(10) = f(2)f(5) \leq 4a - 2$$

$$f(9) \leq 4a - 3$$

$$f(18) = f(2)f(9) \leq 8a - 6.$$

So, comparing the two values of $f(18)$, we have

$$a^2 + 2a + 3 \leq 8a - 6, \quad \text{i.e., } (a - 3)^2 \leq 0,$$

so $f(3) = a = 3$.

Now if $f(p) = p$ for some integer p, then $f(n) = n$ for all integers $n \leq p$ because of the strict monotonicity property.

We will establish

$$f(2^k + 1) = 2^k + 1$$

by induction. We have already shown this for $k = 1$. Suppose it is true for some integer t. Then

$$f(2^{t+1} + 2) = f(2)f(2^t + 1) = 2(2^t + 1) = 2^{t+1} + 2;$$

hence $f(2^{t+1} + 1) = 2^{t+1} + 1$. This completes the induction. Thus $f(p) = p$ for arbitrarily large integers p, and hence $f(n) = n$ for all n.

---

3. Find an integral formula for the solution of the differential equation

$$\delta(\delta - 1)(\delta - 2) \cdots (\delta - n + 1)y = f(x), \qquad x \geq 1,$$

for y as a function of x satisfying the initial conditions $y(1) = y'(1) = \cdots = y^{(n-1)}(1) = 0$, where f is continuous and

$$\delta \equiv x \frac{d}{dx}.$$

Solution. We first show that

$$\delta(\delta - 1)(\delta - 2) \cdots (\delta - n + 1)y = x^n \frac{d^n y}{dx^n}. \tag{1}$$

We prove this by induction on n. It is true for $n = 1$. Assume (1) is true for $n = k$. Since polynomials in the operator δ commute with one another, we have

$$\begin{aligned}\delta(\delta - 1)(\delta - 2) \cdots (\delta - k + 1)(\delta - k)y \\ &= (\delta - k)\delta(\delta - 1)(\delta - 2) \cdots (\delta - k + 1)y \\ &= \left(x \frac{d}{dx} - k\right) x^k \frac{d^k y}{dx^k} \\ &= x^{k+1} \frac{d^{k+1} y}{dx^{k+1}}.\end{aligned}$$

Thus (1) is proved by induction for all n.

The differential equation is thus

$$x^n y^{(n)} = f(x), \quad x \geq 1, \tag{2}$$

and the solution can obviously be obtained by applying the integral operator $\int_1^x$ n times to the function $f(x)x^{-n}$. However, it is possible to collapse the n-fold integration to a single integration as follows:

One of the standard forms of Taylor's theorem is

$$y(x) = y(a) + (x - a)y'(a) + \cdots + \frac{(x-a)^{n-1}}{(n-1)!}y^{(n-1)}(a)$$
$$+ \int_a^x \frac{(x-t)^{n-1}}{(n-1)!}y^{(n)}(t)\,dt,$$

if $y^{(n)}$ is continuous. (See, for example, Thomas, *Calculus and Analytic Geometry*, alternate ed., Addison-Wesley, Reading, Mass., 1972, page 814.) In this case, taking $a = 1$ and using (2) and the initial conditions, we have

$$y(x) = \int_1^x \frac{(x-t)^{n-1}}{(n-1)!}\cdot\frac{f(t)}{t^n}\,dt.$$

It is easy to check by direct differentiation that the function defined by this integral does indeed satisfy the differential equation (2) and the initial conditions.

---

4. Let $\{a_n\}$ be a sequence of positive real numbers. Show that

$$\limsup_{n\to\infty} n\left(\frac{1 + a_{n+1}}{a_n} - 1\right) \geq 1.$$

Show that the number 1 on the right-hand side of this inequality cannot be replaced by any larger number. (The symbol lim sup is sometimes written $\overline{\lim}$.)

Solution. Suppose that for some fixed integer k,

$$n\left(\frac{1 + a_{n+1}}{a_n} - 1\right) \leq 1$$

for all $n \geq k$. Then

$$1 + a_{n+1} \leq \frac{n+1}{n}a_n,$$

$$\frac{a_n}{n} \geq \frac{1}{n+1} + \frac{a_{n+1}}{n+1}$$

for $n \geq k$. Accordingly we have

$$\frac{a_k}{k} \geq \frac{1}{k+1} + \frac{a_{k+1}}{k+1} \geq \frac{1}{k+1} + \frac{1}{k+2} + \frac{a_{k+2}}{k+2}$$

$$\geq \frac{1}{k+1} + \frac{1}{k+2} + \cdots + \frac{1}{k+p} + \frac{a_{k+p}}{k+p}$$

$$\geq \frac{1}{k+1} + \frac{1}{k+2} + \cdots + \frac{1}{k+p}$$

for each p. However, this is impossible since the harmonic series diverges.

Thus for any k there exists an $n \geq k$ such that

$$n\left(\frac{1 + a_{n+1}}{a_n} - 1\right) > 1.$$

Therefore

$$\limsup_{n\to\infty} n\left(\frac{1 + a_{n+1}}{a_n} - 1\right) \geq 1.$$

If we take $a_n = n \log n$, then

$$n\left(\frac{1 + a_{n+1}}{a_n} - 1\right) = \frac{1 + (n+1)\log(n+1) - n\log n}{\log n}$$

$$= \frac{1}{\log n}\left[1 + n \log \frac{n+1}{n} + \log(n+1)\right]$$

$$\leq \frac{1}{\log n}\,[2 + \log(n+1)]$$

since $\log(1 + x) \leq x$ for all $x > -1$. Since the right side has limit 1, we have, for the particular sequence,

$$\limsup_{n\to\infty} n\left(\frac{1 + a_{n+1}}{a_n} - 1\right) \leq 1.$$

Thus we cannot increase the bound 1.

We could also take $a_n = n^{1+\epsilon}$ where $\epsilon > 0$ and obtain

$$\limsup_{n\to\infty} n\left(\frac{1 + a_{n+1}}{a_n} - 1\right) = 1 + \epsilon.$$

5. (i) Prove that if a function f is continuous on the closed interval $[0, \pi]$ and if

$$\int_0^\pi f(\theta) \cos \theta \, d\theta = \int_0^\pi f(\theta) \sin \theta \, d\theta = 0$$

then there exist points α and β such that

$$0 < \alpha < \beta < \pi \quad \text{and} \quad f(\alpha) = f(\beta) = 0.$$

(ii) Let R be any bounded convex open region in the Euclidean plane (that is, R is a connected open set contained in some circular disk, and the line segment joining any two points of R lies entirely in R). Prove with the help of part (i) that the centroid (center of gravity) of R bisects at least three distinct chords of the boundary of R.

Solution. (i) We may assume that $f \not\equiv 0$. Then since $\int_0^\pi f(\theta) \sin \theta \, d\theta = 0$ and $\sin \theta > 0$ for $0 < \theta < \pi$, f must change sign somewhere on $(0, \pi)$, say at α. If α is the only zero of f on $(0, \pi)$, then f has one sign on $(0, \alpha)$ and the opposite sign on (α, π). In the latter event,

$$\int_0^\pi f(\theta) \sin (\theta - \alpha) \, d\theta \neq 0.$$

But

$$\int_0^\pi f(\theta) \sin (\theta - \alpha) \, d\theta = \cos \alpha \int_0^\pi f(\theta) \sin \theta \, d\theta - \sin \alpha \int_0^\pi f(\theta) \cos \theta \, d\theta = 0.$$

This contradiction implies the existence of a second point β with $f(\beta) = 0$.

(ii) Take polar coordinates with pole at the centroid P of the bounded convex open region R, and write the equation of the bounding curve Γ of R as $\rho = g(\theta)$. [Since R is convex, each ray from P meets Γ just once. Thus Γ has such an equation.]

Γ is compact and the mapping $(\rho, \theta) \to (1, \theta)$ (polar coordinates) of Γ into the unit circle is continuous and bijective; therefore, the inverse mapping $(1, \theta) \to (g(\theta), \theta)$ is continuous. So g is continuous.

The moments of the region about the lines $\theta = 0$ and $\theta = \pi/2$ are given by

$$\iint_R \rho \cos \theta \, \rho \, d\rho d\theta \quad \text{and} \quad \iint_R \rho \sin \theta \, \rho \, d\rho d\theta.$$

These are both zero since we selected the origin at the centroid. Integration with respect to ρ gives

$$\frac{1}{3}\int_0^{2\pi}[g(\theta)]^3\cos\theta\, d\theta = 0 = \frac{1}{3}\int_0^{2\pi}[g(\theta)]^3\sin\theta\, d\theta.$$

Since $\cos(\theta+\pi) = -\cos\theta$, $\sin(\theta+\pi) = -\sin\theta$ this gives

$$0 = \int_0^{\pi}[g(\theta)]^3\cos\theta\, d\theta - \int_0^{\pi}[g(\theta+\pi)]^3\cos\theta\, d\theta$$

$$= \int_0^{\pi}\{[g(\theta)]^3 - [g(\theta+\pi)]^3\}\cos\theta\, d\theta = 0,$$

and similarly

$$0 = \int_0^{\pi}\{[g(\theta)]^3 - [g(\theta+\pi)]^3\}\sin\theta\, d\theta = 0.$$

Now according to part (i),

$$[g(\theta)]^3 - [g(\theta+\pi)]^3 = 0$$

holds for at least two values of θ in $(0, \pi)$. For these values

$$g(\theta) = g(\theta+\pi),$$

which means that the centroid bisects the chords having these two directions.

Now since we know that P must bisect at least one chord, we may as well assume that the polar axis was chosen in such a direction. Then the argument above shows that P also bisects at least two other chords.

---

6. Let U and V be any two distinct points on an ellipse, let M be the midpoint of the chord UV, and let AB and CD be any two other chords through M. If the line UV meets the line AC in the point P and the line BD in the point Q, prove that M is the midpoint of the segment PQ.

First Solution. Let UV be the x-axis of an oblique coordinate system with M the origin and the y axis distinct from lines AB and CD. Suppose $y = mx$ and $y = nx$ are the equations of lines AB and CD respectively. Let $ax^2 + by^2 + cxy + dx + ey + f = 0$ be the equation of the ellipse. Since the ellipse passes through $V(t, 0)$ and $U(-t, 0)$, it follows that $d = 0$.

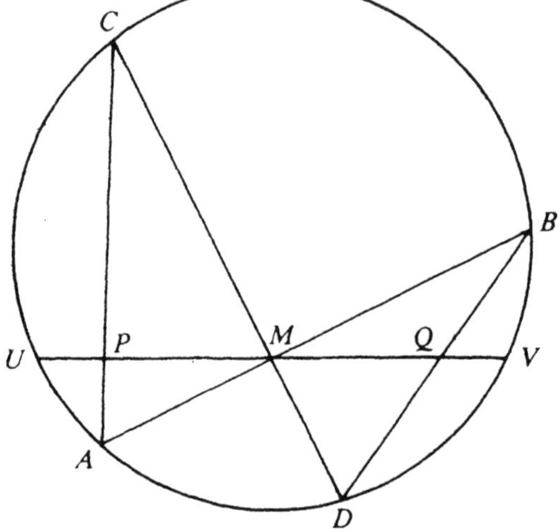

Now any conic through the four points A, B, C, D can be represented by

$$k_1(ax^2 + by^2 + cxy + ey + f) + k_2(y - mx)(y - nx) = 0$$

for suitable choice of k_1, k_2. Such a conic intersects the x-axis in two points whose x coordinates satisfy $k_1(ax^2 + f) + k_2 mn\, x^2 = 0$, i.e., in two points symmetric in M. In particular the degenerate conic consisting of the lines CA and DB intersects the x-axis in points P and Q which are symmetric relative to M: i.e., M is the midpoint of PQ.

Second Solution. We treat the problem projectively. We consider all conics through the four points A, B, C, D. Let $R = AC \cap BD$ and $S = AD \cap BC$. Then the polar of the point M with respect to any of the conics is the line RS.

Let $T = RS \cap UV$. Considering the original ellipse we see that U, V are divided harmonically by M and T. (Since any chord of a conic is divided harmonically by any point on it and the polar of that point.) Since M is the midpoint of UV, it follows that T is on the line at infinity.

Let $P = AC \cap UV = AC \cap MT$ and $Q = BD \cap UV = BD \cap MT$. Then PQ is a chord of the degenerate conic $AC \cup BD$ so it is divided harmonically by M and T. Since T is at infinity, M is the midpoint of PQ.

Third Solution. The following argument avoids explicit consideration of degenerate conics. Let $X \mapsto X'$ be a projective transformation that carries the given ellipse E into a circle with center M'. This is possible since we can first arrange that the polar m of M is transformed into the line at infinity. Then since the line m does not intersect E (M is inside of E), E' is a bounded central conic (i.e., an ellipse) with center M'.

A further transformation (stretching in one direction) will take E' into a circle.

Suppose T is the point at infinity on UV. Then $U, V; M, T$ is a harmonic quadruple. Hence $U', V'; M', T'$ is also a harmonic quadruple. Since M' bisects $U'V'$, it follows that T' is at infinity.

Now $A'C'B'D'$ is a rectangle, since $A'B'$ and $C'D'$ are diameters of the circle E', so it is immediate that M' bisects $P'Q'$. Hence $P', Q'; M', T'$ is harmonic, so $P, Q; M, T$ is harmonic. Since T is at infinity, M bisects PQ.

REMARKS. These proofs show, of course, much more than was asked for. The ellipse can be any conic and the two lines AD and CB can be replaced by any conic through A, B, C, D.

This theorem is well known in the literature of conic sections as the "butterfly theorem." See, for example, *A Survey of Geometry*, H. Eves, Boston, 1972; also M. S. Klamkin, "An Extension of the Butterfly Problem," *Mathematics Magazine*, vol. 38 (1965), pages 206–208, and C. D. Chakerian, G. T. Sallee, and M. S. Klamkin, "The Butterfly Property," *Mathematics Magazine*, 42 (1969).

Afternoon Session

1. For what integer a does $x^2 - x + a$ divide $x^{13} + x + 90$?

Solution. Suppose $x^{13} + x + 90 = (x^2 - x + a)q(x)$ where a is an integer. Then q is a polynomial with integer coefficients. If $a \leq 0$, then $x^2 - x + a$, and hence also $x^{13} + x + 90$, would have a non-negative zero, which is impossible. So $a > 0$.

Substituting $x = -1, 0, 1$, we find

$$(a + 2)q(-1) = 88,$$
$$aq(0) = 90,$$
$$aq(1) = 92.$$

The last two equations show that a divides 2, so $a = 1$ or 2. But if $a = 1$, then 3 would divide 88. So $a = 2$ is the only possibility and, in fact,

$$x^{13} + x + 90 = (x^2 - x + 2)(x^{11} + x^{10} - x^9 - 3x^8 - x^7 + 5x^6 + 7x^5 - 3x^4 - 7x^3 - 11x^2 + 23x + 45).$$

---

2. Let S be the set of all numbers of the form $2^m 3^n$, where m and n are integers, and let P be the set of all positive real numbers. Is S dense in P?

Solution. Because the logarithm function and its inverse, the exponential function, are continuous, the proposed question amounts to asking whether numbers of the form

$$m \log 2 + n \log 3, \qquad m \text{ and } n \text{ integers},$$

are dense in all of $\mathbf{R}$. We use the following theorem.

THEOREM 1. *If $\alpha, \beta \in \mathbf{R}$, then the numbers of the form $m\alpha + n\beta$, m and n integers, are dense in $\mathbf{R}$ unless there are integers p and q not both zero such that*

$$p\alpha + q\beta = 0. \tag{1}$$

(A proof of this theorem is given below.)

In our particular case, $\alpha = \log 2$ and $\beta = \log 3$ and, on taking exponentials, (1) becomes

$$2^p 3^q = 1$$

for integers p and q not both zero. This is clearly impossible by the unique factorization theorem. So we conclude that S is dense in P.

To prove the theorem, we start with another theorem.

THEOREM 2. *If T is a subgroup of the additive group of $\mathbf{R}$, then either $T = \{0\}$, T consists of all multiples of some positive number, or T is dense in $\mathbf{R}$.*

Proof of Theorem 2. If T contains no positive numbers, then $T = \{0\}$.

Suppose T contains a least positive number, say x. We shall prove that T consists of all multiples of x. Clearly, T contains all multiples of x. Suppose $y \in T$. There is an integer n such that $n \le y/x < n + 1$. Then $z = y - nx$ is in T and $0 \le z < x$. Since x is the least positive element of T, $z = 0$; thus, $y = nx$. This proves that T consists of all multiples of x.

Suppose T contains a positive number x, but no least positive number. Let $I = (a, a + \delta)$ be an open interval in $\mathbf{R}$. Since T has infinitely many elements in $(0, x)$, two of them, say t_1 and t_2, are within δ of one another. We may assume $t_1 < t_2$. Then $s = t_2 - t_1$ is in T and $0 < s < \delta$. All multiples of s are in T, and some multiple of s lies in the interval I, so $T \cap I \neq \emptyset$. Since I was arbitrary, T is dense in this case. ∎

Proof of Theorem 1. The numbers of the form $m\alpha + n\beta$ are a subgroup T of $\mathbf{R}$, and $\alpha \in T$ and $\beta \in T$. Suppose T is not dense in $\mathbf{R}$. Then, by the preceding theorem, T consists of all multiples of some non-negative number x. Say $\alpha = qx$, $\beta = -px$, where p and q are integers. Then

$$p\alpha + q\beta = 0.$$

If here $p = q = 0$, then $\alpha = \beta = 0$ and (1) can be satisfied trivially. ■

---

3. Find every twice-differentiable real-valued function f with domain the set of all real numbers and satisfying the functional equation

$$(f(x))^2 - (f(y))^2 = f(x + y)f(x - y)$$

for all real numbers x and y.

Solution. If we put $x = y = 0$ in the given relation, we find $f(0) = 0$. Then differentiating the given relation, first with respect to x and then with respect to y we get

$$\begin{aligned} 2f(x)f'(x) &= f'(x + y)f(x - y) + f(x + y)f'(x - y) \\ 0 &= f''(x + y)f(x - y) - f(x + y)f''(x - y). \end{aligned}$$

Now for any u and v we can choose $x = \frac{1}{2}(u + v)$, $y = \frac{1}{2}(u - v)$; then

$$f''(u)f(v) = f(u)f''(v). \tag{1}$$

If there is a point v_0 such that $f(v_0) \neq 0$ and if $c = f''(v_0)/f(v_0)$, then we can write (1) as

$$f''(u) = cf(u).$$

This differential equation along with the initial condition $f(0) = 0$ has three types of solutions

$$\begin{aligned} f(u) &= A \sinh ku && \text{if } c > 0, \quad c = k^2 \\ &= Au && \text{if } c = 0 \\ &= A \sin ku && \text{if } c < 0, \quad c = -k^2. \end{aligned}$$

We note that the possibility $f(v) \equiv 0$ is included in these cases when $A = 0$.

All three types are indeed solutions of the given functional equation. For example

$$\sin x + \sin y = 2 \sin \frac{1}{2}(x + y) \cos \frac{1}{2}(x - y)$$

$$\sin x - \sin y = 2 \sin \frac{1}{2}(x - y) \cos \frac{1}{2}(x + y)$$

so that

$$\begin{aligned}(A \sin x)^2 - (A \sin y)^2 &= \left(2A \sin \frac{1}{2}(x + y) \cos \frac{1}{2}(x + y)\right) \\ &\quad \times \left(2A \sin \frac{1}{2}(x - y) \cos \frac{1}{2}(x - y)\right) \\ &= (A \sin (x + y))(A \sin (x - y)).\end{aligned}$$

The check for the hyperbolic sine can be carried out similarly, or by noting that it follows by replacing x by ix and y by iy in the sine solution above. The remaining case, $f(u) = Au$, is trivial.

---

4. Let C be a closed plane curve that has a continuously turning tangent and bounds a convex region. If T is a triangle inscribed in C with maximum perimeter, show that the normal to C at each vertex of T bisects the angle of T at that vertex. If a triangle T has the property just described, does it necessarily have maximum perimeter? What is the situation if C is a circle? (A convex region is a connected open set such that the line segment joining any two points of the set lies entirely in the set.)

Solution. As in Problem A.M. 2 of the Third Competition, many of the hypotheses in this problem tend to obscure the generality and the simplicity of the result. C need not bound a convex region and the continuous turning of the tangent is irrelevant.

We first prove the following general fact, from which the assertion of the problem will quickly follow.

THEOREM. *If C is any set of points in the plane and T is a triangle with vertices in C whose perimeter is maximal among all such triangles, then C has at each vertex of T a support line perpendicular to the angle bisector of T at that vertex.*

Proof. Suppose P, Q, R are the vertices of T and let l be the line through P perpendicular to the bisector of $\angle QPR$. If Q^* is the reflection of Q in l then it is immediate that R, P, Q^* are collinear.

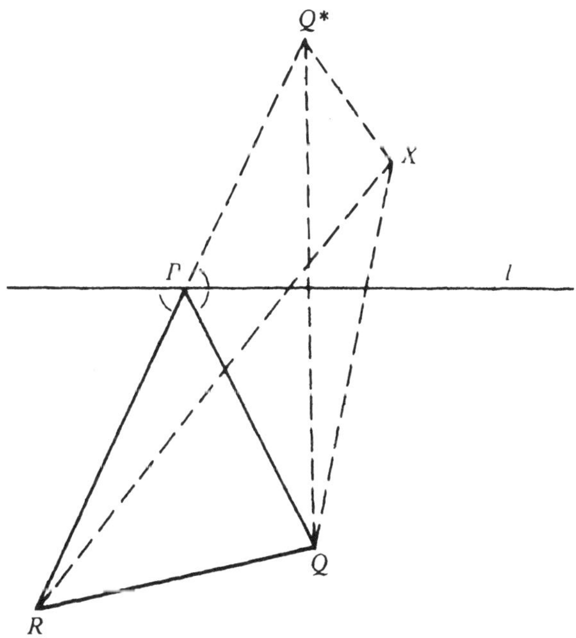

If X is any point on the Q^*-side of l, then $RQ + RX + XQ > RQ + RX + XQ^* \geq RQ + RQ^* = RQ + RP + PQ$. Thus ΔRXQ has greater perimeter than ΔPQR and X is not in C. By definition, then, l is a support line of C as claimed. ■

In the case where C is a differentiable closed curve, the support lines of C are precisely the tangents to C. (This is clear, but more details of this argument are given in the solution to A.M. 2, Third Competition.)

If C is approximately an equilateral triangle with slightly rounded corners, then the triangle with its vertices at the midpoint of the sides of C has the property that the normals to C bisect its angles, but it is clearly not of maximum perimeter.

If C is a circle, then standard compactness arguments show that there is a triangle inscribed in C having largest perimeter. This triangle has its angle bisectors concurrent at the center O of C. Hence, in the figure, $\widehat{QS} = \widehat{RS}$, so $\widehat{PQ} = \widehat{PR}$. By symmetry $\widehat{PQ} = \widehat{QR} = \widehat{RP}$, so ΔPQR is equilateral. Therefore:

Of all triangles inscribed in a given circle the equilateral triangles (and no others) have maximum perimeter.

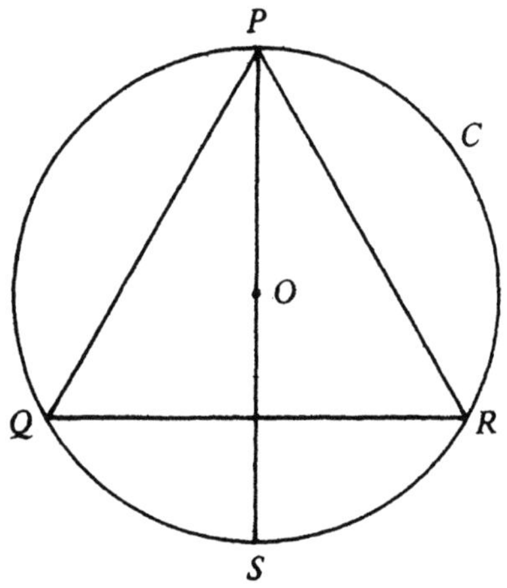

---

5. Let $\{a_n\}$ be a sequence of real numbers satisfying the inequalities

$$0 \leq a_k \leq 100a_n \text{ for } n \leq k \leq 2n \quad \text{and} \quad n = 1, 2, \ldots,$$

and such that the series

$$\sum_{n=0}^{\infty} a_n$$

converges. Prove that

$$\lim_{n \to \infty} na_n = 0.$$

Solution. For each positive integer k, let

$$S_k = \sum_{n \geq k/2}^{k} a_n .$$

Since $\Sigma\, a_n$ converges, we have $S_k \to 0$ as $k \to \infty$.

Rewriting the given condition slightly we have

$$0 \leq a_k \leq 100\, a_n \quad \text{for} \quad \frac{1}{2}k \leq n \leq k.$$

For each k there are at least $k/2$ integers n satisfying this double inequality.

Adding these inequalities we have, therefore,

$$\frac{1}{2} k a_k \leq 100 S_k .$$

Hence

$$\limsup k a_k \leq 200 \lim S_k = 0.$$

Since $a_k \geq 0$, we conclude $\lim k a_k = 0$.

---

6. Let E be a Euclidean space of at most three dimensions. If A is a nonempty subset of E, define $S(A)$ to be the set of all points that lie on closed segments joining pairs of points of A. For a given nonempty set A_0, define $A_n \equiv S(A_{n-1})$ for $n = 1, 2, \ldots$. Prove that $A_2 = A_3 = \cdots$. (A one-point set should be considered to be a special case of a closed segment.)

Solution. We consider E as a vector space as usual. If $a_1, a_2, \ldots, a_n$ are in E, then any point (vector) p which can be written in the form

$$p = \sum_{i=1}^{n} \lambda_i a_i ,$$

where $\lambda_i \geq 0$ and $\Sigma \lambda_i = 1$, is called a *convex combination* of the a's.

Convex combination is transitive; that is, if $b_1, b_2, \ldots, b_j$ are each convex combinations of $a_1, a_2, \ldots, a_n$, and c is a convex combination of $b_1, b_2, \ldots, b_j$, then c is a convex combination of $a_1, a_2, \ldots, a_n$. It follows that, if A is a given set in E, the set K of all convex combinations of elements of A is a convex set; indeed, it is the smallest convex set containing A. K is called the *convex hull* of A.

The essence of the problem lies in the following theorem.

THEOREM. *Suppose E has dimension n, and A is a subset of E. Then every point in the convex hull K of A can be written as a convex combination of at most $n + 1$ points of A.*

Proof. Suppose $p \in K$, but p cannot be written as a convex combination of fewer than $n + 2$ points of A. Since $p \in K$, we can write

$$p = \sum_{i=1}^{q} \lambda_i a_i , \tag{1}$$

where $a_i \in A$, $\lambda_i \geq 0$, and $\Sigma\, \lambda_i = 1$. Of all such representations we choose one with q as small as possible. Then $q > n + 1$.

There exist numbers $\mu_1, \mu_2, \ldots, \mu_q$, not all zero, such that

$$\sum_{i=1}^{q} \mu_i = 0$$

(2)

$$\sum_{i=1}^{q} \mu_i a_i = \underset{\cdot}{0},$$

since (2) can be regarded as a system of $n + 1$ linear homogeneous equations in more than $n + 1$ unknowns μ_i.

From (1) and (2) we have

$$p = \sum_{i=1}^{q} (\lambda_i + \sigma\mu_i)a_i$$

for any $\sigma \in \mathbf{R}$. Here the coefficient sum is always 1. We can choose σ so that *one* of the new coefficients $\lambda_i + \sigma\mu_i = 0$ while the rest are non-negative. Indeed, let σ be the largest of the (negative) numbers $-\lambda_i/\mu_i$, considering only i's for which $\mu_i > 0$, of which there must be at least one. But this gives a representation of p as a convex combination of fewer than q elements of A, which is impossible because of our choice of q. This contradiction proves the theorem. ■

Returning to the problem, let K be the convex hull of A_0. If X is any subset of K, then $X \subseteq S(X) \subseteq K$. (The first inclusion requires that degenerate segments be allowed; x is on the "segment" from x to x.) It follows that

(3) $$A_0 \subseteq A_1 \subseteq A_2 \subseteq A_3 \subseteq \cdots \subseteq K.$$

Suppose $p \in K$. By the theorem we can write

$$p = \lambda_1 a_1 + \lambda_2 a_2 + \lambda_3 a_3 + \lambda_4 a_4,$$

where $a_i \in A_0$, $\Sigma\, \lambda_i = 1$, and $\lambda_i > 0$. (If perchance p can be represented as a convex combination of fewer than four points of A_0, we can allow several of the a_i's to be the same and thus arrange that all the coefficients are positive.) Put

$$q = \frac{\lambda_1}{\lambda_1 + \lambda_2} a_1 + \frac{\lambda_2}{\lambda_1 + \lambda_2} a_2$$

$$r = \frac{\lambda_3}{\lambda_3 + \lambda_4} a_3 + \frac{\lambda_4}{\lambda_3 + \lambda_4} a_4.$$

Then

$$q \in S(A_0) = A_1, \qquad r \in A_1, \quad \text{and}$$
$$p = (\lambda_1 + \lambda_2)q + (\lambda_3 + \lambda_4)r \in S(A_1) = A_2.$$

This proves that $K \subseteq A_2$. Then from (3) it follows that

$$A_2 = A_3 = \cdots = K.$$

THE TWENTY-FIFTH WILLIAM LOWELL PUTNAM MATHEMATICAL COMPETITION

December 5, 1964

Morning Session

1. Given a set of 6 points in the plane, prove that the ratio of the longest distance between any pair to the shortest is at least $\sqrt{3}$.

Solution. Suppose three of the points, say A, B, C, are collinear in that order. Then either $|AC| \geq 2|BC|$ or $|AC| \geq 2|AB|$. Hence a ratio of at least two occurs in this case, so we assume from now on that no three of the given points are collinear.

Next we prove that some three of the points form a triangle with one angle at least $120°$.

If the convex hull of the given points is a hexagon, then the angle sum is $720°$, and one of the angles must be at least $120°$, so some three consecutive vertices of the hexagon form the required triangle.

If the convex hull of the given points has fewer than six vertices, then one of the points, say P, is in the interior of the convex hull of the others. P must be in the interior of a triangle spanned by some other three of the given points (since we have disposed of the case where three points are collinear). Say P lies inside the triangle QRS, where Q, R, S are among the given points. Then one of the angles QPR, RPS, SPQ must be at least $120°$, since their sum is $360°$.

Finally, we show that in a triangle in which one angle is at least $120°$, the ratio of the longest side to the shortest side is at least $\sqrt{3}$. Let ABC be a triangle with $\angle A \geq 120°$. Then $-\cos A \geq \frac{1}{2}$. Suppose $b \geq c$.

By the law of cosines

$$a^2 = b^2 + c^2 - 2bc \cos A \geq b^2 + c^2 + bc \geq 3c^2,$$

so $a \geq \sqrt{3}c$, as claimed.

Remarks. This is not the best possible result. It can be shown that for six points in the plane the minimum ratio of longest distance to shortest is $2 \sin 72° \sim 1.902$, which is attained by the vertices of a regular pentagon and its center.

The minimum ratio of longest distance to shortest for sets of a fixed size in a specified space has been called the *critical ratio*. Thus we claim that the critical ratio for sextuples in the plane is 2 sin 72°.

For sets of $n + 2$ points in E^n the critical ratio has been determined and all sets attaining this ratio are known to be similar. (See J. J. Seidel, "Quasi-regular Distance Sets," *Nederl. Akad. Wetensch. Proc. Ser. A*, 72 = *Indag. Math.* 31 (1969), pp. 64–70.)

A somewhat related question is considered in Problem E 2193, *American Mathematical Monthly*, vol. 77 (1970), page 770: "For A, B and C, three points in the Euclidean plane, define B to be '*weakly between*' A and C if and only if angle $ABC \geq 120°$. Determine the minimal number of points required to insure the existence of at least one such weak-betweenness relation." The solution there shows that this minimal number is six, and the proof parallels the proof given above.

---

2. Find all continuous positive functions $f(x)$, for $0 \leq x \leq 1$, such that

$$\int_0^1 f(x)\,dx = 1$$

$$\int_0^1 f(x)x\,dx = \alpha$$

$$\int_0^1 f(x)x^2\,dx = \alpha^2$$

where α is a given real number.

Solution. Multiply the first integral equation by α^2, the second by -2α, the third by 1, and then add to obtain

$$\int_0^1 f(x)(\alpha - x)^2 dx = 0.$$

But this integral is clearly positive for any positive continuous function f. Hence there are no functions satisfying the conditions of the problem.

---

3. Let $P_1, P_2, \ldots$ be a sequence of distinct points which is dense in the interval (0, 1). The points $P_1, P_2, \ldots, P_{n-1}$ decompose the interval into n parts, and P_n decomposes one of these into two parts. Let a_n and b_n be the

lengths of these two intervals. Prove that

$$\sum_{n=1}^{\infty} a_n b_n (a_n + b_n) = 1/3.$$

(A sequence of points in an interval is said to be dense when every subinterval contains at least one point of the sequence.)

Solution. Let S_n be the sum of the cubes of the lengths of the segments formed by the points $P_1, P_2, \ldots, P_{n-1}, P_n$. Take $S_0 = 1$. We can obtain S_n from S_{n-1} by removing the term $(a_n + b_n)^3$ and replacing it by $a_n^3 + b_n^3$, leaving all other terms fixed.

Hence

$$\begin{aligned} S_{n-1} - S_n &= (a_n + b_n)^3 - a_n^3 - b_n^3 \\ &= 3a_n b_n (a_n + b_n). \end{aligned}$$

Therefore

$$1 - S_k = S_0 - S_k = \sum_{n=1}^{k} 3a_n b_n (a_n + b_n).$$

Since, as we shall prove below, $\lim_{k\to\infty} S_k = 0$,

$$\sum_{n=1}^{\infty} 3a_n b_n (a_n + b_n) = 1,$$

and the required result follows immediately.

We now show that $\lim_{k\to\infty} S_k = 0$. Let t be a positive integer. Because the set $\{P_i\}$ is dense, we can choose an integer q so large that $\{P_1, P_2, \ldots, P_q\}$ meets each of the intervals

$$\left[0, \frac{1}{t}\right], \left[\frac{1}{t}, \frac{2}{t}\right], \ldots, \left[\frac{t-1}{t}, 1\right].$$

Suppose $k \geq q$, and let $l_0, l_1, \ldots, l_k$ be the lengths of the intervals determined by the division points $P_1, P_2, \ldots, P_k$. Then each of these intervals has length at most $2/t$, so

$$S_k = \sum_{i=0}^{k} l_i^3 \leq \left(\frac{2}{t}\right)^2 \sum_{i=0}^{k} l_i = \frac{4}{t^2}.$$

Since t was arbitrary, this proves

$$\lim S_k = 0.$$

---

4. Let p_n $(n = 1, 2, \ldots)$ be a bounded sequence of integers which satisfies the recursion

$$p_n = \frac{p_{n-1} + p_{n-2} + p_{n-3}p_{n-4}}{p_{n-1}p_{n-2} + p_{n-3} + p_{n-4}}.$$

Show that the sequence eventually becomes periodic.

Solution. It is easy to prove a much more general theorem. Suppose f is any function with k arguments and $\{p_n : n = 1, 2, \ldots\}$ is a bounded sequence of integers satisfying the recursion

$$p_{n+k} = f(p_n, p_{n+1}, \ldots, p_{n+k-1})$$

for all $n = 1, 2, \ldots$. Then $\{p_n\}$ is eventually periodic.

Let q_n stand for the k-tuple $(p_n, p_{n+1}, \ldots, p_{n+k-1})$. Let $M = \sup\{|p_n|\}$. Then each p_n is one of the $2M + 1$ integers $-M, -M + 1, \ldots, M$ and there are at most $A = (2M + 1)^k$ possible k-tuples that q_n might be. Hence there must be some duplication in the sequence $q_1, q_2, \ldots, q_{A+1}$. Suppose then that $i < j$ and $q_i = q_j$. This implies that $p_{i+t} = p_{j+t}$ for $0 \le t < k$; then the recursion relation shows that this is true for $t = k$ also, and, by induction, for all $t \ge 0$. Thus the sequence $\{p_n\}$ is periodic with period $j - i$ starting with the ith term.

---

5. Prove that there is a constant K such that the following inequality holds for any sequence of positive numbers $a_1, a_2, a_3, \ldots$:

$$\sum_{n=1}^{\infty} \frac{n}{a_1 + a_2 + \cdots + a_n} \le K \sum_{n=1}^{\infty} \frac{1}{a_n}.$$

Solution. Let $k = 2t$ be some fixed even positive integer. We shall prove

$$\sum_{n=1}^{k} \frac{n}{a_1 + a_2 + \cdots + a_n} \le 4 \sum_{n=1}^{k} \frac{1}{a_n}.$$

From this inequality, the required inequality for infinite sums, with $K = 4$, follows immediately.

Let $b_1, b_2, \ldots, b_k$ be the terms $a_1, a_2, \ldots, a_k$ enumerated in increasing order. For $1 \le p \le t$ we have

$$\begin{aligned} a_1 + a_2 + \cdots + a_{2p} &\ge a_1 + a_2 + \cdots + a_{2p-1} \\ &\ge b_1 + b_2 + \cdots + b_{2p-1} \ge pb_p \end{aligned}$$

since all terms are positive and since the last p terms are at least b_p. Therefore

$$\frac{2p-1}{a_1 + a_2 + \cdots + a_{2p-1}} \le \frac{2p-1}{pb_p} < \frac{2}{b_p}.$$

Also

$$\frac{2p}{a_1 + a_2 + \cdots + a_{2p}} \le \frac{2}{b_p}.$$

Hence

$$\frac{2p-1}{a_1 + a_2 + \cdots + a_{2p-1}} + \frac{2p}{a_1 + a_2 + \cdots + a_{2p}} < \frac{4}{b_p}.$$

Thus

$$\sum_{p=1}^{t} \left[\frac{2p-1}{a_1 + a_2 + \cdots + a_{2p-1}} + \frac{2p}{a_1 + a_2 + \cdots + a_{2p}} \right] < \sum_{p=1}^{t} \frac{4}{b_p}.$$

Rewriting the left-hand sum, we have

$$\sum_{n=1}^{k} \frac{n}{a_1 + a_2 + \cdots + a_n} \le \sum_{p=1}^{t} \frac{4}{b_p} < \sum_{p=1}^{k} \frac{4}{b_p} = 4 \sum_{n=1}^{k} \frac{1}{a_n}.$$

If the series $\sum_{n=1}^{\infty} (1/a_n)$ diverges, the given inequality is deemed to be satisfied automatically. Otherwise we have

$$\sum_{n=1}^{k} \frac{n}{a_1 + \cdots + a_n} \le 4 \sum_{n=1}^{\infty} \frac{1}{a_n}$$

and hence

$$\sum_{n=1}^{\infty} \frac{n}{a_1 + \cdots + a_n} \le 4 \sum_{n=1}^{\infty} \frac{1}{a_n}.$$

REMARK. This inequality was first established by K. Knopp (*Jour. London Math. Soc.*, vol 3 (1928), pp. 205–211). He also proved that the least constant K that will do is 2.

Inspired by this problem, R. M. Redheffer (*Proc. London Math. Soc.*, vol. 17 (1967), pp. 683–699) has investigated a large class of related inequalities. He proves among other things the following inequality that is stronger than Knopp's:

$$\frac{3}{a_1} + \frac{5}{a_1 + a_2} + \frac{7}{a_1 + a_2 + a_3} + \cdots \le 4 \sum \frac{1}{a_n}. \tag{1}$$

In fact he proves the sharp finite version

$$\frac{(k+1)^2}{A_k} + \sum_{n=1}^{k-1} \frac{2n+1}{A_n} \le 4 \sum_{n=1}^{k} \frac{1}{a_n} \tag{2}$$

where $A_n = a_1 + a_2 + \cdots + a_n$, for any positive sequence $\{a_n\}$ with equality if and only if $a_1, a_2, \ldots, a_k$ are proportional to $1, 2, \ldots, k$. Here is his proof of (2).

LEMMA. *Suppose* $\lambda > 0$. *Then*

$$\frac{(\lambda + 2)^2}{1 + x} \le \frac{4}{x} + \lambda^2$$

for all positive x *with equality if and only if* $x = 2/\lambda$.

Proof. Let

$$f(x) = \frac{(\lambda + 2)^2}{1 + x} - \frac{4}{x}.$$

Then

$$f'(x) = -\frac{(\lambda + 2)^2}{(1 + x)^2} + \frac{4}{x^2}$$

and this is zero on $(0, \infty)$ only for $x = 2/\lambda$. This critical point is easily checked to be a maximum point for f and the lemma follows. ∎

Proof of (2). This is clearly true for $k = 1$. Assume it is true for $k = p$. Putting $x = a_{p+1}/A_p$ and applying the lemma with $\lambda = p$, we have

$$\frac{(p+2)^2}{A_{p+1}} = \frac{1}{A_p}\left(\frac{(p+2)^2}{1+x}\right) \leq \frac{1}{A_p}\left(\frac{4}{x} + p^2\right) = \frac{4}{a_{p+1}} + \frac{p^2}{A_p}. \tag{3}$$

Adding this inequality to the assumed inequality for $k = p$, we get

$$\frac{(p+2)^2}{A_{p+1}} + \frac{(p+1)^2}{A_p} + \sum_{n=1}^{p-1} \frac{2n+1}{A_n} \leq 4\sum_{n=1}^{p+1} \frac{1}{a_n} + \frac{p^2}{A_p}.$$

Canceling p^2/A_p from both sides, we obtain (2) for $k = p + 1$. This completes the proof by induction. There can be equality only if there is equality in (3) at each stage, i.e., only if $a_{p+1} = 2A_p/p$ from which follows that $a_p = pa_1$.

Now (1) follows immediately from (2). If the series on the right of (1) is convergent, then the a's are not proportional to the integers, and it follows that the inequality is strict.

---

6. Let S be a finite subset of a straight line. Say that S has the repeated distance property when every value of the distance between pairs of points of S (except for the longest) occurs at least twice. Show that if S has the repeated distance property then the ratio of any two distances between two points of S is a rational number.

Solution. All proofs seem to depend on considering the line as a vector space over $\mathbf{Q}$, the rational field. Once this idea is introduced, it is no harder to prove a more general result.

Suppose $S = \{s_1, s_2, \ldots, s_n\}$ is a finite set in a vector space V over $\mathbf{Q}$. We consider the $n(n-1)$ differences

$$s_1 - s_2, s_1 - s_3, \ldots, s_2 - s_1, s_2 - s_3, \ldots, s_n - s_{n-1}.$$

Some vectors may appear in this list more than once and we refer to them as repeated differences.

THEOREM. *The linear span of the non-repeated differences is the linear span of all the differences.*

Proof. Suppose this theorem is false. Then there is a linear functional f: $V \to \mathbf{Q}$ that annihilates all non-repeated differences but not all differences, (since we can replace V by the linear span of S, we may assume that V is finite-dimensional; then the existence of f follows from basic linear theory.) Let $M = f(s_i)$ and $m = f(s_j)$ be, respectively, the largest and smallest numbers in $f(S)$; then $M \neq m$. The linear functionals that map the finite set S injectively to $\mathbf{Q}$ (i.e., those that annihilate no differences) are dense, so there is such a linear functional g that satisfies

$$|f(s) - g(s)| \leq \frac{1}{5}(M - m)$$

for all $s \in S$. If $g(s_p)$ and $g(s_q)$ are the largest and smallest numbers in $g(S)$, then $s_p - s_q$ is certainly not a repeated difference, so $f(s_p - s_q) = 0$. Therefore

$$g(s_i) - g(s_j) \leq g(s_p) - g(s_q) \leq f(s_p) - f(s_q) + \frac{2}{5}(M - m) = \frac{2}{5}(M - m).$$

But also

$$g(s_i) - g(s_j) \geq f(s_i) - \frac{1}{5}(M - m) - f(s_j) - \frac{1}{5}(M - m) = \frac{3}{5}(M - m).$$

This contradiction proves the theorem. ■

Returning to the problem, let S be a set on a line with the repeated distance property. We identify the line with $\mathbf{R}$ (i.e., introduce a coordinate) so that 0 and 1 are the extreme members of S. We regard $\mathbf{R}$ as a vector space over $\mathbf{Q}$. The repeated distance property shows that 1 and -1 are the only non-repeated differences of S, so by the theorem all differences are in the linear span of 1 (over $\mathbf{Q}$). Hence all differences in S are rational numbers and all distances in S are rational multiples of the largest distance and hence have rational ratios to one another.

REMARKS. The result was first published by Mikusinski and Schinzel (*Acta Arithmetica*, vol. 9 (1964), pp 91-95) in connection with a problem in polynomial factorization. The more general result proved above was discovered by a group of UCLA undergraduates and published by E. G. Straus ("Rational Dependence in Finite Sets of Numbers," *Acta Arithmetica*, vol. 11 (1965), pp. 203-204.) Compare the date of the first paper with the date of this contest.

Afternoon Session

1. Let u_k ($k = 1, 2, \ldots$) be a sequence of integers, and let V_n be the

number of those which are less than or equal to n. Show that if

$$\sum_{k=1}^{\infty} 1/u_k < \infty,$$

then

$$\lim_{n\to\infty} V_n/n = 0.$$

Solution. It must have been intended that the u's be positive, since otherwise we could have $\Sigma\, 1/u_k$ convergent but all the V's infinite. Hence we assume that $\Sigma\, 1/u_k$ is a convergent series of positive terms. Then $u_k \to \infty$, and the V's are finite. Since the convergence of a series of positive terms is not affected by rearrangement, we can assume that $u_1 \le u_2 \le u_3 \le \cdots$. Then, for a fixed p and any n so large that $V_n > p$ we have

$$\sum_{k=p+1}^{V_n} 1/u_k \ge \frac{V_n - p}{n},$$

because there are $V_n - p$ terms in the sum and each is at least $1/n$.

Therefore

$$\limsup_{n\to\infty} \frac{V_n}{n} \le \sum_{k=p+1}^{\infty} 1/u_k.$$

Since this is true for any p we have

$$\limsup_{n\to\infty} \frac{V_n}{n} \le \lim_{p\to\infty} \sum_{k=p+1}^{\infty} 1/u_k = 0.$$

But $V_n/n \ge 0$ for all n, and therefore $\lim_{n\to\infty} V_n/n = 0$.

REMARKS. We need not assume the u's are integers.

We might consider u's of mixed sign and let V_n be the number of indices j for which $|u_j| \le n$, but the convergence of $\Sigma\, 1/u_k$ does not then imply that $V_n/n \to 0$, as we see by considering $u_k = (-1)^k k$.

---

2. Let S be a set of $n > 0$ elements, and let $A_1, A_2, \ldots, A_k$ be a family of distinct subsets, with the property that any two of these subsets meet. Assume that no other subset of S meets all of the A_i.

Prove that $k = 2^{n-1}$.

Solution. There are 2^n distinct subsets of S which we may consider as being arranged in 2^{n-1} complementary pairs. If k, the number of subsets in the family, is greater than 2^{n-1}, then two of these subsets would be complementary, and these two subsets would not intersect. On the other hand, if $k < 2^{n-1}$, we can find two complementary sets X and Y, neither of which is among the given sets $A_1, A_2, \ldots, A_k$. But no set not among the A's meets all the A's, so there is a set A_i with $A_i \cap X = \emptyset$, i.e., $A_i \subseteq Y$. Similarly there is a set A_j with $A_j \subseteq X$. But then $A_i \cap A_j \subseteq X \cap Y = \emptyset$, which contradicts the fact that any two of the A's meet. So the only possibility is $k = 2^{n-1}$.

REMARK. Evidently, such a family is given by the collection of all subsets of S which contain a fixed element of S. There are, however, other possibilities; for example, if $S = \{1, 2, 3\}$, then S, $\{1, 2\}$, $\{1, 3\}$, $\{2, 3\}$ is such a family.

---

3. Let $f(x)$ be a real continuous function defined for all real x. Assume that for every $\epsilon > 0$

$$\lim_{n\to\infty} f(n\epsilon) = 0, \qquad \text{(where } n \text{ is a positive integer).}$$

Prove that

$$\lim_{x\to\infty} f(x) = 0.$$

Solution. We begin by proving the following fact.

LEMMA. *If $0 < a < b$ and k is any positive integer, then*

$$\bigcup_{n=k}^{\infty} [na, nb]$$

contains the ray $[c, \infty)$ for some c.

Proof. Suppose t is an integer such that $t \geq k$, $t \geq a/(b-a)$. If $n \geq t$, then $n(b-a) \geq a$ so $nb \geq (n+1)a$. Hence the intervals $[na, nb]$ and $[(n+1)a, (n+1)b]$ overlap. Therefore

$$\bigcup_{n=k}^{\infty} [na, nb] \supseteq [ta, \infty). \quad \blacksquare$$

We return to the problem. Let $\alpha > 0$ be given. Define

$$F_n = \{x: |f(nx)| \leq \alpha\}$$

and

$$E_k = \{x: (\forall n \geq k) \, |f(nx)| \leq \alpha\}.$$

Then $E_k = \bigcap_{n \geq k} F_n$. Because f is continuous, each F_n is closed, and therefore each E_k is closed. If $y \in (0, \infty)$, then $\lim_{n\to\infty} f(ny) = 0$; hence for some k and all $n \geq k$, $|f(ny)| \leq \alpha$; that is, $y \in E_k$. Thus $(0, \infty) \subseteq \bigcup E_k$. By the Baire category theorem (proof below) one of the E's, say E_m, contains an interval $[a, b]$. This means

$$(\forall x \in [a, b])(\forall n \geq m) \, |f(nx)| \leq \alpha.$$

Therefore

$$\left(\forall y \in \bigcup_{n=m}^{\infty} [na, nb]\right) |f(y)| \leq \alpha.$$

Choose c so that $\bigcup_{n=m}^{\infty} [na, nb] \supseteq [c, \infty)$. Then

$$(\forall y \geq c) \, |f(y)| \leq \alpha.$$

Since α was arbitrary, this proves that $\lim_{y\to\infty} f(y) = 0$, as required.

BAIRE CATEGORY THEOREM. *Suppose $\{E_k\}$ is a sequence of closed subsets of $\mathbf{R}$ such that $\bigcup_{k=1}^{\infty} E_k$ contains an interval. Then at least one of the sets E_k contains an interval.*

Proof. Let I_0 be a bounded closed interval in $\bigcup_{k=1}^{\infty} E_k$. Assuming that the conclusion of the theorem is false, we shall construct inductively a decreasing sequence of closed intervals $I_0, I_1, I_2, \ldots$ such that $(\forall n > 1)$, $I_n \cap E_n = \emptyset$.

Since $I_0 \not\subseteq E_1$, there is a point $x_1 \in I_0 - E_1$. Since E_1 is closed, there is an interval about x_1 which does not meet E_1 and in this interval we can choose a closed interval $I_1 \subseteq I_0$. Continuing inductively, if we have chosen I_{n-1}, there is a point $x_n \in I_{n-1} - E_n$, and we can find a closed interval $I_n \subseteq I_{n-1}$ such that $I_n \cap E_n = \emptyset$.

Now the intersection of a nested sequence of bounded closed intervals cannot be void, so there exists a point x such that $x \in \bigcap_{n=0}^{\infty} I_n$. Then $x \in I_0$ but $x \notin \bigcup_{k=1}^{\infty} E_k$, and this contradicts the fact that $\bigcup_{k=1}^{\infty} E_k \supseteq I_0$. ∎

Note: The Baire Category Theorem is usually stated in the more general context of complete metric spaces, for example: Suppose $\{E_k\}$ is a sequence of closed subsets of a complete metric space such that $\bigcup_{k=1}^{\infty} E_k$ contains a non-void open set G. Then at least one of the sets E_k contains a non-void open subset of G.

---

4. Into how many regions do n great circles (no three concurrent) decompose the surface of the sphere on which they lie?

First Solution. Let $f(n)$ be the number of regions on the surface of a sphere formed by n great circles of which no three are concurrent. Clearly $f(1) = 2$, $f(2) = 4$. Suppose n circles have been drawn and an $(n + 1)$st circle is added. The new circle meets each of the old ones in two points, making $2n$ points of intersection, and these $2n$ points are all different since no three circles are concurrent. The $2n$ points divide the new circle into $2n$ arcs. Each of these arcs divides one of the old regions into two parts. Thus there are $2n + f(n)$ regions formed by the $(n + 1)$ circles. Hence we have

$$f(n+1) = 2n + f(n), \quad \text{for } n \geq 1. \tag{1}$$

It follows easily by induction that

$$f(n) = n^2 - n + 2 \quad \text{for } n \geq 1.$$

Obviously, $f(0) = 1$. Note that the argument leading to (1) breaks down if $n = 0$

Second Solution. Suppose n is at least two and consider the subdivision of the sphere given by n great circles. Let V be the number of vertices, E the number of edges, and F the number of faces (i.e., regions) in the subdivision. By Euler's formula

$$V - E + F = 2.$$

Now there are $2 \cdot n(n - 1)/2$ vertices, since each two great circles cross twice to make two vertices. Since there are four edges terminating at each vertex, $2E = 4V$. Hence for $n \geq 2$ we have

$$F = 2 + E - V = n^2 - n + 2. \tag{2}$$

For $n = 0$ there is just one region, and for $n = 1$, there are two, so (2) holds for $n = 1$, but not for $n = 0$.

Euler's formula for networks on a sphere is valid when the edges are topologically segments and the regions are topologically disks. It fails therefore when $n = 0$, because the one region is not a disk. It fails when $n = 1$, because the edge is not a segment; if we add two vertices on the circle, however, we have $V = 2$, $E = 2$, $F = 2$, and the formula is again valid.

---

5. Let u_n $(n = 1, 2, 3, \ldots)$ denote the least common multiple of the first n terms of a strictly increasing sequence of positive integers (for example, the sequence 1, 2, 3, 4, 5, 6, 10, 12, ...). Prove that the series

$$\sum_{n=1}^{\infty} 1/u_n$$

is convergent.

First Solution. Let u be a positive integer. For each divisor d of u exceeding $\sqrt{u}$ there is another u/d that is less than $\sqrt{u}$. Hence at least half of the positive divisors of u are less than $\sqrt{u}$, so the number of positive divisors is at most $2\sqrt{u}$.

Now u_n, being the least common multiple of n distinct positive integers, has at least n positive divisors, so $2\sqrt{u_n} \geq n$. Therefore, $\Sigma\ 1/u_n$ is dominated by the convergent series $\Sigma\ 4/n^2$ and is itself convergent.

Second Solution. The following proof, contributed by R. L. Graham, shows that the upper bound for such sums is 2.

Let $a_1, a_2, \ldots$ be a strictly increasing sequence of positive integers. Then

$$\frac{1}{a_1} = \sum_{n=2}^{\infty}\left(\frac{1}{a_{n-1}} - \frac{1}{a_n}\right) = \sum_{n=2}^{\infty} \frac{a_n - a_{n-1}}{a_{n-1}a_n}$$

$$\geq \sum_{n=2}^{\infty} \frac{\text{g.c.d.}\{a_{n-1}, a_n\}}{a_{n-1}a_n} = \sum_{n=2}^{\infty} \frac{1}{\text{l.c.m.}\{a_{n-1}, a_n\}}$$

$$\geq \sum_{n=2}^{\infty} \frac{1}{\text{l.c.m.}\{a_1, a_2, \ldots, a_n\}} = \sum_{n=2}^{\infty} \frac{1}{u_n}.$$

(Here g.c.d. stands for greatest common divisor and l.c.m. for least common multiple.) Since $u_1 = a_1$, we have

$$\sum_{n=1}^{\infty} \frac{1}{u_n} \leq \frac{2}{a_1}.$$

There are many a-sequences for which $\Sigma\, 1/u_n = 2$; for example, 1, 2, 4, 8, ... or 1, 2, 3, 6, 12, ..., $3 \cdot 2^{n-3}$, ...

---

6. Show that the unit disk in the plane cannot be partitioned into two disjoint congruent subsets.

Solution. Let D be the unit disk, p its center and ρ the usual distance function.

Suppose D is the union of two disjoint congruent sets A, B, where the notation is so chosen that $p \in A$. Let $x \to x^*$ be the congruence map from A to B. Then p^* is in B. Let r and s be the endpoints of the diameter of D perpendicular to pp^*.

Since $\rho(p, a) \leq 1$ for every point $a \in A$, $\rho(p^*, b) \leq 1$ for every $b \in B$. Clearly $\rho(p^*, r) = \rho(p^*, s) > 1$; so we conclude that r and s belong to A. Then $\rho(r^*, s^*) = \rho(r, s) = 2$. So r^* and s^* are the endpoints of another diameter of D. Then p^* must be the midpoint of this diameter since $\rho(p^*, r^*) = \rho(p, r) = 1$ and similarly $\rho(p^*, s^*) = 1$. Thus $p^* = p$, a contradiction, since $A \cap B = \emptyset$. Therefore D is not the union of two disjoint congruent sets.

APPENDIX

The following two articles are reprinted from the *American Mathematical Monthly*, vol. 72 (1965), pages 469–483.

THE WILLIAM LOWELL PUTNAM MATHEMATICAL COMPETITION: EARLY HISTORY

G. BIRKHOFF, Harvard University

The basic idea underlying the Putnam Competition was expressed by William Lowell Putnam in a three-page article in the Harvard Graduates' Magazine of December, 1921 (reprinted here as an Appendix) from which the following excerpt is taken: "... it is a curious fact that no effort has ever been made to organize contesting teams in regular college studies. All rewards for scholarship are strictly individual and are given in money, or in prizes or in honorable mention. No opportunity is offered a student by diligence and high marks in examinations to win or help in winning honor for his college. All that is offered to him is the chance of personal reward. Little appeal is made to high ideals or to unselfish motives.

Is not this one of the reasons why the effort to interest the great bulk of the undergraduate body in their studies is such an uphill task?"

These words expressed Mr. Putnam's profound conviction in the value of intellectual competition, especially by *teams* whose members took pride in the achievements of their team as a whole and the standing of the institution which it represented. This conviction was shared by Mrs. Putnam, and by her brother Abbott Lawrence Lowell, then President of Harvard. In a letter about the competition some years later, Mr. Lowell referred to the importance of "the promotion of scholarship by contest, as athletics have been," and of "the spirit of intercollegiate emulation in scholarship."

To give substance to the ideas expressed above, Mrs. Putnam established a $125,000 trust in her will, written in 1927. She directed that the trustees "shall have always in mind the purpose of this trust, as set out in my husband's own words," and incorporated a copy of his article in her will.

It was some years after Mrs. Putnam wrote her will, however, before the Putnam Competition assumed its present form. In the meantime, various shortlived experiments were tried.

The first experiment was a competition in 1928 in the field of English between Harvard and Yale. The Yale and Harvard teams each contained ten senior concentrators in English. One member of the Harvard team was Nathan Marsh Pusey, now president of that institution. It was won by

Harvard, a prize of $5,000 going to the winning institution. Both Yale and Princeton declined to repeat the contest, and an offer to compete in economics with Cambridge University was also declined, partly because of a fear of undesirable publicity.

The second experiment was a similar mathematical competition in 1933 between Harvard and West Point juniors. The examination stressed ability to solve problems in the calculus, analytic geometry, and elementary differential equations. This close competition was won by the somewhat more mature and (according to local tradition) better trained West Point cadets. Again, the competition was not repeated, though all ten West Point participants expressed enthusiasm in personal letters to Mrs. Putnam.

Mr. Lowell having retired as President of Harvard in 1933, no further contest was held until after the death of Mrs. Putnam in 1935. In her will, she appointed her sons, George Putnam and August Lowell Putnam, as trustees for the William Lowell Putnam Memorial Fund. These two brothers then consulted George David Birkhoff at Harvard as to the best use to be made of the bequest, a question which the latter had often discussed with Mrs. Putnam. It seemed natural to have the competition in mathematics, both because mathematical ability can better be tested by a set examination than ability in most subjects, and because of the strong mathematical traditions of the Putnam family.

The possibility of discovering outstanding mathematical ability at an early age by set examinations had, indeed, been demonstrated conclusively in Hungary, where a competitive examination (named after the physicist Roland Eötvös) had been given annually (since 1894) to young people entering the University. It certainly contributed to the development of Hungarian mathematics; among the winners were such internationally known names as Leopold Fejér, Theodore von Kármán, Dénes König, Alfred Haar, Marcel Riesz, Gabor Szegö, Tibor Radó, and Edward Teller.

Somewhat similarly, the Cambridge University mathematical examinations were used in the nineteenth century, as a means of detecting both mathematical and legal ability, the Senior Wrangler being considered a man "most likely to succeed" in either profession.

The mathematical interests of the Putnam family were unusual, to say the least. Mrs. Putnam's brother, former President Abbott Lawrence Lowell of Harvard, received in 1877 his A.B. *summa cum laude* in Mathematics at Harvard, and his undergraduate thesis was published in volume 13 (1877), of the Proceedings of the American Academy of Arts and Sciences, pages 222-250. Mr. William Lowell Putnam received in 1882 an A.B. *magna cum laude* in Mathematics at Harvard, and was for many years Chairman of the Visiting Committee of the Mathematics Department there. George Putnam, a trustee of the Putnam Fund, had also majored in mathematics and was active in his turn on the same Visiting Committee, while the present form of the Putnam Competition was being planned. Roger Lowell Putnam also majored in mathematics at Harvard, graduating *magna cum laude* in 1915, while William Lowell Putnam's grandson, McGeorge Bundy (later Dean of the Faculty at Harvard), and his brother Harvey majored in mathematics at Yale.

George David Birkhoff saw in these various facts an important opportunity to stimulate interest in mathematics, to establish national standards of undergraduate mathematical achievement, and to discover and encourage mathematical talent at a time when good fellowships for first-year graduate students were few and far between. With these objectives in mind he proposed, in essence, the principles governing the present Putnam Competition.

The first principle was that the competition should be open to three-man teams selected by the faculty of, *and* to individuals from, *any* American or Canadian college. This was to encourage participation by smaller colleges, who might have only one or two unusually able mathematical concentrators in any given year. The second principle was to have the competition administered by the Mathematical Association of America, which is the professional organization representing college mathematics teachers. The third principle was to distribute prizes and honorable mention to several teams and individuals, so that distinguished performance would be broadly recognized. (To be one of the top 25 or even 50 competitors in any one year, out of 1500 competitors selected from many thousands of mathematics concentrators, is no mean achievement.) The fourth principle was to offer a graduate fellowship at Harvard to *one* of the *five* top competitors. This choice was allowed because Birkhoff recognized that other evidence of talent—especially evidence of creative originality—should be weighed in evaluating men having nearly equal examination scores.

Birkhoff took a very active part in making up the first examination, which was made up by the Harvard Mathematics Department in 1938. So as to avoid favoring the few schools where advanced mathematical courses were then open to able undergraduates, a deliberate effort was made to stress questions covered in standard courses in the calculus, analytic geometry, and elementary mechanics. The emphasis was correspondingly on thoroughness, accuracy, and a clear command of detail.

The first competition was won by the University of Toronto, which was asked to set the next examination and disqualify itself from competing the following year. A similar procedure was followed for about five years, the winning teams being alternately from Toronto and Brooklyn College. These victories were attributed partly to the greater maturity and more intensive concentration program of Canadian mathematical concentrators, partly to the British tradition of problem-solving, and partly to the ability and devotion of Professor Harris MacNeish of Brooklyn in training students how to diagnose and solve set problems.

During the years 1943-5 the Putnam competition was not held because of war conditions. After the war, it was renewed with a few minor modifications. Thoroughness and facility in handling standard course material was given less emphasis, ingenuity in devising and using algorithms and in logical analysis being emphasized in their place. To construct an examination testing these qualities, a special committee was appointed consisting of the distinguished Hungarian-born mathematicians George Pólya and Tibor Radó, and the first Putnam Scholar, Irving Kaplansky. Since that time, membership in the committee making out the examination has rotated, and institutions have been free to win the competition in successive years whenever the ability of their teams made this possible!

APPENDIX

A SUGGESTION FOR INCREASING THE UNDERGRADUATE INTEREST IN STUDIES

WILLIAM LOWELL PUTNAM

[The following article, which appeared in the December, 1921, issue of the *Harvard Graduates' Magazine*, and is reprinted with permission, was the genesis of the William Lowell Putnam Mathematical Competition.]

The idealism of the undergraduate student, his eagerness to achieve something for his college, for his country or for any cause which fills him with enthusiasm is constantly referred to with admiration by those in charge of universities. This unselfish impulse is recognized as one of the strongest forces in a student's life, and great results have been and are being accomplished by appealing to it.

The most telling of these appeals at present are those made in behalf of athletics, where boys are asked and expected to work hard to win laurels for the college on the gridiron or the river or the track. They have to train regularly and monotonously for long periods, to give up smoking, drinking, and late entertainments, to keep off probation, which often requires laborious and irksome study, and they make all these sacrifices willingly and even eagerly for the sake of their college.

In none of these cases is the undergraduate primarily interested in winning honor for himself. He is anxious to have his Alma Mater win, and very glad to play a useful even if an inconspicuous part in the preparation of the team by which her victory is secured. A Harvard man prefers, for example, to be a member of a tennis team to play against Yale and to win a match in that contest rather than to be champion of Harvard, which is only an individual honor. In short, the undergraduate likes to work for the success of his college and particularly likes to work for it as one of a team.

These intercollegiate contests are not confined to athletics. There are competing debating teams which attract some attention from the newspapers and there are chess teams which bring victories that are noticed by those who are interested in this game. But it is a curious fact that no effort has ever been made to organize contesting teams in regular college studies. All rewards for scholarship are strictly individual and are given in money, or in prizes or in honorable mention. No opportunity is offered a student by diligence and high marks in examinations to win or help in winning honor for his college. All that is offered to him is the chance of personal reward. Little appeal is made to high ideals or to unselfish motives.

Is not this one of the reasons why the effort to interest the great bulk of the undergraduate body in their studies is such an uphill task? Some few of them may be wise enough to realize the truth of what they are told often; namely, that if they work hard at their books they will lay the foundation for a future career which will reflect credit on their college; but this is a rather conceited view and the prospect is too remote to appeal to most young men. Those who distinguish themselves as students usually do so either from a desire to improve their own chances of obtaining assistance in college or good employment after graduation, or else from a wish to keep up some family tradition of scholarship or to please some inspiring instructor or from some other similar incentive.

Probably one reason why undergraduate teams do not contest in scholarship is the difficulty of fixing the terms and conditions of such a contest and arranging the details, but in these days, when all contests are fought in the presence of umpires and under highly artificial and technical rules, is there any more difficulty in fixing the terms of a contest of scholars than in arranging for the International Olympic Games or other similar events?

I want to lay particular stress on the point that the competition to be valuable should be between teams and not individuals. Undoubtedly, one reason for the surpassing interest in football games is that the victory is the result of an immense amount of cooperative work, much of it done by people who are never seen by the public. Some twenty-five or more men actually take part in the crowning event of the season, but the efficiency of the team is the result of the coordinate efforts of three times as many players and coaches, each of whom contributes an essential part to the final result. It is a very inspiring thing in itself to form part of such a large body working harmoniously to achieve a great end.

The absence of the opportunity for team-work in undergraduate studies is the more noteworthy from the fact that both before entering college and after graduation most boys do their work as members of a competing team or organization. The preparatory schools take a definite pride in entering their boys without conditions and in a creditable manner and a prize is given to the school whose candidates do best. Every boy as he prepares for his examination feels that he is a part of the school team competing for this prize.

After graduation men usually get into the employ of a firm or corporation or similar body and become intensely loyal to its interests and eager to work hard for its success. Even those who continue their studies at the

Law School get an opportunity to become editors of the *Law Review*, or to become members of law clubs which have interclub contests, so that students are able to do work directly connected with their studies as part of a team which wins credit for the club or the *Review*.

But the regular work of a man's college years—those four most impressionable years—must be done individually and for his own selfish advantage, and one can but ask whether the effort to induce the bulk of the students and their parents to take more interest in their studies would not accomplish more if an appeal were made to higher motives and an opportunity offered for teamwork and for winning laurels for the college.

It seems probable that the competition which has inspired young men to undertake and undergo so much for the sake of athletic victories might accomplish some result in academic fields.

Editorial Note: Professor L. E. Bush, the author of the following paper, which originally appeared in the *American Mathematical Monthly* in 1965, has kindly given us permission to augment and update his study. Accordingly, the tabular data and the list of addresses have been revised.

THE WILLIAM LOWELL PUTNAM COMPETITION: LATER HISTORY AND SUMMARY OF RESULTS

L. E. BUSH, Kent State University

It was most fortunate for undergraduate mathematics in America that the Birkhoffs, father (George David) and son (Garrett), were close friends of the Putnam family. As such, they had the opportunity of formulating the principles of the Putnam Competition in accordance with Mr. Putnam's ideas as expressed in his article. In formulating these principles and in advising the Putnam Trustees on all changes in the regulations governing the Competition, they have always held in mind Mr. Putnam's belief in the value of intellectual competition and his belief that this competition should not be by individuals but by pre-selected teams whose members take pride in the achievement of their team as a whole.

In accordance with Professor Birkhoff's second principle, the competition was placed under the administration of the Mathematical Association of America in the person of its secretary, Professor W. D. Cairns of Oberlin College. Professor Cairns published the general regulations governing the contest and award in this Monthly, 45 (1938) 64–66. These regulations have served as a model for all later revisions. The regulations provided that: "The president of the Association will appoint a committee of three (not necessarily members of the Association) which will select two examiners, mathematicians qualified and willing to undertake the preparation of the examination, and a reader qualified to grade the examination books; in view of the special relation of the donor of the Fund to Harvard University, one member of the committee shall be a member of the Division of Mathematics at Harvard University." In fact, the first examination was prepared by the Harvard Department of Mathematics, and Harvard did not enter a team. The examination was held on April 16, 1938. This competition was won by the University of Toronto, which alternated with Brooklyn College in winning the first five competitions.

The second, third, and fourth examinations were each prepared by the department of mathematics of the institution winning the previous competition, and it appears that these institutions disqualified themselves from the competition for which their department had prepared the examination. The examination for the fifth competition (1942) seems, however, to have been prepared by a committee consisting of Professor B. H. Brown of Dartmouth College and Professor Marie Litzinger of Mount Holyoke College.

The original three team prizes were \$500, \$300, and \$200, but these were changed for the fifth competition to four prizes of \$400, \$300, \$200, and \$100, respectively. In 1958 the number of team prizes was increased to five, with values of \$500, \$400, \$300, \$200, and \$100, respectively; in addition, the members of each team received prizes of \$50, \$40, \$30, \$20, and \$10, respectively, while the five highest ranking individuals received \$75 and the next five highest \$35 each.

The William Lowell Putnam Fellowship had a value during the first five years of \$1000. The value has been gradually increased until it is now worth \$2500 plus tuition at Harvard (\$1520 for 1963-64), or a total monetary value of \$4020.

In the October, 1942 issue of the Monthly it was announced that the Putnam Competition was being postponed indefinitely by agreement of the Putnam Trustees and the Board of Governors, "mainly because of the preoccupation of both teachers and students with war courses in mathematics." When the Competition was renewed after the war the Association was not able to resume its administration immediately, although the President of the Association appointed a committee of three persons to prepare the examination. Members of the Department of Mathematics of Harvard University were asked to assume the responsibility of administration. The first post-war competition, the sixth, was administered by Professor Garrett Birkhoff, and the examination was held June 1, 1946. The seventh and eighth competitions were held May 24, 1947, and March 20, 1948, under the direction of Professor George Mackey.

Beginning with the first post-war competition, the examination has been prepared by a committee appointed by the President of the Association. The members of this committee are appointed for three-year terms, one member retiring each year. The third-year-member of the committee serves as Chairman. Since the resumption of the Competition after the war, every otherwise eligible institution has been free to compete regardless of whether one of their staff members is on the examination committee or is a paper grader. It has been assumed that persons officially connected with the Competition would not take part in organizing or coaching a team. Some of the most distinguished mathematicians of the United States and Canada have served on the committee which makes up the examination, including Professors Tibor Radó, George Pólya, Mark Kac, Irving Kaplansky, Orrin Frink, Jr., Bancroft H. Brown, Emory P. Starke, Ralph G. Sanger, Andrew M. Gleason, L. M. Kelly, R. E. Greenwood, Leo Moser, W. R. Scott, R. E. Bellman, Ivan Niven, D. E. Richmond, John M. H. Olmsted, Gian-Carlo Rota, H. S. M. Coxeter, and A. M. Garsia.

At the September 6, 1948, meeting of the Board of Governors it was agreed that the Association should resume the direction of the Putnam Competition and that the President should be authorized to appoint a director of the competition for a term of five years. Professor Lester R. Ford, Sr., President of the Association, appointed this writer, then of the College of Saint Thomas, as director. I was reappointed in 1953, 1958 and 1963. In 1964 I asked to be relieved after the completion of the twenty-fifth competition.

My first task was to propose a revision of the regulations of the competition for the approval of the Board of Governors and the Putnam Trustees. This revision was necessitated by the new administrative arrangements. It was patterned after Professor Cairns' original general regulations and diverged from that pattern only where the new structure made it necessary. These regulations were approved and, except for several amendments to take care of special situations, they are the present governing instrument of the competition.

That there has been a steady increase in interest in the competition is evidenced by the steadily increasing number of inquiries, both written and oral, about such matters as the level of grades on the examination and the mechanics of operation. The increase in interest also is shown by the growth of participation, as indicated in Table I. The figures given for the first eight competitions may not be exactly comparable with those given for the succeeding competitions.

The first twenty-five competitions show twenty-four institutions whose teams have shared in the prizes and eighteen additional ones which have won at least one honorable mention. Table II shows the distribution of the team honors. One year there was a tie for second place, and both teams were counted as second-place winners; subsequently, ties for lower positions were treated similarly. Fourth-place winners were not named until the fifth competition; fifth-place winners were not named until the nineteenth competition.

Like the Eötvos Competition in Hungary, the Putnam Competition lists among its early winners a large number of the most creative mathematicians of the present day. The number of these is even more impressive when it is realized that the winners of the very first competition are probably still well under sixty-two years old and that the winners of the first post-war competition, the sixth, probably are under fifty-two.

There have been five contestants who have been among the five highest individual contestants three times. They are Edward L. Kaplan, Carnegie Institute of Technology; Andrew M. Gleason, Yale University; D. J. Newman, College of the City of New York; J. B. Herreshoff IV, University of California at Berkeley; and Samuel J. Klein, College of the City of New York.

There have been fifteen contestants who have been among the five highest individual contestants twice. They are Bernard Sherman, Brooklyn

Table I
Participation in the Putnam Competition

Competition Number	*Date of Examination*	*Teams Participating*	*Individuals Participating*	*Institutions Registered*
1	4/16/38	42	163	67
2	3/ 4/39	41	200	69
3	3/ 2/40	45	208	68
4	3/ 1/41	28	146	44
5	3/17/42	—	114	31
6	6/ 1/46	14	—	17
7	5/24/47	32	145	36
8	3/20/48	24	120	29
9	3/26/49	33	155	51
10	3/25/50	41	223	56
11	3/31/51	42	209	56
12	3/22/52	52	295	76
13	3/23/53	49	256	63
14	3/ 6/54	44	231	67
15	3/ 5/55	48	256	70
16	3/ 3/56	54	291	78
17	3/ 2/57	65	378	103
18	2/ 8/58	71	430	96
19	11/22/58	86	506	119
20	11/21/59	95	633	141
21	12/ 3/60	128	867	166
22	12/ 2/61	165	1094	197
23	12/ 1/62	157	1187	192
24	12/ 7/63	170	1260	205
25	12/ 5/64	—	1439	219

College; Maxwell Rosenlicht, Columbia University; W. F. Stinespring, Harvard University; Eoin L. Whitney, University of Alberta; J. W. Milnor, Princeton University; Kenneth G. Wilson, Harvard University; Trevor Barker, Kenyon College; Everett C. Dade, Harvard University; David Mumford, Harvard University; David Bloom, Columbia University; Joseph Lipman, University of Toronto; Alfred W. Hales, California Institute of Technology; Edward Bender, California Institute of Technology; John Hathaway Lindsey, California Institute of Technology; and William C. Waterhouse, Harvard University.

The five highest ranking individuals in each of the first twenty-five competitions are listed in Table III by years in alphabetical order. Where there were ties for fifth place, all those tying are listed. In parentheses following the name of the contestant is the institution which he represented in the competition. Next is given his present position and his present address (whenever we were able to obtain this information). Those awarded the William Lowell Putnam Fellowship are marked by an asterisk (*); although awarded the Fellowship, Mr. Gleason (1940) and Mr. Shepp (1958) never actually held it.

Table III
Winners of the Putnam Competitions

Abbreviations

S = State	C = College	I = Institute	T = Technology
U = University	a = Assistant	A = Associate	P = Professor

Academic subject is mathematics unless noted otherwise.

First Competition — 1938

Robert W. Gibson (Fort Hays Kansas S.C.)—Civil Engineering—AP, Auburn U.
*Irving Kaplansky (U. of Toronto)—P, U. of Chicago.
George W. Mackey (Rice U.)—P, Harvard U.
Michael J. Norris (C. of Saint Thomas)—Member, Technical Staff, Sandia Laboratories. Home address: 12412 Casa Grande, N.E., Albuquerque, N.M. 87112.
Bernard Sherman (Brooklyn C.)—P, New Mexico T., Socorro.

Second Competition — 1939

Richard P. Feynman (M.I.T.)—Physics—P, Cal. I. of T.
Abraham Hillman (Brooklyn C.)—P, U. of New Mexico.
*Edward L. Kaplan (Carnegie I. of T.)—P, Oregon S.U.
William Nierenberg (CCNY)—Physics—Director, Scripps I. of Oceanography, U. of Cal. at San Diego, La Jolla, Cal. 92093.
Bernard Sherman (see First Competition).

Third Competition — 1940

W. J. R. Crosby (U. of Toronto)—AP, U. of Toronto.
*Andrew M. Gleason (Yale U.)—P, Harvard U.
Edward L. Kaplan (see Second Competition).
John Cotton Maynard (U. of Toronto)—Senior Vice-Pres. and Chief Actuary, Canada Life Assurance Co., 330 University Ave., Toronto, Ontario, Canada, M5G 1R8.
Robert Maughan Snow (George Washington U.)—U.S. Department of Transportation. Home address: 222 Marlborough St., Apt. 41, Boston, Mass. 02116.

Table II
Winning Teams in the First Twenty-Five Competitions

Name of Institution	*First Place*	*Second Place*	*Third Place*	*Fourth Place*	*Fifth Place*	*Honorable Mention*
Harvard University	7	5	4	2		2
University of Toronto	4	4	2	2		4
Brooklyn College	3	2	1			3
California Institute of Technology	3	1	2	2		6
Cornell University	2	1	1	1	1	5
Michigan State University	2			1		1
Polytechnic Institute of Brooklyn	2				1	3
University of California at Berkeley	1	1		1	1	7
Queen's University (Kingston)	1		1	1		1
Massachusetts Institute of Technology		4	4	2	1	5
Yale University		3	1			5
Columbia University		2	3			6
University of Pennsylvania		1	1	1		2
College of the City of New York		1		4		5
Dartmouth College		1			1	1
Cooper Union			2			3
Carnegie Institute of Technology			1	1		8

Table II
Winning Teams in the First Twenty-Five Competitions

Name of Institution	First Place	Second Place	Third Place	Fourth Place	Fifth Place	Honorable Mention
New York University			1			3
Mississippi Women's College			1			
Kenyon College				1		2
McGill University				1		1
University of Manitoba				1		1
University of California at Los Angeles						3
Case Institute of Technology				1	1	1
Swarthmore College, Yeshiva University						3 each
Oberlin College, University of British Columbia, Princeton University						2 each
Brown University, Northwestern University, Reed College, Rice University, University of Alberta, University of Colorado, University of Michigan, University of Montreal, University of Notre Dame, University of Rochester, Wesleyan University, University of California at Davis, and University of Waterloo						1 each
42 Institutions	25	26	25	22	7	103

Fourth Competition — 1941

*Richard F. Arens (U. of Cal. at L.A.)—P. U. of Cal. at L.A.

Samuel I. Askovitz (U. of Penn.)—Medicine (Ophthalmology)—A. in Ophthalmology, U. of Penn. Hospital.

Andrew M. Gleason (see Third Competition).

Edward L. Kaplan (see Second Competition).

Paul C. Rosenbloom (U. of Penn.)—P. Teachers C., Columbia U.

Fifth Competition — 1942

*Harvey Cohn (CCNY)—P, CUNY.

Andrew M. Gleason (see Third Competition).

Warren S. Loud (M.I.T.)—P, U. of Minn.

Harold Victor Lyons (U. of Toronto)—Actuary, New York Life Insurance Co., 51 Madison Ave. New York, N.Y. 10010.

Melvin A. Preston (U. of Toronto)—Physics—P of Physics and Vice-Pres. (Academic), U. of Saskatchewan, Saskatoon, Saskatchewan, Canada S7N 0W0.

Sixth Competition — 1946

Felix Browder (M.I.T.)—P, U. of Chicago.

Eugenio Calabi (M.I.T.)—P, U. of Penn.

Donald A. Fraser (U. of Toronto)—Statistics—P, U. of Toronto.

J. Arthur Greenwood (Harvard U.)—Last known address: 115 East 39th St., New York, N.Y. 10016.

*Maxwell A. Rosenlicht (Columbia U.)—P, U. of Cal. at Berkeley.

Seventh Competition — 1947

Clarence Wilson Hewlett, Jr. (Harvard U.)—Elec. Engineering—Last known address: 920 Northgate Ave., Waynesboro, Va. 22980.

Maxwell Rosenlicht (see Sixth Competition).

W. Forrest Stinespring (Harvard U.)—P, U. of Illinois at Chicago Circle, Chicago, Ill. 60680.

*William Turanski (U. of Penn.)—Died in automobile accident.

Eoin L. Whitney (U. of Alberta)— was AP, U. of Alberta, died in 1966.

Eighth Competition — 1948

George F. D. Duff (U. of Toronto)—P. U. of Toronto.

Leonard Geller (Brooklyn C.)—Chemistry—Vice Pres., G. M. Stoller Corp., New York, N.Y. Home address: 336 DeMott Ave., Rockville Centre, N.Y. 11570.

Harry Gonshor (McGill U.)—AP, Rutgers U.
Robert L. Mills (Columbia U.)—Physics—P, Ohio State U.
Donald J. Newman (CCNY)—P, Yeshiva U.
*Eoin L. Whitney (see Seventh Competition).

Ninth Competition — 1949

J. W. Milnor (Princeton U.)—I. for Advanced Study.
*Donald J. Newman (see Eighth Competition).
W. Forest Stinespring (see Seventh Competition).
David L. Yarmush (Harvard U.)—Biology-Res. Assoc., Columbia U.
Ariel Zemach (Harvard U.)—Physics—Los Alamos Scientific Laboratory, Los Alamos, N.M. 87544.

Tenth Competition — 1950

John P. Mayberry (U. of Toronto)—P, Brock U., St. Catharines, Ontario, Canada.
*Z. Alexander Melzak (U. of British Columbia)—P, U. of British Columbia.
J. W. Milnor (see Ninth Competition).
Donald J. Newman (see Eighth Competition).
Richard J. Semple (U. of Toronto)—AP, Carleton U., Ottawa, Ontario, Canada.

Eleventh Competition — 1951

Arthur P. Dempster (U. of Toronto)—Statistics—P and Dept. Chairman, Harvard U.
*James B. Herreshoff IV (U. of Cal., Berkeley)—1708 Euclid Ave., Berkeley, Cal. 94709.
Herbert C. Kranzer (NYU)—P, Adelphi College, Garden City, N.Y. 11530.
Peter John Redmond (Cooper Union)—Physics—General Research Corp., 5383 Hollister Ave., Santa Barbara, Cal. 93111.
Harold Widom (CCNY)—P, Natural Science Division, U. of Cal., Santa Cruz, Cal. 95060.

Twelfth Competition — 1952

Walter L. Bailey Jr. (M.I.T.)—P, U. of Chicago.
James B. Herreshoff IV (see Eleventh Competition).
Gerhard Rayna (Harvard U.)—AP, Lehigh U.
*Eugene R. Rodemich (Washington U., St. Louis)—Group Supvr. Jet Propulsion Lab., Cal. I. of T.
Richard G. Swan (Princeton U.)—P, U. of Chicago.

Thirteenth Competition — 1953

Norman Bauman (Harvard U.)—Medicine—Senior Research Scientist, Lederle Labs. Home address: 99 Briar Rd., Nanuet, N.Y. 10954.

Marshall L. Freimer (Harvard U.)—AP, Business Administration—U. of Rochester.

James B. Herreshoff IV (see Eleventh Competition).

Samuel Jacob Klein (CCNY)—473 Hamilton St., Apt. #12, Geneva, N.Y. 14456.

*Tai Tsun Wu (U. of Minnesota)—Physics—Gordon McKay P. of Applied Physics, Harvard U.

Fourteenth Competition — 1954

*James Daniel Bjorken (M.I.T.)—Physics—P, Stanford Linear Accelerator Center, Stanford U.

Leonard Evens (Cornell U.)—P, Northwestern U.

William P. Hanf (U. of Cal. at Berkeley)—P, U. of Hawaii.

Benjamin Muckenhoupt (Harvard U.)—P, Rutgers U.

Kenneth G. Wilson (Harvard U.)—Physics—P, Cornell U.

Fifteenth Competition — 1955

*Trevor Barker (Kenyon C.)—Retired from SUNY at Buffalo. Home address: Apt 1, 2640 Windsor St., (840 E) Salt Lake City, Utah 84106.

Everett C. Dade (Harvard U.)—P, U. of Illinois.

David B. Mumford (Harvard U.)—P, Harvard U.

Howard C. Rumsey, Jr. (Cal. I. of T.)—Member, Technical Staff, Cal. I. of T. Home address: 151 Linda Vista, Pasadena, Cal. 91105.

Jack Towber (Brooklyn C.)—Last known address: 1234 Stratford Ave., Bronx, N.Y. 10072.

Sixteenth Competition — 1956

Trevor Barker (see Fifteenth Competition).

David M. Bloom (Columbia U.)—AP, Brooklyn C. (CCNY).

*Richard Michael Friedberg (Harvard U.)—Physics—P, Barnard C. of Columbia U.

David B. Mumford (see Fifteenth Competition).

Kenneth G. Wilson (see Fourteenth Competition).

Seventeenth Competition — 1957

David M. Bloom (see Sixteenth Competition).

Richard T. Bumby (M.I.T.)—AP, Rutgers U.

Everett C. Dade (see Fifteenth Competition).
*Rohit J. Parikh (Harvard U.)—P, Boston U.
J. Ian Richards (U. of Minnesota)—AP, U. of Minnesota.

Eighteenth Competition — 1958 (Spring)

David R. Brillinger (U. of Toronto)—Statistics—P, U. of Cal. at Berkeley.
Donald J. C. Bureś (Queens U.)—P, U. of British Columbia.
Richard M. Dudley (Harvard U.)—P, M.I.T.
Joseph Lipman (U. of Toronto)—P, Purdue U.
*Lawrence A. Shepp (Polytechnic I. of Brooklyn)—Member, Technical Staff, Bell Telephone Labs., Murray Hill, N.J. 07974.

Nineteenth Competition — 1958 (Fall)

Alfred W. Hales (Cal. I. of T.)—P, U. of Cal. at Los Angeles.
Robert C. Hartshorne (Harvard U.)—P, U. of Cal. at Berkeley.
John Rex Forrester Hewett (U. of Toronto)—Last known address: 500 Avenue Rd., Toronto, Ontario, Canada.
Joseph Lipman (see Eighteenth Competition).
*Alan Gaisford Waterman (San Diego S.C.)—Home address: 5020 Porter Hill Road, La Mesa, Cal. 92041.

Twentieth Competition — 1959

Stephen L. Adler (Harvard U.)—Physics—P, Princeton U.
Donald S. Gorman (Harvard U.)—Home address: 355 Grove St., Half Moon Bay, Cal. 94019.
Alfred W. Hales (see Nineteenth Competition).
Martin Isaacs (Polytechnic I. of Brooklyn)—P, U. of Wisconsin, Madison.
*Stephen Lichtenbaum (Harvard U.)—P, Cornell U.
Donald Passman (Polytechnic I. of Brooklyn)—P, U. of Wisconsin, Madison.
Daniel G. Quillen (Harvard U.)—P, M.I.T.

Twenty-first Competition — 1960

William R. Emerson (Cal. I. of T.)—P, Queens C., Flushing, N.Y. 11367.
Jon H. Folkman (U. of Cal. at Berkeley)—Died January 1969.
Melvin Hochster (Harvard U.)—P, U. of Michigan.
*Louis Jaeckel (U. of Cal. at Los Angeles)—Self-employed Statistical Consultant. Home address: 1507 Belvedere Ave., Berkeley, Cal. 94702.
Samuel Jacob Klein (see Thirteenth Competition).

Twenty-second Competition — 1961

Edward Anton Bender (Cal. I. of T.)—AP, U. of Cal. at San Diego, La Jolla, Cal. 92037.

Elwyn R. Berlekamp (M.I.T.)—Elec. Engineering—P, Computer Science and Mathematics, U. of Cal. at Berkeley.

*John Hathaway Lindsey (Cal. I. of T.)—Last known address: 92 Field Rock Road, Southport, Conn. 06490.

William C. Waterhouse (Harvard U.)—AP, Penn. S.U.

Barry Wolk (U. of Manitoba)—Research Associate, Mathematics and Physics, U. of Manitoba.

Twenty-third Competition — 1962

Edward Anton Bender (see Twenty-second Competition).

John Hathaway Lindsey (see Twenty-second Competition).

Robert S. Strichartz (Dartmouth C.)—P, Cornell U.

*William C. Waterhouse (see Twenty-second Competition).

John William Wood (Harvard U.)—AP, U. of Illinois at Chicago Circle, Chicago, Ill. 60680.

Twenty-fourth Competition — 1963

Lawrence J. Corwin (Harvard U.)—AP, Rutgers U.

*Stephen E. Crick, Jr. (Mich. S.U.)—aP, U. of Pittsburgh (died April 1979).

Robert E. Greene (Mich. S.U.)—P, U. of Cal. at Los Angeles.

Joel H. Spencer (M.I.T.)—AP, SUNY at Stony Brook, Stony Brook, N.Y. 11790.

Lawrence A. Zalcman (Dartmouth C.)—AP, U. of Maryland.

Twenty-fifth Competition — 1964

Rufus Bowen (U. of Cal. at Berkeley)—P, U. of Cal. at Berkeley.

*Roger E. Howe (Harvard U.)—P, Yale U.

Barry B. MacKichan (Harvard U.)—AP, New Mexico S.U., Las Cruces, N.M. 88001.

Vern Sheridan Poythress (Cal. I. of T.)—aP, Westminster Theological Seminary, Chestnut Hill, Philadelphia, Pa. 19118.

Fred William Roush (U. of North Carolina)—aP, Alabama S.U.

Students and faculty of the competing institutions have expressed great interest in the details of the scoring of the examination papers. Since the ninth competition, the examination has always consisted of twelve questions (six questions in the morning and six in the afternoon), each of which counted ten points. The graders have given partial credit for partially answered questions and for imaginative attempts whenever these seemed to deserve recognition. Thus the maximum score on each examination was 120 points. Information relative to the grading of the first eight examinations is not available, but since the ninth competition no paper has been judged worthy of the full 120 points. The lowest score on every one of these sixteen examinations has been zero, and on six of them more than ten per cent of the papers had a score of zero. Table IV gives for each of these examinations the scores, to the nearest integer, of contestants falling in the tenth, twenty-fifth, fiftieth, seventy-fifth, and ninetieth percentiles. It also shows the highest score on each examination.

Some readers may wish to make deductions about the relative difficulty of the examinations and the relative number of easy and difficult questions in each examination. Since any such deductions must depend also on the grader, the following information is pertinent. There have been four graders involved in the above sixteen competitions, and since they must remain anonymous, I shall call them *A*, *B*, *C*, and *D*. Where two graders graded a competition, one graded the morning papers and one the afternoon papers. Competitions nine through nineteen were graded by *A* alone, competition twenty by *A* and *B*, competition twenty-one and twenty-two by *B* and *C*, and the last two competitions by *C* and *D*.

Table IV
Scores of Contestants in the Ninth Through the Twenty-fifth Competition

	Percentile					
Competition	10	25	50	75	90	*Highest Score*
9	1	4	9	20	32	74
10	0	2	8	22	42	100
11	6	14	29	49	65	103
12	6	16	31	51	69	117
13	4	11	21	38	53	114
14	0	5	16	31	53	86
15	1	5	16	33	51	91
16	1	6	15	26	45	107
17	0	2	9	22	34	89
18	0	1	5	22	41	97
19	0	3	10	25	43	106
20	3	10	21	36	55	96
21	5	10	18	28	40	87
22	0	6	14	26	44	107
23	2	6	13	25	41	90
24	5	9	14	21	28	62
25	0	5	11	18	26	56

Many examination supervisors have written me apologetically regarding the showing made by their contestants. They can be encouraged by the fact that on seven of the last seventeen competitions at least 10 per cent of the scores were zero and that a score of six is the highest tenth percentile rank on any of these seventeen examinations. A contestant does not have to earn a high score to feel that it has been worth his while to spend six hours on the examination. He may derive great benefit from seeing the kind of problems that the ablest undergraduates are interested in. It is my hope that he may thus gain encouragement to delve more deeply into this kind of mathematical activity.

The following two papers, one by the late Professor L. J. Mordell and the other by Professor L. M. Kelly, are reprinted from the May 1963 issue of the *American Mathematical Monthly*. Together, they serve as an examination in depth of both some of the weaknesses and some of the strengths of the Putnam Competition.

THE PUTNAM COMPETITION

L. J. MORDELL, University of Arizona and St. John's College, Cambridge

Editorial Note. We are all aware of Professor Mordell's distinguished career at Cambridge University, but his extensive activities in this country may be less familiar. He went to high school in Philadelphia, belongs to both the MAA and AMS, and has published in most of our journals. He has given talks in forty universities in this country and seven in Canada. He has been Visiting Professor at well-known universities for five years in this country and three years in Canada. He has participated in coaching students for the Putnam Contest. On the basis of this experience, he is evidently well qualified to offer an enlightened opinion on the examinations.

The twenty-second competition was held in December 1961, and it may be assumed that the Putnam competition is well established on a permanent basis. It has been a factor of the utmost importance in arousing and stimulating interest in mathematics in the Colleges and Universities of the United States and Canada. Increasing numbers of students are entering the contest, and 1109 students from 166 institutions took part in the twenty-first competition in December 1960. The competition has undoubtedly played no small part in raising the status, the levels and standards of mathematical education. It is therefore worth while to review recent competitions, and to pay some attention to the kind of pattern that seems to be emerging. Suggestions will be made for increasing the usefulness of the competition and its services to mathematics.

It was customary some years ago to publish in this Monthly a syllabus of the subjects of the examination. Thus, for competition 18 held in February 1958, it was stated "The questions will be taken from the fields of calculus (elementary and advanced) with applications to geometry and mechanics not involving techniques beyond the usual applications, higher algebra (determinants and theory of equations), elementary differential equations and geometry (advanced plane and solid analytical geometry)."

It was stated in this notice that the requirements for the examination would be met by two years (12 semester hours) in the calculus, a half year's work (3 semester hours) each in higher algebra and differential equations and a year's work (6 semester hours) in analytic geometry.

The syllabus would not seem a very formidable one to most students interested in mathematics, and they might well think that the subjects mentioned would give ample opportunity for testing their knowledge, ability and skill. They might hope there was a reasonable chance of not doing too badly in the examination independently of where they may have studied.

They might not have been aware that the syllabus was not always adhered to. Thus, in competition 14 (1954), Part I, Question 7 is: "Prove that there are no integers x and y for which $x^2 + 3xy - 2y^2 = 122$." The published solution reduces the equation to the form $x^2 - 17y^2 = 488$ whence $x^2 \equiv 12 \pmod{17}$. It is then stated that by examining a complete residue class or by the theory of quadratic residues, that this congruence is impossible. This question seems to be outside the scope laid down.

It can also be said that the expression "not involving techniques beyond the usual applications" should be taken with a grain of salt. In competition 17 (1957) Part I, Question 2 is: "A uniform wire is bent into a form coinciding with the portion of the curve $y = e^x$, $0 \le x \le a$, $a > 1$, and the line segment $a - 1 \le x \le a$, $y = e^a$. The wire is then suspended from the point $(a - 1, e^a)$ and a horizontal force F is applied at the point $(0, 1)$ to hold the wire in coincidence with the curve and segment. Assuming the x axis is horizontal, show that the force F is directed to the right." It is easy to show that this requires that if $a > 1$, then

$$\int_0^a (x - a + 1)\sqrt{(1 + e^{2x})}dx + \frac{1}{2} > 0.$$

It requires more than a knowledge of the usual techniques to prove this though it is not too difficult when one knows how.

The procedure for the eighteenth competition held in February 1958 and the later ones was entirely different from the preceding ones. No syllabus was published in the Monthly but it was presumably given in the notices circulated to the various institutions. The syllabus and the requirements for the examination seem to have been fundamentally changed without any public announcement. The syllabus for the twenty-second competition, held in December 1961, is as follows:

"The examinations will be constructed to test originality as well as technical competence. It is expected that the contestant will be familiar with the formal theories embodied in undergraduate mathematics through differential equations. It is assumed that such training, designed for mathematics and physical science majors, will include somewhat more sophisticated mathematical concepts than is the case in minimal courses. Thus the differential equations course is proposed to include some references to qualitative existence theorems and subtleties beyond the routine solution devices. Questions will be included which cut across the bounds of various disciplines and self-contained questions which do not fit into any of the usual categories may be included. It will be assumed that the contestant has acquired a familiarity with the body of mathematical lore commonly discussed in mathematics clubs or in courses with such titles as 'survey of the foundations of mathematics.' It is also expected that the self-contained questions involving elementary concepts from group theory, set theory, graph theory, lattice theory, number theory, and cardinal arithmetic, will not be entirely foreign to the contestant's experience."

The new syllabus is open to serious criticism. Emphasis is now laid upon acquiring what can only be described as a smattering of knowledge of various topics, when a student would be better employed in acquiring, say, a greater knowledge of analysis in its various aspects. There is no reason why attendance at mathematical clubs should give one a working knowledge of a particular subject. A survey course in foundations need not add to technical or manipulative skill. How can one expect that a student at this stage of his career should have a working knowledge of graphs and lattice theory? He can still be a good and promising student without any such knowledge.

The syllabus assumes a very sophisticated type of student, but will handicap very seriously contestants from a large number of institutions who have neither the opportunity nor time for coping with the new syllabus. (Why the special emphasis on differential equations?) It is inevitable that certain institutions will do better than others in any kind of examination. The better students will naturally gravitate to them. These institutions may also devote considerably more time to the study of mathematics than others do, and they may also pursue subjects to a much higher level. It seems undesirable that the examination should accentuate those features in which these institutions are sure to have too great an advantage over others.

It would have been much better to have introduced specifically into the syllabus a subject such as, say, modern algebra. In due course, this might become more and more a subject taught in the earlier years of a mathematics course. In ways of this kind, the competition could exercise a healthy and decisive influence on mathematical curricula, and far more so than by specifying a miscellaneous collection of topics.

There are two very important facts a student would like to know. Does the syllabus include all those topics dealt with in the examination? This is not so since there is no mention of probability though questions are often set on it. It is true that the syllabus says "Questions will be included which cut across the bounds of various disciplines and self-contained questions which do not fit into any of the usual categories may be included." Can one take this to include a reference to probability? Much more important, however, is how much of a subject is the student expected to know? Without some specifications, the word "elementary" is too vague to convey a clear meaning. In fact, questions are set in the examination which by no stretch of the imagination can be described as elementary, and which would be more suitable for graduate students of some maturity. Perhaps one might accept that a rough definition of "elementary" would be the contents of the usual books read by undergraduate students, but not topics found in only one or two books on a subject.

An obvious suggestion is that the content and extent of the syllabus should be referred periodically to a committee representative of every kind of College and University, and also that this committee should draw up a wide list of books useful in preparing for the examination.

Let us now consider in some detail the questions set in the examination. One has every sympathy with the examiners, since setting the questions is not a simple matter. They must not be too easy but also not so difficult as to induce a feeling of dismay and frustration. It must be very discouraging and disheartening to a candidate when he realizes that his unexpected lack of knowledge makes it completely impossible for him to do a number of questions and so to do himself justice in the examination.

It is sometimes rather hard to decide whether a question is easy. It may appear simple enough when the solution is known. To find this, however, may have taken the proposer considerable time, and even then he may have been in the favorable position of having relevant thoughts and ideas present in his mind. Without these, there is often no reason why one should find the solution in a short time; and it is generally most unlikely that he will do so under the stress of examination conditions. In any case, a question really requiring a new idea will rarely be done in an examination. It can be said, however, that the papers do contain questions which would be considered as easy or difficult by any standard.

It may be remarked that the published solutions are often of a synthetic character, i.e., they show how the question was made up, but of course there is very little reason for a student to know this, and so the given solution may not be representative of what is done in the examination or of the difficulty of the question.

First of all, some general remarks about the examination. It should not be of such a nature as to encourage desultory reading as opposed to reading to obtain a thorough knowledge of the subjects of an undergraduate course. This does not mean, however, that a student should not read widely if he is disposed to do so and can profit by it. The examination should not give him a one-sided view of the scope of mathematics or a false idea of what is worth while. It should not suggest that he spend a great deal of time in acquiring dexterity in topics not really important in mathematics.

Questions are sometimes set which have appeared as problems in journals and were not likely to be done without considerable thought and time. Others are taken from research papers or are really of a research character. The examination room is not the place for research, and such questions are more suitable for contemplation in the quiet of one's study.

One instance may suffice. In competition 19 (1958), Part I, Question 2 is: "Given a set of $n + 1$ positive integers none of which exceeds $2n$, show that at least one number of the set must divide another member of the set." This was proposed by Erdös as Problem 3739 in this Monthly. A beautifully simple solution appeared in the February issue (1937), and is much nicer than that published for the competition. The simplicity of a solution may have no relation to the difficulty of a question. Thus the very next Monthly Problem 3740, also due to Erdös was: "From a point O inside a given triangle ABC, the perpendiculars OP, OQ, OR, are drawn to its sides. Prove that

$$OA + OB + OC \geq 2(OP + OQ + OR)."$$

This was considered a very difficult problem, indeed, and was communicated by Erdös to many people who could not find a solution. However, Mordell gave a simple solution which appeared in this Monthly for April 1937. This so-called Erdös-Mordell theorem has been ever since the subject of many papers.

Most important is what types of question should be set? Obviously, they should be reasonable in character and this theme will now be developed. Enunciation of questions should be of a helpful kind. There are often questions set in which an answer is asked for. There seems no reason why answers should not be given more frequently, especially in complicated questions. It saves a student time when he sees he has arrived at the correct answer, and he is saved the trouble of verifying his work. It also conduces to a feeling of confidence and improves his morale when he has good reason to think his work is sound. If the answer is too involved to be given, this is perhaps a good reason for either modifying the question or omitting it. The published answers of some questions suggest most emphatically that the questions ought never to have been set in the form in which they appeared.

Questions are frequently set in which one is asked to consider all possible cases or to verify that all solutions have been obtained. It would be desirable sometimes to word a question in such a way as to impress this upon the student and to keep him from going astray. Take competition 21 (1960), Part II Question 1, as a slight instance. "Find all solutions of $n^m = m^n$ in integers n and $m (n \neq m)$. Prove that you have obtained all of them." The form of the equation might easily lead one to believe that only positive integers are required. It would have been better to add after m, "positive or negative" and omit $n \neq m$.

Those who set the papers should have a clear picture in their mind of the qualities they are testing and what they are looking for in the student. Are they, for example, more concerned with the few students who have acquired knowledge far beyond the usual courses, and do they consequently set problems that only these exceptional students can attempt?

Sometimes one is under the impression that they attach undue weight to the ability to carry out troublesome arithmetical calculations in an examination, especially when these add no point to the question, but only lead to spending more time on it with more opportunities for making arithmetical mistakes.

Some of the questions remind one of the proverbial problem of finding a needle in a haystack. If one has some idea of its location or an infallible tool e.g., an electro-magnet, there will be no difficulty. Without some such indications, one will be helpless. Take competition 20 (1959), Part II, Question 3: "Give an example of a continuous real valued function from [0, 1] to [0, 1] which takes every value in [0, 1] an infinite number of times." The given solution notes that this is really a question on finding a Peano space-filling curve and then utilizes the Cantor set C of those numbers $0 \leq x \leq 1$ which can be written in the ternary scale using only 0's and 2's. How can any one who has never seen something of functions of this kind before, and probably this applies to most undergraduates, be expected to do such a question? It seems almost impossible that this could be discovered in an examination.

The questions seem to suggest that the examiners are more concerned with a knowledge of the morbid and pathological aspects of mathematics, especially of sets and functions. These of course have led to important mathematical advances but surely the respectable or normal parts of mathematics are worth more attention. The examiners often emphasize the bizarre, the whimsical and the exotic aspects of mathematics and questions often seem artificial in character. They tend to stress modes of thought which are not possessed by many good mathematicians, (and they are not necessarily the worse for that) and also aspects of mathematics which are not very important nor significant for an undergraduate. This applies in particular to the more difficult questions dealing with the distribution of lines and points, the distances between points in various sets, or covering of sets by others. Of course, such questions may require considerable ingenuity of a sort, and are suitable for a kind of research, but how relevant are they to sound mathematical knowledge and ability? They can easily induce a feeling of repulsion in many people when such emphasis is laid upon them. In some topics, such as combinatorial analysis, there may be considerable difficulty in translating the statement of the question into a mathematical argument. The bizarre aspects of questions occur in those dealing with towers of exponents and an infinity of radicals. There is a real danger that students will be spending too much time on such questions when they would be better employed in studying what is more worth while.

There are many reasons why questions might be called undesirable or unreasonable and a number are now considered in detail.

Questions should not be set depending upon advanced knowledge of which most of the candidates are sure to be ignorant. In competition 18 (1958), Part II, Question 7 is: "Prove that if $f(x)$ is continuous for $a \le x \le b$ and $\int_a^b x^n f(x)dx = 0$ for $n = 0, 1, 2, \ldots$, then $f(x)$ is identically zero on $a \le x \le b$."

The published proof of this theorem of Lerch is simple since it uses, but only quotes, the well-known result of Weierstrass on the approximation of continuous functions by polynomials (as is also done in Ford's "Differential Equations"). This result may be found in books dealing with real variables or Fourier series. The question places a great premium on knowledge far beyond what most undergraduates know, for I cannot believe that a student is likely to find a proof during the examination. It might be mentioned that the solution tacitly assumes that a and b are finite. In view of the frequent emphasis in the examination to consider all cases, it might have been mentioned in the question that a and b are finite.

Take competition 17 (1957), Part II, Question 6: "The curve $y = f(x)$ passes through the origin with a slope of 1. It satisfies the differential equation $(x^2 + 9)y'' + (x^2 + 4)y = 0$. Show that it crosses the x-axis between $x = 3\pi/2$ and $\pi\sqrt{63}/\sqrt{53}$."

This is a not too difficult application of Sturm's theorem on results of this kind, but involves a little unpleasant arithmetic. The theorem is not found in the usual books on differential equations. It is given, however, in Ford's book on the chapter dealing with Bessel functions—a natural place for such theorems. How much better it would have been for the student to know of this book, but even then the topic might not have been discussed in the course.

Problems should not involve an inordinate amount of arithmetic especially when this adds no value to the question or gives it any special significance. The mere need for such calculations is the best reason for either rejecting the question or for radical modification. Take competition 20 (1959), Part II, Question 5: "Find the equation of the smallest sphere which is tangent to both of the lines (i) $x = t + 1$, $y = 2t + 4$, $z = -3t + 5$, and (ii) $x = 4t - 12$, $y = -t + 8$, $z = t + 17$."

The published solution is

$$(502x + 915)^2 + (502y - 791)^2 + (502z - 8525)^2 = 5{,}423{,}859.$$

It seems almost cruel to set questions with so many opportunities of making numerical errors. How much better it would have been to present the question rather differently. The point of the question is that the centre of the sphere is at the midpoint of the common perpendicular of the two lines. Call the two lines $x = l_1t + a_1, \ldots,$ and $x = l_2t + a_2 \ldots$. The perpendicular has direction cosines $L, M, N,$ proportional to $m_1n_2 - m_2n_1, \ldots$. Hence its length i.e., the diameter of the sphere, is the projection of the line joining the points a_1, b_1, c_1 and a_2, b_2, c_2 on the perpendicular and so is easily written down. An elementary knowledge of solid analytical geometry gives at once the equation of the perpendicular, then its intersection with the two given lines. More simply if A, B, C is the centre of the sphere of radius r, the points $A \pm Lr$, $B \pm Mr$, $C \pm Nr$ are on the lines and so A, B, C can be found. Surely the question set is of a type which should be set only in a form requiring no arithmetic.

Sometimes a question seems to involve considerable arithmetic sufficient to discourage a student, and is made worse by requiring technical knowledge. Take competition 21 (1960) Part I, Question 6: "A player throwing a die scores as many points as on the top face of the die and is to play until his score reaches or passes a total n. Denote by $p(n)$ the probability of making exactly the total n, and find the value of $\lim_{n\to\infty} p(n)$."

Here $6p(n) = p(n-1) + p(n-2) + \cdots + p(n-6)$, and it requires some knowledge of probability to see that this is a method of attack. The student might feel unhappy at the prospect of dealing with this recurrence formula of the sixth degree. The solution given is as follows. Write

$$P(s) = 6/6 - s - s^2 \cdots - s^6 = 6/(1-s)Q.$$

Then

$$P(s) = \sum_{n=0}^{\infty} p(n)s^n.$$

Also

$$P(s) = \frac{\frac{2}{7}}{1-s} + \frac{2(15 + 10s + 6s^2 + 3s^3 + s^4)}{7(6 + 5s + 4s^2 + 3s^3 + 2s^4 + s^5)}.$$

Expanding the two fractions into power series, we obtain from the coefficients of $P(s)$ that $p(n) = \frac{2}{7} + f_n$ where the f_n are coefficients in the expansion of the second fraction. It can be shown that $f_n \to 0$ and therefore that $\lim p(n) = \frac{2}{7}$. What had the examiner in mind for a proof when he states that $f_n \to 0$?

In fact, a very bright student could have done the question with practically no arithmetic. He might have argued that it suffices to find only the term $\frac{2}{7}/(1-s)$ in the decomposition of $P(s)$ and to show that the roots of $Q = 0$ have moduli >1. This is obvious since if $|s| \leq 1$, $|s + \cdots + s^6| \leq 6$ with equality only when $s = 1$.

If such a probability question were thought to be within a candidate's powers, the idea might have retained by using a die with each of the numbers 1, 2, 3 occurring twice. Then if he adopted the published form of solution, $3p(n) = p(n-1) + p(n-2) + p(n-3)$, and now $3 - s - s^2 - s^3 = (1-s)(3 + 2s + s^2)$, and it is obvious that the roots of $3 + 2s + s^2 = 0$ have moduli >1.

Unnecessary complication should be avoided in proposing a question and it is often worth while stripping a question of unessential parts which add no value to the questions. Take competition 20 (1959), Part I, Question 5: "A sparrow, flying horizontally in a straight line, is 50 feet directly below an eagle and 100 feet directly above a hawk. Both hawk and eagle fly directly toward the sparrow, reaching it simultaneously. How far does each bird fly? At what rate does the eagle fly?"

Curves of pursuit lead to differential equations and there is already some work involved in finding the solution. Two pursuers and numerical details are of no special mathematical significance, and merely make the solution more troublesome. How much simpler the question would have been with only one pursuer and asking, say, only for the distance travelled by the prey.

Problems should be stated in such a way that there is no doubt about the meaning of the question or what is required of the candidate. Take competition 22 (1961), Part I, Question 6: "If $J_2 = \{0, 1\}$ is the field of integers modulo 2, and if $J_2[x]$ is the integral domain of polynomials in one indeterminate with coefficients in J_2, prove that $p(x) = 1 + x + \cdots + x^n$ is reducible (factorizable) in case $n + 1$ is composite. Is the converse true? That is, if $n + 1$ is prime, is $p(x)$ irreducible?"

The first part of the question is simple but what is expected of the candidate in the second part? If he is familiar with the theory of finite fields, he will answer that $f(x)$ is irreducible if and only if 2 is a primitive root of $n + 1$, and may wonder if this will be accepted as an answer. As he is unlikely to know this, he might try to see whether $p(x)$ is reducible, for example, to the product of two polynomials $g(x)$, $h(x)$ each of degree $\frac{1}{2}(n + 1)$. On changing x into $1/x$, he is led to conditions on the coefficients of $g(x)$, $h(x)$. These suggest that when $n = 6$, a possible decomposition may be $(1 + x + x^3)(1 + x^2 + x^3)$, and it is easily verified that these are the factors of $f(x)$. Trial will show that $f(x)$ is not reducible when $n = 4$. Would it not have been better to ask to prove the results for $n = 4, 6$?

Take competition 21 (1960), Part I, Question 4: "Given two points in the plane P and Q at fixed distances from a line L and on the same side of the line, as indicated, the problem is to find a third point R so that $PR + RQ + RS$ is a minimum where RS is perpendicular to L. Consider all cases."

A diagram is given in which R lies between the perpendiculars from P and Q on L, and also between L and the line through P parallel to L. What is the significance of the diagram and the phrase "consider all cases?" Does it mean that R is to be restricted to the region in which it seems to be? Apparently not. The published solution takes L to be the x-axis and the points P, Q to be $(-a, p)$, (a, q), and we may take $0 < p < q$. If R is (x, y), the given solution is an attempt to minimize

$$F(x, y) = |y| + \sqrt{(x + a)^2 + (y - p)^2} + \sqrt{(x - a)^2 + (y - q)^2}.$$

The presence of $|y|$ here means that the phrase "all cases" is meant to include the one when R is below L. Since $|y|$ is not differentiable at $y = 0$, it may be remarked that $y = 0$ must be considered separately.

The diagram serves no purpose except possibly to mislead candidates. It should have been omitted since the exceptional case may occur when R is at P.

The question would read much better as this: "Given in the plane a line L and two points P and Q on the same side of L. Find a third point R such that the sum of the distances of R from P, Q and the line is a minimum."

Questions dealing with elementary parts of mathematics become difficult if they assume the knowledge of some elementary facts with which the student is not likely to be familiar and which he is not likely to discover in the examination. Take competition 17 (1957), Part I, Question 3: "If A and B are real numbers and k is a positive integer, show that

$$\left(\frac{\cos kB \cos A - \cos kA \cos B}{\cos B - \cos A}\right) < k^2 - 1$$

whenever the left hand side is defined."

It is noted in the published solution that $<$ should be $\leq$. The result can be deduced easily from the result $|\sin kA/\sin A| \leq k$, but it seems unlikely that many candidates would know this. It would have been much better if they had been asked first to prove this and then the main part. The published solution starts from $|\sin kA/\sin A| \leq k$ and shows how the given question arose—a synthesis which makes sense for a student only when the solution has been found otherwise. In fact, it is stated in the results of the competition that no candidate did the question completely.

There is undue emphasis on problems of combinatory analysis. Proficiency in these is not easily developed under examination conditions. Take competition 20 (1959), Part II, Question 4: "Given the following matrix of 25 elements

$$\begin{pmatrix} 11, & 17, & 25, & 19, & 16 \\ 24, & 10, & 13, & 15, & 3 \\ 12, & 5, & 14, & 2, & 18 \\ 23, & 4, & 1, & 8, & 22 \\ 6, & 20, & 7, & 21, & 9 \end{pmatrix},$$

choose five of these elements, no two coming from the same row or column, in such a way that the minimum of these five elements is as large as possible. Prove that your answer is correct."

Probably thoughts about such matrices have not been present in the minds of many contestants. Confronted with an apparently meaningless collection of numbers, whose significance or relation to each other is not easily grasped, most students would be helpless. It seems most undesirable that a mathematician's competence should be tested or judged by this kind of question.

Too much emphasis is sometimes placed upon topics not justified by the role they play in undergraduate work, or their importance in mathematics as a whole. It has already been stated that nowhere is probability mentioned as a subject of the examination. When or where is the student to acquire a knowledge of this? He may acquire a little from some books on algebra or calculus. There are institutions in which probability appears as an undergraduate course, but there may have been no reason for a student to attend this if other courses seemed more attractive to him. It may be remarked that almost every subject will be taught in undergraduate courses in some place or other. Time spent in designing a well balanced curriculum would be worth while.

If questions on probability are to be set, it seems undesirable to assume more than a simple common-sense knowledge. This is done for Question 3 in Part I of Competition 18 (1958): "Real numbers are chosen at random from the interval $0 \leq x \leq 1$. If after choosing the nth number, the sum of the numbers so chosen first exceeds 1, show that the expected or average value of n is e."

Question 4 in Part II of the same competition is: "What is the average straight line distance between two points on a sphere of radius 1?" Two questions on average values are too many.

What is to be the future of this competition which has to cater for increasing numbers of students of all kinds? Most of them cannot expect to profit financially or to do brilliantly in the examination. They probably regard it as a kind of sporting, free-for-all contest, and think the effort worth while. They might be happier if the competition was split into two distinct ones, say an A and a B competition, with adjustments in the value of the prizes.

The B competition should have a syllabus comparable with the previous one prior to the change. Most of the entries for this would come from the smaller institutions and others in which mathematics is not taught on a high level. These institutions might feel that honor is satisfied if their students feel they have made a creditable showing. The A competition would be primarily for those from the larger institutions and others in which mathematics is intensively studied. It should, however, be open to all who wish to enter independently of any place of study. The A syllabus should cover a greater range than the B one and make wider demands on the contestants. Both examinations should be such as not to give the student a jaundiced mathematical outlook. They should give him the feeling that mathematics is real, mathematics is earnest.

THE PUTNAM COMPETITION—REJOINDER

L. M. KELLY, Michigan State University

The very searching analysis of the Putnam examinations of the past few years by such an eminent mathematician as L. J. Mordell warrants the interest and concern of all who feel, as I do, that this distinguished institution has played, and should continue to play, an important role in undergraduate mathematical training in this country and Canada. My interest in the competition dates from my undergraduate days and became quite intense, of course, when I served on the examination committee some years ago. The examination committee consists of three members appointed by the president of the Association for three year terms. The appointments are on a rotating schedule, one member retiring and one new member joining each year. During my tenure the revision of what Professor Mordell refers to as the syllabus was suggested and I actively participated in drawing up the new version. Before I go into a detailed discussion of this and other matters it may be in order to point out that these opinions are entirely my own and are not to be given any official status. I know that numerous colleagues who have served on previous examination committees share many of the sentiments which are expressed herein but I claim only to speak for myself unless specific quotations are employed. In particular I will make liberal use of a letter which Professor Ivan Niven has made available.

Professor Mordell's criticisms seem to center on four main points:

1. The present syllabus is not well conceived.
2. The examinations do not always adhere to the syllabus.
3. There is excessive emphasis on the bizarre, the exotic and the pathological.
4. Some problems are poorly constructed, are occasionally ambiguous, and involve tedious calculations to no good end.

In order to maintain some perspective on the effect of the appearance of an occasional unfortunate problem perhaps the following observations may be useful. Experience has shown that in order to discriminate among the top ten or so contestants the examination must be of the order of difficulty of those which have been given. As those who have worked with competitions know, however, it is very difficult to devise an examination which will provide a significant linear ordering very far down the list. In these examinations it has been the custom to present the contestant with 14 problems from which they were to select 12. As a matter of fact the successful solving of 6 or 7 of the 14 quite usually placed the contestant among the top 30 or 40. Thus the inclusion of as many as 2 or 3 "bad" problems on a single examination will not seriously affect the chances of success of the vast majority of those participating. I am not contending that we should not be concerned about such a state of affairs, that it may not have some psychological effect, and that it would not have been better to have more suitable problems in their stead, but I think it is unrealistic to make the claim that the presence of an occasional unreasonable problem is a decisive factor in the performance of students from the small or less distinguished schools whose fate seems to be one of Professor Mordell's principal concerns.

My principal disagreement with the contention of point 4 is the extent to which such problems occur and the degree of their effect on the performance of most of the contestants. A similar observation holds for point 2 as well. Many people will take issue with some of the examples which Professor Mordell has chosen to illustrate his points. For example, Professor Niven notes that the problem concerning the sphere tangent to the skew lines was not a calculated form of cruelty but became such because of a misprint in one of the signs. These things happen, of course, and should not form the basis for serious criticism. My own perusal of old examinations reveals relatively few problems involving onerous calculations. Professor Niven further observes that he had in mind a much simpler solution of the probability problem concerning the die than that envisaged by either Professor Mordell or the graders. He suggests that it might be a good idea to have the examination committee submit its solutions for publication rather than leaving this entirely in the hands of the graders as is now done. It is my understanding that Professor Bush intends to follow this procedure in the future.

I also would disagree with some of the illustrations. The problem concerning the 5×5 matrix is quite a trivial problem and has been so evaluated by several contestants. I have frequently presented this to high school students who were neither "helpless" nor dismayed. In fact they found the problem rather entertaining and its solution not too difficult. Or again, the problem involving the force on the bent wire amounts to establishing the inequality

$$\int_0^a x\sqrt{1 + e^{2x}}\, dx > (a - 1) \int_0^a \sqrt{1 + e^{2x}}\, dx, \qquad a > 1.$$

While this may involve a degree of ingenuity it hardly qualifies as "outside the range of the usual applications."

But it is not my intention to take issue with each of Professor Mordell's examples of problems he considers inappropriate. I readily agree that there have been many such examples in the 23 year history of the competition and not all in the past few years. It would, in fact, be a striking achievement if a committee effort over such an extended period had not produced such instances. I do claim, however, that on the whole the examinations have been remarkably good, have stimulated interest, and have evoked the respect of the mathematics community in the competition.

Professor Mordell seems to suggest that something slightly irregular happened in 1958. "The syllabus and the requirements for the examination seem to have been fundamentally changed without any public announcement," he writes. The announcement was given, of course, in the brochure announcing the forthcoming examination but apparently Professor Mordell feels that there should have been more publicity signalling the event. The fact of the matter is that those of us involved in the revision did not regard this as an effort to change the pattern and nature of subsequent examinations but rather to recognize and describe properly what had been going on for some years. It was apparent to us that almost from the outset, problems had been included which would fit within the announced scope only by the most flexible interpretation.

On the ninth competition we find: "Assume the complex numbers a_1, $a_2, a_3, \ldots, a_n, \ldots$ are all different from zero and that $|a_r - a_s| > 1$ for $r \neq s$. Show that $\Sigma 1/a_n^3$ converges." Does this fall within the range of advanced calculus, vintage 1949?

Or again in the 11th competition this problem appears: "Find the volume of a four dimensional hypersphere $x^2 + y^2 + z^2 + t^2 = r^2$ and also the hypervolume of its interior $x^2 + y^2 + z^2 + t^2 < r^2$." What in the syllabus covers this? A 1950 course in advanced calculus or a course in solid analytics?

As a final example consider PM 3 again from the 9th competition: "Let K be a closed plane curve such that the distance between any two points of K is less than 1. Show that K lies inside a circle of radius $1/\sqrt{3}$." I will comment more on this problem shortly. For the moment I merely observe that it is hard to justify this as being in the realm of the original examination syllabus.

In the light of such examples the committee sought advice from several competent sources including Professor Bush and drafted the current version to which Professor Mordell objects. In a general way, the committee was attempting to introduce more flexibility into the description, realizing that such descriptions are necessarily vague and should serve only as a broad suggestion of what to expect on any given examination and not as some sort of legal straight jacket within which the committee must operate. Some effort was made to anticipate slightly possible desirable trends but largely the motive was to describe the state of affairs at that time.

No doubt an improvement in the present language can be effected but whatever is done should not rule out such elegant and suitable problems as AM 4 of the 15th competition, PM 5 of the 16th competition, and PM 5 of the 17th (see note at end of this paper) which can only be classified as combinatorial analysis, the theory of linear graphs, and set (or lattice) theory respectively. Most committees, I am sure, make an effort to conform to the tradition of the examination while introducing some element of novelty. Thus the best syllabus is a reference to recent past examinations.

I am unable to subscribe to the picture of frustration painted by Professor Mordell, of the contestant who innocently reads the syllabus and decides on the basis that he is adequately prepared, only to be rudely shocked when confronted with the actual examination. Surely any serious contestant would glance over some previous examinations and would in addition have the advice and assistance of a faculty member of some experience.

I close consideration of the syllabus by remarking that Professor Mordell weakens his case when he asks where a student is supposed to acquire a "working knowledge" of graph theory. This is grossly unfair reading of the syllabus. The syllabus reads, "It is also expected that *self-contained* questions involving elementary concepts from group theory, set theory, graph theory, lattice theory, number theory, and cardinal arithmetic will not be *entirely foreign* to the contestant's experience."

The emphasis here is on the *entirely foreign* aspect. Surely no essential fault can be found with a problem such as PM 5 from the 17th competition. It involves no technical knowledge not possessed by a good freshman. Yet if the contestant has no *prior experience* with this type of problem, he is at a definite disadvantage. The committee was merely trying to suggest to the faculty advisers that the prospective contestants should be prepared, psychologically at least, for questions of this character. Again we remark that such broad descriptions take on more meaning by reference to a few past examinations.

The objection to the inclusion of *research* problems as illustrated by the Erdös-Anning theorem (see note) is well taken but such problems are hardly characteristic of recent examinations. PM 3 from the ninth competition, referred to above, is essentially the celebrated theorem of Jung and in some respects more difficult to prove than the Erdös-Anning result. Certainly knowing the very simple and elegant solution of the Erdös-Anning theorem a committee would be hard put to decide whether this should be regarded as suitable. The history of the result suggests not, since Erdös and Anning first published a rather involved solution before Erdös observed the almost trivial argument that he eventually published. In addition, the problem has had quite a wide circulation and it is quite possible that a number of the contestants had already seen the solution.

In the final analysis the way to produce a good examination is to have a good committee. The composition of the committees these past few years has been in my opinion quite good and quite representative. The present committee, for example, has members representing Williams College, M.I.T., and Southern Illinois University. Among the types of schools usually participating in the examination can one imagine a better cross sectional representation?

In criticising problems as being "bizarre, whimsical, and exotic," we enter a very subjective area and one which is certainly hard to debate. As Professor Niven puts it, "It seems to me that Mordell, in criticising the selection of problems used in Putnam examinations, is really criticising the current interests, fashions, outlooks and standpoints of contemporary American mathematics. The difficulty is that what is respectable and normal to one is exotic to another." Keeping in mind that the examination committee changes each year and the president of the association who appoints them, every two years, certainly any "trend" which Professor Mordell has noted over the past, say ten years' represents the judgment of a large number of competent mathematicians.

As a partial remedy for some of the faults which he sees in the present arrangements, Prof. Mordell suggests a second and simpler examination for the less talented or less sophisticated. This may be an attractive "solution" to the progressive educator who visualizes all examinations as a form of group therapy but I am surprised to find it advocated by a member of the old school, particularly one trained in the British tradition. Students, I believe, are proud to participate in this competition even if they do not do too well because they respect the examination and they recognize that doing well in it represents an achievement of high order. Resorting to some classification or handicapping scheme will, in my view, cheapen the entire program and undermine the respect which this splendid competition has generated over the past 25 years. I have no objection to some sort of minor league competition if some other group wishes to sponsor it but let's leave the Putnam competition simple and unencumbered, the winning of which represents one of the most significant achievements of American undergraduate mathematics.

Note. For the convenience of the reader we include here the statements of the problems referred to in the rejoinder.

AM 4, 15th competition. On a circle n points are selected and the chords joining them in pairs are drawn. Assuming that no three of these chords are concurrent (except at the end points) how many points of intersection are there?

PM 5, 16th competition. Consider $2n$ points in space, $n > 1$. Suppose they are joined by at least $n^2 + 1$ segments. Show that at least one triangle is formed. Show that for each n it is possible to have $2n$ points joined by n^2 segments without any triangle being formed.

PM 5, 17th competition. With each subset X of a set is associated a second subset $f(X)$. The association is such that whenever X contains Y then $f(X)$ contains $f(Y)$. Show that for some set A, $f(A) = A$.

The Anning-Erdös Theorem If the distances between pairs of points of an infinite plane set are all integers, the set is linear.

INDEX

References are to page numbers. An ordered number pair, such as (a, b), indicates that the item referenced appears in the statement of the problem on page a and again in the solution to the problem, beginning on page b.